# THE WORLD TREASURY OF PHYSICS,
# ASTRONOMY, AND MATHEMATICS

# The World Treasury of

# PHYSICS, ASTRONOMY, AND MATHEMATICS

■

*Edited by* TIMOTHY FERRIS

*With a Foreword by* Clifton Fadiman, General Editor

LITTLE, BROWN AND COMPANY

*Boston   New York   Toronto   London*

**FIRST PAPERBACK EDITION**

Acknowledgments of permission to reprint previously copyrighted material
appear on pages 837–847.

Library of Congress Cataloging-in-Publication Data

The World treasury of physics, astronomy, and mathematics / edited by
    Timothy Ferris; with a foreword by Clifton Fadiman, general editor. — 1st
    ed.
        p.   cm.
Includes bibliographical references and index.
ISBN 0-316-28133-6
1. Physics.   2. Astronomy.   3. Mathematics.   4. Science — Philosophy.
5. Physicists — Biography.   6. Astronomers — Biography.
7. Mathematicians — Biography.   I. Ferris, Timothy.
    II. Fadiman, Clifton, 1904–
QC71.W67   1991
500.2 — dc20                                                            90-45693

10  9  8  7  6  5  4  3  2
MV-PA

Designed by Barbara Werden
Published simultaneously in Canada by Little, Brown & Company
(Canada) Limited

PRINTED IN THE UNITED STATES OF AMERICA

# CONTENTS

# Two: The Wider Universe

## THE SUN AND BEYOND

## THE STRUCTURE OF THE UNIVERSE

## BEGINNINGS AND ENDINGS

# Three: The Cosmos of Numbers

## ABOUT MATHEMATICS

Contents

## ARTIFICIAL INTELLIGENCE AND ALL THAT

## MATH ANGST

# Four: The Ways of Science

## SCIENTISTS' LIVES AND WORKS

## Contents

### THE POETRY OF SCIENCE

### PHILOSOPHY AND SCIENCE

# FOREWORD

THIS VOLUME forms one of a continuing series of World Treasuries published in cooperation with the Book-of-the-Month Club. Each of these Treasuries gathers, from as many literatures as possible, previously published, high-order writing of our time, fiction or nonfiction. Each covers a specific field or genre of general interest to the intelligent reader. Each has been constructed and edited by a recognized authority in the field.

The underlying purpose of the series is to meet the expectations of the thoughtful modern reader, one conscious of the striking change that has taken place in our view of the world. That change has run counter to parochialism. It recognizes the contribution made by thinkers and prose artists of the Orient as well as the Occident, of those who use languages other than English. Hence *World* Treasuries.

The series is planned to cover a great variety of fields, from the life sciences to religious thought, from science fiction to love stories to mystery and detection, and will be issued over the next several years.

CLIFTON FADIMAN
*General Editor*

# PREFACE

*Anthology:* From the Greek *anthologia*,
literally, flower gathering.
— *Webster's New International Dictionary,* Second Edition

An anthology is like all the plums and orange peel
picked out of a cake.
— Sir Walter Raleigh (1861–1922)

THE DELIGHTS of science and mathematics — their revelations of natural beauty and harmony, their visions of things to come, and the joy of discovery in itself, the light and shadow it casts on the mystery dance of mind and nature — are too profound, and too important, to be left to scientists and mathematicians alone. They belong to the cultural heritage of the entire world, and to know something about them is to be acquainted with the finest new achievements of the human mind. Science and mathematics are to our technically inclined societies what the composition of epic poetry was to the Homeric Greeks, or shipbuilding to the ninth-century Norsemen, or landscape painting to the Sung Dynasty Chinese; they are what we do best.

To familiarize ourselves with a development in science or mathematics, however, is for most of us rather more complicated than to read a poem or view a painting. The reason, of course, is that technical matters customarily are announced in technical terms — in words and equations that make sense chiefly to initiates.* If the com-

*This said, it should be added that the difficulty of understanding a work of science, as opposed to a work of art, is frequently exaggerated. I suspect that the reasons for this have to do with the way we are taught in school. Virtually all students everywhere learn to read and write and are exposed to at least a little painting and sculpture, but relatively few learn much mathematics or science. Consequently, a high-school graduate may feel emboldened to try reading Tolstoy but is far less likely to try reading Heisenberg or Bohr. The intellectual challenges posed by all three authors are comparable, however; if the scientists seem more forbidding it is because they, and their tradition, are less familiar to most teachers and students.

munication of science went no further than publication in the scientific journals, the general public would be left out of the process. That would be a shame; it would be as if Tolstoy's novels and Rembrandt's portraits were kept locked in a room to which only a tiny elite were granted admittance. That this is *not* the case, that the work of the astronomers, quantum physicists, and set theorists is to a considerable extent accessible to us all, is due in large measure to the efforts of the "science writers" — those scientists, mathematicians, journalists, and essayists who explain and interpret the work in language that all educated people can understand.

The sciences as they have developed have gone beyond narrowly technical concerns and have begun to touch on questions that have deep and lasting resonances for every thinking person. Astronomers and physicists today find themselves obliged to contemplate such grand matters as the origin of the universe and to consider that the very act of observation disrupts the collection of atoms in the laboratory (or photons coming from a distant star) that they are trying to observe. Mathematicians have learned that their equations are both more limited than they had supposed (they cannot, it seems, predict everything in perfect detail) and more sweeping in scope (they can chart the shoals where determinism and freedom meet, even plumb the enigma of human intelligence). As such questions enter into the province of science, the subject matter the science writer confronts expands, to include themes whose artistic weight echoes that of poetry, music, and painting.

This anthology represents an attempt to collect some of the best modern science writing — the most elegant and intelligible nontechnical writing on astronomy, physics, and mathematics published in the twentieth century. Many a pious rationale might be proffered in defense of its publication. One might, for instance, argue that such a book could help drive back the darkness of scientific illiteracy. That scientific illiteracy exists is not in doubt. A survey conducted for the National Science Foundation in the mid 1980s found that only one-third of the American public understands what a molecule is, that nearly half reject the theory of evolution, and that only one in ten can distinguish astronomy from astrology. Far less certain, though, is the assertion that any one book can do much to improve the situation. I am more inclined to hope that this anthology makes for a good read, and a little education into the bargain, and that it encourages its readers to keep learning.

An editor, naturally enough, hopes that an anthology will do justice to the etymology of its name — that it will, as the Greeks envisioned it, be as satisfying as a bouquet of flowers — but fears that, as in Sir Walter Raleigh's characterization, it may instead amount to plucking out the plums and orange peels while discarding most of the cake. Still, plums and orange peels are better than nothing, and if the whole cake had been included — the whole corpus of twentieth-century astronomy, physics, and mathematics, that uniquely grand and glorious attainment of our times — this book would consist of many weighty volumes rather than just one; it would comfort its editor but exhaust its readers. What we have, then, is a collection of generally nontechnical writing about the noble sciences that investigate the foundations of the material world from atoms to stars, and about the mathematical relationships that illuminate the laws by which their marvelously diverse empires behave.

This book is *not* a great many things. It is not a history of science, nor a comprehensive introduction, nor a textbook; more subjects have been omitted here than could be included. It *is*, optimistically, a compendium of pieces, written by both scientists and lay authors, that are distinguished both by the quality and significance of the ideas expressed and by their stylistic, even literary, merit. It aspires to act as a finger pointing at the moon but cannot stand in for the moon itself.

Most of the writing in this book is nontechnical, meaning that it can be understood by the 99.7 percent of the public who are trained in neither science nor engineering. The difficulty of the selections varies considerably, however. To have taken an alternative course, limiting the collection to writing of one particular level of difficulty, would have meant excluding many readers. Instead, I have tried to present many sorts of writing, on many levels, in the desire that novices and science buffs alike might find something of value here.

There is necessarily some overlap in the material covered between one article and another; because I anticipate that most readers will want to read *in* this book, rather than reading it linearly from front to back, it seemed better to permit a bit of repetition than to refer the reader constantly to passages elsewhere in the book. On the other hand, some worthy subjects have been omitted. There is, for instance, little about the planets in this book, though a great deal has been learned about the planets in recent years, thanks principally to the success of unmanned space probes dispatched throughout most of the solar system. But the quality of writing about the new planetology has not (yet) been very high, owing no doubt to the fact that many discoveries in this field are new and have not yet been long thought about.

More difficult to justify — though not, I think, less unavoidable — are the effects of personal taste and bias. I do not find it difficult (and critics may find it easier still!) to see that the colors of my individual predilections stain this book. I am interested in the philosophy of science; those who are less so, many excellent working scientists among them, may feel there is too much philosophy here. In selecting pieces for inclusion I have stressed literary quality; some doubtless would have placed more stress on content and less on felicities of style. A good case could be made for an encyclopedic approach, in order to cover recent scientific advances more thoroughly. Nevertheless, I think there is something to be said for a personal point of view, however flawed, if only because a book is a highly personal medium and seldom fares well under the evenhanded manipulations of a committee.

The scientific organism is, happily, alive and thriving, and even the most competent writing of only a decade or so ago may, by today's lights, prove to contain misconceptions and technical errors. Nevertheless I have constrained myself to a minimum of correcting and updating. The worldview of each generation is of a piece, and to start correcting every outdated reference can mean shredding a perfectly coherent piece of writing into tatters. I have, though, corrected an occasional figure or term, where the danger of being misled seemed especially great and where the correction could be made with reasonably little damage; such insertions appear in brackets. In addition, some technical references have been deleted. So have footnotes citing the sources of direct quotations and the like; readers interested in full citations are referred to the original articles, which are listed in the bibliography. Brief deletions are indicated by ellipses points, longer ones by five asterisks. Not every technical term is explained in every instance; readers encountering unfamiliar names or terms are urged to consult the index for other references.

In making selections from the vast literature of modern astronomy, physics, and mathematics, I have chosen to emphasize their intellectual core — what the physicist Victor Weisskopf calls the "intensive" rather than "extensive" areas of these sciences. "Intensive research goes for the fundamental laws, extensive research goes for the *explanation* of phenomena in terms of known fundamental laws," Weisskopf writes. "Solid state physics, plasma physics, and perhaps also biology are extensive. High energy physics and a good part of nuclear physics are intensive."

To concentrate on intensive science means leaving out a number of things. For instance, this book contains nothing on superconductivity, or the latest parallel-processor computer architecture, or the design of the Galileo space probe. I would seek to justify such choices in terms not only of brevity but of currency. Extensive fields tend to change more quickly; a piece on so-called room-temperature superconductivity written in 1987, when expectations were running highest, would read quite differently than one written in 1990, when growing evidence had begun to suggest that practical applications of the new superconductors were to be rather restricted. The same is true for the concept of "cold fusion," which created a global flap in March through May of 1989, only to collapse once further experiments failed to confirm the evidences of nuclear fusion originally reported. Intensive research is more centered, hangs on fewer basic principles, and stays closer to them, so a book that concentrates on such work may instruct better, and last longer, than would one that pays closer heed to the newspaper headlines.

In compiling this anthology I have benefited from the aid and advice of more friends and colleagues than I can properly thank. I would, though, like to express my particular gratitude to Isaac Asimov, Andrew Fraknoi, Owen Gingerich, Alan Guth, Dennis Overbye, Allan Sandage, and Kip Thorne, and, for assistance with the preliminary research, to Marjorie Dobkin, Steve Merryman, Julia Rechter, and Eric Smith. As always, I have been sustained in my work by the affection, forbearance, and good counsel of my wife, Carolyn Zecca Ferris, our children, and my mother, Jean Baird Ferris.

Science has been called the twentieth century's greatest art form, and it is natural to expect that writers will increasingly find inspiration in the grandeur, rigor, and elegance of science. Surely something of the sort has been going on at least sporadically since the days of the Greek and Latin poets. We may be a long way from producing a modern-day Lucretius who can weave real poetry out of cosmology or quantum physics. But this book will have served its function if its readers find, here and there within its pages, a few sparks of the illumination with which art and science alike serve to brighten our darkling times.

# PART ONE

# *The Realm of the Atom*

As physicists probe nature on scales a billion times smaller than the unaided eye can see, they find much that is strange, to be sure, but also ample evidence of connections linking the tiny precincts of the subatomic world to the wider universe. The electrons orbiting each atomic nucleus obey weird rules — performing quantum leaps, for instance, which means disappearing from one spot and appearing at another without having traversed the space in between — but it is thanks to this odd behavior that atoms can link up in fantastic combinations, generating the chemistry of everything from rocks to living cells. Radioactive atoms decay in ways that are impossible to predict in detail, yet large collections of such atoms decay at rates statistically so reliable that they provide scientists with natural clocks that can be consulted for information about the ages of the earth, sun, and stars.

Science often stands accused of shattering nature's wholeness into bite-sized bits for analysis, but the results of such analysis have a way of resurrecting the whole. The laws of thermodynamics, originally investigated in order to design more efficient steam engines, are currently being applied to the study of black holes. Einstein's special theory of relativity, which he composed in an effort to understand better the infinitesimal particles of which light is composed, turned out to be relevant to such wildly divergent concerns as the longevity of interstellar astronauts (the faster you go, the more slowly you age) to how the stars shine (by releasing a little of the energy locked up inside atoms). Even Heisenberg's uncertainty principle, which revealed that there is a fundamental limit to our ability to predict quantum events, has implications that work more to involve us in nature than to alienate us from it: If, as the uncertainty principle implies, no atom can be observed without disturbing it, then the illusion of the passive observer has been shattered, and we are left with no choice but to recognize ourselves as active participants in the atomic world, as cosmic meddlers who of necessity leave our fingerprints everywhere.

# Atoms and Quarks

## RICHARD P. FEYNMAN

■

Richard Feynman (1918–1988), regarded as one of the most innovative physicists of the twentieth century, was known for his lively and unpretentious way of talking about science. This excerpt comes from one of the introductory lectures he gave to a class of freshmen and sophomores at the California Institute of Technology in 1961–1962.

## Atoms in Motion

... IF, IN SOME cataclysm, all of scientific knowledge were to be destroyed, and only one sentence passed on to the next generations of creatures, what statement would contain the most information in the fewest words? I believe it is the *atomic hypothesis* (or the atomic *fact*, or whatever you wish to call it) that *all things are made of atoms — little particles that move around in perpetual motion, attracting each other when they are a little distance apart, but repelling upon being squeezed into one another*. In that one sentence, you will see, there is an *enormous* amount of information about the world, if just a little imagination and thinking are applied.

To illustrate the power of the atomic idea, suppose that we have a drop of water a quarter of an inch on the side. If we look at it very closely we see nothing but water — smooth, continuous water. Even if we magnify it with the best optical microscope available — roughly two thousand times — then the water drop will be roughly forty feet across, about as big as a large room, and if we looked rather closely, we would *still* see relatively smooth water — but here and there small football-shaped things swimming back and forth. Very interesting.

3

These are paramecia. You may stop at this point and get so curious about the paramecia with their wiggling cilia and twisting bodies that you go no further, except perhaps to magnify the paramecia still more and see inside. This, of course, is a subject for biology, but for the present we pass on and look still more closely at the water material itself, magnifying it two thousand times again. Now the drop of water extends about fifteen miles across, and if we look very closely at it we see a kind of teeming, something which no longer has a smooth appearance — it looks something like a crowd at a football game as seen from a very great distance. In order to see what this teeming is about, we will magnify it another two hundred and fifty times and we will see something similar to what is shown in Figure 1. This is a picture of water magnified a billion times, but idealized in several ways. In the first place, the particles are drawn in a simple manner with sharp edges, which is inaccurate. Secondly, for simplicity, they are sketched almost schematically in a two-dimensional arrangement, but of course they are moving around in three dimensions. Notice that there are two kinds of "blobs" or circles to represent the atoms of oxygen (black) and hydrogen (white), and that each oxygen has two hydrogens tied to it. (Each little group of an oxygen with its two hydrogens is called a molecule.) The picture is idealized further in that the real particles in nature are continually jiggling and bouncing, turning and twisting around one another. You will have to imagine this as a dynamic rather than a static picture. Another thing that cannot be illustrated in a drawing is the fact that the particles are "stuck together" — that they attract each other, this one pulled by that one, etc. The whole group is "glued together," so to speak. On

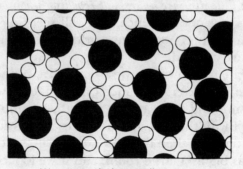

Water Magnified One Billion Times

Figure 1

the other hand, the particles do not squeeze through each other. If you try to squeeze two of them too close together, they repel.

The atoms are 1 or $2 \times 10^{-8}$ cm in radius. Now $10^{-8}$ cm is called an *angstrom* (just as another name), so we say they are 1 or 2 angstroms (A) in radius. Another way to remember their size is this: if an apple is magnified to the size of the earth, then the atoms in the apple are approximately the size of the original apple.

Now imagine this great drop of water with all of these jiggling particles stuck together and tagging along with each other. The water keeps its volume; it does not fall apart, because of the attraction of the molecules for each other. If the drop is on a slope, where it can move from one place to another, the water will flow, but it does not just disappear — things do not just fly apart — because of the molecular attraction. Now the jiggling motion is what we represent as *heat:* when we increase the temperature, we increase the motion. If we heat the water, the jiggling increases and the volume between the atoms increases, and if the heating continues there comes a time when the pull between the molecules is not enough to hold them together and they *do* fly apart and become separated from one another. Of course, this is how we manufacture steam out of water — by increasing the temperature; the particles fly apart because of the increased motion.

In Figure 2 we have a picture of steam. This picture of steam fails in one respect: at ordinary atmospheric pressure there might be only a few molecules in a whole room, and there certainly would not be as many as three in this figure. Most squares this size would contain none — but we accidentally have two and a half or three in the picture (just so it would not be completely blank). Now in the case of steam we see the characteristic molecules more clearly than in the

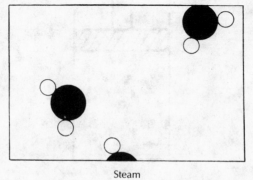

Steam

Figure 2

case of water. For simplicity, the molecules are drawn so that there is a 120° angle between them. In actual fact the angle is 105°3', and the distance between the center of a hydrogen and the center of the oxygen is 0.957 A, so we know this molecule very well.

Let us see what some of the properties of steam vapor or any other gas are. The molecules, being separated from one another, will bounce against the walls. Imagine a room with a number of tennis balls (a hundred or so) bouncing around in perpetual motion. When they bombard the wall, this pushes the wall away. (Of course we would have to push the wall back.) This means that the gas exerts a jittery force which our coarse senses (not being ourselves magnified a billion times) feel only as an *average push*. In order to confine a gas we must apply a pressure. Figure 3 shows a standard vessel for holding gases (used in all textbooks), a cylinder with a piston in it. Now, it makes no difference what the shapes of water molecules are, so for simplicity we shall draw them as tennis balls or little dots. These things are in perpetual motion in all directions. So many of them are hitting the top piston all the time that to keep it from being patiently knocked out of the tank by this continuous banging, we shall have to hold the piston down by a certain force, which we call the *pressure* (really, the pressure times the area is the force). Clearly, the force is proportional to the area, for if we increase the area but keep the number of molecules per cubic centimeter the same, we increase the number of collisions with the piston in the same proportion as the area was increased.

Now let us put twice as many molecules in this tank, so as to double the density, and let them have the same speed, i.e., the same

Gas

Figure 3

temperature. Then, to a close approximation, the number of collisions will be doubled, and since each will be just as "energetic" as before, the pressure is proportional to the density. If we consider the true nature of the forces between the atoms, we would expect a slight decrease in pressure because of the attraction between the atoms, and a slight increase because of the finite volume they occupy. Nevertheless, to an excellent approximation, if the density is low enough that there are not many atoms, *the pressure is proportional to the density*.

We can also see something else: If we increase the temperature without changing the density of the gas, i.e., if we increase the speed of the atoms, what is going to happen to the pressure? Well, the atoms hit harder because they are moving faster, and in addition they hit more often, so the pressure increases. You see how simple the ideas of atomic theory are.

Let us consider another situation. Suppose that the piston moves inward, so that the atoms are slowly compressed into a smaller space. What happens when an atom hits the moving piston? Evidently it picks up speed from the collision. You can try it by bouncing a Ping-Pong ball from a forward-moving paddle, for example, and you will find that it comes off with more speed than that with which it struck. (Special example: if an atom happens to be standing still and the piston hits it, it will certainly move.) So the atoms are "hotter" when they come away from the piston than they were before they struck it. Therefore all the atoms which are in the vessel will have picked up speed. This means that *when we compress a gas slowly, the temperature of the gas increases*. So, under slow *compression*, a gas will *increase* in temperature, and under slow *expansion* it will *decrease* in temperature.

We now return to our drop of water and look in another direction. Suppose that we decrease the temperature of our drop of water. Suppose that the jiggling of the molecules of the atoms in the water is steadily decreasing. We know that there are forces of attraction between the atoms, so that after a while they will not be able to jiggle so well. What will happen at very low temperatures is indicated in Figure 4: the molecules lock into a new pattern which is *ice*. This particular schematic diagram of ice is wrong because it is in two dimensions, but it is right qualitatively. The interesting point is that the material has a *definite place for every atom*, and you can easily appreciate that if somehow or other we were to hold all the atoms at one end of the drop in a certain arrangement, each atom in a certain place, then because of the structure of interconnections, which is rigid, the other end miles away (at our magnified scale) will have a definite location. So if we

7

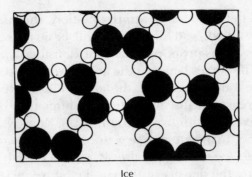

Ice

Figure 4

hold a needle of ice at one end, the other end resists our pushing it aside, unlike the case of water, in which the structure is broken down because of the increased jiggling so that the atoms all move around in different ways. The difference between solids and liquids is, then, that in a solid the atoms are arranged in some kind of an array, called a *crystalline array,* and they do not have a random position at long distances; the position of the atoms on one side of the crystal is determined by that of other atoms millions of atoms away on the other side of the crystal. Figure 4 is an invented arrangement for ice, and although it contains many of the correct features of ice, it is not the true arrangement. One of the correct features is that there is a part of the symmetry that is hexagonal. You can see that if we turn the picture around an axis by 120°, the picture returns to itself. So there is a *symmetry* in the ice which accounts for the six-sided appearance of snowflakes. Another thing we can see from Figure 4 is why ice shrinks when it melts. The particular crystal pattern of ice shown here has many "holes" in it, as does the true ice structure. When the organization breaks down, these holes can be occupied by molecules. Most simple substances, with the exception of water and type metal, *expand* upon melting, because the atoms are closely packed in the solid crystal and upon melting need more room to jiggle around, but an open structure collapses, as in the case of water.

Now although ice has a "rigid" crystalline form, its temperature can change — ice has heat. If we wish, we can change the amount of heat. What is the heat in the case of ice? The atoms are not standing still. They are jiggling and vibrating. So even though there is a definite order to the crystal — a definite structure — all of the atoms are vibrating "in place." As we increase the temperature, they vibrate

8

with greater and greater amplitude, until they shake themselves out of place. We call this *melting*. As we decrease the temperature, the vibration decreases and decreases until, at absolute zero, there is a minimum amount of vibration that the atoms can have, but *not zero*. This minimum amount of motion that atoms can have is not enough to melt a substance, with one exception: helium. Helium merely decreases the atomic motions as much as it can, but even at absolute zero there is still enough motion to keep it from freezing. Helium, even at absolute zero, does not freeze, unless the pressure is made so great as to make the atoms squash together. If we increase the pressure, we *can* make it solidify.

So much for the description of solids, liquids, and gases from the atomic point of view. However, the atomic hypothesis also describes *processes*, and so we shall now look at a number of processes from an atomic standpoint. The first process that we shall look at is associated with the surface of the water. What happens at the surface of the water? We shall now make the picture more complicated — and more realistic — by imagining that the surface is in air. Figure 5 shows the surface of water in air. We see the water molecules as before, forming a body of liquid water, but now we also see the surface of the water. Above the surface we find a number of things: First of all there are water molecules, as in steam. This is *water vapor,* which is always found above liquid water. (There is an equilibrium between the steam vapor and the water which will be described later.) In addition we find some other molecules — here two oxygen atoms stuck together by them-selves, forming an *oxygen molecule,* there two nitrogen atoms also stuck together to make a nitrogen molecule. Air consists almost entirely of nitrogen, oxygen, some water vapor, and lesser amounts of carbon

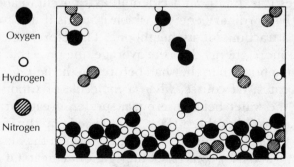

Water Evaporating in Air

Figure 5

dioxide, argon, and other things. So above the water surface is the air, a gas, containing some water vapor. Now what is happening in this picture? The molecules in the water are always jiggling around. From time to time, one on the surface happens to be hit a little harder than usual, and gets knocked away. It is hard to see that happening in the picture because it is a *still* picture. But we can imagine that one molecule near the surface has just been hit and is flying out, or perhaps another one has been hit and is flying out. Thus, molecule by molecule, the water disappears — it evaporates. But if we *close* the vessel above, after a while we shall find a large number of molecules of water amongst the air molecules. From time to time, one of these vapor molecules comes flying down to the water and gets stuck again. So we see that what looks like a dead, uninteresting thing — a glass of water with a cover, that has been sitting there for perhaps twenty years — really contains a dynamic and interesting phenomenon which is going on all the time. To our eyes, our crude eyes, nothing is changing, but if we could see it a billion times magnified, we would see that from its own point of view it is always changing: molecules are leaving the surface, molecules are coming back.

Why do *we* see *no change*? Because just as many molecules are leaving as are coming back! In the long run "nothing happens." If we then take the top of the vessel off and blow the moist air away, replacing it with dry air, then the number of molecules leaving is just the same as it was before, because this depends on the jiggling of the water, but the number coming back is greatly reduced because there are so many fewer water molecules above the water. Therefore there are more going out than coming in, and the water evaporates. Hence, if you wish to evaporate water turn on the fan!

Here is something else: Which molecules leave? When a molecule leaves it is due to an accidental, extra accumulation of a little bit more than ordinary energy, which it needs if it is to break away from the attractions of its neighbors. Therefore, since those that leave have more energy than the average, the ones that are left have *less* average motion than they had before. So the liquid gradually *cools* if it evaporates. Of course, when a molecule of vapor comes from the air to the water below there is a sudden great attraction as the molecule approaches the surface. This speeds up the incoming molecule and results in generation of heat. So when they leave they take away heat; when they come back they generate heat. Of course when there is no net evaporation the result is nothing — the water is not changing temperature. If we blow on the water so as to maintain a continuous preponderance in the number evaporating, then the water is cooled. Hence, blow on soup to cool it!

Of course you should realize that the processes just described are more complicated than we have indicated. Not only does the water go into the air, but also, from time to time, one of the oxygen or nitrogen molecules will come in and "get lost" in the mass of water molecules, and work its way into the water. Thus the air dissolves in the water; oxygen and nitrogen molecules will work their way into the water and the water will contain air. If we suddenly take the air away from the vessel, then the air molecules will leave more rapidly than they come in, and in doing so will make bubbles. This is very bad for divers, as you may know.

Now we go on to another process. In Figure 6 we see, from an atomic point of view, a solid dissolving in water. If we put a crystal of salt in the water, what will happen? Salt is a solid, a crystal, an organized arrangement of "salt atoms." Figure 7 is an illustration of the three-dimensional structure of common salt, sodium chloride. Strictly speaking, the crystal is not made of atoms, but of what we call *ions*. An ion is an atom which either has a few extra electrons or has lost a few electrons. In a salt crystal we find chlorine ions (chlorine atoms with an extra electron) and sodium ions (sodium atoms with one electron missing). The ions all stick together by electrical attraction in the solid salt, but when we put them in the water we find, because of the attractions of the negative oxygen and positive hydrogen for the ions, that some of the ions jiggle loose. In Figure 6 we see a chlorine ion getting loose, and other atoms floating in the water in the form of ions. This picture was made with some care. Notice, for example, that the hydrogen ends of the water molecules are more likely to be near the chlorine ion, while near the sodium ion we are more likely to find the

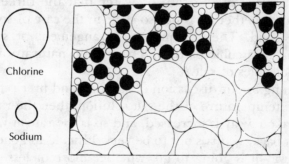

Chlorine

Sodium

Salt Dissolving in Water

Figure 6

11

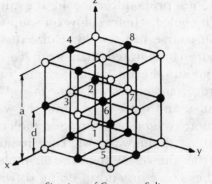

| Crystal | ● | ○ | a(A) |
|---------|-----|-----|------|
| Rocksalt | Na | Cl | 5.64 |
| Sylvine | K | Cl | 6.28 |
| | Ag | Cl | 5.54 |
| | Mg | O | 4.20 |
| Galena | Pb | S | 5.97 |
| | Pb | Se | 6.14 |
| | Pb | Te | 6.34 |

Structure of Common Salt

Nearest neighbor
distance d = a/2

Figure 7

oxygen end, because the sodium is positive and the oxygen end of the water is negative, and they attract electrically. Can we tell from this picture whether the salt is *dissolving in* water or *crystallizing out* of water? Of course we *cannot* tell, because while some of the atoms are leaving the crystal other atoms are rejoining it. The process is a *dynamic* one, just as in the case of evaporation, and it depends on whether there is more or less salt in the water than the amount needed for equilibrium. By equilibrium we mean that situation in which the rate at which atoms are leaving just matches the rate at which they are coming back. If there is almost no salt in the water, more atoms leave than return, and the salt dissolves. If, on the other hand, there are too many "salt atoms," more return than leave, and the salt is crystallizing.

In passing, we mention that the concept of a *molecule* of a substance is only approximate and exists only for a certain class of substances. It is clear in the case of water that the three atoms are actually stuck together. It is not so clear in the case of sodium chloride in the solid. There is just an arrangement of sodium and chlorine ions in a cubic pattern. There is no natural way to group them as "molecules of salt."

Returning to our discussion of solution and precipitation, if we increase the temperature of the salt solution, then the rate at which atoms are taken away is increased, and so is the rate at which atoms are brought back. It turns out to be very difficult, in general, to predict which way it is going to go, whether more or less of the solid will dissolve. Most substances dissolve more, but some substances dissolve less, as the temperature increases.

12

In all of the processes which have been described so far, the atoms and the ions have not changed partners, but of course there are circumstances in which the atoms do change combinations, forming new molecules. This is illustrated in Figure 8. A process in which the rearrangement of the atomic partners occurs is what we call a *chemical reaction*. The other processes so far described are called physical processes, but there is no sharp distinction between the two. (Nature does not care what we call it, she just keeps on doing it.) This figure is supposed to represent carbon burning in oxygen. In the case of oxygen, *two* oxygen atoms stick together very strongly. (Why do not *three* or even *four* stick together? That is one of the very peculiar characteristics of such atomic processes. Atoms are very special: they like certain particular partners, certain particular directions, and so on. It is the job of physics to analyze why each one wants what it wants. At any rate, two oxygen atoms form, saturated and happy, a molecule.)

The carbon atoms are supposed to be in a solid crystal (which could be graphite or diamond[1]). Now, for example, one of the oxygen molecules can come over to the carbon, and each atom can pick up a carbon atom and go flying off in a new combination — "carbon-oxygen" — which is a molecule of the gas called carbon monoxide. It is given the chemical name CO. It is very simple: the letters "CO" are practically a picture of that molecule. But carbon attracts oxygen much more than oxygen attracts oxygen or carbon attracts carbon. Therefore in this process the oxygen may arrive with only a little energy, but the oxygen and carbon will snap together with a tremendous vengeance and commotion, and every-

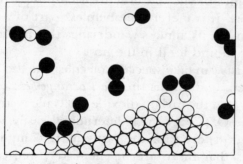

Carbon Burning in Oxygen

Figure 8

1. One *can* burn a diamond in air.

13

thing near them will pick up the energy. A large amount of motion energy, kinetic energy, is thus generated. This of course is *burning;* we are getting *heat* from the combination of oxygen and carbon. The heat is ordinarily in the form of the molecular motion of the hot gas, but in certain circumstances it can be so enormous that it generates *light.* That is how one gets *flames.*

In addition, the carbon monoxide is not quite satisfied. It is possible for it to attach another oxygen, so that we might have a much more complicated reaction in which the oxygen is combining with the carbon, while at the same time there happens to be a collision with a carbon monoxide molecule. One oxygen atom could attach itself to the CO and ultimately form a molecule, composed of one carbon and two oxygens, which is designated $CO_2$ and called carbon dioxide. If we burn the carbon with very little oxygen in a very rapid reaction (for example, in an automobile engine, where the explosion is so fast that there is not time for it to make carbon dioxide) a considerable amount of carbon monoxide is formed. In many such rearrangements, a very large amount of energy is released, forming explosions, flames, etc., depending on the reactions. Chemists have studied these arrangements of the atoms, and found that every substance is some type of *arrangement of atoms.*

To illustrate this idea, let us consider another example. If we go into a field of small violets, we know what "that smell" is. It is some kind of *molecule,* or arrangement of atoms, that has worked its way into our noses. First of all, *how* did it work its way in? That is rather easy. If the smell is some kind of molecule in the air, jiggling around and being knocked every which way, it might have *accidentally* worked its way into the nose. Certainly it has no particular desire to get into our nose. It is merely one helpless part of a jostling crowd of molecules, and in its aimless wanderings this particular chunk of matter happens to find itself in the nose.

Now chemists can take special molecules like the odor of violets, and analyze them and tell us the *exact arrangement* of the atoms in space. We know that the carbon dioxide molecule is straight and symmetrical: O—C—O. (That can be determined easily, too, by physical methods.) However, even for the vastly more complicated arrangements of atoms that there are in chemistry, one can, by a long, remarkable process of detective work, find the arrangements of the atoms. Figure 9 is a picture of the air in the neighborhood of a violet; again we find nitrogen and oxygen in the air, and water vapor. (Why is there water vapor? Because the violet is *wet.* All plants transpire.)

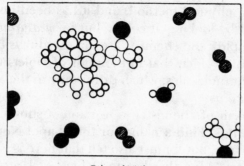

Odor of Violets

Figure 9

However, we also see a "monster" composed of carbon atoms, hydrogen atoms, and oxygen atoms, which have picked a certain particular pattern in which to be arranged. It is a much more complicated arrangement than that of carbon dioxide; in fact, it is an enormously complicated arrangement. Unfortunately, we cannot picture all that is really known about it chemically, because the precise arrangement of all the atoms is actually known in three dimensions, while our picture is in only two dimensions. The six carbons which form a ring do not form a flat ring, but a kind of "puckered" ring. All of the angles and distances are known. So a chemical *formula* is merely a picture of such a molecule. When the chemist writes such a thing on the blackboard, he is trying to "draw," roughly speaking, in two dimensions. For example, we see a "ring" of six carbons, and a "chain" of carbons hanging on the end, with an oxygen second from the end, three hydrogens tied to that carbon, two carbons and three hydrogens sticking up here, etc.

How does the chemist find what the arrangement is? He mixes bottles full of stuff together, and if it turns red, it tells him that it consists of one hydrogen and two carbons tied on here; if it turns blue, on the other hand, that is not the way it is at all. This is one of the most fantastic pieces of detective work that has ever been done — organic chemistry. To discover the arrangement of the atoms in these enormously complicated arrays the chemist looks at what happens when he mixes two different substances together. The physicist could never quite believe that the chemist knew what he was talking about when he described the arrangement of the atoms. For about twenty years it has been possible, in some cases, to look at such molecules (not quite as complicated as this one, but some which contain

15

parts of it) by a physical method, and it has been possible to locate every atom, not by looking at colors, but by *measuring where they are.* And lo and behold!, the chemists are almost always correct.

It turns out, in fact, that in the odor of violets there are three slightly different molecules, which differ only in the arrangement of the hydrogen atoms.

One problem of chemistry is to name a substance, so that we will know what it is. Find a name for this shape! Not only must the name tell the shape, but it must also tell that here is an oxygen atom, there a hydrogen — exactly what and where each atom is. So we can appreciate that the chemical names must be complex in order to be complete. You see that the name of this thing [Figure 10] in the more complete form that will tell you the structure of it is 4-(2, 2, 3, 6 tetramethyl-5-cyclohexanyl)-3-buten-2-one, and that tells you that this is the arrangement. We can appreciate the difficulties that the chemists have, and also appreciate the reason for such long names. It is not that they wish to be obscure, but they have an extremely difficult problem in trying to describe the molecules in words!

How do we *know* that there are atoms? By one of the tricks mentioned earlier: we make the *hypothesis* that there are atoms, and one after the other results come out the way we predict, as they ought to if things *are* made of atoms. There is also somewhat more direct evidence, a good example of which is the following: the atoms are so small that you cannot see them with a light microscope — in fact, not even with an *electron* microscope. (With a light microscope you can only see things which are much bigger.) Now if the atoms are always in motion, say in water, and we put a big ball of something in the water, a ball much bigger than the atoms, the ball will jiggle around — much as in a push ball game, where a great big ball is

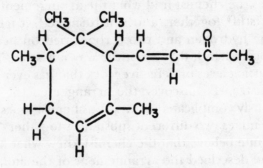

The α-irone Molecule

Figure 10

pushed around by a lot of people. The people are pushing in various directions, and the ball moves around the field in an irregular fashion. So, in the same way, the "large ball" will move because of the inequalities of the collisions on one side to the other, from one moment to the next. Therefore, if we look at very tiny particles (colloids) in water through an excellent microscope, we see a perpetual jiggling of the particles, which is the result of the bombardment of the atoms. This is called the *Brownian motion.*

We can see further evidence for atoms in the structure of crystals. In many cases the structures deduced by x-ray analysis agree in their spatial "shapes" with the forms actually exhibited by crystals as they occur in nature. The angles between the various "faces" of a crystal agree, within seconds of arc, with angles deduced on the assumption that a crystal is made of many "layers'" of atoms.

*Everything is made of atoms.* That is the key hypothesis. The most important hypothesis in all of biology, for example, is that *everything that animals do, atoms do.* In other words, *there is nothing that living things do that cannot be understood from the point of view that they are made of atoms acting according to the laws of physics.* This was not known from the beginning: it took some experimenting and theorizing to suggest this hypothesis, but now it is accepted, and it is the most useful theory for producing new ideas in the field of biology.

If a piece of steel or a piece of salt, consisting of atoms one next to the other, can have such interesting properties; if water — which is nothing but these little blobs, mile upon mile of the same thing over the earth — can form waves and foam, and make rushing noises and strange patterns as it runs over cement; if all of this, all the life of a stream of water, can be nothing but a pile of atoms, *how much more is possible?* If instead of arranging the atoms in some definite pattern, again and again repeated, on and on, or even forming little lumps of complexity like the odor of violets, we make an arrangement which is *always different* from place to place, with different kinds of atoms arranged in many ways, continually changing, not repeating, how much more marvelously is it possible that this thing might behave? Is it possible that that "thing" walking back and forth in front of you, talking to you, is a great glob of these atoms in a very complex arrangement, such that the sheer complexity of it staggers the imagination as to what it can do? When we say we are a pile of atoms, we do not mean we are *merely* a pile of atoms, because a pile of atoms which is not repeated from one to the other might well have the possibilities which you see before you in the mirror.

KENNETH W. FORD

■

Kenneth W. Ford (b. 1926), a nuclear physicist and professor at
Brandeis University, guides us through the atomic landscape in
terms of fundamental units of space, time, and energy. This
introduction originally appeared in his book *The World of Elementary Particles*, published in 1963.

# The Large and the Small

IT IS EASY to talk about the "incredibly short" lifetime of an elementary particle or about the "fantastically small" size of an atomic nucleus. It is not so easy to visualize these things. On the submicroscopic
frontier of science (as well as on the cosmological frontier) man has
proceeded so far away from the familiar scale of the world encom-
passed by his senses, that he must make a real effort of the imagination to relate these new frontiers to the ordinary world. But the
reward of being able to think pictorially over the whole panorama,
from infinitesimal to enormous, adequately repays the effort.

In order to describe nature, the scientist needs a number of
concepts which are so well defined that they are not merely descriptive ideas but are measurable quantities. The simplest of these are
the ideas of size (length measurement) and duration (time measurement). Each property of an elementary particle — its mass, electric
charge, energy, spin, angular momentum — is such a measurable
concept. Roughly, a measurable, or quantitative, concept is anything
to which both a name and a number can be attached, for example,
6 inches, 90 miles per hour, 30 minutes, 110 volts. (Soldiers, sailors,
and prisoners, although they have both a name and a number, do
not qualify. The number must be an indication of size or quantity,
not a serial number.)

For each of the concepts used to describe nature, a unit of measurement is introduced to allow meaningful comparisons of mea-

surements in different places or at different times. This is a long way of saying something we all know from daily life. Our height is expressed in terms of feet and inches, our weight in pounds, our age in years. Each quantity needs a unit in terms of which it can be expressed. Unfortunately there is no international agreement about units (although the situation is much less chaotic within science than in the everyday world) but this circumstance of a multiplicity of units serves all the better to demonstrate the need for units. An American, asked his weight, might answer, "I weigh 154." An Englishman of the same size weighs 11. A Frenchman of similar build might claim to weigh 70. Their weights are all the same, of course, but the American is reckoning in pounds, the Englishman in stone, and the Frenchman in kilograms. A statement of the number without the unit is quite meaningless, although in a sufficiently provincial gathering the unit may, by common agreement, be understood and therefore not stated.

There are in science some "pure" numbers, also called dimensionless numbers. These numbers in fact hold a special fascination for scientists just because they are independent of any set of units. If we say, "Table 1 lists twelve kinds of particles," the "twelve" is a pure number, the result of counting, and refers to no particular unit. But in the main the quantities needed to describe the particles do require units.

The normal scientific units, such as the centimeter of length (frequently abbreviated cm) and the second of time, are "man-sized" units adopted for convenience in our macroscopic world. The centimeter, about half an inch, is roughly the thickness of a man's finger; in one second a man can blink a few times, say "one thousand and one," or stroll about a hundred centimeters. These units are handy and easy to visualize. But in the worlds of the large and the small they become ridiculously inappropriate. The distance from earth to sun is an enormous number of centimeters; the size of a hydrogen atom, a tiny fraction of one centimeter. The age of the earth is a vast number of seconds, the lifetime of a pion an imperceptible part of a second. Journalists are fond of writing out numbers from the cosmological or submicroscopic world in their full glory, with a string of zeros before or after the decimal point: The pion lives 0.000000026 second; the number of hydrogen atoms in a quart of water is 60,000,000,000,000,000,000,000,000. These numbers are impressive, but a bit confusing and not very instructive.

Scientists have done two things about this situation. First, and most simply, they have replaced the lengthy newspaper notation for large and small numbers by what is called the exponential notation. In this notation one hundred is written $10^2$, one thousand is $10^3$, one million is $10^6$, three million is $3 \times 10^6$, forty billion dollars . . . is $4 \times 10^{10}$ dollars. The long number in the preceding paragraph is $6 \times 10^{25}$. The exponent, or power of ten, may be looked upon as an instruction to shift the decimal point so many places to the right. Two million is $2 \times 10^6$, that is, two followed by six zeros, 2,000,000. This notation is so simple and convenient that everyone should know it.

Multiplication is carried out by *adding* exponents. One billion is one thousand millions: $10^6$ multiplied by $10^3$ is $10^9$. . . . If light travels at $3 \times 10^{10}$ cm per second, and there are $3 \times 10^7$ seconds in one year, then one light-year, the distance traveled by light in one year, is $3 \times 10^{10}$ multiplied by $3 \times 10^7$, that is, $9 \times 10^{17}$ cm. This is a large number; it is about 300 million times around the earth. Returning to dollars, if $9 \times 10^{17}$ dollars were divided equally among the world's population, every man, woman, and child would control a fortune of nearly half a billion dollars.

The exponential notation is also used for small numbers. One tenth is $10^{-1}$; one millionth is $10^{-6}$; three billionths is $3 \times 10^{-9}$. The rule for adding exponents still holds in multiplication. For example, $10^{-3} \times 10^9$ equals $10^6$, recalling that adding a negative number is the same as subtracting a positive number. In words, one thousandth of one billion is one million. The lifetime of a muon is $2 \times 10^{-6}$ seconds, that is, two millionths of a second. The size of an atom is about $10^{-8}$ cm, or one one-hundred millionth of a centimeter.

The second approach to dealing with large and small quantities is to invent new units more appropriate to the domain being considered. Thus, for cosmological purposes, the light-year (which we calculated above to be $9 \times 10^{17}$ cm) is a convenient unit of length. For dealing with atoms, the Angstrom unit, equal to $10^{-8}$ cm, is frequently used; for nuclear and particle phenomena the fermi ($10^{-13}$ cm), one hundred thousand times smaller, is a more suitable unit. Earlier we followed this approach, adopting the spin of the photon as the unit spin, the charge of the proton as the unit charge, and the mass of the electron as the unit mass. It is, of course, still necessary to know how to convert these units back to the conventional units, just as it is necessary to know how to convert among centimeters, inches, feet, yards, miles, and light-years.

The table summarizes some of the quantities used to characterize the world of elementary particles.

## TABLE OF MEASUREMENTS

| Physical Quantity | Common Unit in the Large-Scale World | Scale of the Submicroscopic World |
|---|---|---|
| Length | Centimeter (about half an inch) | Size of atom about $10^{-8}$ cm = 1 Angstrom unit<br><br>Size of particle about $10^{-13}$ cm = 1 fermi |
| Speed | Centimeter per second (speed of a snail) | Speed of light = $3 \times 10^{10}$ cm/sec |
| Time | Second (the swing of a pendulum) | Natural time unit of particle about $10^{-23}$ sec<br><br>Typical lifetime of a "long-lived" particle about $10^{-10}$ sec |
| Mass | Gram (mass of one cubic centimeter of water) | Mass of electron = $9 \times 10^{-28}$ gm |
| Energy | Erg (energy of a lazy bug)<br>Food calorie (40 billion ergs) | 1 ev (electron volt) = $1.6 \times 10^{-12}$ erg<br><br>Air molecule has about one fortieth ev<br><br>Proton in largest accelerator about 30 billion ev |
| Charge | Coulomb (lights a lamp for one second) | Electron charge = $1.6 \times 10^{-19}$ coulomb |
| Spin | Gm × cm × cm/sec (grasshopper turning around) | Spin of photon = $h = 10^{-27}$ gm × cm × cm/sec |

## Length

One of the best ways to try to visualize the very great or the very small is by analogy. For example, to picture the nucleus, whose size is about $10^{-4}$ to $10^{-5}$ of the size of an atom, one may imagine the atom expanded to, say, 10,000 feet ($10^4$ feet) or nearly two miles. This is about the length of a runway at a large air terminal such as

New York International Airport. A fraction $10^{-4}$ of this is one foot, or about the diameter of a basketball. A fraction $10^{-5}$ is ten times smaller, or about the diameter of a golf ball.

A golf ball in the middle of New York International Airport is about as lonely as the proton at the center of a hydrogen atom. The basketball would correspond to a heavy nucleus such as uranium. On this scale, one centimeter would be expanded to $2 \times 10^8$ (200 million) miles, or about twice the distance from earth to sun.

To arrive at the number of atoms in a cubic centimeter of water (a few drops), first cover the earth with airports, one against the other. Then go up a mile or so and build another solid layer of airports. Do this 100 million times. The last layer will have reached out to the sun and will contain some $10^{16}$ airports (ten million billion). The number of atoms in a few drops of water will be the number of airports filling up this substantial part of the solar system. If the airport-construction rate were *one million* each second, the job could just have been finished in the known lifetime of the universe (something over 10 billion years).

The size of a proton is about $10^{-13}$ cm; this distance has been given the name fermi, in honor of Enrico Fermi, who pioneered studies of the nuclear particles in the 1930's. The smallest distance probed in any experiment so far conducted is about a tenth of a fermi, or $10^{-14}$ cm. Most of the elementary particles have a size of about 1 fermi, but a few, such as the electron, may be much smaller.

It is of some interest to compare the astronomical scale of size with the submicroscopic. The known part of the universe extends out to about $10^{10}$ light-years, or $10^{28}$ cm. Man, a creature about $10^2$ cm high, is thus smaller than the universe by a factor of $10^{26}$, but larger than a proton by a factor of "only" $10^{15}$. The universe is about as much larger than the whole solar system as man is larger than the proton. From the smallest known distance ($10^{-14}$ cm) to the largest ($10^{28}$ cm), man has spanned a factor of $10^{42}$ in size. The number $10^{42}$ is so enormous that not even analogy is of much help in comprehending it. Suppose we wanted to let the population explosion run its course until there were $10^{42}$ people. The earth itself can only accommodate a bit over $10^{15}$ people standing shoulder to shoulder. A million earths similarly packed could handle $10^{21}$. To reach our goal of $10^{42}$ people, we would have to make each couple on these million earths personally responsible for finding and peopling to the limit a million new earths. This would require $10^{27}$ earths altogether, a substantial number. There are only about $10^{23}$ stars in the universe.

If every star had ten planets, and every planet was packed with people like sardines in a can, the universe would still not hold as many as $10^{42}$ people.

Thinking up ways to picture large and small numbers can be a fascinating game. It is recommended that the reader try it himself.

### Speed

A snail in a hurry can travel at a speed of about 1 centimeter per second (or, in briefer scientific notation, 1 cm/sec). A man strolls at about $10^2$ cm/sec, drives a car at $3 \times 10^3$ cm/sec, and rides in a jet plane at near the speed of sound, which is $3 \times 10^4$ cm/sec (about 700 miles per hour).

In distance, man has encountered no limit, large or small. But, in speed, nature seems to have established a very definite limit, the speed of light, $3 \times 10^{10}$ cm/sec. This is exactly a million times the speed of sound in air, an easy ratio to remember. Even an astronaut falls short of the speed of light by a factor of forty thousand. He needs an hour and a half to get once around the earth, while a photon (if it could be caused to travel in a curved path) could complete the trip in a tenth of a second. Still, man is not so far removed from nature's top speed as he is from the frontiers in space and time.

Atoms and molecules, executing their continual restless motion in solids, liquids, and gases on earth, move sluggishly about at only one to ten times the speed of sound in air, a factor $10^5$ to $10^6$ short of the speed of light. But for elementary particles, speeds near the speed of light are common. The photon and neutrinos have no choice, of course, and travel at exactly the speed of light (as does the graviton). In all of the larger modern accelerators, particles with mass — usually electrons or protons — are pushed very near to the speed of light, and the unstable particles formed in nuclear collisions also frequently emerge with speeds near the speed of light.

Astronauts in science fiction frequently shift into "superdrive" and scoot about the galaxy above the speed of light. Is there any chance that this will become reality? It is extremely unlikely, and for a very simple reason. The lighter an object is, the more easily it can be accelerated. Freight trains lumber slowly up to speed, automobiles more quickly, and protons in a cyclotron still more quickly. A particle with no mass whatever should be the easiest to accelerate; indeed, the massless photon jumps instantaneously to the speed of light when it is created. But not beyond. If anything at all were able to go faster

23

than light, then light itself, being composed of massless photons, should go faster.

### Time

In dealing with the time scale of elementary particles, it is essential first to get rid of preconceived notions about what is a "short" time or a "long" time. A millionth of a second certainly seems to qualify as a short time, yet for an elementary particle it is an exceedingly long time. On the other hand, a million years is the mere blink of an eye for the stately cosmological march of events.

If an automobile had its last bolt tightened at the end of the assembly line, then was driven a hundred feet away, where it promptly collapsed in a heap, we should say that it had had a short lifetime. If it covered some 20 billion miles before collapsing, we should say that it had had an amazingly long lifetime, indeed, that it was the most extraordinarily long-lived car ever built. Let us translate these distances to the elementary-particle world. The size of a particle is about $10^{-13}$ cm; it travels typically at about $10^{10}$ cm/sec. Thus, to cover a distance ten times its own size (comparable to the car moving 100 feet), the particle needs only $10^{-22}$ sec. The duration, $10^{-23}$ sec, is a kind of natural time unit for a particle. Yet almost every particle lives at least $10^{-10}$ sec, an enormously long time compared to $10^{-23}$ sec. In $10^{-10}$ sec the particle can cover a whole centimeter, more than a million million times its own size. A particle moving one centimeter is comparable to a car going 20 billion miles. Any particle that can move one centimeter away from its birthplace before dying deserves to be called long-lived. The pion and muon, with lifetimes of $10^{-8}$ sec and $10^{-6}$ sec, respectively, can move much farther even than 1 cm. The neutron is a strange special case. Its lifetime of 17 minutes is practically infinite for the elementary-particle world.

The new particles, or "resonances," now being discovered have lifetimes of $10^{-20}$ sec or less. They are indeed short lived and perhaps they do not even deserve to be called particles. They are like the car which collapses before it gets out of the factory gate. The manufacturer might be tempted to say, "That was no car; it was just an unstable phenomenon with a transitory existence" (for which the physicist uses the word "resonance").

Since the shortest distance probed experimentally is about $10^{-14}$ cm, it is fair to say that the shortest known time interval is $10^{-24}$ sec (although direct time measurements are still very far from reaching this short an interval). The longest known time is the "lifetime of the universe," that is, the apparent duration of the expansion of the universe, which is also a few times the age of the earth. This amounts to about $10^{18}$ sec (30 billion years). The ratio of these is $10^{42}$, the same enormous number as the ratio of the largest and smallest distances. This is not a coincidence. The outermost reaches of the universe are moving away from us at a speed near the speed of light and the particles used to probe the submicroscopic world are moving at speeds near the speed of light. On both the cosmological and submicroscopic frontiers, the speed of light appears to be the natural link between distance and time measurements.

### Mass

The technical definition of mass is rather complicated and we have no need to go into it in detail. It will be adequate for our present purposes to think of the mass of an object as the amount of material in the object. This is actually a very unscientific definition, but it provides a way of thinking about mass. The idea of mass is confused by the circumstance that the pull of gravity on an object is proportional to the mass in the object. A parcel of large mass on a postal scale is pulled down more strongly than a parcel of small mass and consequently causes the scale to show a higher reading. We say that the more massive object "weighs" more; that is, it is pulled more strongly by the earth, and in fact in exact proportion to its mass.

To understand mass a bit better, imagine some astronauts floating freely inside the cabin of their space ship in a weightless condition. If two of them join hands, then push and let go, they will float apart — in a particular way. The larger man will move away a bit more slowly than the smaller man; we attribute this to his greater mass. The important fact about mass is that it is a measure of resistance to a change of motion, a property usually called inertia. If a single astronaut floating in mid cabin tosses a baseball, he will recoil and drift backward slowly as the ball moves swiftly off in the opposite direction. If the ball leaves the point of separation five hundred times faster than the astronaut, it is because its mass is only one five-hundredth as great as the astronaut's, and it has, therefore, five hundred times less resistance to being set into motion. Anyone who has fired a gun knows about recoil. If a hunter had no more mass than his bullet, he and the bullet would be accelerated to equal speed. Because he has a great deal more mass — hence, more inertia — than the bullet, he has more resistance to being set into motion and is moved much more slowly than the bullet.

On the human scale, the elementary particles have exceedingly small masses. It is therefore relatively easy to set them into high-speed motion and they normally fly about near the speed of light. The massless particles have, in fact, no resistance whatever to being speeded up, and upon being created, move instantly at the speed of light.

The electron is the lightest particle with nonvanishing mass and it is therefore usual to adopt the mass of the electron as a convenient unit in the submicroscopic world. The heaviest particle, the xi, is more than 2,000 times heavier than the electron, yet a million million xi particles would not be able to tip the world's most sensitive balance.

The scientific unit of mass is the gram, about the mass of an average vitamin pill. A quart of water has a mass of about 900 grams. The transatlantic air traveler is permitted a mass of luggage of 20,000 grams (tourist class), or 20 kilograms, which is about 44 pounds. The electron's mass is $9 \times 10^{-28}$ gm, that of the "heavy" xi particle about $2 \times 10^{-24}$ gm.

To end the discussion of mass on a cosmic note we ask, what is the mass of the universe? This is certainly not well known, but a rough estimate can be made. There are about $10^{23}$ stars in the universe (this number is roughly the same as the number of molecules in a gram of water). An average star weighs about $10^{35}$ gm, making the total mass some $10^{58}$ gm. Each gram of matter contains about $10^{24}$ protons, so that what we know of the universe contains (very roughly) $10^{82}$ protons. (The known part of the universe presumably contains a like number of electrons, about $10^{82}$. It contains fewer neutrinos, perhaps about $10^{79}$, but uncountably many photons and gravitons. The unstable particles, on the other hand, are far less numerous than protons.)

### Energy

The most remarkable fact about energy is its diversity. Like a clever actor who can assume many guises, energy appears in a variety of forms, and can shift from one role to another. Because of this richness of form, energy appears in nearly every part of the description of nature and can make a good claim to be the most important single concept in science.

One common form of energy is energy of motion, kinetic

energy, which is a measure of how much force over how great a distance is required to set an object into motion or to bring it to rest. The faster a particle moves, the more kinetic energy it possesses. This is not far removed from our everyday use of the word; for example, in speaking of an "energetic" person, we refer to someone in constant motion, capable of doing a great deal of work. Work itself has a technical definition in physics, force multiplied by distance; and energy is the capacity for doing work. Every form of energy, if suitably transformed, can do work. Heat energy, for example, is the accumulated kinetic energy of the restless random motion of molecules, and a steam engine is a device which derives mechanical work from heat energy.

The great significance of energy springs in part from the variety of its manifestations, in part from the fact that it is conserved — that the total amount of energy remains always the same, with the loss of one kind of energy being compensated by the gain of another kind of energy. Trace, for example, the flow of energy from sun to earth to man; this illustrates both the variety and the conservation of energy. When protons in the sun unite to form helium nuclei, nuclear energy is released. This energy may go first to the kinetic energy of motion of nuclei, which contributes to the sun's heat energy. Some of the energy is then carried away from the sun by photons, the particles which are bundles of electromagnetic energy. The energy content of the photons may be transformed, by the complicated and not yet fully understood process of photosynthesis, into stored chemical energy in plants. Either by eating plants or by eating animals which ate plants, man acquires this solar energy, which is then made available to power his brain and muscles and to keep him warm.

That mass is one of the forms of energy was first realized at the beginning of this century. The energy of mass and the energy of motion are the two forms of energy which dominate the elementary-particle world. Mass energy can be thought of as the "energy of being," matter possessing energy just by virtue of existing. A material particle is nothing more than a highly concentrated and localized bundle of energy. The amount of concentrated energy for a motionless particle is proportional to its mass. If the particle is moving it has still more energy, its kinetic energy. A massless particle such as a photon has only energy of motion (kinetic energy) and no energy of being (mass).

Einstein's most famous equation, $E = mc^2$, provides the relation between the mass, $m$, of a particle and its intrinsic energy, or energy of being, $E$. The quantity $c$ in this formula is the speed of light; $c^2$ is the notation for $c \times c$, the square of the speed of light. The important statement Einstein's equation makes is that energy is proportional to mass. Twice as much mass means twice as much intrinsic energy; no mass means no intrinsic energy. The factor $c^2$ is called a constant of proportionality and does the job of converting from the units in which mass is expressed to the units in which energy is expressed. By analogy, consider the equation giving the cost of filling your car with gasoline, $C = GP$. The cost, $C$, is equal to the number of gallons, $G$, multiplied by the price per gallon, $P$. The cost is proportional to the number of gallons and $P$ is the constant of proportionality which converts the number of gallons to a total cost. In a similar way, $c^2$ is a price. It is energy per unit mass, the price which must be paid in energy to create a unit of mass.

In an isolated nuclear collision, or particle reaction, the total energy remains unchanged. In fact, practically every event in the submicroscopic world *is* isolated, for the distance over which the particles interact with one another is generally exceedingly small compared with the distance between neighboring atoms, about $10^{-8}$ cm. The individual event occurs with the particles effectively unaware of anything else in the universe. In a collision, or a reaction, or a decay process, there are two ways in which energy may be supplied. A particle may be slowed down, thus giving up some of its kinetic energy; or a particle with mass may be destroyed, thus giving up its intrinsic energy. Analogously, energy may be taken up in two ways. A particle may be speeded up or a new particle may be created. The rule of energy conservation can be stated as follows: The total energy supplied must be equal to the total energy taken up; that is, the energy loss must equal the energy gain.

Consider, for example, the decay of a pion at rest. Since it is motionless its only energy is the intrinsic energy associated with its mass. It decays spontaneously into a muon and a neutrino. The vanishing of the pion makes its mass energy available. Some of this is used to create the mass of the lighter muon. The rest is supplied as energy of motion, and the muon and neutrino fly off at high speed with just enough kinetic energy to make up the difference. This example makes clear why a particle can decay spontaneously only into other particles lighter than itself. In a high-energy collision, on the other hand, such as those which occur near a particle accelerator, the projectile particle may be slowed down and some of its kinetic energy made available to create the mass of new particles. This is the way in which antiprotons and the various unstable particles are created in the modern high-energy laboratory.

The average citizen who pays an electric bill and watches his weight is probably aware of at least two of the energy units in common use, the kilowatt-hour and the food calorie. Ten 100-watt light bulbs require a kilowatt; if they all remain lit for one hour, one kilowatt-hour of energy is used up. Reasonably enough, it is electrical energy for which one pays. An overweight person might remark that it is also calories for which one pays. One food calorie is the energy released by a good-sized pinch of sugar when it is oxidized. A thousand or more calories of energy are needed each day to keep the human machine running efficiently.

A more usual scientific unit of energy is called the erg. It is actually quite small, being the energy of motion of a 2-gram bug crawling at 1 cm/sec. A food calorie, for example, is some 40 billion ($4 \times 10^{10}$) ergs, and a kilowatt-hour is nearly a thousand times larger still, being $3.6 \times 10^{13}$ ergs. Nevertheless, the erg is considerably larger than the energies usually encountered in the submicroscopic world, where another energy unit, the electron volt, has been introduced (the last energy unit we shall mention!). The electron volt is about one millionth of one millionth of an erg ($1.6 \times 10^{-12}$ erg, to be exact). (The electron volt has nothing special to do with the electron. It is the energy given to *any* charged elementary particle [since all carry the same size of charge] by an electrical potential of one volt.)

The average kinetic energy of the incessant random motion of molecules and atoms provides a kind of a baseline for energy comparisons in the world of the very small. A single molecule at normal temperatures moves about with an average kinetic energy of about one fortieth of an electron volt. On the much hotter surface of the sun this thermal motion averages about half an electron volt. But in accelerators man can readily push the energies of particles to much higher values. Early cyclotrons accelerated protons to over one million electron volts (1 Mev). Other accelerators completed after World War II pushed the energy of protons toward one billion electron volts (1 Bev). The Bevatron in Berkeley (whose name obviously has something to do with its energy) accelerates protons to 6 Bev. The largest accelerators operational in 1962 are in Brookhaven, New York, and Geneva, Switzerland; as projectiles, both use protons that reach an energy of over 30 Bev. Some of the cosmic-ray particles from outer space arrive with energies even much greater than this. How these particles are accelerated to such enormous energies remains a mystery.

Since mass and energy are equivalent, how much energy is needed to create the mass of an elementary particle? The creation of the lightest particle, the electron, requires 500,000 electron volts. The older accelerators were capable of supplying sufficient energy to create positrons and electrons. But the proton mass is equivalent to nearly 1 billion electron volts. The production of antiprotons had to await the construction of the 6 Bev Bevatron. The new 30 Bev machines can create all of the known particles with energy to spare; it is quite possible that new and still heavier particles may be discovered with the help of these accelerators.

Mass is a very potent and highly concentrated form of energy. The power of the atomic bomb, in which less than a tenth of one per cent of the mass is converted to energy, serves as an eloquent reminder of this fact. To understand this in terms of the energy equivalents given above, recall the typical energies of thermal motion. Even on the white-hot surface of the sun, a proton has a kinetic energy of less than 1 ev, a mere nothing compared with the billion ev locked up in its mass. The mass of a single proton converted to energy on the earth is sufficient to heat up a billion atoms to a temperature higher than that of the sun's surface.

## Charge

Electric charge is best described as being like French perfume. It is that certain something worn by particles which makes them attractive — specifically, attractive to the opposite kind of particle. Particles that do not have it are called neutral and have no influence (at least no electrical influence) on other particles. Charge can lead to pairing of particles; the hydrogen atom, for example, consists of a proton and electron held together by electrical attraction. More energetic particles are not held together by the electric force; it merely causes them to deviate slightly from a straight course.

The charge carried by a particle may be positive or negative. Two charges of the same kind repel each other, two of opposite kinds attract. The protons within a nucleus, for example, repel each other, but the overpowering nuclear force nevertheless holds the nucleus together. Eventually, however, for very heavy nuclei, the electrical repulsion becomes more than the nuclear forces can counteract, and the nucleus flies apart. It is for this reason that no nuclei heavier than uranium exist in nature.

Which charge is called positive and which negative is entirely arbitrary, and is the result of historical accident. The definition that led to electrons being negative and protons positive probably stems from a guess made by Benjamin Franklin about the middle of the eighteenth century. His choice of nomenclature was based on the erroneous supposition that it is positive electricity which flows most readily from one object to another. We now know that it is the negative electrons which are mobile and account for the flow of electric current in metals.

Charge is still very mysterious to the physicist. He understands it little better than he understands the workings of French perfume. If we think of a particle as a small structure spread out over a tiny region of space, it is logical to think of its charge as being spread out too. But if this is so, why do the various bits of charge making up the particle not repel each other and cause the whole particle to disintegrate and fly apart? No one knows a satisfactory answer to this question. It is also perplexing that all particles have exactly the same magnitude of charge. If the charge of the electron is called $-e$ (it is negative) then the charge of every other particle is either $-e$ or $+e$ or zero. No other possibilities are realized in nature. We have no understanding of this fact, and also have no clue as to why the electron has the charge it does have, rather than some other value. The true nature of charge and the reason it comes only in lumps of a certain size are among the most important problems in elementary-particle physics.

The fact that we do not understand charge at a fundamental level has been no hindrance to making extensive use of charge for practical purposes. Electrons can be detached from atoms rather readily — at least in certain metals called conductors — and by means of electrical forces they can be pushed and pulled through wires or sent flying through empty space, as within a vacuum tube or a television tube. Almost all of the fine control and all of the communications in the world are effected by electrons in electronic circuits. A great part of the world's heavy labor is also done by electrons turning motors or supplying heat.

The fact that there is a voltage across the two holes in a household electrical outlet means that an electric force is standing ready to do some work. If an electric light is plugged in, the electrons are sent flying from one prong of the plug, through the light (where they expend some energy which appears as light and heat), then back to the other prong. The number of electrons involved in such a flow is enormous. In a typical household light bulb, about $10^{19}$ electrons flow through the filament each second. In heavy machinery or in the high-tension lines connecting cities, the number is far greater. Even through the tiniest and most delicate electronic circuit, many billions of electrons flow each second.

If a comb is passed through dry hair, perhaps a million million electrons leave the hair and stick on the comb. Nevertheless, the comb is almost neutral. For every extra electron it has acquired the comb has a million million neutral atoms. It is fortunate for us that the objects in our macroscopic world remain always almost neutral. If the comb acquired anything close to an electrification of one extra electron per atom, the consequence would be dire. Either there would be a powerful and deadly bolt of lightning from comb to man as the charge was neutralized or the enormous force of electrical attraction would draw the comb back so violently that it would be a dangerous weapon.

Intrinsically, the electrical force in nature is a great deal stronger than the gravitational force and in the submicroscopic world the gravitational force is usually ignored altogether. The matter in our macroscopic world, however, exists in such a fine state of electrical balance that gravity has a chance to make itself felt. In every object in our world, the number of positive charges is almost precisely equal to the number of negative charges. The effects nearly cancel, and what we regard as marked electrical effects arise from an exceedingly tiny imbalance in positive and negative charge. If a big imbalance were ever realized (there is no chance of this) the disastrous result would make the force of gravity appear to be truly inconsequential.

One of the common scientific units of charge is the coulomb, named for the French scientist Charles A. Coulomb, who discovered the exact law of electrical force in 1785. One coulomb is roughly the amount of charge that moves through a 100-watt light bulb in a second, or through an electric iron in a fifth of a second. (The coulomb is not commonly encountered in the house, but a very closely related unit, the ampere, is. On a household fuse one might see stamped "15 amp." The ampere, or amp for short, is one coulomb per second. If more than 15 coulombs of charge flow through this fuse each second, a wire inside it will melt and stop the flow.) The basic unit of charge in the world of particles is the electron charge,

1.6 × 10⁻¹⁹ coulombs. This is less than one billionth of one billionth of one coulomb.

## *Spin*

Rotational motion seems to be a characteristic of most of the structures in the universe, from neutrinos to galaxies. Our earth rotates on its axis once a day, and rotates once around the sun in a year. The sun itself turns on its axis once in 26 days and, along with the other stars of our galaxy, travels once around the galaxy in 230 million years. It is not known whether larger structures, such as clusters of galaxies, have an over-all rotational motion, but it would be surprising if they do not.

Going down the scale, the atoms that compose molecules can rotate about each other, and do, although at a rate which varies from time to time as the molecule is disturbed by its neighbors. Within the atom, electrons rotate about the nucleus at speeds of from one per cent up to more than ten per cent of the speed of light, thereby giving a kind of solidity to the sphere of empty space in which they move. The remarkable fact that the electron also spins like a top about its own axis was first discovered in 1925 and now we know that many particles have this property of intrinsic spin.

The spin of an elementary particle unlike, say, the rotation of a molecule, is an invariable property of that particle, always fixed at a definite value. The electron can not be stopped from rotating, nor can its rotation be speeded up. The spin of an electron is such an essential feature of the electron that it can be changed only at the expense of destroying the electron altogether. Actually, this is a rather subtle point. It is probably more accurate to say that if the electron is caused to spin more rapidly, this so drastically changes the properties of the electron that it is more convenient to think of the resulting structure as a completely new and distinct particle. To what extent the different particles are really independent and to what extent they are only different states of motion of some common underlying structure is, of course, the great, unsolved problem of particle physics, and it would be idle to speculate further about this point here. It can only be re-emphasized that physicists retain their faith that there exists a simpler structure underlying the particles.

Spin is measured in terms of the quantity called angular momentum, which is a combined measure of the mass, the size, and the speed of the rotating system. The rate of rotation of an electron can not actually be measured but it has to be so great that the charge within the electron is moving at nearly the speed of light. In spite of this frantic rate of rotation, the electron is not able to generate much angular momentum because of its small size and small mass. A man swiveling slowly in his chair to watch a tennis match has at least $10^{33}$ times more angular momentum than a single electron.

As the electron carries nature's smallest unit of electric charge, it also carries the indivisible unit of spin, which, for historical reasons, is denoted by $\frac{1}{2}\hbar$. At the turn of this century, Max Planck discovered the existence of a constant in nature which relates the frequency and the energy of photons. We now call it Planck's constant and write it $h$. About ten years later Niels Bohr discovered that this constant also has something to do with the rotation of electrons about the nucleus in an atom. In their orbits about the nucleus the electrons have an angular momentum which is always equal to $h$ divided by $2\pi$ ($h/2\pi$) or two times $h/2\pi$ or three times $h/2\pi$, and so on, never any value between. Since it is a nuisance to write out the extra factor of $2\pi$ whenever it occurs, the notation $\hbar$ (pronounced "h bar") has come into common use for $h/2\pi$. Finally, when Samuel Goudsmit and George Uhlenbeck discovered in 1925 that the electron spins, they found that its spin angular momentum is not $\hbar$, which had been thought to be the indivisible unit, but only $\frac{1}{2}\hbar$. The quantity $h$ has been adopted as the unit of spin and angular momentum in the submicroscopic world, even though the smallest indivisible unit is only half so great. The leptons and baryons all have spin $\frac{1}{2}$, the photon has spin 1, and the graviton spin 2, in this unit.

For the record, $\hbar$ is equal to $1.0544 \times 10^{-27}$ gm $\times$ cm $\times$ cm/sec (the units are mass $\times$ size $\times$ speed). Recall that $\pi$ is the number $3.14159\cdots$, which occurs in geometry as the ratio of the circumference of a circle to its diameter.

The principle of spin quantization applies in the macroscopic world as well as in the microscopic, but is too subtle an effect ever to be measured for a large object. When the spectator at a tennis tournament turns to follow the ball, his angular momentum in units of $\hbar$ might be $10^{33}$ or $10^{33} + 1$ or $10^{33} + 2$, but never $10^{33} + \frac{1}{3}$. The increment $\hbar$ between allowed values is so infinitesimal that it is entirely beyond hope ever to notice this discreteness in the macroscopic world. A change of one penny in the gross national product of the United States is a disturbance more than a billion billion times greater than a change by one unit of the spectator's angular momentum. It is no wonder that man did not discover the quantum theory of angular momentum until he was able to study in detail the structure of a system as small as an atom.

The units of measurement man normally uses, even in scientific work, have been defined in an arbitrary way, chosen merely for convenience. They have nothing to do with "natural units" and are not in particular harmony with the basic structure of the world. Yet in this century we have learned of the existence of two natural units which are now commonly employed in studying the elementary particles. It seems quite likely that a deeper understanding of the particles will be accompanied by the discovery of a third natural unit.

The meter was originally defined as one ten-millionth of the distance from the earth's pole to the equator. (It is not quite that, since the meter was standardized in the nineteenth century, but the knowledge of the size of the earth has been improved since then.) The centimeter, in turn, is one one-hundredth of a meter. The gram is then defined as the mass of a cube of water one centimeter on a side. Both the centimeter and the gram thus depend on the size of the earth, and there is no reason to believe that there is anything very special about the size of the earth. The third basic unit, the second, also depends on a property of our earth, its rate of rotation — again nothing very special. For no better reasons than that the Egyptians divided the day and the night into twelfths and the Sumerians liked to count in sixties, the hour is one twenty-fourth of a day, the minute a sixtieth of an hour, and the second a sixtieth of a minute.

Within the first five years of this century two natural units were discovered which appeared to be obvious choices for a basis of measurement in the submicroscopic world. These were the speed of light $c$ and Planck's constant $h$. Neither is directly a mass, a length, or a time, but they are simple combinations of these three. If they were joined by a third natural unit, they would form a basis of measurement as complete as, and much more satisfying than, the gram, the centimeter, and the second. (The thoughtful reader might propose the charge on the electron as an obvious candidate for the third natural unit. Unfortunately, it will not serve, for it is not independent of $c$ and $h$ — just as speed is not independent of time and distance.)

The fact that light travels at a fixed speed $c$ has been known for several centuries, but the central significance of this speed in nature could be appreciated only when the theory of relativity was developed. Relativity revealed first of all that $c$ is a natural speed limit, attainable not only by light but by any massless particle. The theory also showed that this constant appeared in various surprising places which had nothing to do with speed, for instance, in the mass-energy relation, $E = mc^2$. Planck's constant was brand new in 1900, but its significance also mounted over the next few decades as it came to be recognized as the fundamental constant of the quantum theory, governing not only the allowed values of spin, but every other quantized quantity.

It has to be recognized that every measurement is really the statement of a ratio. If you say you weigh 151 pounds you are, in effect, saying your weight is 151 times greater than the weight of a standard object (a pint of water), which is arbitrarily called one pound. A 50-minute class is 50 times longer than the arbitrarily defined time unit, the minute. When one uses natural units, the ratio is taken with respect to some physically significant unit rather than an arbitrary one. On the natural scale, a jet-plane speed of $10^{-6}c$ is very slow, a particle speed of $0.99c$ is very fast. An angular momentum of $10,000\hbar$ is large, an angular momentum of $\frac{1}{2}h$ is small.

The difficult point to recognize here is that once the speed of light has been adopted as the unit of speed, it no longer makes any sense to ask how fast light travels. The only answer is: Light goes as fast as it goes. Since every measurement is really a comparison, there must always be at least one standard that can not be compared with anything but itself. This leads to the idea of a "dimensionless physics." Having agreed on a standard of speed, we can say a jet plane travels at a speed of $10^{-6}$, that is, at one one-millionth of the speed of light. The $10^{-6}$ is a pure number to which no unit need be attached; it is the ratio of the speed of the plane to the speed of light. To make possible a dimensionless physics we need one more independent natural unit, and this has not yet emerged. This unit, if it is found, may be a length, and there is much speculation that such a unit will be connected with a whole new view of the nature of space (and of time) in the world of the very small.

It should be added that a dimensionless physics is not so profound as it may sound, nor would it necessarily be a terminus of man's downward probing. Its lack of profundity springs from the fact that it, too, would rest on an arbitrary agreement among men about units. The hope is, however, that all scientists will be led naturally and uniquely to agree that there is only one sensible set of natural units, in contrast with the present situation, where the fact about which we all agree is that there is nothing special whatever about the centimeter, the gram, and the second. Even if the dimensionless physics is realized, there will remain dimensionless, or pure, numbers still to be explained, and that explanation may lie in still deeper layers of the substructure of nature.

# GEORGE GAMOW

■

George Gamow (1904–1968), the Russian-American nuclear physicist who developed the first quantum theory of radioactivity, was an effervescent and outgoing man whose nontechnical books brought science to life for millions of readers. This portrait of electrodynamics is from his *Mr. Tompkins Explores the Atom*.

# *The Gay Tribe of Electrons*

... WHILE FINISHING HIS DINNER, Mr. Tompkins remembered that it was the night of the professor's lecture on the structure of the atom, which he had promised to attend. But he was so fed up with his father-in-law's interminable expositions that he decided to forget the lecture and spend a comfortable evening at home. However, just as he was getting settled with his book, Maud cut off this avenue of escape by looking at the clock and remarking, gently but firmly, that it was almost time for him to leave. So, half an hour later, he found himself on a hard wooden bench in the university auditorium together with a crowd of eager young students.

"Ladies and gentlemen," began the professor, looking at them gravely over his spectacles, "in my last lecture I promised to give you more details concerning the internal structure of the atom, and to explain how the peculiar features of this structure account for its physical and chemical properties. You know, of course, that atoms are no longer considered as elementary indivisible constituent parts of matter, and that this role has passed now to much smaller particles such as electrons, protons, etc.

"The idea of elementary constituent particles of matter, representing the last possible step in divisibility of material bodies, dates back to the ancient Greek philosopher Democritus who lived in the fourth century B.C. Meditating about the hidden nature of things,

38

Democritus came to the problem of the structure of matter and was faced with the question whether or not it can exist in infinitely small portions. Since it was not customary at this epoch to solve any problem by any other method than that of pure thinking, and since, in any case, the question was at that time beyond any possible attack by experimental methods, Democritus searched for the correct answer in the depths of his own mind. On the basis of some obscure philosophical considerations he finally came to the conclusion that it is 'unthinkable' that matter could be divided into smaller and smaller parts without any limit, and that one must assume the existence of 'the smallest particles which cannot be divided any more.' He called such particles 'atoms,' which, as you probably know, means 'indivisibles' in Greek.

"I do not want to minimize the great contribution of Democritus to the progress of natural science, but it is worth keeping in mind that besides Democritus and his followers, there was undoubtedly another school of Greek philosophy the adherents of which maintained that the process of divisibility of matter *could* be carried beyond any limit. Thus, independent of the character of the answer which had to be given in the future by exact science, the philosophy of ancient Greece was well secured with an honorable place in the history of physics. At the time of Democritus, and for centuries later, the existence of such indivisible portions of matter represented a purely philosophical hypothesis, and it was only in the nineteenth century that scientists decided that they had finally found these indivisible building-stones of matter which were foretold by the old Greek philosopher more than two thousand years ago.

"In fact, in the year 1808 an English chemist, John Dalton, showed that the relative proportions . . ."

Almost from the beginning of the lecture Mr. Tompkins had felt an irresistible urge to close his eyes and doze through the rest of the lecture, and it was only the academic hardness of the bench that kept him from doing so. However, Dalton's ideas concerning the law of "relative proportions" proved the last straw, and the hushed auditorium was soon permeated by a gentle wheeze coming from the corner where Mr. Tompkins was sitting.

When Mr. Tompkins dropped off to sleep, the discomfort of the uncompromising bench seemed to melt into the pleasant sensation of floating on air, and opening his eyes he was surprised to find himself dashing through space at what he considered a pretty reckless speed. Looking around he saw that he was not alone on this fantastic trip. Near him a number of vague, misty forms were swooping around a large heavy-looking object in the middle of the crowd. These strange beings were traveling in pairs, gaily chasing each other along circular and elliptic tracks. Suddenly Mr. Tompkins felt very lonely because he realized that he was the only one of the whole group who had no playmate.

"Why didn't I bring Maud along with me?" Mr. Tompkins wondered gloomily. "We could have had a wonderful time with this happy-go-lucky crowd." The track he was moving along was outside all the others, and while he wanted very much to join the party, the uncomfortable feeling of being odd man out kept him from doing so. However, when one of the electrons (for by now Mr. Tompkins realized he had miraculously joined the electronic community of an atom) was passing close by on its elongated track, he decided to complain about the situation.

"Why haven't *I* got anyone to play with?" he shouted across.

"Because this is an odd atom, and you are the valency electr-o-o-on . . ." called the electron as it turned and plunged back into the dancing crowd.

"Valency electrons live alone or find companions in other atoms," squeaked the high pitched soprano of another electron rushing past him.

> "If you want a partner fair,
> Jump into chlorine and find one there,"

chanted another mockingly.

"I see you are quite new here, my son, and very lonely," said a friendly voice above him, and raising his eyes Mr. Tompkins saw the stout figure of a monk clothed in a brown tunic.

"I am Father Paulini,"* went on the monk, moving along the track with Mr. Tompkins, "and it is my mission in life to keep watch over the morals and social life of electrons in atoms and elsewhere. It is my duty to keep these playful electrons properly distributed among the different quantum cells of the beautiful atomic structures erected by our great architect Niels Bohr. To keep order and to preserve the proprieties, I never permit more than two electrons to follow the same track; a *ménage à trois* always gives a lot of trouble, you know. Thus electrons are always grouped in pairs of opposite

*Gamow has in mind the physicist Wolfgang Pauli. [editor's note]

'spin' and no intruder is permitted if the cell is already occupied by a couple. It is a good rule, and I may add that not a single electron has yet broken my commandment."

"Maybe it *is* a good rule," objected Mr. Tompkins, "but it is rather inconvenient for me at the moment."

"I see it is," smiled the monk, "but it is just your bad luck, being a valency electron in an odd atom. The sodium atom to which you belong is entitled by the electric charge of its nucleus (that big dark mass you see in the center) to hold eleven electrons altogether. Well, unfortunately for you, eleven is an odd number, hardly an unusual circumstance when you consider that exactly one half of all numbers are odd, and only the other half even. Thus, as the latecomer you will have to be alone for a while at least."

"You mean there is a chance that I can get in later?" asked Mr. Tompkins eagerly. "Kicking one of the oldtimers out, for example?"

"It isn't exactly done," said the monk wagging a plump finger at him, "but, of course, there is always a chance that some of the inner circle members will be thrown out by an external disturbance, leaving an empty place. However, I wouldn't count on it much, if I were you."

"They told me I'd be better off if I moved into chlorine," said Mr. Tompkins, discouraged by Father Paulini's words. "Can you tell me how to do that?"

"Young man, young man!" exclaimed the monk sorrowfully, "why are you so insistent on finding company? Why can't you appreciate solitude and this Heaven-sent opportunity to contemplate your soul in peace? Why must even electrons lean always to the worldly life? However, if you insist on companionship, I will help you get your wish. If you look where I'm pointing, you will see a chlorine atom approaching us, and even at this distance you can see an unoccupied spot where you would most certainly be welcomed. The empty spot is in the outer group of electrons, the so-called 'M-shell,' which is supposed to be made up of eight electrons grouped in four pairs. But, as you see, there are four electrons spinning in one direction and only three in the other, with one place vacant. The inner shells, known as 'K' and 'L,' are completely filled up, and the atom will be glad to get you and have its outer shell complete. When the two atoms get close together, just jump over, as valency electrons usually do. And may peace be with you, my son!" With these words the impressive figure of the electron priest suddenly faded into thin air.

Feeling considerably more cheerful, Mr. Tompkins gathered his strength for a neckbreaking jump into the orbit of the passing chlorine atom. To his surprise he leapt over with an easy grace and found himself in the congenial surroundings of the members of the chlorine M-shell.

"It was delightful of you to join us!" called his new partner of opposite spin, gliding gracefully along the track. "Now no one can say that our community is not complete. Now we shall all have fun together!"

Mr. Tompkins agreed that it really was fun — lots of fun — but one little worry kept stealing into his mind. "How am I going to explain this to Maud when I see her again?" he thought rather guilt-ily, but not for long. "Surely she won't mind," he decided. "After all, these are only electrons."

"Why doesn't that atom you've left go away now?" asked his companion with a pout. "Does it still hope to get you back?"

And, as a matter of fact, the sodium atom, with its valency elec-tron gone, *was* sticking closely to the chlorine one as if in the hope that Mr. Tompkins would change his mind and jump back to his lonely track.

"Well how do you like that!" said Mr. Tompkins angrily, frowning at the atom which had first received him so coldly. "There's a dog in the manger for you!"

"Oh they always do that," said a more experienced member of the M-shell. "I understand it is not so much the electronic community of the sodium atom which wants you back as the sodium nucleus itself. There is almost always some disagreement between the central nucleus and its electronic escort: the nucleus wants as many electrons around it as it can possibly hold with its electric charge, whereas the electrons themselves prefer to be only enough in number to make the shells complete. There are only a few atomic species, the so-called *rare gases,* or *noble gases* as the German chemists call them, in which the desire of the ruling nucleus and the subordinate electrons are in full harmony. Such atoms as helium, neon and argon, for example, are quite satisfied with themselves and neither expel their number nor invite new ones. They are chemically inert, and keep away from all other atoms. But in all other atoms electronic communities are always ready to change their membership. In the sodium atom, which was your former home, the nucleus is entitled by its electric charge to one more electron than is necessary for harmony in the shells. On the other hand, in our atom the normal contingent of electrons is not enough for complete harmony, and thus we welcome your arrival, in spite of the fact that your presence overloads our nucleus. But as long as you stay here, our atom is not neutral any more, and has an extra electric charge. Thus the sodium atom which you left stands by, held by the force of electric attraction. I once heard our great priest, Father Paulini, say that such atomic communities, with extra electrons or electrons missing, are called negative and positive 'ions.' He also uses the word 'molecule' for groups of two or more atoms bound together by electric force. This particular combination of sodium and chlorine atoms he calls a molecule of 'table salt,' whatever that may be."

"Do you mean to tell me you don't know what table salt is?" said Mr. Tompkins, forgetting to whom he was talking. "Why that's what you put on your scrambled eggs at breakfast."

"What are 'scram bulldeggs' and what is 'break-fust'?" asked the intrigued electron. Mr. Tompkins sputtered and then realized the futility of trying to explain to his companions even the simplest details of the lives of human beings. "That's why I don't get more out of their talk about valency and complete shells," he told himself, deciding to enjoy his visit to this fantastic world without worrying about understanding it. But it was not so easy to get away from the talkative electron who evidently had a great desire to pass on all the knowledge collected during a long electronic life.

"You must not think," he continued, "that the binding of atoms into molecules is always accomplished by one valency electron alone. There are atoms, like oxygen for example, which need two more electrons to complete their shells, and there are also atoms which need three electrons and even more. On the other hand, in some atoms the nucleus holds two or more extra — or valency — electrons. When such atoms meet, there is quite a lot of jumping over and binding to do, as a result of which quite complex molecules, often consisting of thousands of atoms, are formed. There are also the so-called 'homopolar' molecules, that is molecules made up of two identical atoms, but that is a very unpleasant situation."

"Unpleasant, why?" asked Mr. Tompkins, getting interested again.

"Too much work," commented the electron, "to keep them together. Some time ago I happened to get that job and I didn't have a moment to myself all the while I stayed there. Why, it isn't at all the way it is here where the valency electron just enjoys himself and lets the electrically hungry and deserted atom stand by. No sir! In order to keep the two identical atoms together, he has to jump to and fro, from one to the other and back again. My word! One feels like a Ping-Pong ball."

Mr. Tompkins was rather surprised to hear the electron, which did not know what scrambled eggs were, speak so glibly of Ping-Pong, but he let it pass.

"I'll never take on that job again!" grumbled the lazy electron, overwhelmed by a wave of unpleasant memories. "I am quite comfortable where I am now.

"Wait!" he exclaimed suddenly. "I think I see a still better place for me to go. So lo-o-o-ong!" And with a giant leap he rushed toward the interior of the atom.

Looking in the direction in which his interlocutor had gone, Mr. Tompkins understood what had happened. It seems that one of the electrons of the inner circle was thrown clear of the atom by some foreign high-speed electron which had unexpectedly penetrated into their system, and a cozy place in the "K" shell was now wide open. Chiding himself for missing this opportunity to join the inner circle, Mr. Tompkins now watched with great interest the course of the electron he had just been talking to. Deeper and deeper into the atomic interior this happy electron sped, and bright rays of light accompanied his triumphant flight. Only when it finally reached the internal orbit did this almost unbearable radiation finally stop.

"What was that?" asked Mr. Tompkins, his eyes aching from the sight of this unexpected phenomenon. "Why all this brilliance?"

"Oh that's just the X-ray emission connected with the transition," explained his orbit companion, smiling at his embarrassment. "Whenever one of us succeeds in getting deeper into the interior of the atom, the surplus energy must be emitted in the form of radiation. This lucky fellow made quite a big jump and let loose a lot of energy. More often we have to be satisfied with smaller jumps, here in the atomic suburbs, and then our radiation is called 'visible light' — at least that is what Father Paulini calls it."

"But this X-light, or whatever you call it, is also visible," protested Mr. Tompkins. "I should call your terminology rather misleading."

"Well, we are electrons and are susceptible to any kind of radiation. But Father Paulini tells us that there exist gigantic creatures, 'Human Beings' he calls them, who can see light only when it falls within a narrow energy-interval, or wavelength-interval as he puts it. He told us once that it took a great man, Roentgen I think his name was, to discover these X-rays and that now they are largely used in something called 'medicine.'"

"Oh yes. I know quite a lot about that," said Mr. Tompkins, feeling proud that now *he* could show off his knowledge. "Want me to tell you more about it?"

"No thanks," said the electron yawning. "I really don't care. Can't you be happy without talking? Try to catch me!"

For a long time Mr. Tompkins went on enjoying the pleasant sensation of diving through space with the other electrons in a kind of glorified trapeze act. Then, all of a sudden he felt his hair stand on end, an experience he had felt once before during a thunder storm in the mountains. It was clear that a strong electric disturbance was approaching their atom, breaking the harmony of the electronic motion, and forcing the electrons to deviate seriously from their normal tracks. From the point of view of a human physicist, it was only a wave of ultraviolet light passing through the spot where this particular atom happened to be, but to the tiny electrons it was a terrific electric storm.

"Hold on tight!" yelled one of his companions, "or you will be thrown out by photo-effect forces!" But it was already too late. Mr. Tompkins was snatched away from his companions and hurled into space at a terrifying speed, as neatly as if he had been seized by a pair of powerful fingers. Breathlessly he hurtled further and further through space, tearing past all kinds of different atoms so fast he could hardly distinguish the separate electrons. Suddenly a large atom loomed up right in front of him and he knew that a collision was unavoidable.

"Pardon me, but I am photo-effected and cannot . . ." began Mr. Tompkins politely, but the rest of the sentence was lost in an earsplitting crash as he ran head on into one of the outer electrons. The two of them tumbled head over heels off into space. However, Mr. Tompkins had lost most of his speed in the collision and was now able to study his new surroundings somewhat more closely. The atoms which towered around him were much larger than any he had seen before, and he could count as many as twenty-nine electrons in each of them. If he had known his physics better he would have recognized them as atoms of copper, but at these close quarters the group as a whole did not look like copper at all. Also they were spaced rather close to one another forming a regular pattern which extended as far as he could see. But what surprised Mr. Tompkins most was the fact that these atoms did not seem to be very particular about holding on to their quota of electrons, particularly their outer electrons. In fact the outer orbits were mostly empty, and crowds of unattached electrons were drifting lazily through space, stopping from time to time but never for very long, on the outskirts of one atom or another. Rather tired after his breakneck flight through space, Mr. Tompkins tried at first to get a little rest on a steady orbit of one of the copper atoms. However he was soon infected with the prevailing vagabondish feeling of the crowd, and he joined the rest of the electrons in their nowhere-in-particular motion.

"Things are not very well organized here," he commented to himself, "and there are too many electrons not tending to their business. I think Father Paulini should do something about it."

"Why should I?" said the familiar voice of the monk who had suddenly materialized from nowhere. "These electrons are not disobeying my commandments, and besides they are doing a very useful job indeed. You may be interested to know that if all atoms cared as much about holding their electrons as some of them do, there would be no such thing as electric conductivity. Why you wouldn't even be able to have an electric bell in your house, to say nothing of a light or a telephone."

"Oh, you mean these electrons carry electric current?" asked Mr. Tompkins, grasping at the hope that the conversation was turning to a subject more or less familiar to him. "But I don't see that they are moving in any particular direction."

"First of all, my lad," said the monk severely, "do not use the word 'they,' use 'we.' You seem to forget that you are an electron yourself and that the moment someone presses the button to which this copper wire is attached, electric tension will cause you, as well as all the other conductivity electrons, to rush along to call the maid or do whatever else is needed."

"But I don't want to!" said Mr. Tompkins firmly, a note of temper in his voice. "As a matter of fact I am quite tired of being an electron and I don't think it's so much fun any more. What a life, to have to carry out all these electronic duties for ever and ever!"

"Not necessarily forever," countered Father Paulini, who definitely did not like back-talk on the part of plain electrons. "There is always the chance that you will be annihilated and cease to exist."

"B-b-be annihilated?" repeated Mr. Tompkins feeling cold shivers running up and down his spine. "But I always thought electrons were eternal."

"That is what physicists used to believe until comparatively

46

recent times," agreed Father Paulini, amused at the effect produced by his words, "but it isn't exactly correct. Electrons can be born, and die, as well as human beings. There isn't, of course, such a thing as dying of old age; death comes only through collisions."

"Well, I had a collision only a short while ago, and a pretty bad one too," said Mr. Tompkins recovering a little confidence. "And if that one didn't put me out of action, I can't imagine one that would."

"It isn't a question of how forcibly you collide," Father Paulini corrected him, "but of who the other fellow is. In your recent collision you probably ran into another negative electron, very similar to yourself, and there is not the slightest danger in such an encounter. In fact, you could butt into each other like a couple of rams for years and no harm could be done. But there is another breed of electron, the positive ones, which have been discovered only comparatively recently by the physicists. These positive electrons, or positrons, look exactly the way you do, the only difference being that their electric charge is positive instead of negative. When you see such a fellow approaching, you think it is just another innocent member of your tribe and go ahead to greet him. But then you suddenly find that, instead of pushing you away slightly to avoid a collision, as any normal electron would, he pulls you right in. And then it is too late to do anything."

"How terrible!" exclaimed Mr. Tompkins. "And how many poor ordinary electrons can one positron eat up?"

"Fortunately only one, since in destroying a negative electron the positron also destroys itself. One could describe them as members of a suicide club, looking for partners in mutual annihilation. They do not harm one another, but as soon as a negative electron comes their way, it hasn't much chance of surviving."

"Lucky I haven't run into one of these monsters yet," said Mr. Tompkins much impressed by this description. "I hope they are not very numerous. Are they?"

"No, they're not. And for the simple reason that they are always looking for trouble and so vanish very soon after they are born. If you wait a minute, I shall probably be able to show you one."

"Yes, here we are," continued Father Paulini after a short silence. "If you look carefully at that heavy nucleus over there, you will see one of these positrons being born."

The atom at which the monk was pointing was evidently undergoing a strong electromagnetic disturbance owing to some vigorous radiation falling on it from outside. It was a much more violent disturbance than the one which threw Mr. Tompkins out of his chlorine atom, and the family of atomic electrons surrounding the nucleus was being dispersed and blown away like dry leaves in a hurricane.

"Look closely at the nucleus," said Father Paulini, and concentrating his attention Mr. Tompkins saw a most unusual phenomenon taking place in the depths of the destroyed atom. Very close to the nucleus, inside the inner electronic shell, two vague shadows were gradually taking shape, and a second later Mr. Tompkins saw two glittering brand new electrons rushing at great speed away from their birthplace.

"But I see two of them," said Mr. Tompkins, fascinated by the sight.

"That is right," agreed Father Paulini. "Electrons are always born in pairs, otherwise it would contradict the law of conservation of electric charge. One of these two particles, born under the action of a strong gamma ray on the nucleus, is an ordinary negative electron, whereas the other is a positron — the murderer. He is off now to find a victim."

"Well, if the birth of each positron destined to destroy an electron is accompanied by the birth of still another plain electron, then things aren't so bad," commented Mr. Tompkins thoughtfully. "At least it doesn't lead to the extinction of the electronic tribe, and I . . ."

"Look out!" interrupted the monk shoving him aside while the newborn positron whistled by, just an inch away. "You can never be too careful when these murderous particles are around. But I think I'm spending too much time talking to you and I have other business to attend to. I must look for my pet 'neutrino' . . ." And the monk disappeared without letting Mr. Tompkins know what this "neutrino" was and whether or not it was also to be feared. Thus deserted, Mr. Tompkins felt even more lonely than before, and when one or another fellow electron approached him on his journey through space, he even nursed a secret desperate hope that under each innocent exterior might be hidden the heart of a murderer. For a long time, centuries it seemed to him, his fears and hopes were not justified, and he unwillingly bore the dull duties of a conductivity electron.

Then suddenly it happened, and at a moment when he expected it least. Feeling a strong need to talk to somebody, even to a stupid conductivity electron, he approached a particle which was slowly moving by and was evidently a newcomer to this bit of copper wire. Even at a distance, however, he noticed that he had made a bad choice and that an irresistible force of attraction was pulling him along, permitting no retreat. For a second he tried to struggle and tear himself away, but the distance between them was rapidly getting smaller and smaller and it seemed to Mr. Tompkins that he saw a fiendish grin on the face of his captor.

"Let me go! Let me go!" shouted Mr. Tompkins at the top of his voice, struggling with his arms and kicking his legs. "I don't want to be annihilated; I'll conduct electric current for the rest of eternity!" But it was all in vain, and the surrounding space was suddenly illuminated by a blinding flash of intensive radiation.

"Well, I am no more," thought Mr. Tompkins, "but how is it I can still think? Has my body only been annihilated, and my soul gone to a quantum heaven?" Then he felt a new force, more gentle this time, shaking him firmly and resolutely, and opening his eyes he recognized the university janitor.

"I'm sorry, Sir," he said, "but the lecture was over some time ago and we gotta close the hall up now." Mr. Tompkins stifled a yawn and looked sheepish.

"Good night, Sir," said the janitor with a sympathetic smile.

# PIERRE CURIE

■

The French physicist Pierre Curie (1859–1906) concluded his 1903 Nobel Prize address with a prescient warning of the dangers that nuclear energy could pose for all civilization. He was little heeded, though the Curie name carried considerable authority: He, his wife, Marie, and their daughter Irene all won Nobel Prizes for their research in radioactivity.

# *Radioactive Substances*

... I HAVE TO SPEAK to you today on the properties of the *radioactive substances,* and in particular of those of *radium.* I shall not be able to mention exclusively our own investigations. At the beginning of our studies on this subject in 1898 we were the only ones, together with Becquerel, interested in this question; but since then much more work has been done, and today it is no longer possible to speak of radioactivity without quoting the results of investigations by a large number of physicists such as Rutherford, Debierne, Elster and Geitel, Giesel, Kauffmann, Crookes, Ramsay and Soddy, to mention only a few of those who have made important progress in our knowledge of radioactive properties.

I shall give you only a rapid account of the discovery of radium and a brief summary of its properties, and then I shall speak to you of the consequences of the new knowledge which radioactivity gives us in the various branches of science.

Becquerel discovered in 1896 the special radiating properties of *uranium* and its compounds. Uranium emits very weak rays which leave an impression on photographic plates. These rays pass through black paper and metals; they make air electrically conductive. The radiation does not vary with time, and the cause of its production is unknown.

Mme. Curie in France and Schmidt in Germany have shown

that thorium and its compounds possess the same properties. Mme. Curie also showed in 1898 that of all the chemical substances prepared or used in the laboratory, only those containing uranium or thorium were capable of emitting a substantial amount of the Becquerel rays. We have called such substances *radioactive*.

Radioactivity, therefore, presented itself as an atomic property of uranium and thorium, a substance being all the more radioactive as it was richer in uranium or thorium.

Mme. Curie has studied the minerals containing uranium or thorium, and in accordance with the views just stated, these minerals are all radioactive. But in making the measurements, she found that certain of these were more active than they should have been according to the content of uranium or thorium. Mme. Curie then made the assumption that these substances contained radioactive chemical elements which were as yet unknown. We, Mme. Curie and I, have sought to find these new hypothetical substances in a uranium ore, *pitchblende*. By carrying out the chemical analysis of this mineral and assaying the radioactivity of each batch separated in the treatment, we have, first of all, found a highly radioactive substance with chemical properties close to bismuth which we have called *polonium,* and then (in collaboration with Bémont) a second highly radioactive substance close to barium which we called *radium*. Finally, Debierne has since separated a third radioactive substance belonging to the group of the rare earths, *actinium*.

These substances exist in pitchblende only in the form of traces, but they have an enormous radioactivity of an order of magnitude 2 million times greater than that of uranium. After treating an enormous amount of material, we succeeded in obtaining a sufficient quantity of radiferous barium salt to be able to extract from it radium in the form of a pure salt by a method of fractionation. Radium is the higher homologue of barium in the series of the alkaline earth metals. Its atomic weight as determined by Mme. Curie is 225. Radium is characterized by a distinct spectrum which was first discovered and studied by Demarçay, and then by Crookes and Runge and Precht, Exner, and Haschek. The spectrum reaction of radium is very sensitive, but it is much less sensitive than radioactivity for revealing the presence of traces of radium.

The general effects of the radiations from radium are intense and very varied. . . .

A radioactive substance such as radium constitutes a continuous source of energy. This energy is manifested by the emission of the radiation. I have also shown in collaboration with Laborde that radium releases heat continuously to the extent of approximately 100 calories per gram of radium and per hour. Rutherford and Soddy, Runge and Precht, and Knut Angström have also measured the release of heat by radium; this release seems to be constant after several years, and the total energy release by radium in this way is considerable.

The work of a large number of physicists (Meyer and Schweidler, Giesel, Becquerel, P. Curie, Mme. Curie, Rutherford, Villard, etc.) shows that the radioactive substances can emit rays of the three different varieties designated by Rutherford as α-, β- and γ-rays. They differ from one another by the action of a magnetic field and of an electric field which modify the trajectory of the α- and β-rays.

The β-rays, similar to cathode rays, behave like negatively charged projectiles of a mass 2000 times smaller than that of a hydrogen atom (electron). We have verified, Mme. Curie and I, that the β-rays carry with them negative electricity. The x-rays, similar to the Goldstein's canal rays, behave like projectiles 1000 times heavier and charged with positive electricity. The γ-rays are similar to Röntgen rays.

Several radioactive substances such as radium, actinium, and thorium also act otherwise than through their direct radiation; the surrounding air becomes radioactive, and Rutherford assumes that each of these substances emits an unstable radioactive gas which he calls *emanation* and which spreads in the air surrounding the radioactive substance.

The activity of the gases which are thus made radioactive disappears spontaneously according to an exponential law with a time constant which is characteristic for each active substance. It can, therefore, be stated that the emanation from radium diminishes by one-half every 4 days, that from thorium by one-half every 55 seconds, and that from actinium by one-half every 3 seconds.

Solid substances which are placed in the presence of the active air surrounding the radioactive substances themselves become temporarily radioactive. This is the phenomenon of *induced radioactivity* which Mme. Curie and I have discovered. The induced radioactivities, like the emanations, are equally unstable and are destroyed spontaneously according to exponential laws characteristic of each of them. . . .

Finally, according to Ramsay and Soddy, radium is the seat of a continuous and spontaneous production of helium.

The radioactivity of uranium, thorium, radium and actinium seems to be invariable over a period of several years; on the other hand, that of polonium diminishes according to an exponential law; it diminishes by one-half in 140 days, and after several years it has almost completely disappeared.

These are all the most important facts which have been established by the efforts of a large number of physicists. Several phenomena have already been extensively studied by them.

The consequences of these facts are making themselves felt in all branches of science:

The importance of these phenomena for *physics* is evident. Radium constitutes in laboratories a new research tool, a source of new radiations. The study of the β-rays has already been very fruitful. It has been found that this study confirms the theory of J. J. Thomson and Heaviside on the mass of particles in motion, charged with electricity; according to this theory, part of the mass results from the electromagnetic reactions of the ether of the vacuum. The experiments of Kauffmann on the β-rays of radium lead to the assumption that certain particles have a velocity very slightly below that of light, that according to the theory the mass of the particle increases with the velocity for velocities close to that of light, and that the whole mass of the particle is of an electromagnetic nature. If the hypothesis is also made that material substances are constituted by an agglomeration of electrified particles, it is seen that the *fundamental principles of mechanics* will have to be profoundly modified.

The consequences for *chemistry* of our knowledge of the properties of the radioactive substances are perhaps even more important. And this leads us to speak of the source of energy which maintains the radioactive phenomena.

At the beginning of our investigations we stated, Mme. Curie and I, that the phenomena could be explained by two distinct and very general hypotheses which were described by Mme. Curie in 1899 and 1900.

1. In the first hypothesis it can be supposed that the radioactive substances borrow from an external radiation the energy which they release, and their radiation would then be a secondary radiation. It is not absurd to suppose that space is constantly traversed by very penetrating radiations which certain substances would be capable of capturing in flight. According to the recent work of Rutherford, Cooke and McLennan, this hypothesis seems to be useful for

explaining part of the extremely weak radiation which emanates from most of the substances.

2. In the second hypothesis it can be supposed that the radioactive substances draw from themselves the energy which they release. The radioactive substances would then be in course of evolution, and they would be transforming themselves progressively and slowly in spite of the apparent invariability of the state of certain of them. The quantity of heat released by radium in several years is enormous if it is compared with the heat released in any chemical reaction with the same weight of matter. This released heat would only represent, however, the energy involved in a transformation of a quantity of radium so small that it cannot be appreciated even after years. This leads to the supposition that the transformation is more far-reaching than the ordinary chemical transformations, that the existence of the atom is even at stake, and that one is in the presence of a transformation of the elements.

The second hypothesis has shown itself the more fertile in explaining the properties of the radioactive substances properly so called. It gives, in particular, an immediate explanation for the spontaneous disappearance of polonium and the production of helium by radium. This theory of the transformation of the elements has been developed and formulated with great boldness by Rutherford and Soddy who state that there is a continuous and irreversible disaggregation of the atoms of the radioactive elements. In the theory of Rutherford the disaggregation products would be, on the one hand, the projectile rays, and on the other hand, the emanations and the induced radioactivities. The latter would be new gaseous or solid radioactive substances frequently with a rapid evolution and having atomic weights lower than that of the original element from which they are derived. Seen in this way the life of radium would be necessarily limited when this substance is separated from the other elements. In Nature radium is always found in association with uranium, and it can be assumed that it is produced by the latter.

This, therefore, is a veritable theory of the transmutation of elements, although not as the alchemists understood it. The inorganic matter would necessarily evolve through the ages and in accordance with immutable laws.

Through an unexpected consequence, the radioactive phenomena can be important in *geology*. It has been found, for example, that radium always accompanies uranium in minerals. And it has even been found that the ratio of radium to uranium is constant in all minerals (Boltwood). This confirms the idea of the creation of radium from uranium. This theory can be extended to try to explain also other associations of elements which occur so frequently in minerals. It can be imagined that certain elements have been formed on the spot on the surface of the Earth or that they stem from other elements in a time which may be of the order of magnitude of geological periods. This is a new point of view which the geologists will have to take into account.

Elster and Geitel have shown that the emanation of radium is very widespread in Nature and that radioactivity probably plays an important part in *meteorology,* with the ionization of the air provoking the condensation of water vapour.

Finally, in the *biological sciences* the rays of radium and its emanation produce interesting effects which are being studied at present. Radium rays have been used in the treatment of certain diseases (lupus, cancer, nervous diseases). In certain cases their action may become dangerous. If one leaves a wooden or cardboard box containing a small glass ampulla with several centigrams of a radium salt in one's pocket for a few hours, one will feel absolutely nothing. But 15 days afterwards a redness will appear on the epidermis, and then a sore which will be very difficult to heal. A more prolonged action could lead to paralysis and death. Radium must be transported in a thick box of lead.

It can even be thought that radium could become very dangerous in criminal hands, and here the question can be raised whether mankind benefits from knowing the secrets of Nature, whether it is ready to profit from it or whether this knowledge will not be harmful for it. The example of the discoveries of Nobel is characteristic, as powerful explosives have enabled man to do wonderful work. They are also a terrible means of destruction in the hands of great criminals who are leading the peoples towards war. I am one of those who believe with Nobel that mankind will derive more good than harm from the new discoveries.

# ALBERT EINSTEIN

■

When Albert Einstein (1879–1955) discovered, in the first
decade of the twentieth century, that enormous reservoirs of
energy are latent in all matter, he at first rejected the suggestion
that it might ever be possible to release that energy in a nuclear
explosion. This essay was written in 1946, by which time he and
the world had become sadder and wiser.

$$E = mc^2$$

IN ORDER TO UNDERSTAND the law of the equivalence of mass and
energy, we must go back to two conservation or "balance" principles
which, independent of each other, held a high place in pre-relativity
physics. These were the principle of the conservation of energy and
the principle of the conservation of mass. The first of these,
advanced by Leibniz as long ago as the seventeenth century, was
developed in the nineteenth century essentially as a corollary of a
principle of mechanics.

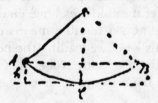

Drawing from Dr. Einstein's Manuscript

Consider, for example, a pendulum whose mass swings back
and forth between the points $A$ and $B$. At these points the mass $m$ is
higher by the amount $h$ than it is at $C$, the lowest point of the path
(see drawing). At $C$, on the other hand, the lifting height has disappeared and instead of it the mass has a velocity $v$. It is as though

the lifting height could be converted entirely into velocity, and vice versa. The exact relation would be expressed as

$$mgh = \frac{m}{2} v^2,$$

with $g$ representing the acceleration of gravity. What is interesting here is that this relation is independent of both the length of the pendulum and the form of the path through which the mass moves.

The significance is that something remains constant throughout the process, and that something is energy. At $A$ and at $B$ it is an energy of position, or "potential" energy; at $C$ it is an energy of motion, or "kinetic" energy. If this concept is correct, then the sum

$$mgh + m\frac{v^2}{2}$$

must have the same value for any position of the pendulum, if $h$ is understood to represent the height above $C$, and $v$ the velocity at that point in the pendulum's path. And such is found to be actually the case. The generalization of this principle gives us the law of the conservation of mechanical energy. But what happens when friction stops the pendulum?

The answer to that was found in the study of heat phenomena. This study, based on the assumption that heat is an indestructible substance which flows from a warmer to a colder object, seemed to give us a principle of the "conservation of heat." On the other hand, from time immemorial it has been known that heat could be produced by friction, as in the fire-making drills of the Indians. The physicists were for long unable to account for this kind of heat "production." Their difficulties were overcome only when it was successfully established that, for any given amount of heat produced by friction, an exactly proportional amount of energy had to be expended. Thus did we arrive at a principle of the "equivalence of work and heat." With our pendulum, for example, mechanical energy is gradually converted by friction into heat.

In such fashion the principles of the conservation of mechanical and thermal energies were merged into one. The physicists were thereupon persuaded that the conservation principle could be further extended to take in chemical and electromagnetic processes — in short, could be applied to all fields. It appeared that in our physical system there was a sum total of energies that remained constant through all changes that might occur.

Now for the principle of the conservation of mass. Mass is defined by the resistance that a body opposes to its acceleration (inert mass). It is also measured by the weight of the body (heavy mass). That these two radically different definitions lead to the same value for the mass of a body is, in itself, an astonishing fact. According to the principle — namely, that masses remain unchanged under any physical or chemical changes — the mass appeared to be the essential (because unvarying) quality of matter. Heating, melting, vaporization, or combining into chemical compounds would not change the total mass.

Physicists accepted this principle up to a few decades ago. But it proved inadequate in the face of the special theory of relativity. It was therefore merged with the energy principle — just as, about 60 years before, the principle of the conservation of mechanical energy had been combined with the principle of the conservation of heat. We might say that the principle of the conservation of energy, having previously swallowed up that of the conservation of heat, now proceeded to swallow that of the conservation of mass — and holds the field alone.

It is customary to express the equivalence of mass and energy (though somewhat inexactly) by the formula $E = mc^2$, in which $c$ represents the velocity of light, about 186,000 miles per second. $E$ is the energy that is contained in a stationary body; $m$ is its mass. The energy that belongs to the mass $m$ is equal to this mass, multiplied by the square of the enormous speed of light — which is to say, a vast amount of energy for every unit of mass.

But if every gram of material contains this tremendous energy, why did it go so long unnoticed? The answer is simple enough: so long as none of the energy is given off externally, it cannot be observed. It is as though a man who is fabulously rich should never spend or give away a cent; no one could tell how rich he was.

Now we can reverse the relation and say that an increase of $E$ in the amount of energy must be accompanied by an increase of $E/c^2$ in the mass. I can easily supply energy to the mass — for instance, if I heat it by 10 degrees. So why not measure the mass increase, or weight increase, connected with this change? The trouble here is that in the mass increase the enormous factor $c^2$ occurs in the denominator of the fraction. In such a case the increase is too small to be measured directly; even with the most sensitive balance.

For a mass increase to be measurable, the change of energy per mass unit must be enormously large. We know of only one sphere in which such amounts of energy per mass unit are released: namely, radioactive disintegration. Schematically, the process goes like this: An atom of the mass $M$ splits into two atoms of the mass $M'$ and $M''$, which separate with tremendous kinetic energy. If we imagine these two masses as brought to rest — that is, if we take this energy of motion from them — then, considered together, they are essentially poorer in energy than was the original atom. According to the equivalence principle, the mass sum $M' + M''$ of the disintegration products must also be somewhat smaller than the original mass $M$ of the disintegrating atom — in contradiction to the old principle of the conservation of mass. The relative difference of the two is on the order of $\frac{1}{10}$ of one percent.

Now, we cannot actually weigh the atoms individually. However, there are indirect methods for measuring their weights exactly. We can likewise determine the kinetic energies that are transferred to the disintegration products $M'$ and $M''$. Thus it has become possible to test and confirm the equivalence formula. Also, the law permits us to calculate in advance, from precisely determined atom weights, just how much energy will be released with any atom disintegration we have in mind. The law says nothing, of course, as to whether — or how — the disintegration reaction can be brought about.

What takes place can be illustrated with the help of our rich man. The atom $M$ is a rich miser who, during his life, gives away no money (*energy*). But in his will he bequeaths his fortune to his sons $M'$ and $M''$, on condition that they give to the community a small amount, less than one thousandth of the whole estate (*energy or mass*). The sons together have somewhat less than the father had (*the mass sum $M' + M''$ is somewhat smaller than the mass $M$ of the radioactive atom*). But the part given to the community, though relatively small, is still so enormously large (*considered as kinetic energy*) that it brings with it a great threat of evil. Averting that threat has become the most urgent problem of our time.

# ROBERT P. CREASE
# AND CHARLES C. MANN

■

In the 1980s Robert Crease, then a graduate student at Columbia University, and the science writer Charles Mann conducted extensive interviews and research in order to compile a history of how physicists had approached a unified theory of matter and energy. The result was their far-reaching book *The Second Creation*. In this excerpt, Crease and Mann describe how the English physicist Paul Dirac derived a quantum theory of the electron and was led by it, albeit unwillingly, to predict the existence of antimatter.

# *The Man Who Listened*

THE STANDARD MODEL of elementary particle interactions was pieced together by three distinct intellectual generations of twentieth-century physicists. Each arrived suddenly, of a piece, in the course of two or three years, a group of young men — women as well, in the case of the third — which emerged with their style of play fully developed, in confident command of the tools of the trade. Like the abstract expressionists, whose intemperate urgency and immediate prominence in postwar New York City stunned their elders, the new physicists startled their contemporaries with the sweep and precision of their attack and the fierceness with which they demanded to be heard. Although, like the expressionist Willem de Kooning, the new workers may actually have labored unrecognized for years, it seems to the community at large that a movement has abruptly formed, fast and bright as a stroke of lightning, and that everything has changed. The construction of quantum mechanics, in the mid-1920s, was the work of the first *nouvelle vague;* names like Heisenberg, Pauli, and Oppenheimer, Rabi, Schrödinger, and Jordan, suddenly appeared on papers that had to be read by every serious practitioner in the field. Meanwhile, the old hands, the Rutherfords and Bohrs, were pushed to new accomplishments. (Some never adjusted; J. J. Thomson was one, and — sadly, grandly — Einstein was another.) The years immediately after the Second World War marked the entrance of another group of scientists — young, hungry, predominantly American; the beginning of the 1970s has witnessed the rise of a third.

In each of these brilliant, collective entrances, there has been one player whose superior insight into the structure of field theory has led his confreres to adjudge him, rightly or wrongly, the most brilliant of the lot. These men are not well-known outside the field — in a related discipline, how many members of the public know of David Hilbert, arguably the finest mathematician this century has produced? — and they may not produce as much physics as others of their generation, but nonetheless the acuity of their vision and the formal tools under their control have awed their contemporaries. Physicists have as much difficulty with complex mathematics as anyone else, and it is little wonder that the legends should accrue around mathematical physicists. In the recent rush to assemble the standard model and move toward unification, the name of Gerard 't Hooft stands out; in 1948, the honor went to Julian Schwinger. During the short, happy heyday of quantum mechanics, the reputation went to Paul Adrien Maurice Dirac.

At a time when young physicists created stirs with their first papers and every dissertation opened a new field, it was Dirac who most shaped the resulting science. If Einstein, with his disdain for the quantum theory, was truly the last classical physicist, then P.A.M. Dirac, as he always signed his work, was the first completely modern one. In an article written soon before Dirac's death in 1984, the physicist Silvan Schweber remarked, "Dirac is not only one of the chief authors of quantum mechanics, but he is also the creator of quantum electrodynamics and one of the principal architects of quantum field theory. All the major developments in quantum field theory in the thirties and forties have as their point of departure some work of Dirac's."

Circumstances seem to have conspired to make Dirac painfully shy and taciturn. He was born on August 8, 1902, in Bristol, England, then, as now, a commercial town on the confluence of the Avon and Frome rivers. His father, a Swiss émigré, was retiring to the point of being antisocial; the family never had guests, and the Diracs never went out. Dirac ate with his father in the dining room, while the rest of the family had dinner in the kitchen; the boy would have preferred to be with his mother, sister, and brother, but there were not enough chairs in the kitchen. At home, Dirac once recalled, "My father made the rule that I should only talk to him in French. He thought it would be good for me to learn French in that way. Since I found that I couldn't express myself in French, it was better for me to stay silent than to talk in English. So I became very silent at that time." Quiet and introverted, Dirac spent much of his time outdoors, taking solitary walks through the English countryside. He liked order and symmetry, the meticulous compression of mathematical relationships. "A great deal of my work is just playing with equations and seeing what they give," Dirac said later. "I don't suppose that applies so much to other physicists; I think it's a peculiarity of myself that I like to play about with equations, just looking for beautiful mathematical relations which maybe don't have any physical meaning at all. Sometimes they do."

Although Dirac's father had no appreciation for the importance of social contact, he did realize the use of a good education, and encouraged his son's mathematical bent. Moreover, this proclivity was fostered by an accident of history: As a teenager, Dirac was pushed into a higher level than was normal for his age, because the more advanced classes had emptied out when older students left to go to war. He liked the Merchant Venturer's School, which shared quarters with the engineering college of Bristol University, partly because the curriculum deemphasized philosophy and the arts, subjects that he found nearly incomprehensible for most of his life. Afraid that there would be no jobs in mathematics, Dirac chose to specialize in engineering when he attended the university. He was a good student, but only the theoretical aspects of the field really interested him; his early on-the-job training ended disastrously when his employer found that Dirac "lacked keenness, and was slovenly." Although his only brother was working in the same factory, the two boys never exchanged a word.

In the fall of 1921, having completed an engineering degree, Dirac found himself unable to obtain employment. The Bristol mathematics professors, who had made known their disappointment that a brilliant mathematician was taking engineering courses, offered Dirac free tuition; because he had nothing else to do, he accepted. The only other student in the honors program of mathematics was a woman firmly intent on studying applied mathematics, especially that which could be used in physics. Because he did not have any firm convictions, Dirac went along with her wishes; in this way the career of one of the great physicists of the century began.

From the haphazard beginning of his career to the end of his days, Dirac was convinced that mathematics is the key to progress in physics. In one of his last addresses, Dirac explained his credo: "[O]ne should allow oneself to be led in the direction which the mathematics suggests . . . [o]ne must follow up [a] mathematical idea and see what its consequences are, even though one gets led to a domain which is completely foreign to what one started with. . . . Mathematics can lead us in a direction we would not take if we only followed up physical ideas by themselves."

At Bristol, Dirac became acquainted with relativity, which electrified him. He won a B.Sc. and, buttressed by two grants, entered Saint John's College, Cambridge, in 1925. By 1927, at the age of twenty-five, his contributions to quantum mechanics had ensured that he was one of the most important physicists in the world.

It should be said that he was not changed unduly by the growth of his reputation; he continued to be so laconic that people who met him often thought him rude. Although an honored member of the Cambridge physics group, he had few students, established no school, and talked rarely to experimenters. Samuel Devons, who spent the late 1930s at the laboratory, once told us, "There was a Cavendish Physical Society meeting, a sort of semiformal gathering once every two weeks. Some lecturer would come in, and Dirac would sit in the front row and listen. Very rarely would he open his mouth. Sometimes he would be prodded by Rutherford, who'd say, 'Now what do you theoretical people think?' Rutherford had this notion that theory was some sort of speculation — the real facts were in the experiments. And Dirac would sit and say nothing."

Dirac spoke so precisely and carefully that he approached the Delphic; when he taught quantum mechanics, he stood behind the podium and read to the class from the book he had written on the subject, believing that he had set down his point of view there as well as he could. In 1928, he gave a series of talks at Leiden, where Paul Ehrenfest, the Dutch theoretician who had quickly submitted the idea of spin before its creators could get scared, was frustrated by the Olympian manner of Dirac's presentation. H.B.G. Casimir was in the audience. Each lecture, he recollected later, "was presented in perfect form. You know Dirac's habit — if you didn't understand things, he would not offer any explanations but would very patiently repeat exactly the same thing. Usually it worked, but it wasn't quite

Ehrenfest's way of doing things." Ehrenfest wanted always to see the human being behind the work. "I" — Casimir again — "remember that once Ehrenfest put a question to Dirac to which Dirac had no immediate answer. And so Dirac began to work it out on the blackboard. He covered the entire blackboard with very small [writing]. And Ehrenfest was [standing] right behind him, trying to see what he did, and exclaiming, *'Kinder, Kinder! Schaut jetzt zu! Jetzt kann man sehen, wie er es macht!'*" Kids, kids — look at this! *Now* we can see what he's up to!

In August of 1925, R. H. Fowler of the Cavendish received the proofs of an as-yet unpublished paper by Werner Heisenberg that, in the opinion of its author, created "some new quantum mechanical relations." It was Heisenberg's first paper on quantum mechanics, the paper that inspired Born to think of matrices. Fowler gave the proofs to his young research assistant, Dirac. Dirac's interest vanished as soon as he saw that Heisenberg had prefaced his work with several paragraphs of philosophical musings about the importance of considering observable quantities. Noticing that the principal example of the new methods Heisenberg had provided was of no real import, Dirac decided that there was no reason to pay much attention. He came back to it a week or so later, and this time realized that Heisenberg's quantum-theoretical series, the variables with which the equations were formulated, were noncommutative, that is, $A \times B$ did not equal $B \times A$. Dirac saw that this was the key to Heisenberg's entire scheme, and that the reason he had not appreciated the article's significance was that Heisenberg in his anxiety had chosen to demonstrate his idea with a sample calculation in which the noncommutativity didn't show; because noncommutativity was central to the whole conception, he had been forced to provide a trivial example. Dirac remarked later that Heisenberg "was afraid this [noncommutativity] was a fundamental blemish in his theory and that probably the whole beautiful idea would have to be given up. . . . At this stage, you see, I had an advantage over Heisenberg because I did not have his fears. I was not afraid of *Heisenberg's* theory collapsing. It would not have affected me as it would have affected Heisenberg. It would not have meant that I would have had to start again from the beginning." In Dirac's opinion, the originators of new ideas are too protective of their brainchildren to develop them properly; creativity implies a corresponding lack of perspective.

Like Born, Dirac quickly realized that $pq - qp = \hbar/i$. But unlike

Born, Dirac did not see at first that $p$ and $q$ were associated with matrices. He regarded them simply as strange versions of position and momentum, and looked for some means to connect Heisenberg's new quantum mechanics with the classical mechanics Dirac felt he understood.

> At this time [September of 1925] I used to take long walks on Sundays alone, thinking about these problems, and it was during one such walk that the idea occurred to me that the commutator $A$ times $B$ minus $B$ times $A$ was very similar to the Poisson bracket which one had in classical mechanics when one formulates the equations in the Hamiltonian form. That was an idea that I just jumped at as soon as it occurred to me. But then I was held back by the fact that I did not know very well what was a Poisson bracket. It was something which I had read about in advanced books of dynamics, but there was not really very much use for it, and after reading about it, it had slipped out of my mind and I did not very well remember what the situation was.

The French mathematician Siméon-Denis Poisson had introduced this idea in 1809, as a curiosity, to help him make some calculations about the motion of a planet in its orbit. Although Poisson brackets had drawn occasional interest from physicists, until Dirac's insight they were rarely used.

> Well, I hurried home and looked through all my books and papers and could not find any reference in them to Poisson brackets. The books that I had were all too elementary. It was a Sunday. I could not go to a library then; I just had to wait impatiently through the night and then the next morning early, when the libraries opened, I went and checked what a Poisson bracket really is and found that it was as I had thought. . . . This provided a very close connection between the ordinary classical mechanics which people were used to and the new mechanics involving the noncommutating quantities which had been introduced by Heisenberg.

A lengthy letter in Dirac's minute, fussy handwriting arrived in Göttingen on November 20, 1925. Written in English, it explained to the astonished Heisenberg how in a few concise steps his version of quantum mechanics could be reformulated in classical terms using a mathematical device he had never heard of. Moreover, working alone, Dirac had come up with a version of quantum mechanics much more general and complete than that produced by Heisenberg, Born, and Jordan in their just-completed joint paper. Stunned at this authoritative communiqué from a complete stranger, Heisenberg replied immediately — in German, a language that Dirac was barely able to follow:

> I have read your extraordinarily beautiful paper on quantum mechanics with the greatest interest, and there can be no doubt that all your results are correct as far as one believes at all in the newly proposed theory. . . . [Moreover, your paper is] really better written and more concentrated than our attempts here.

Casually, Heisenberg mentioned that he and others have been working along some of the same lines.

> I hope you are not disturbed by the fact that part of your results have already been found here some time ago and are being published independently in two papers, one by Born and Jordan, the other by Born, Jordan and me. [He didn't mention that "some time ago" meant about a month.] Your results, in particular . . . the connection of the quantum conditions with the Poisson bracket, go considerably further.

He then peppered Dirac with questions about the applicability of Poisson brackets to quantum theory. In the next ten days, Heisenberg sent off a postcard and two more letters, all with further questions.

Unfortunately, Dirac's replies are lost. They were among the papers that American military authorities confiscated from Heisenberg at the end of World War II, and, despite all pleas, never returned. But Dirac recalled later that, in one way or another, he was able to answer all of Heisenberg's objections and show that Poisson brackets do, indeed, allow quantum mechanics to be cast neatly into Hamiltonian form by simpler and more classical methods than those Heisenberg used. "That," Dirac said, "was the beginning of quantum mechanics so far as I was concerned."

Three months later, in March of 1926, Schrödinger's wave equation appeared. Like Heisenberg, Dirac was annoyed, but for entirely different reasons. Dirac was hot in pursuit of the mathematical analogies he had unearthed between Heisenberg's quantum mechanics and classical mechanics, and he resented having to turn his attention to a formidable-looking new set of basic ideas whose value was not clear. But he, too, was eventually unable to avoid the wave function.

By the time Dirac had completed his Ph.D. in the spring of 1926, he had completely reorganized quantum mechanics, and was ready to begin pushing its frontier forward. Following graduation, Dirac had the chance to travel to further his studies. His first thought was, unsurprisingly, Göttingen, the home of matrix mechanics. But Fowler, who was fond of Bohr, strongly encouraged him to go to Copenhagen. Torn, Dirac decided to spend several months at each place, going first to Copenhagen. He arrived in the second week of September 1926.

Despite Bohr's garrulity, his overwhelming concern with the philosophical implications of physics, and his notorious difficulties with precise expression, he hit it off immediately with Dirac. "I think mostly Bohr was talking and I was listening," Dirac recalled. "That rather suits me because I'm not very fond of talking." At late-night talkfests in Copenhagen taverns, Dirac met Heisenberg, Pauli, and the rest of the "quantum mechanics" for the first time — a loquacious, snobbish, simpatico bunch that appears to have gotten him to open up a little. Like many of the shy, smart people in that time and place, Dirac had violently colored political views; with passionately lofty detachment, he told his Continental colleagues that there was no reason for the poor to suffer, that he saw little purpose in rewarding the greedy with wealth, and that organized religion was a ludicrous sham. After one such disquisition, Wolfgang Pauli, who had mystical leanings, is supposed to have remarked, "Dirac has a new religion — there is no God, and Dirac is His prophet!"

In Copenhagen, Dirac began to work on some ideas that he hoped would reconcile quantum mechanics and relativity. Quantum mechanics had much to say about the principles of action in the subatomic domain, but it did not take into account the special effects that occur at speeds near that of light. What happens when a flashlight beam hits a wall? The electrons in the wall absorb and emit photons, in the process moving at the enormous velocities where relativity is important. A proper theory of light and matter thus had to describe quantized, relativistic behavior. Dirac set out to create such a theory during the last months of 1926 and the first weeks of 1927, the same period in which Heisenberg wrestled with the uncertainty principle. In the optimism of the time, Dirac had little notion of what a tangle he was to create, and how many years — decades, in fact — it would take to unravel.

Dirac's principal tool was an old idea, that of a *field*. A field is a region of space in which particular quantities are precisely defined at every point. For instance, the pattern of arcs that iron filings form around a magnet illustrates the magnet's field. Each point along each arc is subject to a force of particular strength and direction. The alignment of each filing indicates the direction of the field at that point, and the density of the filings indicates the field's strength. Fields can be defined for such diverse domains as temperature, sound, and matter. The idea of a field was slowly developed by nineteenth-century physicists, and culminated in James Clerk Maxwell's demonstration, in 1861, that light of all varieties could be described as a pattern of electric and magnetic fields. (For this reason, visible light, radar waves, X rays, and infrared beams are all called electromagnetic radiation.) In 1905, Einstein showed that electromagnetic fields are associated with quanta — photons — that act as the agents of the field. If the field is like the domain of a feudal overlord, then photons can be thought of as the soldiers, tax collectors, magistrates, and factota who make the wishes of the ruler felt in that area. By the time Dirac considered the problem of the interaction of matter and electromagnetic radiation, physicists fully realized that to describe the comings and goings of electrons and photons it would be necessary to treat the field associated with the photons. And this is exactly what he set out to do.

Like Planck, Heisenberg, and Schrödinger before, Dirac had recourse to systems of oscillators. But unlike his predecessors, he thought of the atom *and* the field in these terms. Using Heisenberg's quantum mechanics, Dirac was able to come up with a Hamiltonian for the field that was fully compatible with the Hamiltonian for the atom from quantum mechanics. Dirac was thus able to say that the Hamiltonian for the entire process could be found by adding up the separate Hamiltonians for the atom, the field, and the interaction. Moreover, Dirac showed that by juggling the Hamiltonians through an appropriate mathematical procedure, he could prove a law that Einstein had discovered, which gave the probability that a given atom in a given state that sat in a particular field in a particular configuration would absorb or release a photon.

The result was the first real quantum field theory. Because it linked quantum theory with the dynamics of electromagnetic fields,

Dirac called it *quantum electrodynamics*. Another way of describing Dirac's accomplishment might be to say that in the beginning of 1927, human beings understood fairly accurately for the first time what happens on the atomic level when someone shines a light in a mirror and the glow is reflected back.

Highly satisfied, Dirac submitted his paper to the *Proceedings of the Royal Society* in the last days of January of 1927, just three weeks before Heisenberg wrote his long letter to Pauli about the uncertainty principle. One of the most influential papers in the history of twentieth-century physics, "The Quantum Theory of the Emission and Absorption of Radiation" begins with the prescient assertion that there is little further work to do in quantum mechanics.

> The new quantum theory, based on the assumption that the dynamical variables do not obey the commutative law of multiplication, has by now been developed sufficiently to form a fairly complete theory of dynamics. . . . On the other hand, hardly anything has been done up to the present on quantum electrodynamics.

"Hardly anything" is a classic bit of Oxbridgian understatement. *Nothing* had been done before on quantum electrodynamics.

> The questions of the correct treatment of a system in which the forces are propagated with the velocity of light instead of instantaneously, of the production of an electromagnetic field by a moving electron, and the reaction of this field on the electron have not yet been touched.

Dirac said often that he was beset by fears when he came up with his new theory. Perhaps. The tone of the article doesn't betray it.

> [I]t will be impossible to answer any one question completely without at the same time answering them all.

In any case, he soon discovered that his fear was justified. Quantum electrodynamics was indeed a great step forward, but it came at a great price. Dirac had set down the beginnings of the modern theory of electromagnetism — the first solid piece of the standard model — but he had also unwittingly let loose an onslaught of conceptual demons that would change our views of space and matter. As a step to quantizing the electromagnetic field, Dirac hypothesized that his oscillators did not disappear when there was no field. Rather, they went into a "zero state," in which they existed but could not be detected. Associated with the zero state oscillators were zero state photons; these, too, could not be detected. It didn't trouble Dirac that his mathematical scheme implied that empty space should contain billions of invisible photons. As long as they didn't show up, their presence made no difference.

But when Dirac's theory was interpreted in light of the uncertainty principle, it turned out that in some sense the zero state photons *do* show up. According to the uncertainty principle, the energy of a field over any given time cannot be determined exactly. (The indeterminacy relations apply equally to particles and fields.) Paradoxically, the less time one spends measuring, the greater this margin of error becomes. It is therefore conceivable that within a time interval on the order of trillionths of a trillionth of a second the zero state oscillators of a field actually might not be at the zero state. In fact, they might have a vast amount of energy that escapes detection. As Einstein's equation $E = mc^2$ demonstrated, mass and energy are two forms of the same thing. Thus the uncertainty principle dictates that if any small area can contain undetectable energy, then, according to Einstein's equation, it can contain undetectable matter. This basic uncertainty is not just a lacuna in our knowledge. Mathematically, there is no difference between this uncertainty and actual random fluctuations in the energy (or matter) measured. Therefore, at least in theory, because any space *might* harbor particles for a short time, it *must* do so.

As strained through the uncertainty principle, quantum field theory exposed a frightful chaos on the lowest order of matter. The spaces around and within atoms, previously thought to be empty, were now supposed to be filled with a boiling soup of ghostly particles. From the perspective of quantum field theory, the vacuum contains random eddies in space-time: tidal whirlpools that occasionally hurl up bits of matter, only to suck them down again. Like the strange virtual images produced by lenses, these particles are present, but out of sight; they have been named *virtual particles*. Far from being an anomaly, virtual particles are a central feature of quantum field theory, as Dirac himself was soon to demonstrate.

At first, however, he does not seem to have realized the full implications of quantum electrodynamics. Instead, he worried about how to make his work jibe with that of his colleagues. There were several major difficulties. First, quantum electrodynamics was based on Dirac's formulation of quantum mechanics, which was itself not entirely in accord with the dictates of relativity. (Dirac's electromagnetic field was relativistic, but his matter field was not.) Deeply sympathetic to the urge for consistency and order that was one of the wellsprings of Einstein's thought, Dirac thought the discrepancy was terribly bothersome. Moreover, in recent months two German physicists, Walter Gordon and Oskar Klein, had produced a relativistic version of Schrödinger's equation for the electron. A little while before his death, Dirac said that the Klein-Gordon equation "could not be interpreted in terms of my general [version] of quantum mechanics and was therefore unacceptable to me. Other physicists with whom I talked at the time were not so obsessed with the need for having quantum theory agreeing [*sic*] with the general . . . theory, and they were rather inclined to let it go as it was. But I just stuck to this problem." He stuck to his last until the end of 1928, when he returned to England. By then, he had developed yet another equation for the electron's motion — the one used today. To reconcile his own version of quantum mechanics, relativity, Klein-Gordon, and the experimental data, Dirac produced a single equation for an electron traveling through space that had four components, two associated with the spin, and two with the energy of the particle.

The Dirac equation, as it is called, suddenly explained a host of puzzles about the electron. To give one example, it confirmed Bohr's old prediction that the problem of the velocity of electron spin would vanish when the *real* quantum theory was found. According to Dirac's theory, an electron is not localized in a particular place, but has a set of probable locations that are scattered around a point of maximum probability like the cluster of holes around the bull's-eye of a sharpshooter's target. Moreover, these locations circulate around the center as if the target were spinning; according to Dirac's equation, this motion is the spin of the electron. Lorentz had envisioned a little ball rotating when he calculated the electron spin velocity, and found that it turned faster than the speed of light — a flat impossibility. But Dirac said that you should use the average radius of the cluster instead, which is more than a hundred times bigger than the electron envisioned by Lorentz. This made the spin velocity much slower, and removed the contradiction with relativity. Not only that, Dirac's theory predicted that this circulation would create a tiny magnetic field around each electron, and accurately predicted its strength, something previous descriptions of spin had been unable to do.

It was difficult for physicists not to be impressed, even awed, by the Dirac equation. Like children shaking fruit from a tree, they extracted equations describing the collision of two electrons, the interaction of photons and free electrons, and the correct formulas for the spectral lines of the hydrogen atom. The results came so quickly that some members of the field were sure that it was only a matter of time — months, perhaps — before Dirac or someone else would come up with an equation for the last remaining piece of the puzzle, the proton. With photons, electrons, and protons disposed of, scientists would soon knock off the nucleus, and then, except for filling in a few loose ends, all of matter and energy would be explained. Unification would occur, and physics would be over.

"There was one of the regular conferences in Copenhagen," the physicist Rudolf Peierls has recalled. "It may have been the conference of thirty-two or thirty-one; I don't know. The interesting point is that there was a general feeling among some people there, not everybody, that physics was almost finished. This looks ridiculous looking back, but if you look at it from the point of view of the time, practically all of the mysteries had resolved themselves — nearly all. Everything that had bothered one about the atom and molecules and solids and so on, had suddenly fallen into place. . . . Now, I'm not saying this was the common view. I don't think I ever shared it, really. I don't think Niels Bohr, for example, would ever have had any such illusions. But there were sort of over-lunch — or sometimes quite serious — discussions about what we would do when physics was finished. By 'finished' was meant the basic structure; of course, there are all the applications. The majority of people said that that would be the time to turn to biology. Only one person really took that seriously and did turn to biology, and that was Max Delbrück, who certainly was present at these discussions." Delbrück did well in molecular biology, winning the Nobel Prize for medicine in 1969 for beginning the chain of investigations that led to the discovery of the doubled helix of DNA. On the other hand, physics was anything but over.

In the meantime, a small physics community had begun to form in the United States, but it was as yet a paltry thing compared to its equivalent on the other side of the Atlantic. America had the resources to do experimental work, but except for isolated savants like Josiah Gibbs of Yale, the theory came entirely from Europe. More important, the context of physics was set by Continental researchers. Classes at Caltech and Columbia University were assigned textbooks written in Göttingen and Leiden, and graduate students in Cambridge, Massachusetts, angled for the opportunity to travel to Cambridge, England. Aided by grants from the newly formed National Research Council, a pool of European-trained physicists slowly accumulated in this country; from them would later come the leaders of the next generation, one dominated by Americans. I. I. Rabi was among them.

Rabi got his dissertation in 1927, and promptly left for Europe with his wife, Helen. "I found I was actually better prepared than most Europeans of the same level," he told us. "What we lacked — and my generation was to supply it in this country — was a kind of understanding and feeling for the subject, which is hard to get if you're not in contact somehow with a verbal tradition, with people who are making the subject. [You have] to see the living tradition." After a month in Munich and two in Copenhagen, Rabi settled down in Hamburg to work with Wolfgang Pauli. There he met all the creators of quantum mechanics and quantum field theory. He took Dirac to Hamburg's well-known Hagenbeck Zoo, and was startled to discover that the Englishman's love for order extended to an insistence that they see the exhibits *in seriatum*. For Rabi's benefit, Max Born listed all the accomplishments of the past few years. "It was heady stuff," Rabi admitted, a half-smile playing on his lips. "Born said, 'We *have* all that. I think in six months we'll have the proton, and physics as we know it will be over.' There would be a lot to do, but the heroic part was finished. I was just a fresh postdoc [a postdoctoral fellow], and he was foreclosing on the field!"

Not everyone was as sanguine as Born. Heisenberg and Pauli, for example, were troubled by some of Dirac's assumptions, and spent 1928 and 1929 working on their own version of quantum electrodynamics. (To the relief of other physicists, the two versions were subsequently shown to be identical, much as Schrödinger wave mechanics and Heisenberg matrix mechanics were discovered to be different ways of describing the same thing.) Dirac was perhaps the least satisfied of all. It pleased him that two of the four components of his electron equation corresponded to the spin of the particle. But the two other components, those corresponding to the energy of the electron, were more difficult to interpret. When the Dirac equation is solved for the energy of an individual electron, there are two answers, one positive, one negative, in a manner analogous to the way that the square root of a number can be either positive or negative. (The roots of 49, for instance, are 7 and $-7$, because both $7 \times 7$ and $-7 \times -7$ equal 49.) The negative answer was trouble: negative energy, itself a puzzling idea, implied, by $E = mc^2$, negative mass — an absurd impossibility. Ordinarily, physicists would assume that the world had started off with all electrons in positive energy states, in which they would remain, and no harm would come from the theoretical existence of negative energy. The problem was that a positive energy electron should be able to emit a photon with enough energy to drop into a negative energy state. In fact, *most* electrons should end up with negative energy.

Alas, "negative energy" is not an easily understood concept. Nobody knew what negative energy electrons could be. One theorist called them "donkey electrons," because they always do precisely the opposite of what you want. Since energy and mass are equivalent, negative energy implies negative mass. What does it mean to say that a subatomic particle can have less than zero mass? As Dirac admitted, the whole subject "bothered me very much." He wrestled with negative energies through all of 1929, trying to see if there was some way to keep the Dirac equation but get rid of the negative energy states.

> And then [Dirac recounted later] I got the idea that because the negative energy states cannot be avoided, one must accommodate them in the theory. One can do that by setting up a new picture of the vacuum. Suppose that in the vacuum all the negative energy states are filled up. . . . We then have a sea of negative energy electrons, one electron in each of these states. It is a bottomless sea, but we do not have to worry about that. The picture of a bottomless sea is not so disturbing, really. We just have to think about the situation near the surface, and there we have some electrons lying above the sea that cannot fall into it because there is no room for them.

In other words, we don't notice the negative energy electrons because they are omnipresent — as undetectable to us as water is to a fish. But wait:

> There is, then, the possibility that holes may appear in the sea. Such holes would be places where there is an extra energy, because one would need a negative energy to make such a hole disappear.

Because of quantum randomness, in other words, photons should hit some of the electrons in the sea with enough energy to make them jump out of it — that is, become positive energy electrons — leaving

74

a hole in their former location. This hole would appear as a sort of "opposite" electron: positively charged, because it corresponds to an absence of negative charge. Therefore, Dirac realized, even if he accommodated the negative energy electrons, he was forced to predict the existence of particles just like the electron, except with a positive electric charge.

Here Dirac's courage failed him for one of the few times in his career. He refused to follow the naked mathematics. Although logically the hole particles must have the same mass as electrons, because they are created by them, Dirac's paper on the subject said they must somehow be protons — which are almost two thousand times heavier. He was also attracted by the hope of thus explaining both known elementary particles with the same theory. (The neutron would not be discovered for another two years.) Dirac's suggestion was promptly flattened by, among others, the mathematician Hermann Weyl, who reported from Göttingen that whatever the holes might be, they could not be protons. Unafraid of *Dirac's* theory collapsing, Weyl said that if Dirac wanted to salvage his quantum electrodynamics, he was going to have to predict a previously unheard-of type of matter with the same mass but the opposite charge as an electron. Moreover, these new particles should be all over the place. Not one had ever been noticed.

No matter how good a theory might look on paper, Dirac thought in dismay, if it predicts particles that don't exist there is something terribly wrong. Discouraged, he even wondered if all of quantum field theory might have to be scrapped. "I thought it was rather sick," he said afterward. "I didn't see any chance of making further progress."

It is pleasing to report that the answer came quite literally from the clouds. Decades before, Charles Thomson Rees Wilson, a young and impoverished Scotsman with a scholarship at Cambridge, had gone to the highest mountain in the Highlands, Ben Nevis, where there is a meteorological observatory. He had been entranced with the beauty of the clouds surrounding the hilltop, the shimmering rings that suddenly burst into colored existence when the sun burst through the cumuli, or the great shadows cast by the peaks on the foggy masses below. When Wilson returned to the Cavendish with the announced intention of studying the process of cloud formation, J. J. Thomson was willing to let him try to make clouds himself in

75

the Cavendish. The fruit of these studies, made between 1896 and 1912, was the cloud chamber, which for decades was the primary means of studying subatomic particles.

Clouds, Wilson knew, had something to do with water vapor in the air. The amount of water vapor that a given volume of air can hold depends on its temperature; it can contain more when hot than when cold. Wilson also knew that if a volume of air suddenly expands, its temperature drops. When the compressed gas in an aerosol can is released, for instance, it expands out the nozzle and chills immediately — the reason that spray deodorant feels cool on the hottest of summer days. While Thomson and others at the Cavendish puzzled over radioactivity and X rays, Wilson constructed glass tanks with ingenious pistons tightly fitted into the inside walls. Wilson discovered that if he sprayed the air inside his tank with mist until it could absorb no more water, then suddenly pulled out the piston, expanding and cooling the air, the surplus water vapor "fell out" and formed a miniature cloud.

The condensation was started by droplets of water collecting around dust particles in the air. Wilson tried cleaning the particles out of the chamber to see if that prevented clouds from forming. It did not, although the clouds emerged only if the piston were withdrawn a lot more. Now what was the water coagulating around? In February of 1896, less than two months after the discovery of X rays, Wilson shot a beam of them into his glass tank. It was much easier to make clouds. Wilson was sure that the X rays smacked into air molecules, creating the charged objects known as ions, and the ions caused the clouds.

A year later, when J. J. Thomson discovered the electron, it would be clear that the X rays knock loose electrons, which are easily swept onto tiny droplets of water in the air. When several electrons collect on one droplet, the net charge begins to exert an effect on nearby molecules of water vapor. Water molecules ($H_2O$) are shaped like a broad **V**, with the oxygen nucleus at the vertex and the two hydrogen nuclei — protons — sticking out the ends. The electrons of the hydrogen atoms are yanked toward the oxygen atom by its much bigger, positively charged nucleus, which means that the tips of the arms of the **V** tend to end up with a net positive charge. The positive charge is easily drawn toward the negative electrons on the droplet, with the result that in rapid order a lot of water molecules are pulled to the charged droplets, gas condenses to form liquid, and the number and size of the droplets grow. A tiny cloud is born.

With the excitement in the Cavendish caused by radioactivity, it was soon discovered that water vapor would also condense around the electrons left behind by alpha particles speeding through the chamber. For a few instants — long enough for a photograph to be taken — the paths of these particles are visible as slender white lines that arc across the chamber like the contrails of a jet squadron. Alpha particles leave thick tracks, for these slow and doubly charged entities dislodge many electrons; the electrons from beta radiation have a smaller charge, are much faster, and leave light, wobbly tracks reminiscent of scratches from a cat's claw. If a magnet is wrapped around the chamber, particle identification is even easier: The magnetic field forces negatively charged electrons to curve in one direction, whereas positive alpha particles and protons are sent off in the other. The degree of curvature earmarks the particle's momentum. Thus Wilson could put a lump of uranium by his cloud chamber, pull out the piston with a thump, quickly click the camera shutter, and produce a photograph of the tracks left by the spray of radiation. Using a ruler and compass, he could discern that such-and-such a thick line that curved across the photographic plate in such-and-such a direction was left by an alpha particle with such-and-such momentum, whereas a light little line was a beta electron with a different momentum. Cloud chambers soon became standard equipment in laboratories where radioactivity was studied, and Wilson was awarded the Nobel for his invention in 1927.

In 1930, Robert Millikan, the head of the laboratory at the California Institute of Technology, asked Carl D. Anderson, a fresh young Caltech Ph.D., to build a new cloud chamber for use in studying cosmic rays, the recently discovered high-energy radiation that bombards the earth from space. With the cloud chamber, Anderson built an electromagnet so powerful that the laboratory lights dimmed when the cloud chamber was in use. For the next two years, Anderson expanded the chamber at random times, photographed the tracks of any cosmic rays that happened to pass through, and studied the results.

The great majority of the photographs were blank — no cosmic rays came through. But from the very beginning Anderson also started to find that a small number of the photographs had something strange, tracks from light particles that could either be negatively charged particles traveling upward or positively charged particles traveling downward. Anderson wrote later:

> In the spirit of scientific conservatism we tended at first toward the former interpretation, i.e., that these particles were upward-moving, negative electrons. This led to frequent and at times somewhat heated discussions between Professor Millikan and myself, in which he repeatedly pointed out that everyone knows that cosmic-ray particles travel downward, and not upward, except in extremely rare instances, and that therefore, these particles must be downward-moving protons. This point of view was very difficult to accept, however, since in nearly all cases the [thickness of the track] was too low for particles of proton mass.

To settle the argument with Millikan, Anderson inserted a metal plate in the center of the chamber. Any particle passing through the plate would lose momentum, slow down, and thus be bent more sharply afterward, thereby indicating its direction. Furthermore, Anderson could calculate the particle's mass from the momentum lost passing through the plate.

On August 2, 1932, Anderson obtained a stunningly clear photograph that shocked both men. Despite Millikan's protestations, a particle had indeed shot up like a Roman candle from the floor of the chamber, slipped through the plate, and fallen off to the left. From the size of the track, the degree of the curvature, and the amount of momentum lost, the particle's mass was obviously near to that of an electron. But the track curved the wrong way. The particle was *positive*. Neither electron, proton, or neutron, the track came from something that had never been discovered before. It was, in fact, a "hole," although Anderson did not realize it for a while.

The identification was made by two chagrined English experimenters — chagrined because they had actually seen the particle in their apparatus before Anderson and had been told by Dirac what it might be, but had waited much too long to be sure. Anderson called the new particle a "positive electron"; *positron* was the name that stuck.

Positrons were the new type of matter — antimatter — Dirac had been forced to predict by his theory. (The equation, he said later, had been smarter than he was.) Physicists soon realized that electrons and positrons would annihilate each other when they met, producing two photons — two flashes of light. Similarly, a photon going through matter could split into a virtual electron and a positron. From an embarrassment the negative energy states were transformed into a triumph for a quantum electrodynamics, the first time in history that the existence of a new state of matter had been predicted on purely theoretical grounds. Dirac won the Nobel Prize in 1933; Anderson went to Sweden three years later.

The canonization of the new developments occurred at the Solvay Conference of 1933, in which the evidence for both the positron and Chadwick's just-discovered neutron was discussed eagerly by Bohr, Curie, Dirac, Heisenberg, Pauli, and the rest of the luminaries of subatomic physics. (The Solvay Conferences were periodic gatherings of prominent physicists that were paid for by a rich Belgian industrial chemist named Ernest Solvay.) With a negative, positive, and neutral particle of real matter and the discovery of antiparticles, the mood was of extraordinary excitement. A form of hydrogen, called deuterium, had been discovered which had a proton *and* a neutron in its nucleus, and scientists were beginning to build machines, called particle accelerators, that promised to unlock even more discoveries. Some physicists thought (again) that the subject was rushing to a close. One of the few sour notes came from Ernest Rutherford, who remarked of the positron that he "would find it more to [his] liking if the theory had appeared *after* the experimental facts were established." . . .

# PAUL A. M. DIRAC

■

In these remarks taken from his 1933 Nobel Prize address, Paul
Dirac (1902–1984) describes how he arrived at his theory of the
electron, which predicted the existence of antimatter and forged
the foundations of quantum electrodynamics.

# *Theory of Electrons*
# *and Positrons*

MATTER HAS BEEN found by experimental physicists to be made up
of small particles of various kinds, the particles of each kind being
all exactly alike. Some of these kinds have definitely been shown
to be composite, that is, to be composed of other particles of a simpler nature. But there are other kinds which have not been shown
to be composite and which one expects will never be shown to be
composite, so that one considers them as elementary and fundamental.

From general philosophical grounds one would at first sight like
to have as few kinds of elementary particles as possible, say only one
kind, or at most two, and to have all matter built up of these elementary kinds. It appears from the experimental results, though,
that there must be more than this. In fact the number of kinds of
elementary particles has shown a rather alarming tendency to increase during recent years.

The situation is perhaps not so bad, though, because on closer
investigation it appears that the distinction between elementary and
composite particles cannot be made rigorous. To get an interpretation of some modern experimental results one must suppose that

particles can be created and annihilated. Thus if a particle is observed to come out from another particle, one can no longer be sure that the latter is composite. The former may have been created. The distinction between elementary particles and composite particles now becomes a matter of convenience. This reason alone is sufficient to compel one to give up the attractive philosophical idea that all matter is made up of one kind, or perhaps two kinds of bricks.

I should like here to discuss the simpler kinds of particles and to consider *what can be inferred about them from purely theoretical arguments*. The simpler kinds of particles are:

(*i*) the photons or light-quanta, of which light is composed;

(*ii*) the electrons, and the recently discovered positrons (which appear to be a sort of mirror image of the electrons, differing from them only in the sign of their electric charge);

(*iii*) the heavier particles — protons and neutrons.

Of these, I shall deal almost entirely with the electrons and the positrons — not because they are the most interesting ones, but because in their case the theory has been developed further. There is, in fact, hardly anything that can be inferred theoretically about the properties of the others. The photons, on the one hand, are so simple that they can easily be fitted into any theoretical scheme, and the theory therefore does not put any restrictions on their properties. The protons and neutrons, on the other hand, seem to be too complicated and no reliable basis for a theory of them has yet been discovered.

The question that we must first consider is how theory can give any information at all about the properties of elementary particles. There exists at the present time a general quantum mechanics which can be used to describe the motion of any kind of particle, no matter what its properties are. The general quantum mechanics, however, is valid only when the particles have small velocities and fails for velocities comparable with the velocity of light, when effects of relativity come in. There exists no relativistic quantum mechanics (that is, one valid for large velocities) which can be applied to particles with arbitrary properties. Thus when one subjects quantum mechanics to relativistic requirements, one imposes restrictions on the properties of the particle. In this way one can deduce information about the particles from purely theoretical considerations, based on general physical principles.

This procedure is successful in the case of electrons and positrons. It is to be hoped that in the future some such procedure will be found for the case of the other particles. I should like here to outline the method for electrons and positrons, showing how one can deduce the spin properties of the electron, and then how one can infer the existence of positrons with similar spin properties and with the possibility of being annihilated in collisions with electrons.

We begin with the equation connecting the kinetic energy $W$ and momentum $p_r$, $(r = 1, 2, 3)$, of a particle in relativistic classical mechanics

$$\frac{W^2}{c^2} - p_r^2 - m^2 c^2 = 0 \tag{1}$$

From this we can get a wave equation of quantum mechanics, by letting the left-hand side operate on the wave function $\psi$ and understanding $W$ and $p_r$ to be the operators $i\hbar \partial/\partial t$ and $- i\hbar \partial/\partial x_r$. With this understanding, the wave equation reads

$$\left[ \frac{W^2}{c^2} - p_r^2 - m^2 c^2 \right] \psi = 0 \tag{2}$$

Now it is a general requirement of quantum mechanics that its wave equations shall be linear in the operator $W$ or $\partial/\partial t$, so this equation will not do. We must replace it by some equation linear in $W$, and in order that this equation may have relativistic invariance it must also be linear in the $p$'s.

We are thus led to consider an equation of the type

$$\left[ \frac{W}{c} - \alpha_r p_r - \alpha_o mc \right] \psi = 0 \tag{3}$$

This involves four new variables $\alpha_r$, and $\alpha_v$, which are operators that can operate on $\psi$. We assume they satisfy the following conditions,

$$\alpha_\mu^2 = 1 \qquad \alpha_\mu \alpha_v + \alpha_v \alpha_\mu = 0$$

for

$$\mu \neq v \text{ and } \mu, v = 0, 1, 2, 3$$

and also the $\alpha$'s commute with the $p$'s and $W$. These special properties for the α's make Eq. (3) to a certain extent equivalent to Eq. (2), since if we then multiply (3) on the left-hand side by $W/c + \alpha_r p_r + \alpha_o mc$ we get exactly (2).

The new variables α, which we have to introduce to get a relativistic wave equation linear in $W$, give rise to the spin of the electron. From the general principles of quantum mechanics one can easily deduce that these variables α give the electron a spin angular momentum of half a quantum and a magnetic moment of one Bohr magneton in the reverse direction to the angular momentum. These results are in agreement with experiment. They were, in fact, first obtained from the experimental evidence provided by spectroscopy and afterwards confirmed by the theory.

The variables α also give rise to some rather unexpected phenomena concerning the motion of the electron. These have been fully worked out by Schrödinger. It is found that an electron which seems to us to be moving slowly, must actually have a very high frequency oscillatory motion of small amplitude superposed on the regular motion which appears to us. As a result of this oscillatory motion, the velocity of the electron at any time equals the velocity of light. This is a prediction which cannot be directly verified by experiment, since the frequency of the oscillatory motion is so high and its amplitude is so small. But one must believe in this consequence of the theory, since other consequences of the theory which are inseparably bound up with this one, such as the law of scattering of light by an electron, are confirmed by experiment.

There is one other feature of these equations which I should now like to discuss, a feature which led to the prediction of the positron. If one looks at Eq. (1), one sees that it allows the kinetic energy $W$ to be either a positive quantity greater than $mc^2$ or a negative quantity less than $-mc^2$. This result is preserved when one passes over to the quantum equation (2) or (3). These quantum equations are such that, when interpreted according to the general scheme of quantum dynamics, they allow as the possible results of a measurement of $W$ either something greater than $mc^2$ or something less than $-mc^2$.

Now in practice the kinetic energy of a particle is always positive. We thus see that our equations allow of two kinds of motion for an electron, only one of which corresponds to what we are familiar with. The other corresponds to electrons with a very peculiar motion such that the faster they move, the less energy they have, and one must put energy into them to bring them to rest.

One would thus be inclined to introduce, as a new assumption of the theory, that only one of the two kinds of motion occurs in practice. But this gives rise to a difficulty, since we find from the theory that if we disturb the electron, we may cause a transition from a positive-energy state of motion to a negative-energy one, so that, even if we suppose all the electrons in the world to be started off in positive-energy states, after a time some of them would be in negative-energy states.

Thus in allowing negative-energy states, the theory gives something which appears not to correspond to anything known experimentally, but which we cannot simply reject by a new assumption. We must find some meaning for these states.

An examination of the behaviour of these states in an electromagnetic field shows that they correspond to the motion of an electron with a positive charge instead of the usual negative one — what the experimenters now call a positron. One might, therefore, be inclined to assume that electrons in negative-energy states are just positrons, but this will not do, because the observed positrons certainly do not have negative energies. We can, however, establish a connection between electrons in negative-energy states and positrons, in a rather more indirect way.

We make use of the exclusion principle of Pauli, according to which there can be only one electron in any state of motion. We now make the assumptions that in the world as we know it, nearly all the states of negative energy for the electrons are occupied, with just one electron in each state, and that a uniform filling of all the negative-energy states is completely unobservable to us. Further, *any unoccupied negative-energy state, being a departure from uniformity, is observable and is just a positron.*

An unoccupied negative-energy state, or *hole,* as we may call it for brevity, will have a positive energy, since it is a place where there is a shortage of negative energy. A hole is, in fact, just like an ordinary particle, and its identification with the positron seems the most reasonable way of getting over the difficulty of the appearance of negative energies in our equations. On this view the positron is just a mirror-image of the electron, having exactly the same mass and opposite charge. This has already been roughly confirmed by experiment. The positron should also have similar spin properties to the electron, but this has not yet been confirmed by experiment.

From our theoretical picture, we should expect an ordinary electron, with positive energy, to be able to drop into a hole and fill up this hole, the energy being liberated in the form of electromagnetic radiation. This would mean a process in which an electron and a positron annihilate one another. The converse process, namely the creation of an electron and a positron from electromagnetic radiation, should also be able to take place. Such processes appear to have been found experimentally, and are at present being more closely investigated by experimenters. . . .

# WERNER HEISENBERG

■

Werner Heisenberg (1901–1976) here describes the principle of quantum uncertainty — the realization that the position and velocity of subatomic particles cannot be known precisely but can only be adduced in terms of probabilities. The uncertainty principle, which Heisenberg discovered in 1927, constitutes a recognition that science can neither scrutinize the present nor predict the future with complete exactitude.

# The Copenhagen Interpretation of Quantum Theory

THE COPENHAGEN INTERPRETATION of quantum theory starts from a paradox. Any experiment in physics, whether it refers to the phenomena of daily life or to atomic events, is to be described in the terms of classical physics. The concepts of classical physics form the language by which we describe the arrangement of our experiments and state the results. We cannot and should not replace these concepts by any others. Still the application of these concepts is limited by the relations of uncertainty. We must keep in mind this limited range of applicability of the classical concepts while using them, but we cannot and should not try to improve them.

For a better understanding of this paradox it is useful to compare the procedure for the theoretical interpretation of an experiment in classical physics and in quantum theory. In Newton's mechanics, for instance, we may start by measuring the position and the velocity of the planet whose motion we are going to study. The result of the observation is translated into mathematics by deriving numbers for the co-ordinates and the momenta of the planet from the observation. Then the equations of motion are used to derive from these values of the co-ordinates and momenta at a given time the values of these co-ordinates or any other properties of the system at a later time, and in this way the astronomer can predict the properties of the system at a later time. He can, for instance, predict the exact time for an eclipse of the moon.

In quantum theory the procedure is slightly different. We could for instance be interested in the motion of an electron through a cloud chamber and could determine by some kind of observation the initial position and velocity of the electron. But this determination will not be accurate; it will at least contain the inaccuracies following from the uncertainty relations and will probably contain still larger errors due to the difficulty of the experiment. It is the first of these inaccuracies which allows us to translate the result of the observation into the mathematical scheme of quantum theory. A probability function is written down which represents the experimental situation at the time of the measurement, including even the possible errors of the measurement.

This probability function represents a mixture of two things, partly a fact and partly our knowledge of a fact. It represents a fact in so far as it assigns at the initial time the probability unity (i.e., complete certainty) to the initial situation: the electron moving with the observed velocity at the observed position; "observed" means observed within the accuracy of the experiment. It represents our knowledge in so far as another observer could perhaps know the position of the electron more accurately. The error in the experiment does — at least to some extent — not represent a property of the electron but a deficiency in our knowledge of the electron. Also this deficiency of knowledge is expressed in the probability function.

In classical physics one should in a careful investigation also consider the error of the observation. As a result one would get a probability distribution for the initial values of the co-ordinates and velocities and therefore something very similar to the probability function in quantum mechanics. Only the necessary uncertainty due to the uncertainty relations is lacking in classical physics.

When the probability function in quantum theory has been determined at the initial time from the observation, one can from the laws of quantum theory calculate the probability function at any later time and can thereby determine the probability for a measurement giving a specified value of the measured quantity. We can, for instance, predict the probability for finding the electron at a later time at a given point in the cloud chamber. It should be emphasized, however, that the probability function does not in itself represent a course of events in the course of time. It represents a tendency for events and our knowledge of events. The probability function can be connected with reality only if one essential condition is fulfilled: if a new measurement is made to determine a certain property of the system. Only then does the probability function allow us to calculate the probable result of the new measurement. The result of the measurement again will be stated in terms of classical physics.

Therefore, the theoretical interpretation of an experiment requires three distinct steps: (1) the translation of the initial experimental situation into a probability function; (2) the following up of this function in the course of time; (3) the statement of a new measurement to be made of the system, the result of which can then be calculated from the probability function. For the first step the fulfillment of the uncertainty relations is a necessary condition. The second step cannot be described in terms of the classical concepts; there is no description of what happens to the system between the initial observation and the next measurement. It is only in the third step that we change over again from the "possible" to the "actual."

Let us illustrate these three steps in a simple ideal experiment. It has been said that the atom consists of a nucleus and electrons moving around the nucleus; it has also been stated that the concept of an electronic orbit is doubtful. One could argue that it should at least in principle be possible to observe the electron in its orbit. One should simply look at the atom through a microscope of a very high resolving power, then one would see the electron moving in its orbit. Such a high resolving power could to be sure not be obtained by a microscope using ordinary light, since the inaccuracy of the measurement of the position can never be smaller than the wave length of the light. But a microscope using γ-rays with a wave length smaller than the size of the atom would do. Such a microscope has not yet been constructed but that should not prevent us from discussing the ideal experiment.

Is the first step, the translation of the result of the observation into a probability function, possible? It is possible only if the uncertainty relation is fulfilled after the observation. The position of the electron will be known with an accuracy given by the wave length of the γ-ray. The electron may have been practically at rest before the observation. But in the act of observation at least one light quantum of the γ-ray must have passed the microscope and must first have been deflected by the electron. Therefore, the electron has been pushed by the light quantum, it has changed its momentum and its velocity, and one can show that the uncertainty of this change is just big enough to guarantee the validity of the uncertainty relations. Therefore, there is no difficulty with the first step.

At the same time one can easily see that there is no way of observing the orbit of the electron around the nucleus. The second step shows a wave packet moving not around the nucleus but away from the atom, because the first light quantum will have knocked the electron out from the atom. The momentum of light quantum of the γ-ray is much bigger than the original momentum of the electron if the wave length of the γ-ray is much smaller than the size of the atom. Therefore, the first light quantum is sufficient to knock the electron out of the atom and one can never observe more than one point in the orbit of the electron; therefore, there is no orbit in the ordinary sense. The next observation — the third step — will show the electron on its path from the atom. Quite generally there is no way of describing what happens between two consecutive observations. It is of course tempting to say that the electron must have been somewhere between the two observations and that therefore the electron must have described some kind of path or orbit even if it may be impossible to know which path. This would be a reasonable argument in classical physics. But in quantum theory it would be a misuse of the language which, as we will see later, cannot be justified. We can leave it open for the moment, whether this warning is a statement about the way in which we should talk about atomic events or a statement about the events themselves, whether it refers to epistemology or to ontology. In any case we have to be very cautious about the wording of any statement concerning the behavior of atomic particles.

Actually we need not speak of particles at all. For many experiments it is more convenient to speak of matter waves; for instance, of stationary matter waves around the atomic nucleus. Such a description would directly contradict the other description if one does not pay attention to the limitations given by the uncertainty relations. Through the limitations the contradiction is avoided. The use of "matter waves" is convenient, for example, when dealing with the radiation emitted by the atom. By means of its frequencies and intensities the radiation gives information about the oscillating charge distribution in the atom, and there the wave picture comes much nearer to the truth than the particle picture. Therefore, Bohr advocated the use of both pictures, which he called "complementary" to each other. The two pictures are of course mutually exclusive, because a certain thing cannot at the same time be a particle (i.e., substance confined to a very small volume) and a wave (i.e., a field spread out over a large space), but the two complement each other. By playing with both pictures, by going from the one picture to the other and back again, we finally get the right impression of the strange kind of reality behind our atomic experiments. Bohr uses the concept of "complementarity" at several places in the interpretation of quantum theory. The knowledge of the position of a particle is complementary to the knowledge of its velocity or momentum. If we know the one with high accuracy we cannot know the other with high accuracy; still we must know both for determining the behavior of the system. The space-time description of the atomic events is complementary to their deterministic description. The probability function obeys an equation of motion as the co-ordinates did in Newtonian mechanics; its change in the course of time is completely determined by the quantum mechanical equation, but it does not allow a description in space and time. The observation, on the other hand, enforces the description in space and time but breaks the determined continuity of the probability function by changing our knowledge of the system.

Generally the dualism between two different descriptions of the same reality is no longer a difficulty since we know from the mathematical formulation of the theory that contradictions cannot arise. The dualism between the two complementary pictures — waves and particles — is also clearly brought out in the flexibility of the mathematical scheme. The formalism is normally written to resemble Newtonian mechanics, with equations of motion for the co-ordinates and the momenta of the particles. But by a simple transformation it can be rewritten to resemble a wave equation for an ordinary three-dimensional matter wave. Therefore, this possibility of playing with different complementary pictures has its analogy in the different transformations of the mathematical scheme; it does not lead to any difficulties in the Copenhagen interpretation of quantum theory.

A real difficulty in the understanding of this interpretation arises, however, when one asks the famous question: But what happens "really" in an atomic event? It has been said before that the mechanism and the results of an observation can always be stated in terms of the classical concepts. But what one deduces from an observation is a probability function, a mathematical expression that combines statements about possibilities or tendencies with statements about our knowledge of facts. So we cannot completely objectify the result of an observation, we cannot describe what "happens" between this observation and the next. This looks as if we had introduced an element of subjectivism into the theory, as if we meant to say: what happens depends on our way of observing it or on the fact that we observe it. Before discussing this problem of subjectivism it is necessary to explain quite clearly why one would get into hopeless difficulties if one tried to describe what happens between two consecutive observations.

For this purpose it is convenient to discuss the following ideal experiment: We assume that a small source of monochromatic light radiates toward a black screen with two small holes in it. The diameter of the holes may be not much bigger than the wave length of the light, but their distance will be very much bigger. At some distance behind the screen a photographic plate registers the incident light. If one describes this experiment in terms of the wave picture, one says that the primary wave penetrates through the two holes; there will be secondary spherical waves starting from the holes that interfere with one another, and the interference will produce a pattern of varying intensity on the photographic plate.

The blackening of the photographic plate is a quantum process, a chemical reaction produced by single light quanta. Therefore, it must also be possible to describe the experiment in terms of light quanta. If it would be permissible to say what happens to the single light quantum between its emission from the light source and its absorption in the photographic plate, one could argue as follows: The single light quantum can come through the first hole or through the second one. If it goes through the first hole and is scattered there, its probability for being absorbed at a certain point of the photographic plate cannot depend upon whether the second hole is closed or open. The probability distribution on the plate will be the same as if only the first hole was open. If the experiment is repeated many times and one takes together all cases in which the light quantum has gone through the first hole, the blackening of the plate due to these cases will correspond to this probability distribution. If one considers only those light quanta that go through the second hole, the blackening should correspond to a probability distribution derived from the assumption that only the second hole is open. The total blackening, therefore, should just be the sum of the blackenings in the two cases; in other words, there should be no interference pattern. But we know this is not correct, and the experiment will show the interference pattern. Therefore, the statement that any light quantum must have gone *either* through the first *or* through the second hole is problematic and leads to contradictions. This example shows clearly that the concept of the probability function does not allow a description of what happens between two observations. Any attempt to find such a description would lead to contradictions; this must mean that the term "happen" is restricted to the observation.

Now, this is a very strange result, since it seems to indicate that the observation plays a decisive role in the event and that the reality varies, depending upon whether we observe it or not. To make this point clearer we have to analyze the process of observation more closely.

To begin with, it is important to remember that in natural science we are not interested in the universe as a whole, including ourselves, but we direct our attention to some part of the universe and make that the object of our studies. In atomic physics this part is usually a very small object, an atomic particle or a group of such particles, sometimes much larger — the size does not matter; but it is important that a large part of the universe, including ourselves, does *not* belong to the object.

Now, the theoretical interpretation of an experiment starts with the two steps that have been discussed. In the first step we have to describe the arrangement of the experiment, eventually combined with the first observation, in terms of classical physics and translate this description into a probability function. This probability function follows the laws of quantum theory, and its change in the course of time, which is continuous, can be calculated from the initial conditions; this is the second step. The probability function combines objective and subjective elements. It contains statements about possibilities or better tendencies ("potentia" in Aristotelian philosophy), and these statements are completely objective, they do not depend on any observer; and it contains statements about our knowledge of the system, which of course are subjective in so far as they may be different for different observers. In ideal cases the subjective element in the probability function may be practically negligible as compared with the objective one. The physicists then speak of a "pure case."

When we now come to the next observation, the result of which should be predicted from the theory, it is very important to realize that our object has to be in contact with the other part of the world, namely, the experimental arrangement, the measuring rod, etc., before or at least at the moment of observation. This means that the equation of motion for the probability function does now contain the influence of the interaction with the measuring device. This influence introduces a new element of uncertainty, since the measuring device is necessarily described in the terms of classical physics; such a description contains all the uncertainties concerning the microscopic structure of the device which we know from thermodynamics, and since the device is connected with the rest of the world, it contains in fact the uncertainties of the microscopic structure of the whole world. These uncertainties may be called objective in so far as they are simply a consequence of the description in the terms of classical physics and do not depend on any observer. They may be called subjective in so far as they refer to our incomplete knowledge of the world.

After this interaction has taken place, the probability function contains the objective element of tendency and the subjective element of incomplete knowledge, even if it has been a "pure case" before. It is for this reason that the result of the observation cannot generally be predicted with certainty; what can be predicted is the probability of a certain result of the observation, and this statement about the probability can be checked by repeating the experiment many times. The probability function does — unlike the common procedure in Newtonian mechanics — not describe a certain event but, at least during the process of observation, a whole ensemble of possible events.

The observation itself changes the probability function discontinuously; it selects of all possible events the actual one that has taken place. Since through the observation our knowledge of the system has changed discontinuously, its mathematical representation also has undergone the discontinuous change and we speak of a "quantum jump." When the old adage "Natura non facit saltus" is used as a basis for criticism of quantum theory, we can reply that certainly our knowledge can change suddenly and that this fact justifies the use of the term "quantum jump."

Therefore, the transition from the "possible" to the "actual" takes place during the act of observation. If we want to describe what happens in an atomic event, we have to realize that the word "happens" can apply only to the observation, not to the state of affairs between two observations. It applies to the physical, not the psychical act of observation, and we may say that the transition from the "possible" to the "actual" takes place as soon as the interaction of the object with the measuring device, and thereby with the rest of the world, has come into play; it is not connected with the act of registration of the result by the mind of the observer. The discontinuous change in the probability function, however, takes place with the act of registration, because it is the discontinuous change of our knowledge in the instant of registration that has its image in the discontinuous change of the probability function.

To what extent, then, have we finally come to an objective description of the world, especially of the atomic world? In classical physics science started from the belief — or should one say from the illusion? — that we could describe the world or at least parts of the world without any reference to ourselves. This is actually possible to a large extent. We know that the city of London exists whether we see it or not. It may be said that classical physics is just that idealization in which we can speak about parts of the world without any reference to ourselves. Its success has led to the general ideal of an objective description of the world. Objectivity has become the first criterion for the value of any scientific result. Does the Copenhagen interpretation of quantum theory still comply with this ideal? One may perhaps say that quantum theory corresponds to this ideal as far as possible. Certainly quantum theory does not contain genuine subjective features, it does not introduce the mind of the physicist as a part of the atomic event. But it starts from the division of the world into the "object" and the rest of the world, and from the fact that at least for the rest of the world we use the classical concepts in our description. This division is arbitrary and historically a direct consequence of our scientific method; the use of the classical concepts is finally a consequence of the general human way of thinking. But this is already a reference to ourselves and in so far our description is not completely objective.

It has been stated in the beginning that the Copenhagen interpretation of quantum theory starts with a paradox. It starts from the fact that we describe our experiments in the terms of classical physics and at the same time from the knowledge that these concepts do not fit nature accurately. The tension between these two starting points is the root of the statistical character of quantum theory. Therefore, it has sometimes been suggested that one should depart from the classical concepts altogether and that a radical change in the concepts used for describing the experiments might possibly lead back to a nonstatistical, completely objective description of nature.

This suggestion, however, rests upon a misunderstanding. The concepts of classical physics are just a refinement of the concepts of daily life and are an essential part of the language which forms the basis of all natural science. Our actual situation in science is such that we *do* use the classical concepts for the description of the experiments, and it was the problem of quantum theory to find theoretical interpretation of the experiments on this basis. There is no use in discussing what could be done if we were other beings than we are. At this point we have to realize, as von Weizsäcker has put it, that "Nature is earlier than man, but man is earlier than natural science." The first part of the sentence justifies classical physics, with its ideal of complete objectivity. The second part tells us why we cannot escape the paradox of quantum theory, namely, the necessity of using the classical concepts.

We have to add some comments on the actual procedure in the quantum-theoretical interpretation of atomic events. It has been said that we always start with a division of the world into an object, which we are going to study, and the rest of the world, and that this division is to some extent arbitrary. It should indeed not make any difference in the final result if we, e.g., add some part of the measuring device or the whole device to the object and apply the laws of quantum theory to this more complicated object. It can be shown that such an alteration of the theoretical treatment would not alter the predictions concerning a given experiment. This follows mathematically from the fact that the laws of quantum theory are for the phenomena in which Planck's constant can be considered as a very small quantity, approximately identical with the classical laws. But it would be a mistake to believe that this application of the quantum-theoretical laws to the measuring device could help to avoid the fundamental paradox of quantum theory.

The measuring device deserves this name only if it is in close contact with the rest of the world, if there is an interaction between the device and the observer. Therefore, the uncertainty with respect to the microscopic behavior of the world will enter into the quantum-theoretical system here just as well as in the first interpretation. If the measuring device would be isolated from the rest of the world, it would be neither a measuring device nor could it be described in the terms of classical physics at all.

With regard to this situation Bohr has emphasized that it is more realistic to state that the division into the object and the rest of the world is not arbitrary. Our actual situation in research work in atomic physics is usually this: we wish to understand a certain phenomenon, we wish to recognize how this phenomenon follows from

95

the general laws of nature. Therefore, that part of matter or radiation which takes part in the phenomenon is the natural "object" in the theoretical treatment and should be separated in this respect from the tools used to study the phenomenon. This again emphasizes a subjective element in the description of atomic events, since the measuring device has been constructed by the observer, and we have to remember that what we observe is not nature in itself but nature exposed to our method of questioning. Our scientific work in physics consists in asking questions about nature in the language that we possess and trying to get an answer from experiment by the means that are at our disposal. In this way quantum theory reminds us, as Bohr has put it, of the old wisdom that when searching for harmony in life one must never forget that in the drama of existence we are ourselves both players and spectators. It is understandable that in our scientific relation to nature our own activity becomes very important when we have to deal with parts of nature into which we can penetrate only by using the most elaborate tools.

# HEINZ R. PAGELS

■

The American physicist Heinz Pagels (1939–1988) served as executive director of the New York Academy of Sciences and wrote astute popular accounts of physics and philosophy. An experienced mountaineer, he was killed at the age of forty-nine in a fall on Pyramid Peak, Colorado. In these two chapters from his book *The Cosmic Code,* Pagels considers some of the implications of quantum uncertainty.

# Uncertainty and Complementarity

DETERMINISM — THE WORLD view that nature and our own life are completely determined from past to future — reflects the human need for certainty in an uncertain world. The projection of that need is the all-knowing God some people find in the Bible, a God who knows the past and future down to the finest detail — like a film that has already been developed. We may not have seen the film, but what it holds for us is already fixed.

Classical physics supported the world view of determinism. According to classical physics the laws of nature completely specify the past and future down to the finest detail. The universe was like a perfect clock: once we knew the position of its parts at one instant, they would be forever specified. Human beings could not, of course, know the positions and velocities of all the particles in the universe at one instant. But invoking the medieval concept of "the mind of God," we could imagine that this perfect mind knows the configuration of all the particles, past and future.

With Max Born's statistical interpretation of the de Broglie–Schrödinger wave function, physicists finally renounced the deterministic world view of nature. The world changed from having the

97

determinism of a clock to having the contingency of a pinball machine. Physicists realized that the concept of the perfect all-knowing mind of God has no support in nature. Quantum theory — the new theory that replaced classical physics — makes only statistical predictions. But is there a possibility that beyond quantum theory there exists a new deterministic physics, described by some kind of subquantum theory, and the all-knowing mind uses this to determine the world? According to the quantum theory this is not possible. Even an all-knowing mind must support its knowledge with experience, and once it tries to experimentally determine one physical quantity the rest of the deck of nature gets randomly shuffled again. The very act of attempting to establish determinism produces indeterminism. There is no randomness like quantum randomness. Like us, God plays dice — He, too, knows only the odds.

It is this very randomness that makes the determinist recoil. Physics, as it was conceived of for centuries, was supposed to predict precisely what can happen in nature. In the quantum theory, only probabilities are precisely determined, and the determinist finds it difficult to renounce the hope that behind quantum reality a deterministic reality exists. But in fact the quantum theory has closed the door on determinism.

The randomness at the foundation of the material world does not mean that knowledge is impossible or that physics has failed. To the contrary, the discovery of the indeterminate universe is a triumph of modern physics and opens a new vision of nature. The new quantum theory makes lots of predictions — all in agreement with experiment. But these predictions are for the distribution of events, not individual events — it is like predicting how many times a specific hand of cards gets dealt on the average. Probability distributions are causally determined, not specific events.

After Born's statistical interpretation, other physicists struggled to deepen the understanding of the new quantum theory. What knowledge of nature was possible in the framework of the new theory? For example, the mathematics of quantum theory permitted both a particlelike and a wavelike representation for the electron. But clearly these two representations were opposed and in conflict with any common-sense ideas. Is the electron a wave or a particle? Bohr, Heisenberg, and Pauli in Copenhagen and many others debated these questions for over a year. Frustration set in, but Bohr's persistent optimism kept up a spirit of inquiry. Finally, by the beginning of February 1927, Bohr was exhausted and needed a break from

Heisenberg, and he took a vacation, collecting his thoughts. While on vacation, Bohr had a primary insight into the meaning of the quantum theory. Likewise Heisenberg, in the absence of Bohr but under the lash of Pauli's criticism, came to his own interpretation of the quantum theory. Bohr and Heisenberg each in his own style had come to new breakthroughs in understanding which were conceptually equivalent. Heisenberg had discovered the uncertainty principle, and Bohr had discovered the principle of complementarity. Together these two principles constituted what became known as the "Copenhagen interpretation" of quantum mechanics — an interpretation that convinced most physicists of the correctness of the new quantum theory. The Copenhagen interpretation magnificently revealed the internal consistency of the quantum theory, a consistency which was purchased at the price of renouncing the determinism and objectivity of the natural world.

Heisenberg's forte was expressing physical intuitions in precise mathematical terms. His discovery of the uncertainty relation was an example of this. It came out of the existing mathematical formalism of quantum mechanics and served profoundly to clarify the meaning of that formalism.

Heisenberg . . . invented the matrix mechanics in which the physical properties of a particle such as its energy, momentum, position, and time were represented by mathematical objects called matrices, generalizations of the idea of simple numbers. Simple numbers obey the commutative law of multiplication — the result of the multiplication does not depend on the order in which it is carried out: $3 \times 6 = 6 \times 3 = 18$. But matrix multiplication *can* depend on the order in which it is carried out. For example, if $A$ and $B$ are matrices, then $A \times B$ does not have to equal $B \times A$.

What Heisenberg showed was that if two matrices representing different physical properties of a particle, like the matrix $q$ for the position of the particle and the matrix $p$ for its momentum, had the property that $p \times q$ did not equal $q \times p$, then one could not simultaneously measure both these properties of the particle with arbitrarily high precision. To illustrate this, suppose that I build an apparatus to measure the position and momentum of a single electron. The readout of the apparatus consists of two sets of numbers, one marked "position," the other "momentum." Every time I push a button the apparatus simultaneously measures the position and momentum of the electron and prints out two long numbers, the result of the measurements. For any single measurement, let's say

99

the first one, the two numbers printed out can be as long as I like and hence highly precise. I might think that I have simultaneously measured both the position and momentum of the electron with incredible accuracy. However, to get an idea of the error or uncertainty in this first measurement, I decide to repeat the measurement and push the button again. Again two long numbers are printed out representing the position and the momentum of the electron. Remarkably, they are not exactly the same as for the first measurement, although perhaps the first several digits of each position and momentum measurement agree. By pushing the button again and again I assemble more and more such measurements. Then I can calculate the uncertainty of the position and momentum of the electron by a statistical averaging procedure over the entire set of measurements, so that the quantity denoted by $\Delta q$ is the spread or uncertainty of position measurements about some average value and likewise $\Delta p$ is the spread of momentum measurements about an average value. The uncertainties $\Delta q$ and $\Delta p$ have meaning only if one does a large set of measurements so that one can compare the differences between individual measurements. What the Heisenberg uncertainty relation asserts is that it is impossible to build an apparatus for which the uncertainties so calculated, over a large series of measurements, fail to obey the requirement that the product of the uncertainties, $(\Delta q) \times (\Delta p)$, is greater than or equal to Planck's constant, $h$. This is expressed mathematically by the relation:

$$(\Delta q) \times (\Delta p) \geq h.$$

A similar uncertainty relation is found for the uncertainty in the energy, $\Delta E$, of a particle and the uncertainty in the elapsed time, $\Delta t$:

$$(\Delta E) \times (\Delta t) \geq h.$$

Heisenberg derived these formulas directly from the new quantum theory.

To see what these relations imply, suppose we try to measure the position of an electron with arbitrarily high accuracy. This means that our uncertainty in the position of the electron is zero, $\Delta q = 0$ — we know its location exactly. But the Heisenberg uncertainty relation says that the product of $\Delta q$ and $\Delta p$, the uncertainty in the momentum, must be greater than a fixed quantity, Planck's constant. If $\Delta q$ is zero, this means $\Delta p$ must be infinite — that is, the uncertainty in our knowledge of the momentum of the particle is infinite. Conversely, if we knew precisely that the electron was at rest, so the uncertainty in the momentum is zero, $\Delta p = 0$, then the position uncertainty, $\Delta q$, must be infinite — we have no idea where the particle is located. As Heisenberg remarked, the uncertainty in position and momentum are like "the man and the woman in the weather house. If one comes out the other goes in." Notice that if Planck's constant $h$ were equal to zero in the real world rather than a tiny number, then we could simultaneously measure both the position and momentum of a particle, because the uncertainty relation would be $(\Delta q) \times (\Delta p) \geq 0$ and both $\Delta q$ and $\Delta p$ could be zero. But because Planck's constant is not zero, this is impossible.

I have always thought that wet seeds from a fresh tomato illustrate the Heisenberg relation. If you look at a tomato seed on your plate you may think that you have established both its position and the fact that it is at rest. But if you try to measure the location of the seed by pressing your finger or a spoon on it the seed will slip away. As soon as you measure its position it begins to move. A similar kind of slipperiness for real quantum particles is expressed mathematically by the Heisenberg uncertainty relations.

An important warning must be stated regarding Heisenberg's uncertainty relation: it does not apply to a single measurement on a single particle, although people often think of it that way. Heisenberg's relation is a statement about a statistical average over lots of measurements of position and momenta. As we showed, the uncertainty $\Delta p$ or $\Delta q$ has meaning only if you repeat measurements. Some people imagine that quantum objects like the electron are "fuzzy" because we cannot measure their position and momentum simultaneously and they therefore lack objectivity, but that way of thinking is inaccurate.

To get a feeling of what the Heisenberg relation implies for various objects, we can compare the product of the size of an object times its typical momentum to Planck's constant, $h$ — a measure of how important quantum effects are. For a flying tennis ball, the uncertainties due to quantum theory are only one part in about ten million billion billion billion ($10^{-34}$). Hence a tennis ball, to a high degree of accuracy, obeys the deterministic rules of classical physics. Even for a bacterium the effects are only about one part in a billion ($10^{-9}$), and it really doesn't experience the quantum world either. For atoms in a crystal we are getting down to the quantum world, and the uncertainties are one part in a hundred ($10^{-2}$). Finally, for electrons moving in an atom the quantum uncertainties completely dominate and we have entered the true quantum world governed by the uncertainty relations and quantum mechanics.

Once I tried to imagine what I would see if I could be shrunk down to the size of atoms. I would fly around the atomic nucleus and see what it was like to be an electron. But as I came to understand the meaning of the Copenhagen interpretation of Bohr and Heisenberg I realized that the quantum theory, with its emphasis on superrealism, explicitly denies such a fantasy. I had been trying to form a mental picture of the atom based on the world of my ordinary visual experience, which obeys the laws of classical physics, and applied it precisely where quantum physics says such a picture cannot be maintained. Bohr would insist that if you want to indulge this fantasy then you must precisely specify how the shrinking down to the size of atoms is to be accomplished. Suppose that instead of shrinking down myself I build a tiny little probe that will go down into the atom and tell me what it finds. But since the probe must also be built out of atoms and particles — there isn't anything else out of which to build it — the probe, too, becomes subject to the uncertainty relations, and then we can't even visualize the probe. You see we are stuck. All we can do is perform experiments on atoms and quantum particles resulting in measurements recorded on macroscopic-sized instruments. The quantum theory describes all possible such measurements; we cannot do better. The fantasy of the shrinking person is just that — a fantasy.

Heisenberg's uncertainty relations tied in nicely with Born's previous discovery of indeterminacy, thus deepening physicists' understanding of the internal consistency of the quantum theory. Physicists found that the uncertainty relations *implied* indeterminacy. We can see this easily if we imagine a standard problem in ballistics. Suppose a gun is fired (in outer space, so we can ignore air resistance) and we know both the position of the bullet as it leaves the barrel and its momentum. Using the laws of classical physics and the knowledge of the bullet's initial position and momentum, we can precisely determine its entire future trajectory; everything is predetermined. If we consider this same problem from the standpoint of quantum theory, we are required to draw a different conclusion. The Heisenberg uncertainty relation implies that we cannot simultaneously establish both the position and momentum of the bullet at the instant it leaves the barrel. Hence to the degree that these initial measurements are uncertain, so also is the future trajectory of the bullet undetermined. All we can do is give a statistical or probabilistic description of the future trajectory of the bullet. For real bullets, like tennis balls, these quantum effects are negligible. But for electrons we are compelled to have a probabilistic description of their future motion. That is why Heisenberg's uncertainty relations imply Born's indeterminacy.

While Heisenberg worked on the uncertainty relations, Bohr, in his very different style, independently developed his own interpretation of the quantum theory. Heisenberg's approach was to use mathematics to extract the meaning of the new theory, while Bohr reflected philosophically on the nature of quantum reality. Each physicist's approach supplemented and enriched the other's, and together they constituted the Copenhagen interpretation.

Bohr wondered how we could even talk about the atomic world — it was so far removed from human experience. He struggled with this problem — how can we use ordinary language developed to cope with everyday events and objects to describe atomic events? Perhaps the logic inherent in our grammar was inadequate for this task. So Bohr focused on the problem of language in his interpretation of quantum mechanics. As he remarked, "It is wrong to think that the task of physics is to find out how Nature *is*. Physics concerns what we can say about Nature."

Bohr emphasized that when we are asking a question of nature we must also specify the experimental apparatus that we will use to determine the answer. For example, suppose we ask, "What is the position of the electron and what is its momentum?" In classical physics we do not have to take into account the fact that in answering the question — doing an experiment — we alter the state of the object. We can ignore the interaction of the apparatus and the object under investigation. For quantum objects like electrons this is no longer the case. The very act of observation changes the state of the electron.

The fact that observation can change what is being observed can be seen from examples drawn from ordinary life. The anthropologist who studies a small village isolated from modern life will by his mere presence alter village life. The object of his knowledge changes as a consequence of examination. The fact that people know they are being observed can alter their behavior.

Nature can be most passively accommodating to the quantum experimenter. If he wants to measure the position of an electron with arbitrarily great precision and sets up an apparatus to do this, no law of the quantum theory prevents a definite answer. By "position" I always mean a statistical averaging of many position measurements. The experimentalist would conclude that the electron is a particle, an object at a definite point in space. On the other hand, if he is interested in measuring the wavelength of the electron and sets up another apparatus to do this, he will also get a definite answer. Doing the experiment this way, he would conclude that the electron is a wave, not a particle. No conflict exists between the particle and the wave concept, because, as Bohr taught us, the outcome of the experiments depends on the experimental arrangement, and different experimental arrangements are needed to measure the position and the wavelength of the electron.

Now the experimenter gets persistent. He is fed up with this wave-particle, momentum-position duality nonsense and decides to settle the question once and for all by setting up an apparatus that will try to measure both the position and the momentum of the electron. Nature now becomes most stubborn, because the experimenter runs into the brick wall of the uncertainty relation. No amount of experimental technique seems to do any good, because it is a question of principle that he is up against. Why can't you measure the position and momentum simultaneously — what prevents it? Born describes it this way: "In order to measure space-coordinates and instants of time, rigid measuring rods and clocks are required. To measure momenta and energies, arrangements with movable parts are needed to take up and indicate the impact of the object. If quantum mechanics describes the interaction of the object and the measuring device, both arrangements are not possible." Born is describing that odd feature of the laws of quantum mechanics which imply that we cannot build an apparatus that measures both position and momentum simultaneously — the experimental arrangements for these two measurements are mutually exclusive. Trying simultaneously to measure both position and momentum precisely is like trying to look at the space both in front and behind your head without using a mirror. As soon as you turn to look behind you the space behind your head also turns. You cannot simultaneously see both the space in front of and behind you.

Particle and wave are what Bohr called complementary concepts, meaning they exclude one another. In the analogy between language and mathematics we gave previously, these complementary concepts are different representations of the same object. Physicists speak of the particle representation or the wave representation.

Bohr's principle of complementarity asserts that there exist complementary properties of the same object of knowledge, one of which if known will exclude knowledge of the other. We may therefore describe an object like an electron in ways which are mutually exclusive — e.g., as wave or particle — without logical contradiction provided we also realize that the experimental arrangements that determine these descriptions are similarly mutually exclusive. Which experiment — and hence which description one chooses — is purely a matter of human choice.

Bohr was a philosopher and liked to extend his complementarity principle to other than problems of atomic physics. For example, "duty to society" and "family commitment" as portrayed in Sophocles' *Antigone* were complementary concepts and mutually exclusive in a moral context. As a good citizen, Antigone should consider her brother, who was killed trying to overthrow the king, a traitor. Her duty to the king and society was to renounce her brother. Yet her commitment to her family requires her to bury his body and honor his memory. Bohr later in his life thought that the principle of complementarity applied to the problem of determining the material structure of living organisms. We could either kill an organism and determine its molecular structure, in which case we would know the structure of a dead thing, or we could have a living organism but sacrifice knowledge of its structure. The experimental act of determining the structure also kills the organism. Of course, this latter idea is completely wrong, as molecular biologists have shown in establishing the molecular basis for life. I cite this example because it shows that even if you are as smart as Bohr, extending principles of science beyond their usual domain of application may lead to spurious conclusions.

We thus come to the two crucial points about quantum reality that emerge from the work of Heisenberg and Bohr, the Copenhagen interpretation. The first point is that quantum reality is statistical, not certain. Even after the experimental arrangement has been specified for measuring some quantum property, it may be necessary to repeat the precise measurement again and again, because individual precise measurements are meaningless. The microworld is given only as a statistical distribution of measurements, and these distributions can be determined by physics. The attempt to form a mental picture of the position and momentum of a single electron consistent with a series of measurements results in the "fuzzy" electron. This is a human construct attempting to fit the quantum world into the limitations of everyday sense awareness. People who engage in such constructions or try to find objective meaning in individual events are really closet determinists.

The second main point is that it is meaningless to talk about the physical properties of quantum objects without precisely specifying the experimental arrangement by which you intend to measure them. Quantum reality is in part an observer-created reality. As the physicist John Wheeler says, "No phenomenon is a *real* phenomenon until it is an *observed* phenomenon." This is radically different from the orientation of classical physics. As Max Born put it, "The generation to which Einstein, Bohr, and I belong was taught that there exists an objective physical world, which unfolds itself according to immutable laws independent of us; we are watching this process as the audience watches a play in a theater. Einstein still believes that this should be the relation between the scientific observer and his subject." But with the quantum theory, human intention influences the structure of the physical world.

In summary, the Copenhagen interpretation of the quantum theory rejected determinism, accepting instead the statistical nature of reality, and it rejected objectivity, accepting instead that material reality depended in part on how we choose to observe it. After hundreds of years the world view of classical physics fell. Here, from the very substance of the universe — the atom — the physicists learned a new lesson about reality.

During 1927, through many discussions with Heisenberg and Pauli, Bohr labored on a paper expressing his ideas about complementarity. He was going to present it first in Como at a meeting in honor of the Italian physicist Alessandro Volta and finally at the fifth Solvay conference in Brussels, where many of the leaders of physics would be present, including Einstein. Einstein and Bohr had first met in 1920, when both were already physicists of international stature. More than a professional relation of friendship emerged, and they developed a deep respect and love for one another. Bohr expressed great expectations that the complementarity principle would convince Einstein of the correctness of the quantum theory. When Bohr presented the Copenhagen interpretation of quantum mechanics at the Solvay conference, most physicists present accepted the new work for the triumph of understanding that it was. Not Einstein. He continued to pick away at Bohr's and Heisenberg's ideas, devising thought experiments that would show some flaw in the Copenhagen interpretation. Every time he thought he found a flaw, Bohr would find an error in his reasoning. Nevertheless, Einstein persisted. Finally Paul Ehrenfest said to him, "Einstein, shame on you! You are beginning to sound like the critics of your own theories of relativity. Again and again your arguments have been refuted, but instead of applying your own rule that physics must be built on measurable relationships and not preconceived notions, you continue to invent arguments based on those same preconceptions." The meeting ended with Einstein unconvinced, and this was a profound disappointment to Bohr.

Three years later at the next Solvay conference, Einstein arrived armed with a new thought experiment — the "clock in the box" experiment. Einstein imagined one had a clock in a box preset so that it would open and close a shutter very quickly on the light-tight box. Inside the box was also a gas of photons. When the shutter opened, a single photon would escape. By weighing the box before and after the shutter opened, one could determine the mass and hence energy of the escaped photon. Consequently, it was possible to determine both the energy and the time of escape of the photon with arbitrary precision. This violated the Heisenberg energy-time uncertainty relation, $(\Delta E) \times (\Delta t) \geq h$, and hence, Einstein concluded, quantum theory must be wrong.

Bohr spent a sleepless night thinking about the problem. If Einstein's reasoning was correct, quantum mechanics must fail. By morning he discovered the flaw in Einstein's reasoning. The photon, when it escapes from the box, imparts an unknown momentum to the box, causing it to move in the gravitational field which is being used to weigh it. However, according to Einstein's own theory of general relativity, the rate of a clock depends on its position in the gravity field. Since the position of the clock is uncertain by a small amount because of the "kick" it gets when the photon escapes, so is the time it measures. Bohr showed that the thought experiment devised by Einstein did not in fact violate the uncertainty relation but rather confirmed it.

After this, Einstein never disputed the consistency of the new quantum theory. He continued to dispute that it gave a complete and objective description of nature. This objection, however, was a philosophical issue and not one of theoretical physics. The debate between Einstein and Bohr continued throughout their lives, but it was never resolved. Nor could it have been. Once the debate left the common assumption that reality is instrumentally determined, and became a difference in the appreciation of the nature of reality, there was no possibility of resolution. Bound together by a mutual love, the two titans, the last of the classical physicists and the leader of the new quantum physicists, debated to the end of their days.

By the late 1920s the interpretation of the new quantum theory was intact. A generation of young physicists grew up with it, but they were less interested in the problems of interpretation than in applications. The new theory emphasized, as never before, the paramount role of mathematics in theoretical physics. Individuals with great technical power in abstract mathematics, and the ability to apply it to physical problems, came to the fore.

The new quantum theory became the most powerful mathematical tool for the explication of natural phenomena that ever fell into human hands, an incomparable achievement in the history of science. The theory released the intellectual energy of thousands of young scientists in the industrial nations of the world. No single set of ideas has ever had a greater impact on technology, and its practical implications will continue to shape the social and political destiny of our civilization. We have made contact with new components of the cosmic code — the immutable laws of the universe — which are now programming our development. Practical devices, such as the transistor, the microchip, lasers, and cryogenic technology, have given rise to entire industries at the vanguard of technical civilization. When the history of this century is written, we shall see that political events — in spite of their immense cost in human lives and money — will not be the most influential events. Instead the main event will be the first human contact with the invisible quantum world and the subsequent biological and computer revolutions.

With the new quantum theory the basis for the periodic table of chemical elements, the nature of the chemical bond, and molecular chemistry became understood. These new theoretical developments, supported by experimental research, gave rise to modern quantum chemistry. Dirac could write in a 1929 paper on quantum mechanics, "The underlying physical laws necessary for the mathematical theory of a large part of physics and the whole of chemistry are thus completely known. . . ."

The first generation of molecular biologists were inspired by Erwin Schrödinger's book *What Is Life?* in which he argued that the genetic stability of living organisms must have a material, molecular basis. These researchers, many of them trained physicists, promoted a new attitude toward genetics and introduced the experimental methods of molecular physics foreign to most biologists of that time.

This new attitude toward the problem of life culminated in the discovery of the molecular structure of DNA and RNA, the physical basis for organic reproduction. It was no accident that this discovery was made in a molecular physics laboratory, a discovery that began another revolution in its own right.

The quantum theory of solids was developed. The theory of electrical conductivity, the band theory of solids, and the theory of magnetic materials were all an outgrowth of the new quantum mechanics. In the 1950s there were major breakthroughs in the theory of superconductivity, the phenomenon of electrical current flow without resistance at very low temperature; and of superfluidity, the motion of liquids without friction. The theory of the phase transitions of matter, such as liquid to gas or solid, was advanced.

The new quantum theory provided the theoretical apparatus for the exploration of the atomic nucleus, and nuclear physics was born. The basis for the enormous energy release in radioactive decay was understood — radioactive decay was a nonclassical process involving quantum mechanical events. Physicists knew for the first time the source of the energy of the stars, and astrophysics became a modern science.

Remarkably, the educated public did not follow these developments. The quantum theory never attracted the attention of the public as did the earlier relativity theory. For this there are several reasons. First, in the early 1930s an economic depression was going on. Second, attention to political ideologies preoccupied many intellectuals. Third, and I think most important, the abstract mathematical character of the quantum theory did not relate to immediate human experience.

The quantum theory is a theory of an instrumentally detected material reality — there is an apparatus between the human observer and the atom. Heisenberg commented, "Progress in science has been bought at the expense of the possibility of making the phenomena of nature immediately and directly comprehensible to our way of thought." Or again, "Science sacrifices more and more the possibility of making 'living' the phenomena immediately perceptible by our senses, but only lays bare the mathematical, formal nucleus of the process."

Heisenberg was interested in the contrast between Goethe, the German romantic poet and dramatist, and Newton, regarding color theory. Goethe was interested in colors as an immediate human experience, and Newton was interested in color as an abstract physical phenomenon. On an experimental, material basis one must side with Newton's conclusions. But Goethe's view — and he was one of the fathers of vitalism — speaks to the immediacy of human experience. Vitalists believe there is a special "life force" in living organisms not subject to physical laws. While this appeals to our experience, there is no material basis for it. Life depends only on how ordinary matter is organized. Life-force vitalists are rare today, but they have been replaced by those who believe that human consciousness has some special property that goes beyond the laws of physics. Such neovitalists, searching for the roots of consciousness beyond material reality, might be in for another disappointment.

Goethe was part of the romantic reaction to classical mechanics and modern science — a reaction that continues to this day. This confrontation between Goethe and Newton revealed a modern humanist critique of science that the abstract explanations of science deny the vital core of human experience. The quantum theory and the sciences that emerged from it are prime examples of such abstract explanations.

Science does not deny the reality of our immediate experience of the world; it begins there. But it does not remain there, because the basis for comprehending our experience is not given with sensual experience. Science shows us that supporting the world of sensual experience there is a conceptual order, a cosmic code which can be discovered by experiment and known by the human mind. The unity of our experience, like the unity of science, is conceptual, not sensual. That is the difference between Newton and Goethe — Newton sought universal concepts in the form of physical laws, while Goethe looked for the unity of nature in immediate experience.

Science is a response to the demand that our experience places upon us, and what we are given in return by science is a new human experience — seeing with our mind the internal logic of the cosmos. The discovery of the indeterminate universe by the quantum physicists is an example. The end of determinism meant not the end of physics but the beginning of a new vision of reality. Here in the atomic core of matter, physicists found randomness. . . .

# Schrödinger's Cat

... A NEW GENERATION OF high-speed computers [soon] will appear with switching devices in the electronic components which are so small they are approaching the molecular microworld in size. Old computers were subject to "hard errors" — a malfunction of a part, like a circuit burning out or a broken wire, which had to be replaced before the computer could work properly. But the new computers are subject to a qualitatively different kind of malfunction called "soft errors" in which a tiny switch fails during only one operation — the next time it works fine again. Engineers cannot repair computers for this kind of malfunction because nothing is actually broken.

What causes the soft errors? They occur because a moderately high-energy quantum particle may fly through one of the microscopic switches, causing it to malfunction — the computer switches are so tiny they are influenced by these particles which don't disturb larger electronic components. The source of these quantum particles is the natural radioactivity in the material out of which the microchips are made or cosmic rays raining down on earth. The soft errors are part of the indeterminate universe; their location and effect are completely random. Could the God that plays dice trigger a nuclear holocaust by a random error in a military computer? By shielding the new computers and reducing their natural radioactivity the probability for such an event can be made extremely small. But this example raises the question of whether the quantum weirdness of the microscopic world can creep into our macroscopic world and influence us. Can quantum indeterminacy affect our lives?

The answer is yes — as the example of the soft errors in computers shows. Another example is the random combining of DNA molecules at the moment of conception of a child, in which quantum features of the chemical bond play a role. Atomic events which are completely unpredictable deeply influence our lives — we are in the hands of the God that plays dice.

Unquestionably, quantum indeterminacy can influence our lives. But now a puzzle arises if we think about the implications of the two-hole experiment. The standard Copenhagen interpretation of this experiment showed that indeterminacy — Born's waves of probability — meant that we had to renounce the objectivity of the world, the idea that the world exists independent of our observing it. For example, the electron exists as a real particle at a point in space only if we observe it directly. The puzzle is that if indeterminacy implies nonobjectivity and if the macroscopic human world is influenced by indeterminant events, does this mean that human-scale events lack objectivity — that they exist only if we observe them directly? Do we have to renounce the objectivity not only of an electron passing through a hole but also of the annihilation of the entire human species?

Remarkably, if we adhere strictly to the Copenhagen interpretation of the quantum theory, then the quantum world's weirdness can creep out into everyday reality — the whole world, not just the atomic world, loses objectivity. Erwin Schrödinger devised a clever thought experiment he called the cat-in-the-box to show how crazy the Copenhagen interpretation really was and that it required the whole world to possess quantum weirdness. Unfortunately his intent in this experiment, which was to criticize the Copenhagen interpretation, has been more often misunderstood than understood. Some people who want to see the weird reality of the quanta manifested in the ordinary world have used Schrödinger's experiment to show that this must be so. But they are mistaken. Mathematical physicists have carefully analyzed the cat-in-the-box experiment, especially the physical nature of observation, and arrived at the conclusion that although the macroworld is indeterminate it need not be nonobjective, unlike the microworld. To understand how this is possible, we will first describe a version of Schrödinger's cat-in-the-box experiment and see how indeed it seems to imply the end of the ordinary world's objectivity. Then we analyze more closely the physical act of observation and arrive at the alternate view that we need not apply the Copenhagen interpretation to the macroscopic world — quantum weirdness is only in the microworld.

Schrödinger suggested that we imagine that a cat is sealed in a box along with a weak radioactive source and a detector of radioactive particles. The detector is turned on only once for one minute; let us suppose that the probability that the radioactive source will emit a detectable particle during this minute is one out of two = $\frac{1}{2}$. Quantum theory does not predict the detection of this radioactive event; it only gives the probability as $\frac{1}{2}$. If a particle is detected, a poison gas is released in the box and kills the cat. The well-sealed box is far away on an earth satellite, so we don't know if the cat is alive or dead.

According to the strict Copenhagen interpretation, even after the crucial minute has passed we cannot speak of the cat as in a definite state — alive or dead — because as earthbound people we have not actually observed if the cat is alive or dead. A way of describing the situation is to assign a probability wave to the physical state of a dead cat and another probability wave to the physical state of the live cat. The cat-in-the-box is then correctly described as a wave superposition state consisting of an equal measure of the wave for the live cat and the wave for the dead cat. This superposition state for the cat in the box is characterized not by actualities but by probabilities — macroscopic quantum weirdness! It is as meaningless to talk about the cat's being alive or dead as it is to talk about which hole the electrons go through in the two-hole experiment. The statement "The electron goes either through hole 1 or hole 2" is also meaningless. The electron, if you do not observe which hole it goes through, exists in a superposition state of equal measure of a probability wave for going through hole 1 and through hole 2. Maybe you can accept that weirdness for electrons. But here we have the same kind of statement, "The cat either is dead or the cat is alive," for a cat, not an electron. Cats, like electrons, can be in a quantum never-never land.

Now let us suppose that a space shuttle with a group of scientists goes out to examine the contents of the orbiting cat-in-the-box and when they open the box they are greeted with a loud meow — the cat is alive. The Copenhagen interpretation of this event is that the scientists by opening the box and performing an observation have now put the cat into a definite quantum state — the live cat. This is analogous to examining with light beams the location of the electron at hole 1 or hole 2. For the scientists in the space shuttle, the state of the cat is no longer a superposition of waves for live and dead cat. But because their telecommunications system has broken down the scientists back on earth don't yet know if the cat is alive or dead. For these earthbound scientists, the cat-in-the-box plus the scientists on board the space shuttle who now know the state of the cat are all still in a probability wave superposition state of live cat and dead cat. The quantum never-never land of the superposition state is getting bigger.

Finally, the scientists on board the space shuttle manage to open a communication link to a computer down on earth. They communicate the information that the cat is alive to the computer, and this is stored in its magnetic memory. After the computer receives the information but before its memory is read by the earthbound scientists, the computer is part of the superposition state for the earthbound scientists. Finally in reading the computer output the earthbound scientists reduce the superposition state to a definite one. Then they tell their friends in the next room, and so on. Reality springs into being only when we observe it. Otherwise it exists in a superposition state like the electron going through the holes. Even the reality of the macroscopic world does not have objectivity until we observe it according to this scenario.

Weird as it seems, this is the standard Copenhagen interpretation of reality. We see that it requires a definite line between the observed and the observer, a split between object and mind. At first this line was between the cat-in-the-box and the space-shuttle scientists. After they opened the box the line moved to between the space-shuttle scientists and the computer, and so on. As information about the state of the cat propagated from place to place, so did the objective reality of the live cat. The Copenhagen interpretation demands that a distinction be made between the observer and observed; it does not say where the line between them is drawn, only that it must be drawn.

Something unsettles us in this account of the cat-in-the-box experiment. Somehow we may feel that the microworld of atoms lacks standard objectivity. But should this weirdness get out into the ordinary world of tables, chairs, and cats? Do they exist in a definite state only if we observe them as the Copenhagen interpretation would have it? The analysis of the cat-in-the-box experiment suggests that an observation requires consciousness. Some physicists are of the opinion that the Copenhagen view actually implies that consciousness must exist — the idea of material reality without consciousness is unthinkable. But if we examine closely what an observation is, we find that this extreme view of reality — that tables, chairs, and cats lack definite existence until observed by a consciousness — need not be maintained. The Copenhagen view, while necessary for the atomic world, does not have to be applied to the world of ordinary objects. Those who do apply it to the macroworld do so gratuitously. Let us now examine what actually happens when we observe.

If we observe something, our eyes are receiving energy from that object. But the important feature of an observation is that we obtain information — we know something about the world we didn't know before the observation. In our study of statistical mechanics we learned it is not possible to obtain information without increasing entropy — the measure of disorganization of physical systems. The price we pay for obtaining information is scrambling up the world somewhere else, thus increasing entropy — an inevitable consequence of the second law of thermodynamics. This increase of entropy implies that time has an arrow — there is temporal irreversibility and physical processes exist which can store information; memory is possible. We conclude that irreversibility in time is the principal feature of observation, not consciousness of the observation, although that, of course, also entails irreversibility because it involves memory. Observations can be carried out by dumb machines or computers, provided they have some primitive memory storage. The main point in this analysis of observation is that once information about the quantum world is irreversibly in the macroscopic world, we can safely attribute objective significance to it — it can't slip back into the quantum never-never land. . . .

# TIMOTHY FERRIS

■

One of the most promising developments in contemporary theoretical physics has been the forging of a new alliance between
quantum physicists, who study subatomic processes, and "early
universe" cosmologists, who try to understand what went on
during the big bang, when *all* events were subatomic. This report
was written for a 1988 anthology entitled *Next: The Coming Era
in Science.*

# *Unified Theories of Physics*

SINCE THE DAWN OF HISTORY, Man has pondered the riddle of the
origin and structure of the universe. Pondering, however, didn't get
Man very far. In the absence of hard facts, pure contemplation
tended less to enlighten the mind about the universe than to turn it
into a jack-o'-lantern, projecting old ideas against the sky rather than
learning new ones from it. It was largely through unfettered contemplation that the ancient Sumerians, living as they did at a confluence of rivers, ascribed creation to a kind of mud-wrestling match
among the gods, and that the early Christians in their enchantment
with heaven assigned to the angels the job of pushing the planets
along in their orbits across the sky. These ideas are not without a
certain charm, but they tell us more about ourselves than about the
outer world.

Science, a relatively recent development in the evolution of
human thought, takes a less ambitious approach; scientists tend less
to pronounce grand answers than to pose small, specific questions.
As the Nobel Prize–winning chemist François Jacob writes in his
book *The Possible and the Actual,* "The beginning of modern science
can be dated from the time when such general questions as 'How was
the universe created?' 'What is matter made of?' 'What is the essence
of life?' were replaced by more modest questions like 'How does a

stone fall?' 'How does water flow in a tube?' 'What is the course of blood in the body?' "

This preoccupation with detail has exposed scientists to a certain amount of bemused belittlement ever since the days when Aristotle spent his honeymoon collecting marine biology specimens, but it gets results. "Curiously enough," Jacob writes, "while asking general questions led to very limited answers, asking limited questions turned out to provide more general answers." Charles Darwin's theory of the origin and evolution of life grew less from first principles than from Darwin's tireless scrutiny of the finches' beaks and horses' teeth that he crammed into the holds of the *Beagle.* Isaac Newton was induced to write the *Principia,* his greatest book, by considering the relatively narrow question — put him by Edmund Halley — of just what, exactly, would be the shape of the orbit of a comet if the force of gravitation pulling on the comet were inversely proportional to the distance from the comet to the sun. Yet the results, in both cases, were genuinely universal: Newton's laws of gravitation describe the orbits of planets and stars and galaxies, and Darwin's evolution is the anchoring thread running through all earthly life.

Nowhere has the efficacy of studying the specific in order to learn about the general been demonstrated more dramatically than in recent years in particle physics, the science concerned with the smallest known structures in nature.

Ever since Leucippus and Democritus propounded the theory of the atom in the fifth century B.C., humanity has dreamt, however fitfully, of finding something fundamental — a variety of particle, perhaps, or a natural law — upon which all the wide world is built. The search quickened early in this century, when quantum mechanics began to discern order in the subatomic realm and accelerators were built with which experimental physicists could probe the structures of matter on scales smaller than the nucleus of an atom.

What was found at first seemed anything but simple. The atom, thought to be the fundamental particle, turned out instead to be composed of electrons orbiting a nucleus, and the nucleus, in turn, to be made of protons and neutrons. By the mid 1930s four elementary particles had been identified — protons, neutrons, electrons, and neutrinos — and hopes ran high that this quadrumvirate might represent the foundations of reality. But those hopes were dashed when a bewildering variety of new particles turned up. The roster of allegedly "fundamental" particles soon climbed to over 100:

There were the bosons and fermions, the hadrons and leptons, the quarks and gluons and the charmed, strange mesons that made up what came to be called, not entirely kindly, the particle "zoo." As theories grew ever more elaborate in order to explain new experimental results, particle physics was denounced by skeptics as the modern equivalent of Ptolemy's earth-centered cosmology, to which ever more complex hierarchies of epicycles had to be added in order to square its erroneous assumptions with the observed motions of the planets. Science seemed far indeed from glimpsing the elegant unity thought to underlie the wild diversity of nature.

But the situation has improved considerably in recent years. New, unified theories have appeared to point the way to a bedrock simplicity beneath the myriad subatomic particles. The theoretical progress is ongoing, but real, and experimental testing of the new ideas challenging, but perhaps possible.

Most important, the new theories have a cosmological connection: Though they are concerned with the very smallest structures in the universe, they may well shed light on the largest question of them all, that of how the universe began.

The key to the confluence of large and small is that the new theories indicate that many of the complexities of nature — perhaps all of them — would disappear under conditions of extremely high energy. The universe is understood to have originated in just such a state — in the "big bang," the explosion that initiated the expansion of the universe. If the unified theories are correct, therefore, the simplicity long sought by science is not merely an abstraction or an aspiration, but a fact of cosmic history. The theories suggest, in other words, that the universe began in a state of simplicity, traces of which may be glimpsed today by replicating or envisioning how matter and energy behave under high-energy conditions like those that ruled the universe at the onset of time.

One way to understand the new unified theories is to approach them by way of the four fundamental forces of nature — gravity, electromagnetism, and the "weak" and "strong" nuclear forces. Each force plays a distinct role in the workings of the world. Gravitation rules on the cosmological scale, binding the stars, planets, and galaxies together. Electromagnetism holds atoms together as molecules, brings us light from the sun and stars, and fires the synapses of the brain. The strong and weak nuclear forces, very short in range, keep protons and neutrons bundled together in the nuclei of atoms and govern the transmutation of unstable atoms via radioactive decay.

For physics to have reduced the workings of the universe to a system of only four forces is cause for celebration in physics, of course, but for frustration as well. The scheme is neat, but in some respects arbitrary. Why are there four forces, rather than a dozen, or two, or one? And why do the forces differ so profoundly in character? Gravitation is weak — a hundred million trillion trillion trillion times weaker than the strong nuclear force — but it affects all forms of matter and energy, while the strong force effects only one class of particles, the hadrons, and is ignored by another, the leptons. Electromagnetism manifests itself as both positive and negative electrical charge, but there appears to be no such thing as negative gravity. And so on; the list of unexplained differences among the forces is long.

The "standard model," as it is called in contemporary physics, permits accurate predictions to be made for the outcome of a great variety of events, from the gravitational interaction of a pair of stars to the workings of the strong force in the thermonuclear reactions that power the stars. But each force must be accounted for by a different set of equations. In all, some twenty different parameters must be introduced into the equations; they work, but nobody is certain *why* they work, and this rankles just about any scientist worth his or her salt. Says Steven Weinberg, a theoretical physicist at the University of Texas at Austin, "Our job in physics is to see things simply, to understand a great many complicated phenomena in a unified way in terms of a few simple principles."

It was Weinberg who came up with a key insight in the search for simplicity. He was thinking not about large but about small things — specifically, about the minuscule particles that carry electromagnetism and the weak nuclear force.

Particle physicists envision matter and energy as composed of particles — hence the name of their discipline. The particles that convey the four forces differ considerably from one another, as do the forces themselves. The differences are particularly acute in the case of electromagnetism and the weak nuclear force. Electromagnetism is carried by photons, which have no rest mass at all, while the weak nuclear force is transmitted by weak bosons, which are relatively massive and ponderous. Nonetheless, one autumn morning in 1967, it occurred to Weinberg while he was driving to his office that despite the obvious differences between photons and weak bosons, an obscure but genuine kinship might link them.

Weinberg's hunch, along with his subsequent work and that of colleagues, did much to bridge the perceptual gulf separating the weak force from electromagnetism, and gave rise to the first of the new unified theories, the "electroweak" theory of particle interactions. The electroweak theory demonstrates that electromagnetism and the weak nuclear force may be regarded as aspects of a single, "electroweak" force. For their contributions to developing the electroweak theory, Weinberg shared the 1979 Nobel Prize in physics with Sheldon Glashow of Harvard University and Abdus Salam of the Imperial College of Science and Technology in London and the International Center for Theoretical Physics in Trieste.

The unification of forces envisioned by the electroweak theory — and by the more ambitious "grand unified" theories that sought to unify the strong nuclear force with the electroweak — would manifest itself only if the environment were hotter than can be attained in the highest-energy particle accelerators on Earth, or, for that matter, anywhere else in the contemporary universe. It was largely for this reason that the theory seemed unrealistic, and was at first ignored by the scientific community. If the energy that could forge the two forces into one was unavailable, what had the theory to do with the real world?

The answer came from the opposite end of the scale of size — from cosmology, the science of the very large, the study of the structure and dynamics of the universe as a whole.

The universe is expanding, the farflung galaxies receding from one another at velocities directly proportional to the distances separating them from one another. Galaxies 30 million light years apart are receding from each other at a velocity of some 500 kilometers per second, galaxies ten times farther apart are receding ten times faster, and so on. This literally universal phenomenon was predicted, implicitly, by Einstein's general theory of relativity, published in 1915, but for lack of observational evidence the prediction was at first discounted by almost everyone, including Einstein himself. Then, in one of the most astonishing coincidences in the history of science, evidence of the expansion of the universe soon was discovered, in the late 1920s, by American astronomers who knew nothing of Einstein's theory.

Another way to say that the universe is expanding is to say that it is thinning out — that the amount of matter in a given volume of space is, on the average, steadily decreasing. If so, the universe must, long ago, have been in a state of much higher density. Theoretical investigations of this prospect were first carried out by the Belgian cosmologist Georges Lemaître and the Russian-American physicist George Gamow, and have since been elaborated upon by other scientists. Their research indicates that the expanding universe began as a dense, hot fireball — the "big bang," as the British astronomer Sir Fred Hoyle waggishly dubbed the theory.

Gamow and his colleagues Ralph Alpher and Robert Herman noted that if the universe once was hot, some of the heat from the fireball should still be around, permeating the universe. Thinned out by the subsequent expansion of the universe, the ubiquitous heat should today amount to only a few degrees above absolute zero. At the time there was no way to test this prediction. But in 1965, the American radio astronomers Robert Wilson and Arno Penzias accidently detected just such residual heat, using a state-of-the-art radio telescope. Known today as the cosmic background radiation, the dying murmur of the big bang has been widely studied, and proves to conform to the characteristics predicted by the big bang theory. Thanks to these and other studies, the reality of the big bang stands as one of the best-established tenets of cosmology today.

Here, then, was a time when the energy levels envisioned by the unified theories would have existed. During the first moments of the big bang, temperatures throughout the universe would have been so high that unification could have pertained, and the four forces have functioned as but three forces, or two, or perhaps but a single primordial force. The big bang lends a historical dimension to the physicists' quest for a unified theory: It implies that the universe began in a state of simplicity which has since been buried beneath the detritus of twenty billion years of cosmic history.

Interesting if true, as Dr. Johnson used to say, but are the theories true? A direct test would require recreating the conditions that existed in the big bang, and that is what particle accelerators do. The energy they generate — albeit for but an instant, and on a very small scale — replicate the environment that once embroiled the entire universe. The more powerful the particle accelerator, the earlier the epoch in the big bang that it can recreate. Although no accelerator has yet been built powerful enough to replicate the conditions in which the weak and electromagnetic forces are thought to have functioned as one force, it was just possible to bring an existing accelerator to a power sufficient to test some of the indirect consequences of the theory.

The electroweak theory had predicted the existence of a previously undetected variety of particles called neutral intermediate vector bosons. Like the salamanders of mythology, intermediate vector bosons can thrive only in fire; the amount of energy needed to flush them out was more than could be attained by any accelerator in existence at the time that the theory first was propounded. But by 1983, the giant particle accelerator at CERN, near Geneva, had been souped up until it could achieve the energy required to test the contention of the electroweak theory that neutral intermediate vector bosons actually exist.

The driving force behind the CERN experiment was Carlo Rubbia, a rotund and convivial physicist who commutes between Harvard and Geneva so frequently that he is estimated by his colleagues to have a lifetime average velocity, day and night, of 40 miles per hour. Rubbia teamed up with Simon van der Meere, a resourceful engineer, to increase the energy of the CERN accelerator by equipping it so that particles of antimatter would collide with particles of ordinary matter.

The origin of this bold idea lay in a theory propounded twenty years earlier by Richard Feynman of the California Institute of Technology. Feynman had noted that particles of antimatter — which are identical to those of matter but have opposite electrical charge — could be regarded as ordinary matter moving in reverse time. Some scoffed at Feynman's theory, but none could find fault with his logic, and eventually physicists grew accustomed to entertaining the idea that antimatter particles wander through the universe in backward time, like actors in a film projected in reverse. The idea seemed amusing, albeit useless.

Rubbia and van der Meere proposed to put it to use. Their plan was to inject antimatter particles into the CERN ring, where magnetic pulses are employed to accelerate particles to nearly the velocity of light. The antimatter particles, sensing the pulses in reverse time, would orbit the ring in the opposite of the normal direction. If they could then be directed into the oncoming stream of particles of ordinary matter, the result would be a headlong collision at extremely high velocity, resulting in a tiny explosion of unprecedented intensity.

Big science at its most daring, the CERN matter-antimatter collider experiment absorbed the efforts of 120 physicists from 11 universities, as well as whole platoons of engineers, and it encountered technical problems that took years to overcome. But by the summer of 1983, thousands of matter-antimatter collisions were taking place in the accelerator, and neutral intermediate vector bosons were detected. A key prediction of the Glashow-Weinberg-Salam theory had been upheld, a feat for which Rubbia shared the 1984 Nobel Prize in physics with van der Meere.

Meanwhile, other ingenious experiments were being mounted to test the claims of the "grand" unified theories. The grand unified theories, or "GUTs," contained the startling implication that matter itself might be but a passing phase in the evolution of the universe. Specifically, they mandated that protons — particles found at the heart of every atom — do not last forever, but must eventually decay. The proton decay rate predicted by the grand unified theories is so slow that enormous quantities of protons — at least ten thousand billion billion billion of them — must be monitored in order to have a fifty percent chance of seeing just one proton decay in the course of a year.

Vast collections of protons accordingly were amassed, in the form of tanks of water or stacks of concrete and steel, and were monitored at test sites buried in the bowels of a salt mine in Ohio, a gold mine in India, a railroad tunnel beneath the Alps, and other subterranean sites. (The experiments are run deep underground in order to minimize contamination by high-energy particles streaming in from space, which can collide with particles in the experimental chamber and mimic the effects of proton decay.) By the late 1980s the results of these long-term proton watches were still inconclusive, though they had set a lower limit on the proton half-life that invalidated at least the simplest version of grand unified theory. Other versions of the theory predict a longer proton half-life, and could not be tested until the experiments had continued to run for years longer.

Meanwhile, considerable excitement was being generated in the scientific community by the development of "supersymmetry" theories that might present a unified account of all four forces. The supersymmetry theories, called "SUSYs" for short, purport that the universe originally was built on more than the four dimensions — the three dimensions of time plus one for space — that characterize nature today. According to the SUSYs, the universe could originally have had, say, ten dimensions, of which all but four rapidly collapsed, at the onset of the big bang, into tiny objects too small to be noticed today. Conceivably, these objects might be the subatomic particles themselves — quanta of matter and energy that originated by condensing out of pure space.

Support for the supersymmetry hypothesis came from "superstring" theory, which portrays the subatomic particles not as infinitesimal spheres but as tiny strings. Physicists found it conceivable that all the varied properties of particles, including their mass, electrical charge, and spin, might be explained in terms of possible configurations of the strings, e.g., as lines, loops, and **X**s.

By the late 1980s the prospects for supersymmetry theories — including their cousins the superstring theories — continued to look bright. Since they are inherently geometrical, SUSYs naturally incorporate relativity, which portrays the force of gravitation as a manifestation of the geometrical curvature of space, into their account of quantum mechanics; indeed, writes John Schwarz of Caltech, one of the pioneers of supersymmetry, "it seems impossible to construct a mathematically consistent theory of strings *without* including gravity." And, where previous unified theories had to be "renormalized" — i.e., subjected to mathematical manipulations to eliminate terms that otherwise went uncontrollably to infinity — SUSYs introduce terms that automatically cancel each other out, eliminating the unwanted infinities.

A price paid for all this is complexity. Physics in a ten-dimensional universe is an intricate business, and by 1987 theorists were settling in for a prolonged period of hard work trying to make sense of it all. "The easy problems were solved in the first few months," sighed an MIT researcher working in the field. "Now we're left with the hard ones." Complexity in itself, however, is not necessarily a bad sign — Einstein's general relativity equations are more complicated than the Newtonian equations they superseded — and in some ways it may amount to a virtue. Supersymmetry implies, for instance, that there are whole families of as yet undiscovered subatomic particles; the number of previously unknown particles envisioned under the SUSYs ranges from enormous to infinite. This could be good news for astronomers who have long wondered why ninety percent of the matter in the universe, its presence betrayed by the dynamical behavior of galaxies, is invisible; this "missing mass" could turn out to consist of clouds of particles whose existence was unsuspected until SUSY came along.

If experiment is to keep pace with rapidly developing new theories like SUSY, new and more powerful accelerators will have to be built to recreate conditions of even earlier cosmic history. In 1987 the president of the United States announced his support for such a machine, the Superconducting Supercollider, a particle accelerator sixty miles in circumference. Its power would be more than twenty times that attained by Rubbia's team at CERN. Its significance, however, is likely to go far beyond confirming or denying supersymmetry, which in any case was in the late 1980s still in too preliminary a state for the nature of its experimental tests to yet be clear. The real function of such a giant accelerator would be to explore realms of high energy, on scales of fine resolution, about which little is yet known. The data generated could as easily prompt fresh theoretical insights as test old ones.

Individual theories may stand or fall, but the unification approach is likely to persist, for aesthetic as well as rational reasons. Supersymmetry in particular depicts the universe as having begun in a state of sublime perfection, the contemplation of which elicits such enthusiasm in the minds of theoretical physicists that some have taken to calling the very early universe "paradise lost." As Stephen Hawking of Cambridge University puts it, "The early universe was simpler, and it was a lot more appealing *because* it was a lot simpler."

The picture that elicits the affection of the theorists is of a universe that began in a state of absolute symmetry, in which there was neither place nor time nor varieties of particles and forces. A cosmic fall from grace came during the first fraction of a second of the expansion of the infant universe, when the symmetries of genesis fractured as the universe cooled. Viewed by these lights, the contemporary universe resembles a jumble of broken symmetries, like the heaps of potsherds that archaeologists must painstakingly piece together if they are to glimpse the beauty of the original pot. "The laws of physics are simpler to discern and understand when one goes to higher energies," says Michael Turner, resident cosmologist at Fermilab, "because at high energies symmetries are manifest, while at low energies they are hidden."

The fall from grace is seen as essential to existence as we know it. Absolute symmetry is beautiful, but it is also sterile. Every motion breaks the symmetry of space, by creating a "here" different from "there." Every event breaks the symmetry of time, by creating a difference between "before" and "after." Life thrives on imperfection, and is full of examples of the cosmological principle that one has to break symmetries to get things done, as surely as one must break eggs to make an omelette. Serve the eggs at a circular table at which, say, six diners are seated: Etiquette aside, it doesn't matter whether the dinner guests choose to use the butter plate on their right or their left side; but the choice has to be made, and once one dinner guest has chosen a plate, the symmetry has been broken and all the others must follow suit. A light bulb may be manufactured so that it screws in clockwise, as most do, or counterclockwise, as do the light bulbs used by the New York City subway system in order to discourage riders from unscrewing them and taking them home. But the symmetry must be broken, one way or another, if any threaded light bulbs are to be manufactured. Similarly, a universe that remained forever in a state of perfect symmetry could not have fragmented into the four forces, the subatomic particles, and stars and planets and living beings. In a very real sense, we may owe our lives to the fact that the universe is less than perfect.

The development of the unified theories promises to write a new chapter in the old dichotomy between the perfect symmetries envisioned by the human mind and the asymmetries that permeate the material world. We can think, for instance, of a perfect sphere, but few if any perfect spheres are found in nature. (Stars and planets look spherical at a glance, but upon closer inspection prove to be lumpy, distended, and flattened at the poles.) Spiral galaxies come in pairs, but one member of the pair is usually much larger than the other; human beings have two hands, but one is usually stronger and more adroit than the other; antimatter is the mirror image of matter — antimatter particles are identical to those of matter, but have reversed charge — yet, for some reason, the universe is made almost exclusively of matter, and contains little antimatter.

Plato wrote of the asymmetry that distinguishes the eternal human soul from the mortal body that houses it, and regarded the imperfect, material world as but a shadow of an ideal realm of perfectly symmetrical geometrical forms. In Plato's *Symposium,* Aristophanes identifies asymmetry with the fall from grace; human beings, he says, originally spherical in shape, were cut in two by Zeus as punishment for the sin of pride.

Weinberg, in accepting the Nobel Prize, struck a Platonic theme. "In the seventh book of the *Republic,*" he recalled, "Plato describes prisoners who are chained in a cave and can see only shadows that things outside cast on the cave wall. When released from the cave at first their eyes hurt, and for a while they think that the shadows they saw in the cave are more real than the objects they now see. But eventually their vision clears, and they can understand how beautiful the real world is.

"We are in such a cave," Weinberg continued, "imprisoned by the limitations on the sorts of experiments we can do. In particular, we can study matter only at relatively low temperatures, where symmetries are likely to be spontaneously broken, so that nature does not appear very simple or unified. We have not been able to get out of this cave, but by looking long and hard at the shadows on the cave wall, we can at least make out the shapes of symmetries, which though broken are exact principles governing all phenomena, expressions of the beauty of the world outside."

Contemporary physics resurrects not only Plato's dream of a symmetrical perfection underlying nature, but also his vision of learning as remembrance. Socrates, coaching the untutored slave boy in the precepts of geometry, claimed to have demonstrated that the "soul has been forever in a state of knowledge," and that all real learning therefore is recollection. Generations of philosophers have remained skeptical on this point, and understandably so; at the least, Socrates underestimated his effectiveness as a teacher. But, if traces of cosmic history are indeed etched within every scrap of matter, it is tempting to speculate about the extent to which cosmic history is woven through the human mind as well. Perhaps the dream of an ultimate unified theory is a form of cosmological nostalgia, and human science is a means by which the universe contemplates its past.

# FREEMAN J. DYSON

■

The popular essays of Freeman Dyson (b. 1923) demonstrate, like his scientific research, a vigorous, wide-ranging curiosity and an elegant style. Here Dyson reflects on the enormous range of questions that can occupy (or preoccupy) a working physicist.

# *Butterflies and Superstrings*

THIS QUICK TOUR of the universe will begin with superstrings and end with butterflies. There will be a couple of intermediate stops on the way. Like Dante on his tour of the Inferno, I find at each level some colorful characters to add human interest to an otherwise intimidating scene. I will not explain what butterflies and superstrings are. To explain butterflies is unnecessary because everyone has seen them. To explain superstrings is impossible because nobody has seen them. But please do not think I am trying to mystify you. Superstrings and butterflies are examples illustrating two different aspects of the universe and two different notions of beauty. Superstrings come at the beginning and butterflies at the end because they are extreme examples. Butterflies are at the extreme of concreteness, superstrings at the extreme of abstraction. They mark the extreme limits of the territory over which science claims jurisdiction. Both are, in their different ways, beautiful. Both are, from a scientific point of view, poorly understood. Scientifically speaking, a butterfly is at least as mysterious as a superstring. When something ceases to be mysterious it ceases to be of absorbing concern to scientists. Almost all the things scientists think and dream about are mysterious.

Each of the stops on this tour is a scene taken from my life as a scientist in Princeton. I begin with the scene of my encounter with superstrings. In the spring of 1985 Ed Witten, one of the most brilliant of the young physicists at Princeton University, announced that he would give a talk. Rumors were buzzing. Some people said they had heard that Witten had a new theory of the universe. Other people were skeptical. Whether you believed the rumors or not, it was clear that this talk would be an extraordinary occasion. Perhaps it would be a historic event. When the appointed time came, our seminar room was packed with people, some old and famous, some young, all eager with expectations.

Witten spoke very fast for an hour and a half without stopping. It was a dazzling display of mathematical virtuosity. It was also, as Witten remarked quietly at the end, a new theory of the universe. While he was talking, I was reminded of a talk by the young Julian Schwinger thirty-five years earlier. Julian Schwinger was the most brilliant student of Robert Oppenheimer. He was the Wunderkind of physics in the 1940s. I remembered Oppenheimer saying on that occasion: "When ordinary people give a talk, it is to tell you how to do it, but when Julian gives a talk, it is to tell you that only he can do it." When Ed Witten came to the end of his hour and a half, there was no Oppenheimer in the audience to make a sharp-tongued comment. The listeners sat silent. After a while the chairman of the session called for questions. The listeners still sat silent. There were no questions. Not one of us was brave enough to stand up and reveal the depths of our ignorance. As we trooped glumly out of the room, I could hear voices asking quietly the questions nobody asked out loud. "Is this just another fashionable fad, or is it the real stuff?" "What is a superstring anyway?" "Does this have any connection with anything real?" Many questions but no answers.

I describe this scene because it gives a picture of what it means to explore the universe at the highest level of abstraction. Ed Witten is taking a big chance. He has moved so far into abstraction that few even of his friends know what he is talking about. He is not alone. A crowd of young people in Princeton, and a few hundred elsewhere, are at home with him in the world of superstrings. He did not invent superstrings. Superstrings were invented by two young physicists in California and in London. Ed Witten's role is to build superstrings into a mathematical structure which reflects to an impressive extent the observed structure of particles and fields in the universe. After they had heard him speak, many members of his audience went back to their desks and did the homework they should have done before, reading his papers and learning his language. The next time he talks, we shall understand him better. Next time, we shall perhaps be brave enough to ask questions. I remember that in 1949, the talk of Julian Schwinger which seemed at the time so incomprehensible turned out later to be right. Ed Witten's theory is more ambitious than Schwinger's, but there is a fighting chance that it will also turn out to be right.

It is time now to try to describe what a superstring really is. Here I run into the same difficulty which the geometer Euclid encountered 2,200 years ago. Euclid was trying to convey to his readers his idea of a geometrical point. For this purpose he gave his famous definition of a point: "A point is that which has no parts, or which has no magnitude." This definition would not be very helpful to somebody who was ignorant of geometry and wanted to understand what a point was. Euclid's notion of a point only becomes clear when one reads beyond the definition and sees how points are related to lines and planes and circles and spheres. A point has no existence by itself. It exists only as a part of the pattern of relationships which constitute the geometry of Euclid. This is what one means when one says that a point is a mathematical abstraction. The question, What is a point? has no satisfactory answer. Euclid's definition certainly does not answer it. The right way to ask the question is: How does the concept of a point fit into the logical structure of Euclid's geometry? This question is answered when one has understood Euclid's axioms and theorems. It cannot be answered by a definition.

If I were to follow the example of Euclid and try to give a definition of a superstring, it would be something like this: "A superstring is a wiggly curve which moves in a ten-dimensional space-time of peculiar symmetry." Like Euclid's definition of a point, this tells us very little. It gives us only a misleading picture of a wiggly curve thrashing about in a dark room all by itself. In fact a superstring is never all by itself, just as a Euclidean point is never all by itself. The superstring theory has the ten-dimensional space-time filled with a seething mass of superstrings. The objects with which the theory deals are not individual superstrings but symmetry-groups of states in which superstrings may be distributed. The symmetry-groups are supposed to be observable. If the theory is successful, the symmetry-groups derived from the mathematics of superstrings will be found to be in correspondence with the symmetry-groups of fields and particles seen in the laboratory. The correspondence is still far from being established in detail. Witten's excitement arose from the fact that the theory passed several crucial tests which other theories had failed. To have found a theory of the universe which is not mathematically self-contradictory is already a considerable achievement.

The name "superstring" grew out of a historical analogy. The greatest achievement of Einstein was his 1915 theory of gravity. Sixty years later, a new version of Einstein's theory was discovered which brought gravity into closer contact with the rest of physics. The new version of gravity was called "supergravity." About the same time that supergravity was invented, another model of particle interactions was proposed. The new particle model was called "String Theory," because it represented particles by one-dimensional curves or strings. Finally, the same mathematical trick which turned gravity into supergravity turned strings into superstrings. That is how superstrings acquired their name. The name, like the superstring itself, is a mathematical abstraction.

Superstrings have one striking characteristic which is easy to express in words. They are small. They are extravagantly small. Their extravagant smallness is one of the main reasons why we can never hope to observe them directly. To give a quantitative idea of their smallness, let me compare them with other things that are not so small. Imagine, if you can, four things that have very different sizes. First, the entire visible universe. Second, the planet Earth. Third, the nucleus of an atom. Fourth, a superstring. The step in size from each of these things to the next is roughly the same. The Earth is smaller than the visible universe by about twenty powers of ten. An atomic nucleus is smaller than the Earth by twenty powers of ten. And a superstring is smaller than a nucleus by twenty powers of ten. That gives you a rough measure of how far we have to go in the domain of the small before we reach superstrings.

The main thing which I am trying to suggest by this discussion is the extreme remoteness of the superstring from any objects which we can see and touch. Even for experts in theoretical physics, superstrings are hard to grasp. Theoretical physicists are accustomed to living in a world which is removed from tangible objects by two levels of abstraction. From tangible atoms we move by one level of abstraction to invisible fields and particles. A second level of abstraction takes us from fields and particles to the symmetry-groups by which fields and particles are related. The superstring theory takes us beyond symmetry-groups to two further levels of abstraction. The third level of abstraction is the interpretation of symmetry-groups in terms of states in ten-dimensional space-time. The fourth level is the world of the superstrings by whose dynamical behavior the states are defined. It is no wonder that most of us had a hard time trying to follow Ed Witten all the way to the fourth level.

What philosophical conclusions should we draw from the abstract style of the superstring theory? We might conclude, as Sir James Jeans concluded long ago, that the Great Architect of the Universe now begins to appear as a Pure Mathematician, and that if we work hard enough at mathematics we shall be able to read His mind. Or we might conclude that our pursuit of abstractions is leading us far away from those parts of the creation which are most interesting from a human point of view. It is too early yet to come to conclusions. We should at least wait until the experts have decided whether the superstring theory has anything to do with the universe we are living in. If the theory is wrong, it should be possible to prove it wrong within a few years. If it is right, to prove it right will take a little longer. While we are waiting, let me pass on to the next stop on the tour. The next stop is black holes.

The scene of my encounter with black holes is a visit to Princeton of the great English physicist Stephen Hawking. This happened a few years ago. Hawking is one of very few scientists who have as many profound and original ideas as Ed Witten. In other respects he is as different from Ed Witten as he possibly could be. He is crippled with a form of sclerosis which keeps him confined to a wheelchair. He talked in those days with great difficulty in a voice which only his friends could understand. Now he cannot talk at all. It comes as a shock when you first meet him to see somebody in such a pitiable condition. But very quickly, after you have been talking with him for a few minutes, the strength of his spirit pushes all feelings of pity aside. You forget the illness and enjoy Stephen Hawking as a human being. He is not only a scientific genius but a shrewd observer of the human scene with a fine sense of humor. After you know him well, you think of him not as a disabled person but as a brilliant scientific colleague. By continuing his active scientific life in spite of his severe disabilities, he is a constant source of strength to thousands of other handicapped people who must struggle with lesser obstacles.

I was apprehensive when Stephen Hawking came to Princeton because he came in February, and Februaries in Princeton are apt to be unpleasant. On the average, February is our coldest month. It sometimes snows heavily, and after the snow we often have periods of alternating freeze and thaw with either bumpy ice or deep slush covering the sidewalks. I worried when I thought of Stephen in his wheelchair trying to negotiate snowdrifts and ice-slopes and slush puddles. Of course I need not have worried. On the day that Stephen arrived, warm breezes from the South dried up the streets. The sun shone and continued to shine every day while Stephen was with us. He hummed around Princeton in his electric wheelchair and scoffed at our anxieties.

He talked to us about the new theory of black holes which he had then just worked out. The idea of a black hole was one of the most dramatic consequences of Einstein's theory of gravity. According to Einstein's equations, a massive star at the end of its life, when it has exhausted its nuclear fuel, continues to contract and grow smaller and denser under the influence of its own gravitation. After the nuclear fuel is used up, the star goes into a state of gravitational collapse. All parts of the star fall more or less freely inward toward the center. Now you would imagine that the free fall could not continue for very long, because the falling material would in a short time arrive at the center of the star. But Einstein's equations have the peculiar consequence that the star can continue to exist forever in a state of permanent free fall without ever reaching the bottom. This state of a star in permanent free fall is what we call a black hole.

How can it happen that a star continues to fall freely without ever hitting bottom? What happens is that the space in the region of the hole is so strongly curved that space and time become interchanged. As you fall into the hole, your time becomes space and your space becomes time. More precisely, if you observe the collapsing star from the outside, you see its motion slow down and stop because the direction of time inside the hole is perpendicular to the direction of time as seen from the outside. This is a difficult state of affairs for us to visualize, but it seems to be the way nature works. The collapsing star can continue to fall freely forever because its sense of the flow of time is perpendicular to ours. This picture of a black hole is what we call the classical theory, the theory which was deduced from Einstein's equations and was generally believed to be true until Stephen Hawking came along. The classical theory is still believed, even after Hawking, to be a good approximation for practical purposes. It says that black holes are absolutely permanent and absolutely black. An ideal classical black hole would swallow without trace everything that falls into it, and would emit from its surface no light or any other kind of radiation. We have fairly good evidence that black holes exist and are even quite common in the parts of the universe which we can observe. So far as the observable effects of black holes are concerned, the classical theory gives a satisfactory account of what is going on.

But Stephen Hawking told us that the classical theory is not good enough. Stephen went beyond the classical theory in several directions. He succeeded in putting together a coherent picture of black holes, including not only the laws of gravitation but the laws of thermodynamics. In his new picture, a black hole is not just a bottomless pit but a physical object. A black hole is not totally black but emits thermal radiation at a certain definite temperature. A black hole is not absolutely permanent but will ultimately evaporate into pure radiation. Thus Stephen brought black holes back out of the domain of mathematical abstraction into the domain of things that we can see and measure.

I do not want to give the impression that Stephen's new theory dispels the mystery and gives us a complete understanding of black holes. The new theory is far from being mathematically precise, and even further from being experimentally verified. It leaves the mystery of black holes deeper than ever. Stephen Hawking is still groping in the dark for concepts which will make his theory fully intelligible. As usual when a profoundly original theory is born, the equations come first and a clear understanding of their physical meaning comes later. Perhaps the best way to look at Hawking's discovery is to use another historical analogy. In the year 1900 Max Planck wrote down an equation, $E = h v$, where $E$ is the energy of a light wave, $v$ is its frequency, and $h$ is a constant which we now call Planck's constant. Planck's equation was the beginning of quantum theory. It said that energy and frequency are the same thing measured in different units. Energy is measured in ergs and frequency in cycles. Planck's constant gives you the rate of exchange for converting frequency into energy, namely, $6 \times 10^{-27}$ ergs per cycle. But in the year 1900 this made no physical sense. Even Planck himself did not understand it. He knew only that his equation gave the right answers to certain questions in the theory of radiation. But what does it mean to say that energy and frequency are the same thing? What this means only began to become clear twenty-five years later, when Planck's equation was built into the theory which we now call quantum mechanics.

Now Hawking has written down an equation which looks rather like Planck's equation. Hawking's equation is $S = kA$, where $S$ is the entropy of a black hole, $A$ is the area of its surface, and $k$ is a constant which I call Hawking's constant. Entropy means roughly the same thing as the heat capacity of an object. It is measured in units of calories per degree. $A$ is measured in square centimeters. Hawking's equation says that entropy is really the same thing as area. The exchange rate between area and entropy is given by Hawking's constant, which is roughly $10^{41}$ calories per degree per square centimeter. This factor of $10^{41}$ shows how remote a black hole is from humanly comprehensible and familiar units. But what does it really mean to say that entropy and area are the same thing? We are as far away from understanding this now as Planck was from understanding quantum mechanics in 1900. All that we can say for certain is that Hawking's equation is a clue to the riddle of black holes. Somehow, we can be sure, this equation will emerge as a central feature of the still unborn theory which will tie together gravitation and quantum mechanics and thermodynamics.

Heisenberg, who invented quantum mechanics at the age of twenty-five, was born in the same year as Planck's equation. Hawking's equation is now only twelve years old. I hope we still have some bright twelve-year-olds who are interested in science. We must be careful not to discourage our twelve-year-olds by making them waste the best years of their lives on preparing for examinations. If we do not stifle their interest in science, one of them has a chance to be the new Heisenberg, building Hawking's equation into a fully coherent mathematical theory of black holes. The twelve-year-olds will be twenty-five in the year 2000, and that would be a good moment for a new theory of the universe to be born.

The third stop on my tour of the universe is closer to home. It is called the Oort Cloud. Jan Oort is a famous Dutch astronomer, now eighty-six-years old, who still visits Princeton regularly. At our Institute in Princeton we sometimes organize meetings which are announced as Shotgun Seminars. A Shotgun Seminar is a talk given by an Institute member to a volunteer audience. The subject of the talk is announced a week in advance, but the name of the speaker is not. Before the talk begins, the names of all people in the room are written on scraps of paper, the scraps are put into a box, the box is ceremoniously shaken and one name is picked out at random. The name picked out is the name of the speaker. The unbreakable rule of the seminar is that nobody whose name is not in the box may listen to the talk. This rule ensures that everybody does the necessary homework. The audience is at least as well prepared as the speaker.

The audience is ready to argue and contradict whenever the speaker flounders. Anybody who has not given serious thought to the subject of the seminar had better not come.

A few years ago, when Jan Oort was only eighty-two, he happened to drop in unexpectedly at the seminar room a few minutes before a Shotgun Seminar was supposed to begin. We were sitting around a table drinking coffee. He helped himself to coffee and sat down with us, eager to hear the latest astronomical gossip. This was an embarrassing situation. On the one hand, Jan Oort was our honored guest and we could not ask him to leave the room. On the other hand, we could not break our unbreakable rule and let him stay without putting his name in the box. So we apologetically explained the situation to him. He asked what was to be the subject of the seminar. We told him it was some recent work on the stability of orbits of stars revolving around the center of a galaxy. He said, "No problem, I stay. Put in my name with the others." So he stayed. His name was not picked out of the box. But he took an active part in the seminar, joining the rest of the audience in heckling the speaker. At the end he stood up at the blackboard and gave us a ten-minute summary of the discussion, considerably more lucid than anything else that we heard that morning.

Jan Oort invented the Oort Cloud about forty years ago. The cloud is a huge swarm of comets, loosely attached to the solar system. Oort's idea was that there must be millions and millions of comets, pieces of rock and ice about as big as a terrestrial mountain, moving slowly and randomly around the Sun in the space beyond the planets. Whenever one of these chunks of ice happens to pass close to the Sun, the heat of the Sun boils off its surface, the boiled-off gas and dust forms a long bright tail, and we see it in the sky as a visible comet. Oort calculated that his cloud would explain the appearance of new comets at the observed average rate of about one per year. We believe that the comets were formed out of a collapsing interstellar dust cloud about 4 billion years ago, at the same time as the Earth and Sun. We know from the abundance of old craters on the Moon that the rate of cometary bombardment was much higher when the Moon was young than it is today. The Oort Cloud must have been much denser in the early days than it is now. But there is still enough of it left after 4 billion years to keep the comets dribbling in toward the Sun, to supply occasional portents of good and evil for superstitious humanity.

Once in a while, a comet will hit the Earth. When a large object impacts the Earth, it makes an impact crater similar to the craters which we see on the Moon. On the Moon the craters are visible from impacts dating almost all the way back to the beginning of the solar system. On the Earth the older craters are worn down by erosion and covered with sediments, but still they leave traces which geologists can identify. About fifty large craters have been found on the Earth, and their ages can be measured by examining the rocks lying below them and the sediments lying above them. When the ages of the big craters are collected, it is found that the craters have a tendency to occur in groups. Their ages are not distributed randomly. We find several craters with ages clumped together within a few million years, then a long gap with few craters, then another clump, and so on.

The nearest big crater to Aberdeen is the Rieskessel in Germany. It is well worth a visit. Right in the middle of it there is a beautiful old German town called Nördlingen. Nördlingen still has its medieval town wall preserved. If you go up to the top of the tower over the old wall, you can see the country surrounding the town on all sides. All the way round, it is gentle rolling country until you come to a ring of hills. The ring of hills goes completely around the horizon about five miles from the town. If you stand on the tower and imagine all the grass and trees and people and houses away, this ring of hills on the horizon looks exactly the way a crater on the Moon would look to somebody standing in the middle. Only, because of the trees and houses, the rim of the crater is not obvious. The reason I was visiting Nördlingen was because my wife had an aunt, our Tante Dora, who lived there. Tante Dora had lived in the middle of the Rieskessel for twenty years and had never noticed that it was there until I pointed it out to her. The Rieskessel has been there for about 15 million years. It belongs to the youngest cluster of crater ages.

The Rieskessel is not a big crater by astronomical standards; some of the older craters, both on the Earth and on the Moon, are much bigger. Roughly speaking, a comet hitting the Earth will make a crater ten times as big as itself. The Rieskessel with a diameter of 10 miles could have been made by a 1-mile comet. A big comet like Halley's, with a diameter of about 10 miles, would make a 100-mile crater if it hit the Earth.

Every comet impact, even at the comparatively modest scale of the Rieskessel, must have been an ecological catastrophe for the creatures unlucky enough to be living at the time it happened. It must have darkened the sky with huge quantities of pulverized rock and dust. It might have caused major changes in weather patterns all over the earth. It would not be surprising if many species of plants and animals, hit suddenly by conditions to which they were not adapted, should have become extinct. In fact we find in the fossil record clear evidence that such sudden extinctions have occurred. Paleontologists speak of a "mass extinction" when they find many species disappearing simultaneously from the record. The most famous mass extinction is the one which happened 65 million years ago, when the dinosaurs disappeared leaving not a single species alive. At about the same time that the dinosaurs were coming to grief, many species of microscopic plants and animals in the oceans suddenly died out. We have evidence that, at least in this case, the marine extinctions occurred at a time when some large object from space hit the earth. The limestones in which the microscopic fossils appear are interrupted by a thin layer of clay. The boundary clay has all the chemical properties to be expected if it is made of dust settling out after a large extraterrestrial impact. The position of the clay coincides exactly with the disappearance of a large fraction of the fossils. Above the clay, new species of fossil replace the old.

The crater associated with the 65-million-year mass extinction has not been found. That is the way it often is in science. The evidence for the most famous of all the events is missing. Perhaps the crater associated with the dinosaur extinction is waiting to be discovered somewhere on the bottom of the ocean. Or perhaps it has disappeared into one of the ocean trenches. We now know that in the course of the 65 million years since the great extinction about half of the ocean floor has been freshly made at the mid-ocean ridges and about half of the old floor has sunk into the ocean trenches. So the failure to find the big 65-million-year-old crater is not fatal to the theory that the dinosaur extinction was caused by an impact. The evidence for causal connection between particular events in the remote past can never be conclusive.

When we look, not at any single event, but at the entire history of biological extinctions and of impact craters upon the Earth, we see an overall pattern emerging. The pattern is seen both in the fossil record and in the layers of debris associated with impacts. In both records we see clusters of events. Mass extinctions occur not instantaneously but spread out over periods 1 or 2 million years long. Debris layers are found, with three or four layers clustered within a million years. Between the clusters are long stretches of time in which major extinctions and craters are rare. And the dates of the extinctions and the crater clusters agree within the limits of error of the measurements. The pattern of these events leads us to the conclusion that the Earth has been bombarded, not by a steady hail of comets but by comet showers. Each comet shower apparently lasts for about a million years and results in several major impacts on the Earth. If we were living during a comet shower, we would see the new comets appear at a rate of ten or a hundred per year instead of the one per year that we see now. And our chance of suffering a direct hit would be greater in the same proportion.

To understand why comet showers occur, we go back to the Oort Cloud. The theory of comet showers was worked out by Jack Hills, an American physicist now at Los Alamos. He realized that the movements of the comets in the Oort Cloud are not entirely random. Comets in the cloud are generally moving in random directions, but if a comet happens to be moving in an orbit almost exactly toward the Sun, it will not survive for long. A comet in an orbit coming close to the Sun may get boiled away and disintegrated by the Sun, or alternatively it may swing by one of the planets and pick up enough extra speed from the encounter to carry it right out of the solar system. In either case the comet is lost from the Oort Cloud. Therefore the Oort Cloud has in it a narrow empty cone called a "loss cone." The velocities in the cloud are random except that those comets with velocities inside the narrow cone are missing. The empty cone contains just those velocities which are pointed almost exactly at the Sun. And the Earth is near enough to the Sun so that any comet which would come close enough to impact the Earth would have to be inside the loss cone. That is how we survive. The Earth is sitting all the time in the middle of this shower of comets, but we do not get hit because we are inside the loss cone. The loss cone is like a small umbrella shielding us from the hail of comets. Jack Hills calculated the size of the loss cone and showed that it could be effective in shielding the Earth.

But the Sun and the Oort Cloud are not sitting all alone in the universe. The Sun moves around the galaxy through a constantly shifting pattern of other stars. Occasionally another star will come close enough to the Sun to disturb the orbits of the comets in the cloud. The alien star does not need to come very close. It does not need to disturb the comets very much. It is enough if it disturbs the orbits by only a small angle, so that the narrow empty cone is shifted away from the Sun. The Sun and the Earth are then out of the loss cone. The umbrella protecting us is gone. The Earth will be exposed to the hail of comets with full intensity until the alien star has passed by and the effects of the Sun's heat and of planetary encounters create a new loss cone. It will take about a million years for the new loss cone to be established. During that million years, the Earth experiences a comet shower. After the new loss cone is in place, the Earth is protected until the next close passage of an alien star. And so the cycle continues.

The theory of the Oort Cloud and the theory of comet showers are firmly based in astronomical facts. The association of comet showers with clusters of craters and with biological extinctions is firmly based in the facts of geology and paleontology. But the story of comet showers has another chapter which is not so firmly based. Some paleontologists claim that mass extinctions, or clusters of extinctions, occur at regular intervals of roughly 26 million years. They find big extinctions, with large numbers of species disappearing, at ages around 13, 39, 65, and 91 million years, with a regular 26-million-year cycle. Other paleontologists, looking at the same evidence, find extinctions occurring irregularly without any well-defined period. The community of geologists is similarly divided. Some geologists find convincing evidence of a regular periodicity in the dates of big craters. Other geologists looking at the same evidence are unconvinced. To summarize a long and complicated argument, one can say that the evidence for a regular periodicity of comet showers is suggestive but not conclusive. We know that comet showers have happened. We do not know for sure whether they come at regular intervals.

Piet Hut, another Dutch astronomer fifty years younger than Jan Oort, decided to take seriously the possibility that comet showers are periodic. If they are periodic, the theory that they are caused by the random passing-by of alien stars cannot be right. If showers are periodic, they must be explained by a different theory. Piet Hut and his friend Rich Muller found an alternative theory to explain the periodicity in case it turns out to be real. The alternative theory is called Nemesis. Nemesis is an imagined star belonging to our own solar system. It is gravitationally bound to the Sun and forms with the Sun a binary star. We know rather accurately how far away it is, if it exists. It is 2.5 light-years away. But we do not know even roughly in which direction it lies, and we do not know whether it is bright enough to be seen with a telescope. Several astronomers are searching for it but nobody has found it. It may turn out, after all, to be a chimera, a mythical creature made out of dreams and travelers' tales.

Piet Hut's theory assumes that Nemesis has an elongated orbit, as many double stars do. Most of the time it will be at a large distance from the Sun and will have no effect on the comets in the Oort Cloud. But once every 26 million years it will come rather quickly through the part of its orbit which is close to the Sun. As it passes by the Sun it will disturb the comet orbits and cause a comet shower, just as an alien star would do. Since the passages of Nemesis occur at regular intervals, the comet showers will be periodic.

Since the last cluster of extinctions and craters occurred about 13 million years ago, we are now halfway through the cycle. This means that Nemesis, if she exists, must be close to the farthest point on her orbit. That is why I was able to say that she is exactly 2.5 light-years away. It happens that the maximum distance between Nemesis and the Sun is known rather accurately, once we know the period of her orbit. The other thing that we know for sure about Nemesis, if she exists, is that she is moving extremely slowly relative to the Sun. Her orbital speed must be a few hundred meters per second, about a hundred times slower than the random motions of stars in our neighborhood.

This is one of the most paradoxical things about the Nemesis theory. We have this double star which is very loosely bound, with a huge orbit, and we have other stars which come zipping right through the orbit at a rate of about one every million years. In one revolution of Nemesis around the Sun, twenty or thirty stranger stars will pass through her orbit. You might think that this rain of stars would totally disrupt the orbit and detach Nemesis from the Sun in less than 26 million years. Many people criticized the Nemesis theory on the grounds that a binary star with such a large orbit could not survive for hundreds of millions of years in the environment of our galaxy. But Piet Hut calculated the effect of passing stars accurately, and he found that the double star system survives amazingly well. Nemesis and the Sun are moving so slowly that they hardly notice the quick gravitational jerks produced by the other stars flying by. It is true that Nemesis will not stay bound to the Sun forever, but the time it will take for other stars to disrupt the system is far longer than 26 million years.

It happened three times in the past that theoretical astronomers invented a new planet on the basis of indirect and circumstantial evidence. The first time was in 1845 when Adams and Leverrier independently deduced the existence of the planet Neptune from the perturbations which it had produced in the motion of Uranus. One year later, Neptune was duly discovered in the predicted region of the sky. The successful prediction of the presence of an unseen planet was one of the great events of nineteenth-century science. It impressed the educated public of that time as a spectacular demonstration of the power of human reason to uncover Nature's secrets. The second prediction of a new planet was in 1859 when the same Leverrier, now world-famous from his prediction of Neptune, turned his attention to Mercury. He looked at the unexplained deviation of the motion of Mercury from its theoretical orbit, and deduced the existence of one or more planets circling the Sun inside the orbit of Mercury. This time the end of the story was not so happy. The name Vulcan was given to the undiscovered planet. Many people observed Vulcan, either as a bright speck near the horizon at sunset or sunrise, or as a dark speck silhouetted against the disk of the daytime Sun. Unfortunately, the various observations could not be reconciled with any consistent Newtonian orbits. As the techniques of observation improved, belief in Vulcan's existence slowly faded. Finally in 1915 Einstein showed that the deviation of Mercury's motion was explained as a consequence of general relativity without any help from Vulcan. The third prediction of a planet was also in 1915, when Lowell deduced from residual perturbations of Uranus the existence of another planet beyond Neptune. This time the planet Pluto was found in the predicted place, but subsequent observations showed the mass of Pluto to be too small to produce measurable perturbations of Uranus. The discovery of Pluto appears to have been a happy accident, less impressive than the discovery of Neptune as a demonstration of the power of human reason. Adding up the score, we may say that out of three theoretical predictions of new planets, one turned out right, one turned out wrong, and one turned out to be a fluke.

And now we have the prediction of Nemesis. Will this turn out to be right like Neptune, wrong like Vulcan, or a fluke like Pluto? Only time, and the patient efforts of observers, will tell. If it should happen that the prediction is right, that Nemesis really exists, this will be a triumph even greater than the discovery of Neptune. Neptune is only one more addition to a big family of planets. Nemesis would be something much weightier, something unique, a sister star to the Sun. Nemesis, if she exists, will change drastically our picture of the structure and history of the solar system. Piet Hut and Rich

Muller suggested the name Nemesis for the companion star. They called her Nemesis "after the Greek goddess who relentlessly persecutes the excessively rich, proud and powerful." Just as in the old Greek myth, Nemesis humbled the lords of creation and made room for new beginnings. It is a beautiful theory. Perhaps it is even true. We will not know for certain unless we find Nemesis, wherever she is hiding in the sky.

So far as the consequences for life on Earth are concerned, it is not important whether Nemesis exists or not. The important thing is that comet showers exist. If it turns out that comet showers have a regular period, then they may be explained by the action of Nemesis. If comet showers do not have a regular period, they may be explained by the action of other stars passing randomly by the Sun. In either case, comet showers will occur at the observed rate, about four major showers in 100 million years. To me, the comet-shower theory is exciting because it is not just an exercise in theoretical astronomy. It is not just a theory about the Sun and the stars. It is also a theory about the history of life on Earth. If the theory is right, it changes the way we think about life and its evolution. The theory implies that life has been exposed, at regular or irregular intervals, to a drastic pruning. Every 26 million years, more or less, there has been an environmental catastrophe severe enough to put down the mighty from their seat and to exalt the humble and meek. Creatures which were too successful in adapting themselves to a stable environment were doomed to perish when the environment suddenly changed. Creatures which were unspecialized and opportunistic in their habits had a better chance when Doomsday struck. We humans are perhaps the most unspecialized and the most opportunistic of all existing species. We thrive on ice ages and environmental catastrophes. Comet showers must have been one of the major forces that drove our evolution and made us what we are.

The last stop on our tour of the universe brings us back to my home in Princeton. We have descended from sky to earth, from abstract and speculative theories to the world of everyday reality. My youngest daughter came back from a music camp in Massachusetts carrying some Monarch caterpillars in a jar. She found them feeding on milkweed near the camp. We also have milkweed growing in Princeton and so she was able to keep the caterpillars alive. After a few days they stopped feeding, hung themselves up by their tails and began to pupate. The process of pupation is delightful to watch. They squeeze themselves up into the skin of the pupa, like a fat boy wriggling into a sleeping bag that is three sizes too small for him. At the beginning you cannot believe that the caterpillar will ever fit inside, and at the end it turns out that the sleeping bag was exactly the right size.

Two or three weeks later the butterflies emerge. The emergence is even more spectacular than the pupation. Out of the sleeping bag crawls the bedraggled remnant of the caterpillar, much reduced in size and with wet black stubs for wings. Then, in a few minutes, the body dries, the legs and antennae stiffen and the wings unfurl. The bedraggled little creature springs to life as a shimmering beauty of orange and white and black. We set her free in a nearby field and she flies high over the trees, disappearing into the sky. We hope that the move from Massachusetts to Princeton will not have disrupted the pattern of her autumn migration. With luck she will find companions to share with her the long journey to the Southwest. She has a long way to go, most of it against the prevailing winds.

The world of biology is full of miracles, but nothing I have seen is as miraculous as this metamorphosis of the Monarch caterpillar. Her brain is a speck of neural tissue a few millimeters long, about a million times smaller than a human brain. With this almost microscopic clump of nerve cells she knows how to manage her new legs and wings, to walk and to fly, to find her way by some unknown means of navigation over thousands of miles from Massachusetts to Mexico. How can all this be done? How are her behavior patterns programmed first into the genes of the caterpillar and then translated into the neural pathways of the butterfly? These are mysteries which our biological colleagues are very far from having understood. And yet, we can be confident that we are on the way toward understanding. Progress is rapid in all the necessary disciplines: biochemistry, genetics, embryology, cytology and neurophysiology. Within twenty or fifty years, we will probably be able to read the message that is written in the DNA of the caterpillar. Then we will see in detail how this message is able to direct the formation of a pupa, of legs and wings, and of a brain capable of long-range navigation. Before long, all these marvels of biochemical technology will be within our grasp. And we shall then be able, if we so choose, to apply the technology of the butterfly to our own purposes.

That is the end of our tour. I have given you brief glimpses of four pieces of the universe with which I have to deal as a scientist. First, the superstrings, our latest attempt to impose a deep mathematical unity on the laws of physics. Second, the black holes, the conceptual laboratories in which we play with the structure of space and time. Third, the Oort Cloud, and the comet showers which we imagine guiding the evolution of life on our planet. Fourth, the Monarch butterfly, which flies up into the summer sky, over the trees and far away, a symbol of evanescent beauty and a living proof that nature's imagination is richer than our own.

RICHARD P. FEYNMAN

∎

Richard Feynman (1918–1988) made important contributions to
contemporary scientific thinking about time. (Among them was
the provocative — to some physicists, maddening — suggestion
that particles of antimatter can be regarded as particles of ordinary matter moving backward in time.) The following is a tran-
scription of a talk Feynman gave at Cornell University in 1964;
it was broadcast by the BBC and published in his book *The Character of Physical Law*.

# The Distinction of Past and Future

IT IS OBVIOUS TO EVERYBODY that the phenomena of the world are
evidently irreversible. I mean things happen that do not happen the
other way. You drop a cup and it breaks, and you can sit there a long
time waiting for the pieces to come together and jump back into your
hand. If you watch the waves of the sea breaking, you can stand there
and wait for the great moment when the foam collects together, rises
up out of the sea, and falls back farther out from the shore — it
would be very pretty!

The demonstration of this in lectures is usually made by having
a section of moving picture in which you take a number of phenomena, and run the film backwards, and then wait for all the
laughter. The laughter just means this would not happen in the real
world. But actually that is a rather weak way to put something which
is as obvious and deep as the difference between the past and the

147

future; because even without an experiment our very experiences inside are completely different for past and future. We remember the past, we do not remember the future. We have a different kind of awareness about what might happen than we have of what probably has happened. The past and the future look completely different psychologically, with concepts like memory and apparent freedom of will, in the sense that we feel that we can do something to affect the future, but none of us, or very few of us, believe that there is anything we can do to affect the past. Remorse and regret and hope and so forth are all words which distinguish perfectly obviously the past and the future.

Now if the world of nature is made of atoms, and we too are made of atoms and obey physical laws, the most obvious interpretation of this evident distinction between past and future, and this irreversibility of all phenomena, would be that some laws, some of the motion laws of the atoms, are going one way — that the atom laws are not such that they can go either way. There should be somewhere in the works some kind of a principle that uxles only make wuxles, and never vice versa, and so the world is turning from uxley character to wuxley character all the time — and this one-way business of the interactions of things should be the thing that makes the whole phenomena of the world seem to go one way.

But we have not found this yet. That is, in all the laws of physics that we have found so far there does not seem to be any distinction between the past and the future. The moving picture should work the same going both ways, and the physicist who looks at it should not laugh.

Let us take the law of gravitation as our standard example. If I have a sun and a planet, and I start the planet off in some direction, going around the sun, and then I take a moving picture, and run the moving picture backwards and look at it, what happens? The planet goes around the sun, the opposite way of course, keeps on going around in an ellipse. The speed of the planet is such that the area swept out by the radius is always the same in equal times. In fact it just goes exactly the way it ought to go. It cannot be distinguished from going the other way. So the law of gravitation is of such a kind that the direction does not make any difference; if you show any phenomenon involving only gravitation running backwards on a film it will look perfectly satisfactory. You can put it more precisely this way. If all the particles in a more complicated system were to have every one of their speeds reversed suddenly, then the thing would just unwind through all the things that it wound up into. If you have a lot of particles doing something, and then you suddenly reverse the speed, they will completely undo what they did before.

This is in the law of gravitation, which says that the velocity changes as a result of the forces. If I reverse the time, the forces are not changed, and so the changes in velocity are not altered at corresponding distances. So each velocity then has a succession of alterations made in exactly the reverse of the way that they were made before, and it is easy to prove that the law of gravitation is time-reversible.

The law of electricity and magnetism? Time reversible. The laws of nuclear interaction? Time reversible as far as we can tell. The laws of beta-decay that we talked about at a previous time? Also time reversible? The difficulty of the experiments of a few months ago, which indicate that there is something the matter, some unknown about the laws, suggests the possibility that in fact beta-decay may not also be time reversible, and we shall have to wait for more experiments to see. But at least the following is true. Beta-decay (which may or may not be time reversible) is a very unimportant phenomenon for most ordinary circumstances. The possibility of my talking to you does not depend upon beta-decay, although it does depend on chemical interactions, it depends on electrical forces, not much on nuclear forces at the moment, but it depends also on gravitation. But I am one-sided — I speak, and a voice goes out into the air, and it does not come sucking back into my mouth when I open it — and this irreversibility cannot be hung on the phenomenon of beta-decay. In other words, we believe that most of the ordinary phenomena in the world, which are produced by atomic motions, are according to laws which can be completely reversed. So we will have to look some more to find the explanation of the irreversibility.

If we look at our planets moving around the sun more carefully, we soon find that all is not quite right. For example, the Earth's rotation on its axis is slightly slowing down. It is due to tidal friction, and you can see that friction is something which is obviously irreversible. If I take a heavy weight on the floor, and push it, it will slide and stop. If I stand and wait, it does not suddenly start up and speed up and come into my hand. So the frictional effect seems to be irreversible. But a frictional effect, as we discussed at another time, is the result of the enormous complexity of the interactions of the weight with the wood, the jiggling of the atoms inside. The organized motion of the weight is changed into disorganized, irregular wiggle-waggles of the atoms in the wood. So therefore we should look at the thing more closely.

As a matter of fact, we have here the clue to the apparent irreversibility. I will take a simple example. Suppose we have blue water, with ink, and white water, that is without ink, in a tank, with a little separation, and then we pull out the separation very delicately. The water starts separate, blue on one side and white on the other side. Wait a while. Gradually the blue mixes up with the white, and after a while the water is "luke blue," I mean it is sort of fifty-fifty, the colour uniformly distributed throughout. Now if we wait and watch this for a long time, it does not by itself separate. (You could *do* something to get the blue separated again. You could evaporate the water and condense it somewhere else, and collect the blue dye and dissolve it in half the water, and put the thing back. But while you were doing all that you yourself would be causing irreversible phenomena somewhere else.) By itself it does not go the other way.

That gives us some clue. Let us look at the molecules. Suppose that we take a moving picture of the blue and white water mixing. It will look funny if we run it backwards, because we shall start with uniform water and gradually the thing will separate — it will be obviously nutty. Now we magnify the picture, so that every physicist can watch, atom by atom, to find out what happens irreversibly — where the laws of balance of forward and backward break down. So you start, and you look at the picture. You have atoms of two different kinds (it's ridiculous, but let's call them blue and white) jiggling all the time in thermal motion. If we were to start at the beginning we should have mostly atoms of one kind on one side, and atoms of the other kind on the other side. Now these atoms are jiggling around, billions and billions of them, and if we start them with one kind all on one side, and the other kind on the other side, we see that in their perpetual irregular motions they will get mixed up, and that is why the water becomes more or less uniformly blue.

Let us watch any one collision selected from that picture, and in the moving picture the atoms come together this way and bounce off that way. Now run that section of the film backwards, and you find the pair of molecules moving together the other way and bouncing off this way. And the physicist looks with his keen eye, and measures everything, and says, "That's all right, that's according to the laws of physics. If two molecules came this way they would bounce this way." It is reversible. The laws of molecular collision are reversible.

So if you watch too carefully you cannot understand it at all, because every one of the collisions is absolutely reversible, and yet the whole moving picture shows something absurd, which is that in the reversed picture the molecules start in the mixed condition — blue, white, blue, white, blue, white — and as time goes on, through all the collisions, the blue separates from the white. But they cannot do that — it is not natural that the accidents of life should be such that the blues will separate themselves from the whites. And yet if you watch this reversed movie very carefully every collision is O.K.

Well you see that all there is to it is that the irreversibility is caused by the general accidents of life. If you start with a thing that is separated and make irregular changes, it does get more uniform. But if it starts uniform and you make irregular changes, it does not get separated. It *could* get separated. It is not against the laws of physics that the molecules bounce around so that they separate. It is just unlikely. It would never happen in a million years. And that is the answer. Things are irreversible only in a sense that going one way is likely, but going the other way, although it is possible and is according to the laws of physics, would not happen in a million years. It is just ridiculous to expect that if you sit there long enough the jiggling of the atoms will separate a uniform mixture of ink and water into ink on one side and water on the other.

Now if I had put a box around my experiment, so that there were only four or five molecules of each kind in the box, as time went on they would get mixed up. But I think you could believe that, if you kept watching, in the perpetual irregular collisions of these molecules, after some time — not necessarily a million years, maybe only a year — you would see that accidentally they would get back more or less to their original state, at least in the sense that if I put a barrier through the middle, all the whites would be on one side and all the blues on the other. It is not impossible. However, the actual objects with which we work have not only four or five blues and whites. They have four or five million, million, million, million, which are all going to get separated like this. And so the apparent irreversibility of nature does not come from the irreversibility of the fundamental physical laws; it comes from the characteristic that if you start with an ordered system, and have the irregularities of nature, the bouncing of molecules, then the thing goes one way.

Therefore the next question is — how did they get ordered in the first place? That is to say, why is it possible to start with the ordered? The difficulty is that we start with an ordered thing, and we do not end with an ordered thing. One of the rules of the world is that the thing goes from an ordered condition to a disordered. Incidentally, this word order, like the word disorder, is another of these terms of physics which are not exactly the same as in ordinary life. The order need not be interesting to you as human beings, it is just that there is a definite situation, all on one side and all on the other, or they are mixed up — and that is ordered and disordered.

The question, then, is how the thing gets ordered in the first place, and why, when we look at any ordinary situation, which is only partly ordered, we can conclude that it probably came from one which was more ordered. If I look at a tank of water, in which the water is very dark blue on one side and very clear white on the other, and a sort of bluish colour in between, and I know that the thing has been left alone for twenty or thirty minutes, then I will guess that it got this way because the separation was more complete in the past. If I wait longer, then the blue and white will get more intermixed, and if I know that this thing has been left alone for a sufficiently long time, I can conclude something about the past condition. The fact that it is "smooth" at the sides can only arise because it was much more satisfactorily separated in the past; because if it were not more satisfactorily separated in the past, in the time since then it would have become more mixed up than it is. It is therefore possible to tell, from the present, something about the past.

In fact physicists do not usually do this much. Physicists like to think that all you have to do is say, "These are the conditions, now what happens next?" But all our sister sciences have a completely different problem: in fact all the other things that are studied — history, geology, astronomical history — have a problem of this other kind. I find they are able to make predictions of a completely different type from those of a physicist. A physicist says, "In this condition I'll tell you what will happen next." But a geologist will say something like this — "I have dug in the ground and I have found certain kinds of bones. I predict that if you dig in the ground you will find a similar kind of bones." The historian, although he talks about the past, can do it by talking about the future. When he says that the French Revolution was in 1789, he means that if you look in another book about the French Revolution you will find the same date. What he does is to make a kind of prediction about something that he has never looked at before, documents that have still to be found. He predicts that the documents in which there is something written about Napoleon will coincide with what is written in the other documents. The question is how that is possible — and the only way that is possible is to suggest that the past of the world was more organized in this sense than the present.

Some people have proposed that the way the world became ordered is this. In the beginning the whole universe was just irregular motions, like the mixed water. We saw that if you waited long enough, with very few atoms, the water could have got separated accidentally. Some physicists (a century ago) suggested that all that has happened is that the world, this system that has been going on and going on, fluctuated. (That is the term used when it gets a little out of the ordinary uniform condition.) It fluctuated, and now we are watching the fluctuation undo itself again. You may say, "But look how long you would have to wait for such a fluctuation." I know, but if it did not fluctuate far enough to be able to produce evolution, to be able to produce an intelligent person, we would not have noticed it. So we had to keep waiting until we were alive to notice it — we had to have at least that big a fluctuation. But I believe this theory to be incorrect. I think it is a ridiculous theory for the following reason. If the world were much bigger, and the atoms were all over the place starting from a completely mixed up condition, then if I happened to look only at the atoms in one place, and I found the atoms there separated, I would have no way to conclude that the atoms anywhere else would be separated. In fact if the thing were a fluctuation, and I noticed something odd, the most likely way that it got there would be that there was nothing odd anywhere else. That is, I would have to borrow odds, so to speak, to get the thing lopsided, and there is no use borrowing too much. In the experiment with the blue and white water, when eventually the few molecules in the box became separated, the most likely condition of the rest of the water would still be mixed up. And therefore, although when we look at the stars and we look at the world we see everything is ordered, if there were a fluctuation, the prediction would be that if we looked at a place where we have not looked before, it would be disordered and a mess. Although the separation of the matter into stars which are hot and space which is cold, which we have seen, could be a fluctuation, then in places where we have not looked we would expect to find that the stars are not separated from space. And since we always make the prediction that in a place where we have not looked we shall see stars in a similar condition, or find the same statement about Napoleon, or that we shall see bones like the bones that we have seen before, the success of all those sciences indicates that the world did not come from a fluctuation, but came from a condition which was more separated, more organized, in the past than at the present time. Therefore I think it necessary to add to the physical laws the hypothesis that in the past the universe was more ordered, in the technical sense, than it is today — I think this is the additional statement that is needed to make sense, and to make an understanding of the irreversibility.

That statement itself is of course lopsided in time; it says that something about the past is different from the future. But it comes outside the province of what we ordinarily call physical laws, because we try today to distinguish between the statement of the physical laws which govern the rules by which the universe develops, and the law which states the condition that the world was in in the past. This is considered to be astronomical history — perhaps some day it will also be a part of physical law.

Now there are a number of interesting features of irreversibility which I would like to illustrate. One of them is to see how, exactly, an irreversible machine really works.

Suppose that we build something that we know ought to work only one way — and what I am going to build is a wheel with a ratchet on it — a saw-toothed wheel, with sharp up notches, and relatively slow down notches, all the way round. The wheel is on a shaft, and then there is a little pawl, which is on a pivot and which is held down by a spring (Figure 1).

Now the wheel can only turn one way. If you try to turn it the other way, the straight-edged parts of the teeth get jammed against the pawl and it does not go, whereas if you turn it the other way it just goes right over the teeth, snap, snap, snap. (You know the sort of thing: they use them in clocks, and a watch has this kind of thing inside so that you can only wind it one way, and after you have

Figure 1

wound it, it holds the spring.) It is completely irreversible in the sense that the wheel can only turn one way.

Now it has been imagined that this irreversible machine, this wheel that can only turn one way, could be used for a very useful and interesting thing. As you know, there is a perpetual irregular motion of molecules, and if you build a very delicate instrument it will always jiggle because it is being bombarded irregularly by the air molecules in the neighbourhood. Well that is very clever, so we will connect the wheel with a shaft that has four vanes, like this (Figure 2). They are in a box of gas, and they are bombarded all the time by the molecules irregularly, so the vanes are pushed sometimes one way, sometimes the other way. But when the vanes are pushed one way the thing gets jammed by the ratchet, and when the vanes are pushed the other way, it goes around, and so we find the wheel perpetually going around, and we have a kind of perpetual motion. That is because the ratchet wheel is irreversible.

But actually we have to look into things in more detail. The way this works is that when the wheel goes one way it lifts the pawl up and then the pawl snaps down against the tooth. Then it will bounce off, and if it is perfectly elastic it will go bounce, bounce, bounce, all the time, and the wheel can just go down and around the other way when the pawl accidentally bounces up. So this will not work unless it is true that when the pawl comes down it sticks, or stops, or bounces and cuts out. If it bounces and cuts out there must be what we call damping, or friction, and in the falling down and bouncing and stopping, which is the only way this will work one-way, heat is generated by the friction, so the wheel will get hotter and hotter. However,

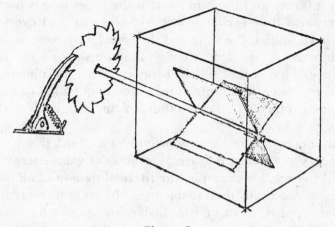

Figure 2

when it begins to get quite warm something else happens. Just as there is Brownian motion, or irregular motions, in the gas round the vanes, so whatever this wheel and pawl are made of, the parts that they are made of are getting hotter and are beginning to move in a more irregular fashion. The time comes when the wheel is so hot that the pawl is simply jiggling because of the molecular motions of the things inside it, and so it bounces up and down on the wheel because of the molecular motion, the same thing as was making the vane turn round. In bouncing up and down on the wheel it is up as much as it is down, and the tooth can go either way. We no longer have a one-way device. As a matter of fact, the thing can be driven backwards! If the wheel is hot and the vane part is cold, the wheel that you thought would go only one way will go the other way, because every time the pawl comes down it comes down on an inclined plane on the toothed wheel, and so pushes the wheel "back-wards." Then it bounces up again, comes down on another inclined plane, and so on. So if the wheel is hotter than the vanes it will go the wrong way.

What has this got to do with the temperature of the gas round the vanes? Suppose we did not have that part at all. Then if the wheel is pushed forward by the pawl falling on an inclined plane, the next thing that will happen is that the straight vertical side of the tooth will bounce against the pawl and the wheel will bounce back. In order to prevent the wheel from bouncing back we put a damper on it and put vanes in the air, so it will be slowed down and not bounce freely. Then it will go only one way, but the wrong way, and so it turns out that no matter how you design it, a wheel like this will go one way if one side is hotter and the other way if the other side is hotter. But after there is a heat exchange between the two, and everything is calmed down, so that the vane and the wheel have come to be at the same temperature, it will neither go the one way nor the other on the average. That is the technical way in which the phenomena of nature will go one way as long as they are out of equilibrium, as long as one side is quieter than the other, or one side is bluer than the other.

The conservation of energy would let us think that we have as much energy as we want. Nature never loses or gains energy. Yet the energy of the sea, for example, the thermal motion of all the atoms in the sea, is practically unavailable to us. In order to get that energy organized, herded, to make it available for use, we have to have a difference in temperature, or else we shall find that although the energy is there we cannot make use of it. There is a great difference between energy and availability of energy. The energy of the sea is a large amount, but it is not available to us.

The conservation of energy means that the total energy in the world is kept the same. But in the irregular jigglings that energy can be spread about so uniformly that, in certain circumstances, there is no way to make more go one way than the other — there is no way to control it any more.

I think that by an analogy I can give some idea of the difficulty, in this way. I do not know if you have ever had the experience — I have — of sitting on the beach with several towels, and suddenly a tremendous downpour comes. You pick up the towels as quickly as you can, and run into the bathhouse. Then you start to dry yourself, and you find that this towel is a little wet, but it is drier than you are. You keep drying with this one until you find it is too wet — it is wetting you as much as drying you — and you try another one; and pretty soon you discover a horrible thing — that all the towels are damp and so are you. There is no way to get any drier, even though you have many towels, because there is no difference in some sense between the wetness of the towels and the wetness of yourself. I could invent a kind of quantity which I could call "ease of removing water." The towel has the same ease of removing water from it as you have, so when you touch yourself with the towel, as much water comes off the towel on to you as comes from you to the towel. It does not mean there is the same amount of water in the towel as there is on you — a big towel will have more water in it than a little towel — but they have the same dampness. When things get to the same dampness then there is nothing you can do any longer.

Now the water is like the energy, because the total amount of water is not changing. (If the bathhouse door is open and you can run into the sun and get dried out, or find another towel, then you're saved, but suppose everything is closed, and you can't get away from these towels or get any new towels.) In the same way if you imagine a part of the world that is closed, and wait long enough, in the accidents of the world the energy, like the water, will be distributed over all of the parts evenly until there is nothing left of one-way-ness, nothing left of the real interest of the world as we experience it.

Thus in the ratchet and pawl and vanes situation, which is a limited one, in which nothing else is involved, the temperatures gradually become equal on both sides, and the wheel does not go round either one way or the other. In the same way the situation is that if you leave any system long enough it gets the energy thoroughly mixed up in it, and no more energy is really available to do anything.

Incidentally, the thing that corresponds to the dampness or the "ease of removing water" is called the temperature, and although I say when two things are at the same temperature things get balanced, it does not mean they have the same energy in them; it means that it is just as easy to pick energy off one as to pick it off the other. Temperature is like "ease of removing energy." So if you sit them next to each other, nothing apparently happens; they pass energy back and forth equally, but the net result is nothing. So when things have become all of the same temperature, there is no more energy available to do anything. The principle of irreversibility is that if things are at different temperatures and are left to themselves, as time goes on they become more and more at the same temperature, and the availability of energy is perpetually decreasing.

This is another name for what is called the entropy law, which says the entropy is always increasing. But never mind the words; stated the other way, the availability of energy is always decreasing. And that is a characteristic of the world, in the sense that it is due to the chaos of molecular irregular motions. Things of different temperature, if left to themselves, tend to become of the same temperature. If you have two things at the same temperature, like water on an ordinary stove without a fire under it, the water is not going to freeze and the stove get hot. But if you have a hot stove with ice, it goes the other way. So the one-way-ness is always to the loss of the availability of energy.

That is all I want to say on the subject, but I want to make a few remarks about some characteristics. Here we have an example in which an obvious effect, the irreversibility, is not an obvious consequence of the laws, but is in fact rather far from the basic laws. It takes a lot of analysis to understand the reason for it. The effect is of first importance in the economy of the world, in the real behaviour of the world in all obvious things. My memory, my characteristics, the difference between past and future, are completely involved in this, and yet the understanding of it is not prima facie available by knowing about the laws. It takes a lot of analysis.

It is often the case that the laws of physics do not have an obvious direct relevance to experience, but that they are abstract from experience to varying degrees. In this particular case, the fact that the laws are reversible although the phenomena are not is an example.

There are often great distances between the detailed laws and the main aspects of real phenomena. For example, if you watch a glacier from a distance, and see the big rocks falling into the sea, and the way the ice moves, and so forth, it is not really essential to remember that it is made out of little hexagonal ice crystals. Yet if understood well enough the motion of the glacier is in fact a consequence of the character of the hexagonal ice crystals. But it takes quite a while to understand all the behaviour of the glacier (in fact nobody knows enough about ice yet, no matter how much they've studied the crystal). However, the hope is that if we do understand the ice crystal we shall ultimately understand the glacier.

In fact, although we have been talking in these lectures about the fundaments of the physical laws, I must say immediately that one does not, by knowing all the fundamental laws as we know them today, immediately obtain an understanding of anything much. It takes a while, and even then it is only partial. Nature, as a matter of fact, seems to be so designed that the most important things in the real world appear to be a kind of complicated accidental result of a lot of laws.

To give an example, nuclei, which involve several nuclear particles, protons and neutrons, are very complicated. They have what we call energy levels, they can sit in states or conditions of different energy values, and various nuclei have various energy levels. And it's a complicated mathematical problem, which we can only partly solve, to find the position of the energy levels. The exact position of the levels is obviously a consequence of an enormous complexity and therefore there is no particular mystery about the fact that nitrogen, with 15 particles inside, happens to have a level at 2.4 million volts, and another level at 7.1, and so on. But the remarkable thing about nature is that the whole universe in its character depends upon precisely the position of one particular level in one particular nucleus. In the carbon$^{12}$ nucleus, it so happens, there is a level at 7.82 million volts. And that makes all the difference in the world.

The situation is the following. If we start with hydrogen, and it appears that at the beginning the world was practically all hydrogen, then as the hydrogen comes together under gravity and gets hotter, nuclear reactions can take place, and it can form helium, and then the helium can combine only partially with the hydrogen and produce a few more elements, a little heavier. But these heavier elements disintegrate right away back into helium. Therefore for a while there was a great mystery about where all the other elements in the world came from, because starting with hydrogen the cooking processes inside the stars would not make much more than helium and less than half a dozen other elements. Faced with this problem, Professors Hoyle and Salpeter[1] said that there is one way out. If three helium atoms could come together to form carbon, we can easily calculate how often that should happen in a star. And it turns out that it should never happen, except for one possible accident — if there happened to be an energy level at 7.82 million volts in carbon, then the three helium atoms would come together and before they came apart, would stay together a little longer on the average than they would do if there were no level at 7.82. And staying there a little longer, there would be enough time for something else to happen, and to make other elements. If there was a level at 7.82 million volts in carbon, then we could understand where all the other elements in the periodic table came from. And so, by a back-handed, upside-down argument, it was predicted that there is in carbon a level at 7.82 million volts; and experiments in the laboratory showed that indeed there is. Therefore the existence in the world of all these other elements is very closely related to the fact that there is this particular level in carbon. But the position of this particular level in carbon seems to us, knowing the physical laws, to be a very complicated accident of 12 complicated particles interacting. This example is an excellent illustration of the fact that an understanding of the physical laws does not necessarily give you an understanding of things of significance in the world in any direct way. The details of real experience are often very far from the fundamental laws.

We have a way of discussing the world, when we talk of it at various hierarchies, or levels. Now I do not mean to be very precise, dividing the world into definite levels, but I will indicate, by describing a set of ideas, what I mean by hierarchies of ideas.

For example, at one end we have the fundamental laws of physics. Then we invent other terms for concepts which are approximate, which have, we believe, their ultimate explanation in terms of the fundamental laws. For instance, "heat." Heat is supposed to be jiggling, and the word for a hot thing is just the word for a mass of atoms which are jiggling. But for a while, if we are talking about heat, we sometimes forget about the atoms jiggling — just as when we talk about the glacier we do not always think of the hexagonal ice and

1. Fred Hoyle, British astronomer, Cambridge. Edwin Salpeter, American physicist, Cornell University.

the snowflakes which originally fell. Another example of the same thing is a salt crystal. Looked at fundamentally it is a lot of protons, neutrons, and electrons; but we have this concept "salt crystal," which carries a whole pattern already of fundamental interactions. An idea like pressure is the same.

Now if we go higher up from this, in another level we have properties of substances — like "refractive index," how light is bent when it goes through something; or "surface tension," the fact that water tends to pull itself together, both of which are described by numbers. I remind you that we have to go through several laws down to find out that it is the pull of the atoms, and so on. But we still say "surface tension," and do not always worry, when discussing surface tension, about the inner workings.

On, up in the hierarchy. With the water we have waves, and we have a thing like a storm, the word "storm" which represents an enormous mass of phenomena, or a "sun spot," or "star," which is an accumulation of things. And it is not worth while always to think of it way back. In fact we cannot, because the higher up we go the more steps we have in between, each one of which is a little weak. We have not thought them all through yet.

As we go up in this hierarchy of complexity, we get to things like muscle twitch, or nerve impulse, which is an enormously complicated thing in the physical world, involving an organization of matter in a very elaborate complexity. Then come things like "frog."

And then we go on, and we come to words and concepts like "man," and "history," or "political expediency," and so forth, a series of concepts which we use to understand things at an ever higher level.

And going on, we come to things like evil, and beauty, and hope . . .

Which end is nearer to God; if I may use a religious metaphor. Beauty and hope, or the fundamental laws? I think that the right way, of course, is to say that what we have to look at is the whole structural interconnection of the thing; and that all the sciences, and not just the sciences but all the efforts of intellectual kinds, are an endeavour to see the connections of the hierarchies, to connect beauty to history, to connect history to man's psychology, man's psychology to the working of the brain, the brain to the neural impulse, the neural impulse to the chemistry, and so forth, up and down, both ways. And today we cannot, and it is no use making believe that we can, draw carefully a line all the way from one end of this thing to

161

the other, because we have only just begun to see that there is this relative hierarchy.

And I do not think either end is nearer to God. To stand at either end, and to walk off that end of the pier only, hoping that out in that direction is the complete understanding, is a mistake. And to stand with evil and beauty and hope, or to stand with the fundamental laws, hoping that way to get a deep understanding of the whole world, with that aspect alone, is a mistake. It is not sensible for the ones who specialize at one end, and the ones who specialize at the other end, to have such disregard for each other. (They don't actually, but people say they do.) The great mass of workers in between, connecting one step to another, are improving all the time our understanding of the world, both from working at the ends and working in the middle, and in that way we are gradually understanding this tremendous world of interconnecting hierarchies.

# MAX PLANCK

∎

It was through his research in thermodynamics, the science of heat, that the German physicist Max Planck (1858–1947) discovered the quantum principle — that energy is emitted not as a continuum but in discrete units, or *quanta*. Here Planck discusses the second law of thermodynamics, which states that the entropy — the disorder — in any system will increase over time unless energy is added to the system.

# *The Second Law of Thermodynamics*

THE SECOND LAW of thermodynamics is essentially different from the first law, since it deals with a question in no way touched upon by the first law, *viz.* the direction in which a process takes place in nature. Not every change which is consistent with the principle of the conservation of energy satisfies also the additional conditions which the second law imposes upon the processes which actually take place in nature. In other words, the principle of the conservation of energy does not suffice for a unique determination of natural processes.

If, for instance, an exchange of heat by conduction takes place between two bodies of different temperature, the first law, or the principle of the conservation of energy, merely demands that the quantity of heat given out by the one body shall be equal to that taken up by the other. Whether the flow of heat, however, takes place from the colder to the hotter body, or *vice versa,* cannot be answered by the energy principle alone. The very notion of temperature is alien to that principle, as can be seen from the fact that it yields no exact definition of temperature. Neither does the first law contain any statement with regard to the direction of the particular process. The special equation, for instance,

$$[H_2] + \frac{1}{2}[O_2] - (H_2O) = 68,400 \text{ cal.,}$$

means only that if hydrogen and oxygen combine under constant pressure to form water, the re-establishment of the initial temperature requires a certain amount of heat to be given up to surrounding bodies; and *vice versa,* that this amount of heat is absorbed when water is decomposed into hydrogen and oxygen. It offers no information, however, as to whether hydrogen and oxygen actually combine to form water, or water decomposes into hydrogen and oxygen, or whether such a process can take place at all in either direction. From the point of view of the first law, the initial and final states of any process are completely equivalent.

In one particular case, however, the principle of the conservation of energy does prescribe a certain direction to a process. This occurs when, in a system, one of the various forms of energy is at an absolute maximum, or absolute minimum. It is evident that, in this case, the direction of the change must be such that the particular form of energy will decrease, or increase. This particular case is realized in mechanics by a system of particles at rest. Here the kinetic energy is at an absolute minimum, and, therefore, any change of the system is accompanied by an increase of the kinetic energy, and, if it be an isolated system, by a decrease of the potential energy. This gives rise to an important proposition in mechanics, which characterizes the direction of possible motion, and lays down, in consequence, the general condition of mechanical equilibrium. It is evident that, if both the kinetic and potential energies be at a minimum, no change can possibly take place, since none of these can increase at the expense of the other. The system must, therefore, remain at rest.

If a heavy liquid be initially at rest at different levels in two communicating tubes, then motion will set in, so as to equalize the levels, for the centre of gravity of the system is thereby lowered, and the potential energy diminished. Equilibrium exists when the centre of gravity is at its lowest, and therefore the potential energy at a minimum, *i.e.* when the liquid stands at the same level in both tubes. If no special assumption be made with regard to the initial velocity of the liquid, the above proposition no longer holds. The potential energy need not decrease, and the higher level might rise or sink according to circumstances.

   If our knowledge of thermal phenomena led us to recognize a
state of minimum energy, a similar proposition would hold for this,
but only for this, particular state. In reality no such minimum has
been detected. It is, therefore, hopeless to seek to reduce the general
laws regarding the direction of thermodynamical changes, as well as
those of thermodynamical equilibrium, to the corresponding propositions in mechanics which hold good only for systems at rest. . . .

   We often find the second law stated as follows: The change of
mechanical work into heat may be complete, but, on the contrary,
that of heat into work must needs be incomplete, since, whenever a
certain quantity of heat is transformed into work, another quantity
of heat must undergo a corresponding and compensating change;
*e.g.* transference from higher to lower temperature. This is quite correct in certain very special cases, but it by no means expresses the
essential feature of the process, as a simple example will show. An
achievement which is closely associated with the discovery of the
principle of energy, and which is one of the most important for the
theory of heat, is the proposition that the total internal energy of a
gas depends only on the temperature, and not on the volume. If a
perfect gas be allowed to expand, doing external work, and be prevented from cooling by connecting it with a heat-reservoir of higher
temperature, the temperature of the gas, and at the same time its
internal energy, remains unchanged, and it may be said that the
amount of heat given out by the reservoir is completely changed into
work without an exchange of energy taking place anywhere. Not the
least objection can be made to this. The proposition of the "incomplete transformability of heat into work" cannot be applied to this
case. . . . When we pass from the consideration of the first law of
thermodynamics to that of the second, we have to deal with a new
fact, and it is evident that no definition, however ingenious, although
it contain no contradiction in itself, will ever permit of the deduction
of a new fact.

   There is but one way of clearly showing the significance of the
second law, and that is to base it on facts by formulating propositions
which may be proved or disproved by experiment. The following
proposition is of this character: It is in no way possible to completely
reverse any process in which heat has been produced by friction. For
the sake of example we shall refer to Joule's experiments on friction,
for the determination of the mechanical equivalent of heat. Applied
to these, our proposition says that, when the falling weights have generated heat in water or mercury by the friction of the paddles, no

process can be invented which will completely restore everywhere the initial state of that experiment, *i.e.* which will raise the weights to their original height, cool the liquid, and otherwise leave no change. The appliances used may be of any kind whatsoever, mechanical, thermal, chemical, electrical, etc., but the condition of *complete* restoration of the initial state renders it necessary that all materials and machines used must ultimately be left exactly in the condition in which they were before their application. Such a proposition cannot be proved *a priori*, neither does it amount to a definition, but it contains a definite assertion, to be stated precisely in each case, which may be verified by actual experiment. The proposition is therefore correct or incorrect.

Another proposition of this kind, and closely connected with the former, is the following: It is in no way possible to completely reverse any process in which a gas expands without performing work or absorbing heat, *i.e.* with constant total energy. The word "completely" again refers to the accurate reproduction of the initial conditions. To test this, the gas, after it had assumed its new state of equilibrium, might first be compressed to its former volume by a weight falling to a lower level. External work is done on the gas, and it is thereby heated. The problem is now to bring the gas to its initial condition, and to raise the weight. The gas kept at constant volume might be reduced to its original temperature by conducting the heat of compression into a colder heat-reservoir. In order that the process may be completely reversed, the reservoir must be deprived of the heat gained thereby, and the weight raised to its original position. This is, however, exactly what was asserted in the preceding paragraph to be impracticable.

A third proposition in point refers to the conduction of heat. Supposing that a body receives a certain quantity of heat from another of higher temperature, the problem is to completely reverse this process, *i.e.* to convey back the heat without leaving any change whatsoever. In the description of Carnot's reversible cycle it has been pointed out, that heat can at any time be drawn from a heat-reservoir and transferred to a hotter reservoir without leaving any change except the expenditure of a certain amount of work, and the transference of an equivalent amount of heat from one reservoir to the other. If this heat could be removed, and the corresponding work recovered without other changes, the process of heat-conduction would be completely reversed. Here, again, we have the problem which was declared earlier to be impracticable.

Further examples of processes to which the same considerations apply are diffusion, the freezing of an overcooled liquid, the condensation of a supersaturated vapour, all explosive reactions, and, in fact, every transformation of a system into a state of greater stability.

*Definition.* A process which can in no way be completely reversed is termed *irreversible,* all other processes *reversible.* That a process may be irreversible, it is not sufficient that it cannot be directly reversed. This is the case with many mechanical processes which are not irreversible. The full requirement is, that it be impossible, even with the assistance of all agents in nature, to restore everywhere the exact initial state when the process has once taken place. The propositions of the three preceding paragraphs, therefore, declare that the generation of heat by friction, the expansion of a gas without the performance of external work and the absorption of external heat, the conduction of heat, etc., are irreversible processes.

We now turn to the question of the actual existence of reversible and irreversible processes. Numerous reversible processes can at least be imagined, as, for instance; those consisting of a succession of states of equilibrium, and, therefore, directly reversible in all their parts. Further, all perfectly periodic processes, *e.g.* an ideal pendulum or planetary motion, are reversible, for, at the end of every period, the initial state is completely restored. Also, all mechanical processes with absolutely rigid bodies and absolutely incompressible liquids, as far as friction can be avoided, are reversible. By the introduction of suitable machines with absolutely unyielding connecting rods, frictionless joints and bearings, inextensible belts, etc., it is always possible to work the machines in such a way as to bring the system completely into its initial state without leaving any change in the machines, for the machines of themselves do not perform work.

If, for instance, a heavy liquid, originally at rest at different levels in two communicating tubes, be set in motion by gravity, it will, in consequence of its kinetic energy, go beyond its position of equilibrium, and, since the tubes are supposed frictionless, again swing back to its exact original position. The process at this point has been completely reversed, and therefore belongs to the class of reversible processes.

As soon as friction is admitted, however, its reversibility is at least questionable. Whether reversible processes exist in nature or not, is not *a priori* evident or demonstrable. There is, however, no purely logical objection to imagining that a means may some day be found of completely reversing some process hitherto considered irreversible: one, for example, in which friction or heat-conduction plays a part. But it can be demonstrated that if, in a single instance, one of the processes here declared to be irreversible should be found to be reversible, then all of these processes must be reversible in all cases. Consequently, either all or none of these processes are irreversible. There is no third possibility. If those processes are not irreversible, the entire edifice of the second law will crumble. None of the numerous relations deduced from it, however many may have been verified by experience, could then be considered as universally proved, and theoretical work would have to start from the beginning. . . . It is this foundation on the physical fact of irreversibility which forms the strength of the second law. If, therefore, it must be admitted that a single experience contradicting that fact would render the law untenable, on the other hand, any confirmation of part supports the whole structure, and gives to deductions, even in seemingly remote regions, the full significance possessed by the law itself.

The significance of the second law of thermodynamics depends on the fact that it supplies a necessary and far-reaching criterion as to whether a definite process which occurs in nature is reversible or irreversible. Since the decision as to whether a particular process is irreversible or reversible depends only on whether the process can in any manner whatsoever be completely reversed or not, the nature of the initial and final states, and not the intermediate steps of the process, entirely settle it. The question is, whether or not it is possible, starting from the final state, to reach the initial one in any way without any other change. The second law, therefore, furnishes a relation between the quantities connected with the initial and final states of any natural process. The final state of an irreversible process is evidently in some way discriminate from the initial state, while in reversible processes the two states are in certain respects equivalent. The second law points out this characteristic property of both states, and also shows, when the two states are given, whether a transformation is possible in nature from the first to the second, or from the second to the first, without leaving changes in other bodies. For this purpose, of course, the two states must be fully characterized. Besides the chemical constitution of the systems in question, the physical conditions — *viz.* the state of aggregation, temperature, and pressure in both states — must be known, as is necessary for the application of the first law.

The relation furnished by the second law will evidently be simpler the nearer the two states are to one another. On this depends the great fertility of the second law in its treatment of cyclic processes, which, however complicated they may be, give rise to a final state only slightly different from the initial state. Since the system, which goes through the cyclic process, returns at the end to exactly the same state as at the beginning, we can leave it entirely out of account on comparing the two states.

If we regard the second law from the mathematical point of view, the distinction between the final and initial states of a process can consist only in an inequality. This means that a certain quantity, which depends on the momentary state of the system, possesses in the final state a greater or smaller value, according to the definition of the sign of that quantity, than in the initial state.

The second law of thermodynamics states that there exists in nature for each system of bodies a quantity, which by all changes of the system either remains constant (in reversible processes) or increases in value (in irreversible processes). This quantity is called, following Clausius, the *entropy* of the system. . . .

Since there exists in nature no process entirely free from friction or heat-conduction, all processes which actually take place in nature, if the second law be correct, are in reality irreversible. Reversible processes form only an ideal limiting case. They are, however, of considerable importance for theoretical demonstration and for application to states of equilibrium.

# DAVID N. SCHRAMM

■

The alliance of physics and astronomy is exemplified in the career of the American physicist David Schramm (b. 1945), who has worked simultaneously as chairman of the astronomy department at the University of Chicago and as a researcher at the Fermilab particle accelerator in Batavia, Illinois. In this essay Schramm examines how nuclear physics can be used to estimate the age of the universe.

# *The Age of the Elements*

FOR AS LONG AS MAN has been able to think abstract thoughts he has wondered about the nature of the universe, including its origin and age. The age can be estimated by studying the process of radioactive decay, which has been going on for billions of years. During this period the mechanisms of the process have not changed; more important, the rate of decay has been constant since the elements were first formed. Today we can be confident that a sample of uranium 238, no matter what its origin, will gradually change into lead, and that this transmutation will occur at a rate such that half of the uranium atoms will have become lead in 4.5 billion years. There is no reason to believe that the nature or the rate of this process was any different in the very remote past, when the universe was new.

The history of these countless nuclear events is written in the chemical elements out of which the earth and the rest of the universe are made. By properly interpreting this history we can assign a date to the formation of those elements. From this date we can infer the age of the universe itself. The scientific discipline that is concerned with these techniques is called nucleocosmochronology.

There are several other ways in which the age of the universe can be determined. One celebrated calculation was made by the Anglican archbishop James Ussher in 1658. By adding up the generations enumerated in the Bible he fixed the time of the Creation at 4004 B.C., which would make the present age of the universe 5,978 years. More recent measurements differ from this figure by six orders of magnitude. The uncertainty of the modern calculations, however, is thought to be a few billion years, whereas Archbishop Ussher considered his date accurate to within one year.

Besides nucleocosmochronology the two principal means of dating cosmological events are measurements of the expansion of the universe and observations of the stars in the globular clusters associated with many galaxies, including our own. Both techniques require precise measurements of astronomical phenomena.

That the universe is expanding was first noted by Edwin P. Hubble in 1929. The expansion is now generally conceded to be explained by some version of the "big bang" theory of cosmology. In this theory the universe is considered to be exploding from a very dense, hot, primordial state. The rival "steady state" theory, which postulates an infinite age, is refuted by the observation of the three-degree black-body background radiation.

The universe did not necessarily begin with the big bang. It is possible it was once a rarefied gas that collapsed, exploded and will now expand forever. It is also possible that it began in the condensed state and will return to it; in fact, it could repeat this cycle over and over again. *Our* universe, however, did begin with the primordial explosion, since we can obtain no information about events that occurred before it. The age of the universe, therefore, is the interval from the big bang to the present.

Allan Sandage, working with the 200-inch Hale telescope on Palomar Mountain, has recently remeasured the rate of expansion. His calculations indicate that if the rate has been constant since the big bang, then the age of the universe is 18 billion years. This is an upper limit of the age, sometimes called the "Hubble time"; it is the correct age only if the rate of expansion has not changed. Since gravitation tends to diminish the rate of expansion, the universe must be somewhat younger than the Hubble time. This deceleration has not been measured, but Sandage has determined its maximum value by noting that if it exceeded that value, it would cause effects that have not been observed. Thus his observations also supply a lower limit to the age of the universe: it probably cannot be younger than about half the Hubble time. The uncertainty of these measurements is considerable: the Hubble time could range from 10 to 20 billion years, and the minimum age as determined by this method could therefore be from five to 10 billion years.

The second astronomical method of dating the universe depends on the singular properties of the globular clusters. Calculations of stellar evolution enable us to estimate the age of a star if we know its mass, its luminosity, its composition and its surface temperature. The composition and surface temperature can be inferred from the star's spectrum; in most cases the mass and luminosity, however, can be determined only if we know the distance to the star. The globular clusters provide a unique set of specimens for these calculations. All the stars in a given cluster were formed at about the same time and out of the same material. Moreover, they are all at approximately the same distance from us.

When the surface temperature of a group of stars is plotted against the luminosity, the relatively young stars fueled by the nuclear fusion of hydrogen form the diagonal band called the main sequence; the more massive stars are in the bright-hot region, the less massive ones in the cool-dim region. This is the Hertzsprung-Russell, or H-R, diagram, named for Ejnar Hertzsprung and Henry Norris Russell. When the hydrogen in the core of a star is exhausted, it moves off the main sequence, soon becoming a red giant. Massive stars leave the main sequence much faster than less massive ones. In a population of stars that were all formed at about the same time one can therefore estimate the age of the stars merely by determining at what point on the H-R diagram the stars are no longer on the main sequence.

Icko Iben, Jr., and Robert Rood have recently calculated that the globular clusters were formed $13 \pm 3$ billion years ago, an age consistent with the nine to 18 billion years given by the rate of expansion of the universe. (Of course, the actual age of the universe must be greater than the age of the globular clusters, but it is thought that the clusters formed much less than a billion years after the big bang.)

As in the determination of the Hubble time, there is considerable uncertainty in these calculations. The principle sources of this uncertainty are in the theoretical calculation of stellar evolution and in the determination of stellar composition. Because of the inherent limitations of these methods a third, independent technique can provide an important means of confirmation. The use of radioactive nuclei to estimate the age of the universe is such a technique.

In the primordial fireball with which we assume the universe began only the lightest elements — hydrogen, helium and possibly some lithium — could have been produced. All the rest of the elements had to have been synthesized later in stars.

The stars of the main sequence, including virtually all the stars formed shortly after the big bang, are fueled by the fusion of hydrogen into helium. As we have seen, it is the exhaustion of the hydrogen in the core that causes a star to leave the main sequence. When the star becomes a red giant, its outer envelope expands, but at the same time the core contracts gravitationally and grows hotter until a new fusion reaction is initiated, burning helium into carbon. Eventually the helium too is exhausted and the star is left with a carbon core.

The subsequent events depend on the mass of the star. A star with a mass less than about four times the mass of the sun will become a planetary nebula and leave behind a white dwarf. A star of from about four to about eight solar masses will probably explode as a supernova when fusion begins in the carbon core. In this case a dense core of neutrons may or may not remain after the explosion. This remnant, if it exists, is a neutron star rather than a white dwarf; the pulsars observed in some supernova remnants are thought to be neutron stars.

A star of more than about eight solar masses probably also becomes a supernova. When such a star explodes, it almost certainly produces a remnant, which may be a neutron star or perhaps a black hole. It is in the explosion of these massive stars that the elements from carbon to iron are formed. Iron is the last element in this sequence, however. Fusion reactions involving iron cannot fuel stellar processes; all such reactions absorb energy rather than release it. Iron and certain closely related elements thus represent the end product of the nuclear process from which stars derive their energy.

It should not seem surprising that all the heavier elements are formed in events that, compared to the number of stars, are rather rare. There have probably been fewer than a billion supernovas among the 100 billion stars in the galaxy. More than 98 percent of the universe is still hydrogen and helium, however. Thus if supernovas are a relatively rare phenomenon, so are the elements they produce.

The elements heavier than iron are formed primarily by the

capture of neutrons by nuclei of iron and its neighboring elements. The capture occurs in two ways: by a slow process (called the *s* process) or a rapid one (the *r* process).

The *s* process takes place in the envelope of red-giant stars. In it neutrons are added to the nuclei one at a time over long periods, so that only relatively stable nuclei can be formed. Isotopes that decay quickly vanish before another neutron can be absorbed. For this reason the *s* process terminates with lead and bismuth; all heavier elements are radioactive to some degree, and those immediately following lead and bismuth are highly unstable.

The *r* process presumably takes place only in supernovas and probably operates in the region of the explosion just outside the neutron-star or black-hole remnant, where the neutron flux is intense. Under these circumstances neutrons can be absorbed by nuclei in rapid succession, so that regions of great nuclear instability are bridged. All the elements heavier than bismuth are believed to be formed in this way.

The nuclei initially produced by the *r* process are far too rich in neutrons to be stable or long-lived. As a result they immediately begin to seek more stable regions by beta decay. In this process a neutron emits an electron (and a neutrino) and becomes a proton; the result is that the atomic mass remains nearly constant but the atomic number increases by one. Beta decay continues until a stable ratio of neutrons to protons is reached. For example, lead 232, formed in the *r* process, with 26 more neutrons than the stable isotope lead 206, decays through eight beta emissions to thorium 232.

The elements that at present make the most suitable nucleochronometers are formed by the *r* process. Since the process presumably operates only in supernovas, a date establishing the origin of these elements in fact dates the supernovas. Most elements from carbon through iron are also formed in supernovas, so that this method truly yields the age of the elements. Moreover, from such calculations one can derive an age for the universe as a whole. Very massive stars that condensed when the galaxies formed must have become supernovas early, shortly after the big bang. A dating of these first supernovas would be approximately equivalent to the age of the galaxy and of the universe. Even if these stars did not evolve as quickly as is assumed, it is clear that the date of the oldest supernova provides a lower limit to the age of the universe.

In order to make these calculations it is not necessary to know the actual abundance of the elements today or at any time in the past. One need only know the ratio in which a suitable pair of elements is found today (the "abundance ratio") and the ratio in which they were found when first formed (the "production ratio").

The abundance ratios are determined by careful experimental measurement. Terrestrial rocks, although they are fundamentally composed of the same material as the rest of the solar system, usually do not provide suitable samples. In the 4.6 billion years since the solar system condensed, rocks on the earth have formed and re-formed many times. Chemical fractionation has altered their composition so that they are no longer representative of "average" solar-system material. Most of the samples are from meteorites, which have been undisturbed during the entire history of the solar system; recently lunar material has also been used.

The samples are first chemically prepared to isolate the appropriate elements. Because the quantities are typically minuscule, care must be taken to avoid contamination. The sample is then analyzed with a mass spectrometer, an instrument that separates elements and isotopes according to the ratio of mass to charge.

The production ratios must be calculated theoretically. In regions between the "magic numbers" of neutrons (2, 8, 20, 28, 50, 82, 126 and probably 184), which yield extraordinarily stable nuclei, the $r$ process gives rise to an approximately equal abundance of each atomic mass. The fact that the initial abundances are roughly equal can be used to advantage in calculating the production ratios of the very heavy nuclei, since in this region more than one mass number contributes to each of the nucleochronometers. The multiple contributions are a result of alpha decay, in which a nucleus emits an alpha particle, or helium nucleus, decreasing its mass by four and its atomic number by two. Numerous heavier elements serve as progenitors, decaying to eventually become the nucleochronometers. One chronometer decays in such a way that it eventually contributes to the abundance of another chronometer. Because several atomic masses contribute to the abundance of each of these nucleochronometers the effect of variations from the average, roughly equal, abundance is reduced.

The chronometers of interest in this mass region are thorium 232, uranium 235, uranium 238 and plutonium 244. Their production rates can be calculated simply by adding up the number of their progenitors. For example, U-238 has progenitors of mass 238, 242 and 246 and receives 10 percent of the decay products of mass 250 (the rest of 250 spontaneously fissions). The total thus is 3.1. Pu-244 has as its progenitors all of atomic mass 244, 92 percent of 248, 89 percent of 252 and 7 percent of 256, for a total of 2.9. Th-232 is produced by nuclei of mass 232, 236 and 240, as well as all the progenitors that contribute to 244, for a total of 5.9. Finally, U-235 has these six progenitors: 235, 239, 243, 247, 251 and 255.

The production ratios of these elements can now be simply expressed. The ratio of Th-232 to U-238 is 5.9 divided by 3.1, or 1.9. The Pu-244 to Th-232 ratio is 2.9 divided by 5.9, or .5. In the case of U-235 to U-238 the ratio calculated by this method is 6 divided by 3.1, or 1.9; because nuclei with odd mass numbers (such as U-235) are more easily destroyed by neutron capture, however, this ratio is adjusted to 1.5.

A simple way to see how these ratios can be used to determine the age of the elements is to assume that all the elements were made in one event. This assumption is known to be wrong, but it provides an idealized model of the nuclear dating processes.

For example, with the chronometer pair Th-232 and U-238 it can be shown by experiment that the abundance ratio today is about 4.0. The calculated production ratio is 1.9; one can determine from the half-lives of the substances that the period required for such a change in ratio is seven billion years. That is the age of the hypothetical single event according to this chronometer pair.

In practice the calculation would not be made in quite this way. Rather than working directly with the present abundance ratio it is more convenient to derive from it the relative abundance at the time the solar system condensed 4.6 billion years ago. In this case the ratio then was 2.4 and the time from the single event to the formation of the solar system is 2.4 billion years. By adding the age of the solar system one arrives at the same age of seven billion years. This procedure becomes important when shorter-lived chronometers are used.

The one-event hypothesis is obviously implausible because it is most unlikely that all the supernovas of the past billions of years exploded at a single moment. The hypothesis is demonstrably wrong because the several nucleochronometer pairs each give a different date for the single event. Indeed, those chronometers with relatively short half-lives would have all but vanished if the only production event had occurred billions of years before the solar system formed. More complicated models are needed.

More than a decade ago William A. Fowler and Fred Hoyle used Th-232, U-235 and U-238 to determine the time scale of nucleosynthesis. With only those chronometers it was not possible to definitely rule out the single-event model. In the past few years, however, techniques have been devised for the use of two new nucleochronometers. They are iodine 129 and plutonium 244, and their observed presence in the solar system cannot be made consistent with a single nucleosynthetic event.

Measurements of the abundance of I-129 and Pu-244 require some subtlety. The half-lives of both substances are much less than the age of the solar system; whatever amount was present at that time, therefore, must by now have been reduced by radioactive decay to virtually zero. (The present abundance would not be zero, although it approaches zero asymptotically and may well be unmeasurable. Investigators at the Los Alamos Scientific Laboratory and the General Electric Company have recently detected traces of Pu-244 in nature.)

Although the plutonium and iodine chronometers themselves are not accessible to us, their decay products are. I-129 decays (by beta emission) to the stable xenon 129. Thus if a given quantity of I-129 was present in the materials that formed a meteorite 4.6 billion years ago, an equal quantity of Xe-129 should be detected in the meteorite today. Many other isotopes of xenon can also be expected. They can be separated by mass spectrometry, however, so that the presence long ago of I-129 in the meteorite would be indicated today by an excess of Xe-129.

John H. Reynolds and his co-workers at the University of California at Berkeley were the first to find this excess of Xe-129 and to prove that it is caused by the decay of I-129. By an ingenious method they were able to determine the ratio of I-129 to the stable isotope I-127 at the time the meteorite was formed, that is, when it first became cool enough to retain xenon gas. To obtain this ratio Reynolds irradiated the sample with neutrons, so that I-127 nuclei would capture a neutron and be converted to I-128, which decays in 25 minutes to Xe-128. Therefore the abundance of I-127 would be related to the excess of Xe-128. He then extracted the xenon by heating the sample in stages, observing the mass spectrum at each stage. At many temperatures only xenon of ordinary origin was driven out. The excess Xe-128 and Xe-129, when they did appear, were driven out at the same time, showing that the original deposits of the two isotopes of iodine were in the same place. It was therefore possible to exclude ordinary xenon from the decay products of iodine. From the ratio of the excess Xe-128 to Xe-129 Reynolds was then able to infer the ratio of the iodine isotopes at the formation of the solar system.

The identification of Pu-244 as a chronometer also involved xenon isotopes. Pu-244 decays by alpha emission with a half-life of 82 million years. One in every 1,000 nuclei, however, instead of following this decay path spontaneously fissions. Among its fission products is xenon, but only the four heavy isotopes Xe-131, Xe-132, Xe-134 and Xe-136. These isotopes are also produced by the fission of other nuclei, but in proportions that are characteristic of each fissionable substance. It is therefore possible by examining the mass spectrum of the xenon isotopes to identify the fissioning nucleus.

Anomalous isotopic spectra of xenon, suggesting an unknown fission component, were discovered in meteorites by P. K. Kuroda of the University of Arkansas. Gerald J. Wasserburg and his co-workers at the California Institute of Technology found in 1969 that the excess heavy xenon was correlated with fossil fission tracks in meteorites. These tracks represent damage done to the crystal structure by the recoiling fragments of the nucleus. At that time the xenon isotope spectrum for Pu-244 fission was unknown, but this nucleus nevertheless became the only reasonable candidate. Several workers, including myself, began to investigate its possible chronometric applications. The issue was not settled until 1971, 10 years after Kuroda's discovery, when Reynolds' group was able to examine the isotopic spectrum of xenon produced by the fission of a known sample of Pu-244.

From the observation of Pu-244 fission products in a meteorite to the formulation of an abundance ratio for Pu-244 and Th-232 is a step fraught with uncertainties. The abundance value is to be representative of the entire solar system, yet possible chemical fractionation of the materials in the meteorite could substantially alter it. The value assigned is .0062, but it should be considered only a best estimate, not a definitive determination.

In the case of I-129 special measures must also be taken in estimating the production ratio. Iodine 129 (53 protons, 76 neutrons) is near the neutron magic number 82, and as a consequence the distribution of isotopes produced by neutron capture is far from uniform. It is necessary to determine the production rate by interpolating from the known production rates of stable isotopes in the immediate vicinity, such as tellurium 128 and 130.

This procedure is generally required to determine the production rates of all the lighter radioactive nuclei. One important exception is a chronology developed for the pair rhenium 187 and osmium 187, the longest-lived of the nucleochronometers, by Donald D. Clayton of Rice University. For these elements it has been shown that the production ratio is related to the ratio of the neutron-capture cross sections of the nuclei. The cross section is a measure of the readiness of the nucleus to absorb a neutron. Unfortunately the cross sections have not yet been experimentally measured, and the theoretical estimates being used provisionally introduce large uncertainties into the chronologies derived from this pair.

Using these several nucleochronometers, one can take a more sophisticated approach than the one-event model to discovering the age of the universe. When many events are assumed, an age can be assigned by determining when the sum of the remaining radioactive products from each supernova satisfies the observed abundance at the time the solar system formed. It can be assumed that the products of each supernova mixed in the interstellar gas as the galaxy rotated and that the solar system condensed from this gas. We do not know, however, how many supernovas exploded at different times throughout the history of the galaxy; an estimate can be supplied only by observations of other galaxies. Variations in the rate at which supernovas explode would obviously have an effect on the age indicated by the nucleochronometers. If, for example, there were an unusually large number of supernovas shortly before the solar system formed, then such long-lived chronometers as U-238 would indicate too young an age for the universe; the observed abundance would be not a result of the rate of decay but an artifact of the rate of production.

One approach to this problem is to devise a model for the relative number of supernovas over the history of the galaxy. The one-event model is one possibility, but there are many other candidates. The model is then tested to see if it can produce the observed abundance ratios of all the chronometer pairs.

Wasserburg and I have developed an approach to nucleocosmochronology that is independent of models. By using the various time scales of the different chronometers to determine the relative number of supernovas in each period, the valid models are derived from the chronometers themselves. For example, any valid model must include some nucleosynthesis within a few hundred million years of the formation of the solar system or it could not explain the presence of I-129 and Pu-244 decay products. This condition alone precludes a single-event model in which the elements were formed seven billion years ago.

By deriving the model from the data we can estimate not only the time scale of the universe but also the rate of supernovas in the galaxy. (Actually we do not specify supernovas per se but merely the rate of the *r* process. As we have seen, however, it is believed the *r* process occurs only in supernovas.)

Since there have probably been almost a billion supernovas in the galaxy, it is easier to use statistical techniques than to sum the effects of each event. Any distribution of supernovas has an average rate and an average age. There is also a time after which supernovas could no longer contribute material to the solar system. Finally, the oldest nuclei must be older than the average age of the elements.

From the data provided by the nucleochronometers it is possible to find the average age and also to find how the supernova rate at certain times compares with the average rate of supernovas. These calculations are made possible by the differences in the lifetimes of the chronometers. A nucleus with any given half-life must have been formed no earlier than a few of its own half-lives before the solar system condensed or it would not have been present in measurable quantities in the solar-system materials. Thus the abundance of the nucleus yields the mean supernova rate averaged over a few half-lives of that nucleus. The average age of the events producing a particular chronometer pair can be shown to be the same as the age of a hypothetical single event producing that pair.

The very long-lived chronometers, such as Re-187, Th-232 and possibly U-238, persist as long as or longer than the entire duration of nucleosynthesis. The average rate of supernovas contributing to these nuclei is therefore the average supernova rate for the galaxy throughout its history. Moreover, the average, or single-event, age for these nuclei is the average age of the elements. Obviously the ages given by the Th-232 and U-238 pair and those given by Re-187 and Os-187 should be the same. The best estimate for the thorium-uranium pair is 2.4 plus 4.6, or seven, billion years, whereas the rhenium-osmium age is 3.5 plus 4.6, or 8.1, billion years. Considering the uncertainties in these calculations, however, the inconsistency is not great.

Variations in the supernova rate can be revealed by the rates given by the shorter-lived chronometers with respect to those of the long-lived, or stable, nuclei. The shortest-lived chronometers convey information about the period just before the solar system formed; by comparing this rate to that given by the longest-lived elements we can find out if the rate of nucleosynthesis was higher or lower than average when the solar system condensed. Because there are two short-lived chronometers (I-129 and Pu-244) we can determine not only the relative rate at this time but also the time between the last nucleosynthetic event and the formation of the objects in the solar system. This interval has been found to be between 100 and 200 million years.

It is a curious fact that this period is comparable to the rotation time of the galaxy. The coincidence has been noted not only by me but also by H. Reeves of Paris and A.G.W. Cameron of Harvard University, who independently suggested that the nucleochronometers may be indirectly measuring the rotation of the galaxy. In the density-wave theory of spiral galaxies stars are thought more likely to form in any given region when the density is high; the period of variation of density would be determined by galactic rotation and would be about 100 million years.

Another possible explanation of this interval is that it represents the time it took the solar system to form after a final supernova triggered the beginning of its condensation. This theory meets with objections, however: all objects in the solar system are known to have solidified within a few million years of one another. Why would they solidify so "suddenly" 100 million years after the event that initiated their formation? It is also thought that the time required for a star such as the sun to form is on the order of 10 million rather than 100 million years. The galactic-rotation theory therefore seems more nearly satisfactory.

By combining the results obtained from the various chronometers one can reconstruct the probable rates of supernova production over the entire duration of nucleosynthesis. The best values available today for the abundance and production ratios imply that the number of supernovas was above average in the early universe and that there may have been a smaller peak in the supernova rate shortly before the solar system was formed. The uncertainties are large enough, however, for a uniform rate of nucleosynthesis to satisfy the requirements of the data.

The second, later peak is very sensitive to the abundance of Pu-244. Unfortunately the value for this abundance is still quite uncertain.

The earlier peak is dependent on the production of U-235 with respect to the longer-lived nuclei. If the average rate of production of U-235 is low, then in an earlier period the supernova rate must have peaked strongly if the U-235 data are to agree with the overall average rate of nucleosynthesis as determined by the longer-lived chronometers. Although the magnitude of this peak is by no means certain, it is probable that the early supernova rate was greater than, or at least equal to, the average rate. A high early rate of supernovas suggests that the galaxy may have been brighter in the past. Perhaps there were so many exploding massive stars in the galactic nucleus that the galaxy was a quasar.

The calculated age of the universe as a whole is given by the longest-lived nuclei. The uncertainties of the data for these chronometers indicate that the age of the elements must be between seven and 15 billion years; the best estimate is approximately 10 billion years. Thus the nucleochronologies yield an age that is quite consistent with the ages calculated from the expansion of the universe and from the stellar populations of the globular clusters.

It should be possible in the future to diminish the uncertainty of these calculations considerably. A large improvement could be achieved by experimentally measuring the neutron-capture cross sections that would allow a better estimate of the production ratio of Re-187 and Os-187. Better determinations of the relative abundances of Pu-244, Th-232 and U-238 would also increase accuracy, as would improved calculations of $r$-process production.

In addition techniques for using other nucleochronometers may be developed. Two promising nuclei are samarium 146 (half-life 100 million years) and lead 205 (half-life 15 million years). These nuclei are not produced in the $r$ process and they might therefore enable us to see if all nucleosynthetic processes yield similar chronologies.

It was once thought that aluminum 26, with a half-life of 740,000 years, could be used as a chronometer. Analysis of meteorites and moon rocks by me and my co-workers at the California Institute of Technology have shown, however, that no significant amounts of aluminum 26 were present when these objects formed.

This result might be expected since the half-life is much less than the 100-million-year interval indicated by the I-129 and Pu-244 chronologies.

These nucleochronologies, coupled with the observed abundances of heavy elements in the stars and theories of star formation, have been used by various investigators in proposing detailed theories of the entire history of the galaxy. Further work on the correlation of nucleochronology with other astronomical information should yield important results in the near future.

It is impressive that three such different techniques as the observation of the recession of the galaxies, the measurement of globular-cluster ages and the calculation of nucleochronologies should yield consistent ages for the universe. If we accept only ages that agree with all three methods, then the universe is from 10 to 15 billion years old. To be able to measure any astronomical number, particularly such an important one, to an accuracy of 33 percent is deeply satisfying.

# ISAAC ASIMOV

■

The prolific science writer Isaac Asimov (b. 1920) is the author of over 450 books on subjects ranging from biochemistry to biblical studies. This discussion of relativity is from his book *The Subatomic Monster,* published in 1985.

# The Two Masses

I SAW Albert Einstein once.

It was on April 10, 1935. I was returning from an interview at Columbia College, an interview on which my permission to enter depended. (It was disastrous, for I was a totally unimpressive fifteen-year-old, and I didn't get in.)

I stopped off in a museum to recover, for I had no illusions as to my chances after that interview, and so confused and upset was I that I've never been able to remember which museum it was. But wandering in a semi-dazed condition through the rooms, I saw Albert Einstein, and wasn't so dead to the world around me that I didn't recognize him at once.

From then on, for half an hour, I followed him patiently from room to room, looking at nothing else, merely staring at him. I wasn't alone; there were others doing the same. No one said a word, no one approached him for an autograph or for any other purpose; everyone merely stared. Einstein paid no attention whatever; I assume he was used to it.

After all, no other scientist, except for Isaac Newton, was so revered in his own lifetime — even by other great scientists, let alone by laymen and adolescents. It is not only that his accomplishments were enormous, but that they are in some respects almost too rarefied to describe, especially in connection with what is generally accepted as his greatest accomplishment: General Relativity.

184

Certainly it's too rarefied for me, since I am only a biochemist (of sorts) and not a theoretical physicist, but in my self-assumed role as busybody know-it-all, I suppose I have to try anyway . . .

In 1905 Einstein had advanced his special theory of relativity (or special relativity for short), which is the more familiar part of his work. Special relativity begins with the assumption that the speed of light in a vacuum will always be measured at the same constant value regardless of the speed of the light source relative to the observer.

From that, an inescapable line of deductions tells us that the speed of light represents the limiting speed for anything in our universe — that if we observe a moving object, we will find its length in the direction of motion and the rate of time passage upon it decreased and its mass increased, as compared with what it would be if the object were at rest. These properties vary with speed in a fixed manner such that at the speed of light, length and time rate would be measured as zero while mass would be infinite. Furthermore, special relativity tells us that energy and mass are related according to the now famous equation $e = mc^2$.

Suppose, though, that the speed of light in a vacuum is *not* unchanging under all conditions. In that case, none of the deductions is valid. How, then, can we decide on this matter of the constancy of the speed of light?

To be sure, the Michelson-Morley experiment had indicated that the speed of light did not change with the motion of the Earth — that is, that it was the same whether the light moved in the direction of Earth's revolution about the Sun or at right angles to it. One might extrapolate the general principle from that, but the Michelson-Morley experiment is capable of other interpretations. (To go to an extreme, it might indicate that the Earth wasn't moving, and that Copernicus was wrong.)

In any case, Einstein insisted later that he had not heard of the Michelson-Morley experiment at the time he conceived of special relativity and that it seemed to him that light's speed must be constant simply because he found himself involved in contradictions if that weren't so.

Actually, the best way to test Einstein's assumption would be to test whether the deductions from that assumption are to be observed in the real universe. If so, then we are driven to the conclusion that the basic assumption must be true, for we would then know of no other way of explaining the truth of the deductions. (The deductions do *not* follow from the earlier Newtonian view of the universe or from any other non-Einsteinian — or nonrelativistic — view.)

It would have been extremely difficult to test special relativity if the state of physical knowledge had been what it was in 1895, ten years before Einstein advanced his theory. The startling changes it predicted in the case of length, mass, and time with speed are perceptible only at great speeds, far beyond those encountered in ordinary life. By a stroke of fortune, however, the world of subatomic particles had opened up in the decade prior to Einstein's announcements. These particles moved at speeds of 15,000 kilometers per second and more, and at *those* speeds relativistic effects are appreciable.

It turned out that the deductions of special relativity were all there, all of them; not only qualitatively but quantitatively as well. Not only did an electron gain mass as it sped by at nine tenths the speed of light, but its mass was multiplied by $3\frac{1}{6}$ times, just as the theory predicted.

Special relativity has been tested an incredible number of times in the last eight decades and it has passed every test. The huge particle accelerators built since World War II would not work if they didn't take into account relativistic effects in precisely the manner required by Einstein's equations. Without the $e = mc^2$ equation, there is no explaining the energy effects of subatomic interactions, the working of nuclear power plants, the shining of the Sun. Consequently, no physicist who is even minimally sane doubts the validity of special relativity.

This is not to say that special relativity necessarily represents ultimate truth. It is quite possible that a broader theory may someday be advanced to explain everything special relativity does and more besides. On the other hand, nothing has so far arisen that seems to require such explanation except for the reported *apparent* separation of quasar components at more than the speed of light, and the betting is that this is probably an optical illusion that can be explained within the limits of special relativity.

Then, too, even if such a broader theory is developed, it would have to work its way down to special relativity within the bounds of present-day experimentation, just as special relativity works itself down to ordinary Newtonian laws of motion, if you stick to the low speeds we use in everyday life.

Why is special relativity called "special"? Because it deals with the special case of constant motion. Special relativity tells you all you need to know if you are dealing with an object moving at constant speed and unchanging direction with respect to yourself.

But what if the speed or direction of movement of an object (or both) is changing with respect to you? In that case, special relativity is insufficient.

Strictly speaking, motion is never constant. There are always present forces that introduce changes in speed, direction, or both, in the case of every moving object. Consequently we might argue that special relativity is always insufficient.

So it is, but the insufficiency can be small enough to ignore. Subatomic particles moving at vast speeds over short distances don't have time to accelerate much, and special relativity can be applied.

In the universe generally, however, where stars and planets are involved, special relativity is grossly insufficient, for there we deal with large accelerations and these are invariably brought about by the existence of vast and ever-present gravitational fields.

At the subatomic level, gravitation is so excessively weak in comparison with other forces that it can be ignored. At the macroscopic level of visible objects, however, it cannot be ignored; in fact, everything *but* gravitation can be ignored.

Near Earth's surface, a falling body speeds up while a rising body slows down, and both are examples of accelerations caused entirely by progress through Earth's gravitational field. The Moon travels in an orbit about the Earth, the Earth about the Sun, the Sun about the galactic center, the galaxy about the local group center and so on, and in every case the orbital motion involves an acceleration since there is a continuing change in direction of motion. Such accelerations are also produced in response to gravitational fields.

Einstein therefore set about applying his relativistic notions to the case of motion *generally,* accelerated as well as constant — in other words, all the real motions in the universe. When worked out, this would be the general theory of relativity, or general relativity. To do this, he had, first and foremost, to consider gravitation.

There is a puzzle about gravitation that dates back to Newton. According to Newton's mathematical formulation of the laws governing the way in which objects move, the strength of the gravitational pull depends upon mass. The Earth pulls on an object with a mass of 2 kilograms exactly twice as hard as it does on an object with a mass of only 1 kilogram. Furthermore, if Earth doubled its own

187

mass it would pull on everything exactly twice as hard as it does now. We can, therefore, measure the mass of the Earth by measuring the intensity of its gravitational pull upon a given object; or we can measure the mass of an object by measuring the force exerted upon it by Earth.

A mass, so determined, is "gravitational mass."

Newton, however, also worked out the laws of motion and maintained that any force exerted upon an object causes that object to undergo an acceleration. The amount of acceleration is in inverse proportion to the mass of the object. In other words, if the same force is exerted on two objects, one which has a mass of 2 kilograms and the other of 1 kilogram, the 2-kilogram object will be accelerated to exactly half the exent of the 1-kilogram object.

The resistance to acceleration is called inertia, and we can say that the larger the mass of an object, the larger its inertia — that is, the less it will accelerate under the push of a given force. We can therefore measure the mass of a body by measuring its inertia — that is, by measuring the acceleration produced upon it by a given force.

A mass so determined is "inertial mass."

All masses that have ever been determined have been measured either through gravitational effects or inertial effects. Either way is taken as valid and they are treated as interchangeable, even though the two masses have no *apparent* connection. Might there not, after all, be some objects, made of certain materials or held under certain conditions, that would show an intense gravitational field but very little inertia, or vice versa? Why not?

Yet whenever one measures the mass of a body gravitationally, then measures the mass of the same body inertially, the two measurements come out to be equal. Yet that may only be appearance. There may be small differences, too small to be noted ordinarily.

In 1909 an important experiment in this connection was performed by a Hungarian physicist, Roland, Baron von Eötvös (the name is pronounced "ut'vush").

What he did was to suspend a horizontal bar from a delicate fiber. At one end of the bar was a ball of one material, and at the other end a ball of another material. The Sun pulls on both balls and forces an acceleration on each. If the balls are of different mass, say 2 kilograms and 1 kilogram, then the 2-kilogram mass is attracted twice as strongly as the 1-kilogram mass, and you might expect it to be accelerated twice as strongly. However, the 2-kilogram mass has twice the inertia of the 1-kilogram mass. For that reason, the

2-kilogram mass accelerates only half as much per kilogram and ends up being made to accelerate only as strongly as the 1-kilogram mass is.

If inertial mass and gravitational mass are *exactly* equal, then the two balls are made to accelerate *exactly* equally, and the horizontal bar may be pulled toward the Sun by an immeasurable amount, but it does not rotate. If inertial mass and gravitational mass are not quite equal, then one ball will accelerate a bit more than the other and the bar will experience a slight turning force. This will twist the fiber, which to a certain extent resists twisting and will only twist so far in response to a given force. From the extent of the twist, one can calculate the amount of difference between the inertial mass and the gravitational mass.

The fiber used was a very thin one so its resistance to twist was very low and yet the horizontal bar showed no measurable turn. Eötvös could calculate that a difference in the two masses of 1 part in 200,000,000 would have produced a measurable twist, so the two masses were identical in amount to within that extent.

(Since then, still more delicate versions of the Eötvös experiment have been carried through and we are now certain by direct observation that inertial mass and gravitational mass are identical in quantity to within 1 part in 1,000,000,000,000.)

Einstein, in working out general relativity, began by assuming that inertial mass and gravitational mass are *exactly* equal because they are, in essence, *the same thing*. This is called "the principal of equivalence" and it plays the same role in general relativity that the constancy of the speed of light plays in special relativity.

It was possible even before Einstein to see that inertially produced acceleration can bring about the same effects as gravitation. Any of us can experience it.

If, for instance, you are in an elevator which starts downward, gaining speed at the start, then during that period of acceleration the floor of the elevator drops out from under you, so to speak, so that you press upon it with less force. You feel your weight decrease as though you were lifting upward. The downward acceleration is equivalent to a lessening of the gravitational pull.

Of course, once the elevator reaches a particular speed and maintains it, there is no longer any acceleration and you feel your normal weight. If the elevator is moving at a constant speed of any amount and in any constant direction, you feel no gravitational effect whatever. In fact, if you are traveling through a vacuum in a totally enclosed box so that you don't see scenery moving, or feel the vibration of air resistance, or hear the whistling of wind, there is absolutely no way you can tell such constant motion from any other (at a different speed or in a different direction) or from being at rest. That is one of the basics of special relativity.

It is because Earth travels through a vacuum at a nearly constant speed and in a nearly constant direction (over short distances) that it is so difficult for people to differentiate the situation from that of Earth being at rest.

On the other hand, if the elevator kept on accelerating downward and moving faster and faster, you would feel as if your weight had decreased permanently. If the elevator accelerated downward at a great enough rate — if it fell at the natural acceleration that gravitational pull would impose upon it ("free fall") — then all sensation of weight would vanish. You would float.

If the elevator accelerated downward at a rate faster than that associated with free fall, you would feel the equivalent of a gravitational pull *upward,* and you would find the ceiling fulfilling the functions of a floor for you.

Naturally you can't expect an elevator to accelerate downward for very long. For one thing, you would need an extraordinarily long shaft within which it might continue moving downward, one that would be light-years long if you want to carry matters to extremes. Then, too, even if you had such an impossibly long shaft, a constant rate of acceleration would soon cause the speed to become a respectable fraction of that of light. That would introduce appreciable relativistic effects and complicate matters.

We can, however, imagine another situation. If an object is in orbit about the Earth it is, in effect, constantly falling toward the Earth at an acceleration imposed upon it by Earth's gravitational pull. However, it is also moving horizontally relative to the Earth's surface, and since the Earth is spherical, that surface curves away from the falling object. Hence the object is always falling, but never reaches the surface. It keeps on falling and falling for billions of years, perhaps. It is in perpetual free fall.

Thus a spaceship that is in coasting orbit about the Earth is held in that orbit by Earth's gravitational pull, but anything on the spaceship falls *with* the spaceship and experiences zero gravity, just as though it were on a perpetually falling elevator. (Actually, astronauts would feel the gravitational pull of the spaceship itself and of each

190

other, to say nothing of the pulls of other planets, and distant stars, but these would all be tiny forces that would be entirely imperceptible.) That is why people on orbiting spaceships float freely.

Again, the Earth is in the grip of the Sun's gravitational pull and that keeps it in orbit about the Sun. So is the Moon. The Earth and the Moon perpetually fall toward the Sun together and, being in free fall, don't feel the Sun's pull as far as their relationship to each other is concerned.

However, the Earth has a gravitational pull of its own which, while much weaker than the Sun's, is nevertheless quite strong. Therefore the Moon, in response to Earth's gravitational pull, moves about the Earth just as though the Sun didn't exist. (Actually, since the Moon is a little removed from the Earth and is sometimes a little closer to the Sun than Earth is, and sometimes a little farther, the Sun's pull is slightly different on the two worlds, and this introduces certain minor "tidal effects" which make evident the reality of the Sun's existence.)

Again, we stand on the Earth and feel only the Earth's pull and not the Sun's at all, since we and the Earth share the free fall with respect to the Sun, and since the tidal effect of the Sun upon ourselves is far too small for us to detect or be conscious of.

Suppose, next, that we are on an elevator accelerating upward. This happens to a minute extent every time we are on an elevator that moves upward from rest. If it is a speedy elevator, then when it starts there is a moment of appreciable acceleration during which the floor moves up toward us and we feel ourselves pressed downward. The acceleration upward produces the sensation of an increased gravitational pull.

Again, the sensation lasts only briefly, for the elevator reaches its maximum speed and then stays there during the course of its trip until it is time for it to stop, when it goes through a momentary slowing and you feel the sensation of a decreased gravitational pull. While the elevator was at maximum speed, and neither speeding up nor slowing down, you would feel perfectly normal.

Well, then, suppose you were in an elevator shaft light-years long and there was an enclosed elevator that could accelerate smoothly upward through a vacuum for an indefinite period, going faster and faster and faster. You would feel an increased gravitational pull indefinitely. (Astronauts feel this for a period of time when a rocket accelerates upward and they are pressed downward uncomfortably. Indeed, there is a limit to how intense an acceleration can be allowed or the additional sensation of gravitational pull can become great enough to press astronauts to death.)

But suppose there is no Earth — just an elevator accelerating upward. If the rate of acceleration were at an appropriate level, you would feel the equivalent of a gravitational pull just like that on Earth's surface. You would walk about perfectly comfortably and could imagine the elevator to be resting motionless on Earth's surface.

Here is where Einstein made his great leap of imagination. By supposing that inertial mass and gravitational mass were identical, he also supposed that there was no way — *no way* — in which you could tell whether you were in an enclosed cubicle moving upward at a steady acceleration of 9.8 meters per second per second, or were in that same enclosed cubicle at rest on the surface of the Earth.

This means that anything that would happen in the accelerating cubicle must also happen at rest on the surface of the Earth.

This is easy to see as far as the falling of ordinary bodies is concerned. An object held out at arm's length in an accelerating cubicle would drop when released and seem to fall at a constantly accelerating rate because the floor of the cubicle would be moving up to meet it at a constantly accelerating rate.

Therefore an object held out on Earth would also fall in the same way. This doesn't mean that the Earth is accelerating upward toward the object. It just means that gravitational pull produces an effect indistinguishable from that of upward acceleration.

Einstein, however, insisted that this included *everything*. If a beam of light were sent horizontally across the upward-accelerating elevator, the elevator would be a little higher when the beam of light finished its journey and therefore the beam of light would seem to curve downward as it crossed the cubicle. Light travels so rapidly, to be sure, that in the time it took for it to cross the cubicle it would have moved downward only imperceptibly — but it would curve just the same; there is no question about that.

Therefore, said Einstein, a beam of light subjected to Earth's gravitational field (or *any* gravitational field) must also travel a curved path. The more intense the gravitational field and the longer the path traveled by the light beam, the more noticeable the curve. This is an example of a deduction that can be drawn from the principle of equivalence that could not be drawn from earlier theories of the structure of the universe. All the deductions put together make up general relativity.

Other deductions include the suggestion that light should take a bit longer to travel from A to B when subjected to a gravitational field, because it follows a curved path; that light loses energy when moving against the pull of a gravitational field and therefore shows a red shift, and so on.

Again, by considering all the deductions, it makes sense to consider space-time to be curved. Everything follows the curve so that gravitational effects are due to the geometry of space-time rather than to a "pull."

It is possible to work up a simple analogy to gravitational effects by imagining an indefinitely large sheet made up of infinitely stretchable rubber extending high above the surface of the Earth. The weight of any mass resting on that sheet pushes down the rubber at that point and creates a "gravity well." The greater the mass and the more compressed it is, the deeper the wall and the steeper the sides. An object rolling across the sheet may skim one edge of the gravity well, sinking down the shallow rim of the well and out again. In this way it will be forced to follow a curved path just as though it had suffered a gravitational pull.

If the rolling object should happen to follow a path that would take it deeper into the well, it might be trapped and made to follow a slanting elliptical path about the walls of the well. If there is friction between the moving object and the walls, the orbit will decay and the object will eventually fall into the greater object at the bottom of the well.

All in all, making use of general relativity, Einstein was able to set up certain "field equations" that applied to the universe as a whole. Those field equations founded the science of cosmology (the study of the properties of the universe as a whole).

# BERTRAND RUSSELL

■

The ABC of Relativity, by the British philosopher and polymath Bertrand Russell (1872–1970), ranks as one of the most readable and insightful popular accounts of Einstein's theories. This selection is Russell's chapter on the general theory of relativity.

# Einstein's Law of Gravitation

BEFORE TACKLING EINSTEIN'S law, it is as well to convince ourselves, on logical grounds, that Newton's law of gravitation cannot be quite right.

Newton said that between any two particles of matter there is a force which is proportional to the product of their masses and inversely proportional to the square of their distance. That is to say, ignoring for the present the question of mass, if there is a certain attraction when the particles are a mile apart, there will be a quarter as much attraction when they are two miles apart, a ninth as much when they are three miles apart, and so on: the attraction diminishes much faster than the distance increases. Now, of course, Newton, when he spoke of the distance, meant the distance at a given time: he thought there could be no ambiguity about time. But we have seen that this was a mistake. What one observer judges to be the same moment on the earth and the sun, another will judge to be two different moments. "Distance at a given moment" is therefore a subjective conception, which can hardly enter into a cosmic law. Of course, we could make our law unambiguous by saying that we are going to estimate times as they are estimated by Greenwich Observatory. But we can hardly believe that the accidental circumstances of the earth deserve to be taken so seriously. And the estimate of distance, also, will vary for different observers. We cannot therefore allow that Newton's form of the law of gravitation can be quite correct, since it will give different results according to which of many equally legitimate conventions we adopt. This is as absurd as it would be if the question whether one person had murdered another were to depend upon whether they were described by their first names or their surnames. It is obvious that physical laws must be the same whether distances are measured in miles or in kilometers, and we are concerned with what is essentially only an extension of the same principle.

Our measurements are conventional to an even greater extent than is admitted by the special theory of relativity. Moreover, every measurement is a physical process carried out with physical material; the result is certainly an experimental datum, but may not be susceptible of the simple interpretation which we ordinarily assign to it. We are, therefore, not going to assume to begin with that we know how to measure anything. We assume that there is a certain physical quantity called "interval," which is a relation between two events that are not widely separated; but we do not assume in advance that we know how to measure it, beyond taking it for granted that it is given by some generalization of the theorem of Pythagoras such as we spoke of [earlier].

We do assume, however, that events have an *order,* and that this order is four-dimensional. We assume, that is to say, that we know what we mean by saying that a certain event is nearer to another than a third, so that before making accurate measurements we can speak of the "neighborhood" of an event; and we assume that, in order to assign the position of an event in space-time, four quantities (coordinates) are necessary — e.g., in our former case of an explosion on an airship, latitude, longitude, altitude, and time. But we assume nothing about the way in which these coordinates are assigned, except that neighboring coordinates are assigned to neighboring events.

The way in which these numbers, called coordinates, are to be assigned is neither wholly arbitrary nor a result of careful measurement — it lies in an intermediate region. While you are making any continuous journey, your coordinates must never alter by sudden jumps. In America one finds that the houses between (say) Fourteenth Street and Fifteenth Street are likely to have numbers between 1400 and 1500, while those between Fifteenth Street and Sixteenth Street have numbers between 1500 and 1600, even if the 1400s were not used up. This would not do for our purposes, because there is a sudden jump when we pass from one block to the next. Or again we might assign the time coordinate in the following way: take the time that elapses between two successive births of people called Smith; an event occurring between the births of the 3000th and the 3001st Smith known to history shall have a coordinate lying between 3000 and 3001; the fractional part of its coordinate shall be the fraction of a year that has elapsed since the birth of the 3000th Smith. (Obviously there could never be as much as a year between two successive additions to the Smith family.) This way of assigning the time coordinate is perfectly definite, but it is not admissible for our purposes, because there will be sudden jumps between events just before the birth of a Smith and events just after, so that in a continuous journey your time coordinate will not change continuously. It is assumed that, independently of measurement, we know what a continuous journey is. And when your position in space-time changes continuously, each of your four coordinates must change continuously. One, two, or three of them may not change at all, but whatever change does occur must be smooth, without sudden jumps. This explains what is *not* allowable in assigning coordinates.

To explain all the changes that are legitimate in your coordinates, suppose you take a large piece of soft india rubber. While it is in an unstretched condition, measure little squares on it, each one-tenth of an inch each way. Put in little tiny pins at the corners of the squares. We can take as two of the coordinates of one of these pins the number of pins passed in going to the right from a given pin until we come just below the pin in question, and then the number of pins we pass on the way up to this pin. In Figure 1, let *0* be the pin we start from and *P* the pin to which we are going to assign coordinates. *P* is in the fifth column and the third row, so its coordinates in the plane of the india rubber are to be 5 and 3.

Now take the india rubber and stretch it and twist it as much as you like. Let the pins now be in the shape they have in Figure 2. The divisions now no longer represent distances according to our usual notions, but they will still do just as well as coordinates. We

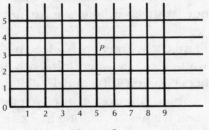

Figure 1

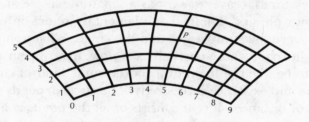

Figure 2

may still take *P* as having the coordinates 5 and 3 in the plane of the india rubber, and we may still regard the india rubber as being in a plane, even if we have twisted it out of what we should ordinarily call a plane. Such continuous distortions do not matter.

To take another illustration: instead of using a steel measuring rod to fix our coordinates, let us use a live eel, which is wriggling all the time. The distance from the tail to the head of the eel is to count as 1 from the point of view of coordinates, whatever shape the creature may be assuming at the moment. The eel is continuous, and its wriggles are continuous, so it may be taken as our unit of distance in assigning coordinates. Beyond the requirement of continuity, the method of assigning coordinates is purely conventional, and therefore a live eel is just as good as a steel rod.

We are apt to think that, for really careful measurements, it is better to use a steel rod than a live eel. This is a mistake, not because the eel tells us what the steel rod was thought to tell, but because the steel rod really tells no more than the eel obviously does. The point is, not that eels are really rigid, but that steel rods really wriggle. To an observer in just one possible state of motion the eel would appear rigid, while the steel rod would seem to wriggle just as the eel does to us. For everybody moving differently both from this observer and ourselves, both the eel and the rod would seem to wriggle. And there is no saying that one observer is right and another wrong. In such matters what is seen does not belong solely to the physical process observed, but also to the standpoint of the observer. Measurements of distances and times do not directly reveal properties of the things measured, but relations of the things to the measurer. What observation can tell us about the physical world is therefore more abstract than we have hitherto believed.

It is important to realize that geometry, as taught in schools since Greek times, ceases to exist as a separate science and becomes merged into physics. The two fundamental notions in elementary geometry were the straight line and the circle. What appears to you as a straight road, whose parts all exist now, may appear to another observer to be like the flight of a rocket, some kind of curve whose parts come into existence successively. The circle depends upon measurement of distance, since it consists of all the points at a given distance from its center. And measurement of distances, as we have seen, is a subjective affair, depending upon the way in which the observer is moving. The failure of the circle to have objective validity was demonstrated by the Michelson-Morley experiment, and is thus, in a sense, the starting point of the whole theory of relativity. Rigid bodies, which we need for measurement, are only rigid for certain observers; for others they will be constantly changing all their dimensions. It is only our obstinately earthbound imagination that makes us suppose a geometry separate from physics to be possible.

That is why we do not trouble to give physical significance to our coordinates from the start. Formerly, the coordinates used in physics were supposed to be carefully measured distances; now we realize that this care at the start is thrown away. It is at a later stage that care is required. Our coordinates now are hardly more than a systematic way of cataloguing events. But mathematics provides, in the method of tensors, such an immensely powerful technique that we can use coordinates assigned in this apparently careless way just as effectively as if we had applied the whole apparatus of minutely accurate measurement in arriving at them. The advantage of being haphazard at the start is that we avoid making surreptitious physical assumptions, which we can hardly help making if we suppose that our coordinates have initially some particular physical significance.

We need not try to proceed in ignorance of all observed physical phenomena. We know certain things. We know that the old Newtonian physics is very nearly accurate when our coordinates have been chosen in a certain way. We know that the special theory of relativity is still more nearly accurate for suitable coordinates. From such facts we can infer certain things about our new coordinates, which, in a logical deduction, appear as postulates of the new theory. As such postulates we take:

1. That the interval between neighboring events takes a general form, like that used by Riemann for distances.
2. That a sufficiently small, light, and symmetrical body

travels on a geodesic in space-time, except in so far as non-gravitational forces act upon it.

3. That a light ray travels on a geodesic which is such that the interval between any two parts of it is zero.

Each of these postulates requires some explanation.

Our first postulate requires that, if two events are close together (but not necessarily otherwise), there is an interval between them which can be calculated from the differences between their coordinates by some such formula as we considered in the preceding chapter. That is to say, we take the squares and products of the differences of coordinates, we multiply them by suitable amounts (which in general will vary from place to place), and we add the results together. The sum obtained is the square of the interval. We do not assume in advance that we know the amounts by which the squares and products must be multiplied; this is going to be discovered by observing physical phenomena. But we do know, because mathematics shows it to be so, that within any small region of space-time we can choose the coordinates so that the interval has almost exactly the special form which we found in the special theory of relativity. It is not necessary for the application of the special theory to a limited region that there should be no gravitation in the region; it is enough if the intensity of gravitation is practically the same throughout the region. This enables us to apply the special theory within any small region. How small it will have to be depends upon the neighborhood. On the surface of the earth, it would have to be small enough for the curvature of the earth to be negligible. In the spaces between the planets, it need only be small enough for the attraction of the sun and the planets to be sensibly constant throughout the region. In the spaces between the stars it might be enormous — say half the distance from one star to the next — without introducing measurable inaccuracies.

Thus, at a great distance from gravitating matter, we can so choose our coordinates as to obtain very nearly a Euclidean space; this is really only another way of saying that the special theory of relativity applies. In the neighborhood of matter, although we can still make our space very nearly Euclidean in a very small region, we cannot do so throughout any region within which gravitation varies sensibly — at least, if we do, we shall have to abandon the view expressed in the second postulate, that bodies moving under gravitational forces only move on geodesics.

We saw that a geodesic on a surface is the shortest line that can be drawn on the surface from one point to another — for example, on the earth the geodesics are great circles. When we come to space-time, the mathematics is the same, but the verbal explanations have to be rather different. In the general theory of relativity, it is only neighboring events that have a definite interval, independently of the route by which we travel from one to the other. The interval between distant events depends upon the route pursued, and has to be calculated by dividing the route into a number of little bits and adding up the intervals for the various little bits. If the interval is spacelike, a body cannot travel from one event to the other; therefore, when we are considering the way bodies move, we are confined to timelike intervals. The interval between neighboring events when it is timelike will appear as the time between them for observers who travel from the one event to the other. And so the whole interval between two events will be judged by people who travel from one to the other to be what their clocks show to be the time that they have taken on the journey. For some routes this time will be longer, for others shorter; the more slowly they travel, the longer they will think they have been on the journey. This must not be taken as a platitude. I am not saying that if you travel from London to Edinburgh you will take longer if you travel more slowly. I am saying something much more odd. I am saying that if you leave London at 10 A.M. and arrive in Edinburgh at 6:30 P.M., Greenwich time, the more slowly you travel, the longer you will take — if the time is judged by your watch. This is a very different statement. From the point of view of a person on the earth, your journey takes eight hours and a half. But if you had been a ray of light traveling around the solar system, starting from London at 10 A.M., reflected from Jupiter to Saturn, and so on, until at last you were reflected back to Edinburgh and arrived there at 6:30 P.M., you would judge that the journey had taken you exactly no time. And if you had gone by any circuitous route, which enabled you to arrive in time by traveling fast, the longer your route, the less time you would judge that you had taken; the diminution of time would be continual as your speed approached that of light. Now I say that when a body travels, if it is left to itself, it chooses the route which makes the time between two stages of the journey as long as possible; if it had traveled from one event to another by any other route, the time, as measured by its own clocks, would have been shorter. This is a way of saying that bodies left to themselves do their journeys as slowly as they can; it is a sort of law of cosmic laziness.

Its mathematical expression is that they travel in geodesics, in which the total interval between any two events on the journey is *greater* than by any alternative route. (The fact that it is greater, not less, is due to the fact that the sort of interval we are considering is more analogous to time than to distance.) For example, if people could leave the earth and travel about for a time and then return, the time between their departure and return would be less by their clocks than by those on the earth: the earth, in its journey around the sun, chooses the route which makes the time of any bit of its course by its clocks longer than the time as judged by clocks which move by a different route. This is what is meant by saying that bodies left to themselves move in geodesics in space-time.

It is important to remember that space-time is not supposed to be Euclidean. As far as the geodesics are concerned, this has the effect that space-time is like a hilly countryside. In the neighborhood of a piece of matter, there is, as it were, a hill in space-time; this hill grows steeper and steeper as it gets nearer the top, like the neck of a bottle. It ends in a sheer precipice. Now, by the law of cosmic laziness which we mentioned earlier, a body coming into the neighborhood of the hill will not attempt to go straight over the top, but will go around. This is the essence of Einstein's view of gravitation. What a body does, it does because of the nature of space-time in its own neighborhood, not because of some mysterious force emanating from a distant body.

An analogy will serve to make the point clear. Suppose that on a dark night a number of people with lanterns were walking in various directions across a huge plain, and suppose that in one part of the plain there was a hill with a flaring beacon on the top. Our hill is to be such as we have described, growing steeper as it goes up and ending in a precipice. I shall suppose that there are villages dotted about the plain, and the people with lanterns are walking to and from these various villages. Paths have been made showing the easiest way from any one village to any other. These paths will all be more or less curved, to avoid going too far up the hill; they will be more sharply curved when they pass near the top of the hill than when they keep some way off from it. Now, suppose that you are observing all this, as best you can, from a place high up in a balloon, so that you cannot see the ground, but only the lanterns and the beacon. You will not know that there is a hill or that the beacon is at the top of it. You will see that people turn out of the straight course when they approach the beacon, and that the nearer they come, the more they turn aside. You will naturally attribute this to an effect of the beacon; you may think that it is very hot and people are afraid of getting burned. But if you wait for daylight you will see the hill, and you will find that the beacon merely marks the top of the hill and does not influence the people with lanterns in any way.

Now in this analogy the beacon corresponds to the sun, the people with lanterns correspond to the planets and comets, the paths correspond to their orbits, and the coming of daylight corresponds to the coming of Einstein. Einstein says that the sun is at the top of a hill, only the hill is in space-time, not in space. (I advise the reader not to try to picture this, because it is impossible.) Each body, at each moment, adopts the easiest course open to it, but owing to the hill, the easiest course is not a straight line. Each little bit of matter is at the top of its own little hill, like the cock on his own dung heap. What we call a big bit of matter is a bit which is the top of a big hill. The hill is what we know about; the bit of matter at the top is assumed for convenience. Perhaps there is really no need to assume it, and we could do with the hill alone, for we can never get to the top of anyone else's hill, any more than the pugnacious cock can fight the peculiarly irritating bird that he sees in the looking glass.

I have given only a qualitative description of Einstein's law of gravitation; to give its exact quantitative formulation is impossible without more mathematics than I am permitting myself. The most interesting point about it is that it makes the law no longer the result of action at a distance; the sun exerts no force on the planets whatever. Just as geometry has become physics, so, in a sense, physics has become geometry. The law of gravitation has become the geometrical law that every body pursues the easiest course from place to place, but this course is affected by the hills and valleys that are encountered on the road. . . .

# ROGER PENROSE

∎

Among the many pregnant predictions of the general theory of relativity was the possibility that objects exist with gravitational fields so intense that nothing — not even their own light — can escape them. These "black holes" (the term was coined by the American relativist John Archibald Wheeler in 1967) are here described by Roger Penrose (b. 1931), an imaginative theoretical physicist and professor of mathematics at Oxford University, in an essay based on a BBC radio lecture he delivered in 1973.

# *Black Holes*

ABOUT 6,000 LIGHT YEARS away, in the constellation of Cygnus the Swan, lies the blue supergiant star HDE 226868. Its mass exceeds that of our sun by a factor of about thirty, and its radius by a factor of nearly twenty-five. This in itself is nothing especially unusual. Many other stars of a similar nature are known. But once every five and one-half days, HDE 226868 circles in orbit about an invisible companion. It is this mysterious companion which concerns us here — with a mass one-half that of HDE 226868 but utterly tiny, its radius apparently being only about thirty miles! The companion of HDE 226868 is now believed by many astronomers to be a *black hole* — a bizarre consequence of the physical laws embodied in Einstein's theory of general relativity. Einstein's theory describes gravitation in terms of "space-time curvature," and in a black hole, the gravitational field has become so strong that this curvature leads to some surprising and weird effects.

The identification of the companion of HDE 226868 as a black hole is not yet quite certain, but looks highly probable at the present time. There are other objects in the heavens which some astronomers claim are likely also to be black holes and it may be that the final definitive judgement on the existence of black holes will be centred instead on one of these. But however it comes, this discovery will be an event of the utmost importance to present-day physical theory. For theory predicts that black holes *should* exist and should occur sometimes as the end-point of stellar evolution. If black holes were found *not* to exist then this would point to some drastic revision necessary in the theory. On the other hand their *existence* will also pose fundamental problems for the theory and I shall attempt to elucidate some of these later in this chapter.

What is a black hole? For astronomical purposes it behaves as a small, highly condensed dark "body." But it is not really a material body in the ordinary sense. It possesses no ponderable surface. A black hole is a region of empty space (albeit a strangely distorted one) which acts as a centre of gravitational attraction. At *one* time a material body *was* there. But the body collapsed inwards under its own gravitational pull. The more the body concentrated itself towards the centre the stronger became its gravitational field and the less was the body able to stop itself from yet further collapse. At a certain stage a point of no return was reached, and the body passed within its "absolute event horizon." I shall say more of this later, but for our present purposes, it is the absolute event horizon which acts as the boundary surface of the black hole. This surface is not material. It is merely a demarcation line drawn in space separating an interior from an exterior region. The interior region — into which the body has fallen — is defined by the fact that no matter, light, or signal of any kind can escape from it, while the exterior region is where it is still possible for signals or material particles to escape to the outside world. The matter which collapsed to form the black hole has fallen deep inside to attain incredible densities, apparently even to be crushed out of existence by reaching what is known as a "space-time singularity" — a place where physical laws, as presently understood, must cease to apply.

Since the black hole acts as a centre of attraction it can draw new material towards it — which once inside can never escape. The material thus swallowed contributes to the effective mass of the black hole. And as its mass increases the black hole grows in size, its linear dimensions being proportional to its mass. Its attractive power likewise increases, so the alarming picture presents itself of an ever-increasing celestial vacuum cleaner — a maelstrom in space which sweeps up all in its path. But things are not quite so bad as this. We are saved by the very minuteness of black holes — a fact which results from the smallness of the gravitational constant.

To see this, let us return to our picture of HDE 226868. Accepting the most recent figures for the dimensions involved, we have a black hole of some thirty miles in radius — in mutual orbit about a giant star whose radius is over 300,000 times larger. Despite its small size, the gravitational influence of the black hole is sufficient to distort the large star considerably out of spherical shape. It becomes rather like an egg whose small end is somewhat pointed in the direction of the black hole. A certain amount of material is dragged from this point and slowly falls inward to the black hole. It does not fall straight in, however. The black hole behaves much like a point mass. Most of the material dragged from the large star will remain circulating about the black hole for a long time. Only gradually, as frictional effects begin to play their part, will the material begin to spiral inwards. Again we must bear in mind the small size of the hole. (Imagine having to drain a normal-sized bath through a plughole a ten-thousandth of an inch across — or a bath the size of Loch Lomond through a normal-sized plughole!) The material can be only very slowly funnelled into the black hole. And as it gets funnelled in it gets compressed and very hot — so hot that the material must be expected to radiate light of very short wavelength, X-rays, in fact. Such X-rays are actually observed coming from the vicinity of HDE 226868. And the source of these X-rays (referred to as Cygnus X-1) appears, on the basis of detailed observations, to be in orbit about the visible component HDE 226868. The observed signals seem to be perfectly consistent with the black-hole picture I have presented. However we should remain cautious about drawing premature conclusions, as it is still conceivable that some alternative explanation of the observations may eventually turn out to be correct. The present evidence seems to be pointing ever more strongly in favor of Cygnus X-1 being a black hole, but even if for some reason this interpretation does turn out to be erroneous after all, it would still be very surprising (on the basis of present theory) if *no* black holes were found to exist. To indicate why, I should explain something of the picture that astronomers and astrophysicists have developed concerning stellar evolution and then indicate some of the theory that lies in support of the black hole picture that I have presented.

Let us consider first what theory and observations tell us to expect for the future of our sun — or of any other normal star of about the same mass. After shining at approximately its present brightness for about 7,000 million years the sun will begin a change which will transform it beyond recognition. According to the well-accepted theory of stellar evolution the sun will grow to an enormous size and become, like stars such as Antares in the constellation Scorpio, a red giant some 200 million miles in diameter. By this time the planets Mercury, Venus and the earth will have been burned away and their former orbits will lie well within the new solar surface. The density of the sun's material will by then have fallen from its present value of a fifth that of the earth to a tenth of the density of air!

As it continues to burn more and more of its nuclear fuel, making heavy inroads on its helium and heavier elements as well as hydrogen, the bloated sun will halt its expansion and begin to contract — down past its present size, smaller and smaller until it stabilises as a white dwarf star perhaps about the size of the earth. At this stage further contraction will be impossible because the electrons of its atoms will be packed together so closely that a law of quantum mechanics known as Pauli's Exclusion Principle will come into play. This principle states that no two electrons in an atom can occupy the same energy state. One can envisage the atoms so closely squeezed together under the dwarf sun's tremendous gravitational field that, pictorially speaking, any closer spacing would force the electrons to get in each other's way so that they cannot, at these temperatures and pressures, be squashed together any further. In this state the density of the solar material will be such that a matchbox full of it would weigh several tons. No material on earth has a density remotely approaching that of a white dwarf but as with red giants, many white dwarfs can be seen in our galaxy. Their ultimate fate is simply to cool off to form black dwarfs and thereafter act merely as centres of strong gravitational attraction. The planets, Jupiter, Saturn, Uranus, Neptune, Pluto and possibly Mars, will continue to circle the ancient sun aeons after it has died.

White dwarfs are part of the normal evolutionary history of average-sized stars like the sun, and astronomical observations show actual stars at each stage of stellar evolution through the stage now reached by our sun, on to the red giant phase and back to white dwarfs. Moreover the theory of stellar physics fits these observations closely, but not all stars can follow this "normal" evolutionary path.

As long ago as 1931, Subrahmanyan Chandrasekhar calculated that there must be a maximum mass above which a white dwarf could not sustain itself against even further gravitational contraction despite Pauli's Exclusion Principle. Furthermore, this mass was not much greater than the mass of the sun — 1.4 times according to Chandrasekhar; a little less according to more recent computations. Many stars have masses considerably more than 1.5 times that of the sun. What is going to happen to them?

The answer depends on just how heavy the star is. Consider a star of twice the mass of the sun. Like the sun it will also expand to an enormous size and then recontract, but being more massive than Chandrasekhar's limit for a white dwarf it will be unable to settle down to final equilibrium in the white dwarf state. To picture what happens it will be useful to consider the giant phase of a star more fully. As soon as the central density of the star reaches that of a white dwarf, the outer layers of the star expand, and they go on expanding as more and more of the central material gets compressed into a white dwarf state. So the giant star develops a growing white dwarf core. In the case of the sun, all the material that remains in the star will eventually become part of this white dwarf. But if the star is too massive, there comes a point at which the core effectively exceeds Chandrasekhar's limit, whereupon it promptly collapses. In the process of collapse there is a tremendous release of energy, much of which is in the form of neutrinos which are absorbed (so it is believed) in the outer regions of the star, heating the envelope to an enormous temperature. A cataclysmic explosion ensues — a supernova explosion — which blows off a considerable proportion of the mass of the star. Many supernovae have been seen, the last one in our galaxy being described by Johannes Kepler in 1604. A supernova will outshine a whole galaxy for a few days and may even be visible by daylight. As much as 90 percent of the star's mass might be thrown off by the cataclysmic explosion to form tenuous debris of heavy elements which may later enrich the stars of a second generation. But especially interesting is the collapsed remnant of the star left behind at the centre of the rapidly expanding cloud of ejected gases. This core is much too compressed to form a white dwarf and it can only find equilibrium as a neutron star.

A neutron star is tiny even by comparison with a white dwarf. Indeed the reduction from a white dwarf to a neutron star is even more than the reduction of 100 to one from the sun to a white dwarf, or probably rather more than the approximate reduction of 250 to one from a red giant to the sun. A neutron star may be only 10 kilometres in radius or only about one seven-hundredth the radius of a white dwarf. The density of a neutron star could be more than a hundred million times the already extraordinary density of a white dwarf.

Our matchbox full of neutron star material would now weigh as much as an asteroid a mile or so in diameter. The star's density would be comparable with the density of the proton or neutron itself — in fact a neutron star could in some ways be regarded as an over-sized atomic nucleus, the only essential difference being that it is bound together by gravitation rather than by nuclear forces. Individual atoms have ceased to exist as such. The nuclei are touching and form one continuous mass. As for the electrons, what has happened is that the stupendous gravitational forces have squeezed the electrons into the only space available to them — that already occupied by the protons, reversing the usual reaction so that the star is now composed mainly of neutrons and it is the Pauli Exclusion Principle acting on these neutrons that supplies the effective forces preventing further collapse.

This picture of a neutron star was predicted theoretically by the Soviet physicist Lev Landau in 1932 and studied in detail by J. Robert Oppenheimer, Robert Serber and George M. Volkoff in 1938 and 1939. For years many astronomers doubted whether neutron stars could actually exist. Since 1967 the observational situation has changed dramatically, because in that year the first pulsars were observed. Since then the theory of pulsars has developed rapidly and it now seems virtually certain that the radio and optical impulses emitted by pulsars owe their energy and extraordinary regularity to the presence of a rotating neutron star. At least two pulsars reside inside supernova remnants, one being the Crab Nebula, and this gives further support to the theory that pulsars are in fact neutron stars.

There is a maximum mass above which a neutron star would not be able to sustain itself against still further gravitational contraction. There is some uncertainty as to the exact value of this maximum-mass limit. The original value given by Oppenheimer and Volkoff in 1939 was about 0.7 times the solar mass. More recently, larger values of up to three solar masses have been suggested. These higher values take into account the idea that the heavy subatomic particles called hyperons may be present in addition to ordinary neutrons and protons. But there are stars whose mass is more than fifty times the mass of the sun. What will happen to these? It seems exceedingly unlikely that all these stars would, as a result of their final collapse phase, or earlier, inevitably throw off so much of their material that their masses would always fall below the limits required for a stable white dwarf or neutron star to be the result. In these the neutron core would be unable to remain in equilibrium and would have to collapse further inwards. But what other forms of condensed matter might be possible considering that here we have densities in excess even of the fantastic value that is maintained inside a neutron star?

In this case, theory tells us a different story: although greater densities can be achieved, it is not possible to obtain any further stable final equilibrium configurations. Instead the gravitational effects become so overwhelming as to dominate everything else. Newtonian gravitation theory becomes quite inadequate to handle the problem, and instead, we must turn to Einstein's theory of general relativity. But in so doing, we are led to a picture so strange that even the phenomenon of the neutron star must seem commonplace by comparison. This new picture is the one which has now earned the description of a black hole.

Briefly, a black hole is a region of space into which a star (or collection of stars or any other bodies) has fallen, but from which no light, matter or signal of any kind can escape. Before examining this picture in some detail, consider the degree of further contraction that would be necessary for a neutron star to be compressed down to the size of a black hole. We have already seen that from the sun's dimension down to that of a neutron star involves a contraction of about 70,000 : 1 in linear size; from a red giant to a neutron star, a linear contraction of about 20,000,000 : 1. In view of this, it is perhaps surprising that a further contraction of only about 3 : 1 is required for a neutron star of one solar mass to become a black hole — in this case, with a radius of about two miles. Larger black holes are also possible, the radius of the hole being proportional to the mass, so they would result from the collapse of a body more massive than the sun. For example, Cygnus X-1 appears to be about 15 times more massive than the sun, so its radius (assuming little rotation) is about 30 miles.

The reason I emphasise the slight nature of this further contraction, only 3 : 1, is that faced with the unsettling nature of the black holes, people have naturally asked whether our physical theories are tenable under these extreme conditions. But these theories seem to have worked well in describing a very large range of stars of enormously different sizes and densities. In any case, the conditions under which a black hole is formed are not so extreme as all that — not necessarily more extreme than the situation of a neutron star. For example, the densities involved as the collapsing star crosses the absolute event horizon are not vastly different from those inside a neutron star. The larger the collapsing mass, the less would be this density — less in inverse proportion to the square of the mass. In the case of Cygnus X-1 this density would be rather less than that of a neutron star. It has often been considered by astronomers that collections of mass of up to 100 million suns or more might be involved in gravitational collapse in galactic centres. The density at the time such a huge mass crosses the event horizon might then be only about that of water. So the local conditions need not be excessive when a black hole is formed and there seems to be no reason to suppose that the black hole condition might render general relativity somehow inapplicable.

On the other hand it must be admitted that general relativity plays anything but an irreplaceable role in observational astronomy. It is still possible that general relativity might be wrong; after all, the experimental tests of general relativity which have been successfully performed are still not very numerous. Although no conflict between the theory and observations exists at present (any apparent conflict being explicable by other means), the observations do not point *conclusively* in the direction of general relativity. There is still scope for alternative theories of gravitation. But no one would deny that general relativity *is* an excellent theory; almost certainly the most satisfactory theory of gravity available to us. Furthermore the theory which at the present time may be regarded as general relativity's most serious rival (namely the Brans-Dicke-Jordan scalar-tensor theory) itself leads to a black hole picture nearly identical to that arising in Einstein's theory. Even in Newtonian theory a phenomenon similar to that of a black hole may be said to occur. As long ago as 1798, Pierre Simon de Laplace used Newtonian theory to predict that a sufficiently massive and concentrated body should be *invisible* because the escape velocity at the surface could be greater than the speed of light. A photon or particle of light emitted from the surface of the body would simply fall back to the surface, it would not escape to be observed at large distances from the body. This description is perhaps arguable, but it shows that there is a situation to be faced, even in Newtonian theory.

I only make these observations to show that if one's sole motive for casting doubts on the tenability of general relativity is to get out of the black hole situation, one might just as well stick with relativity, because its rejection will not necessarily make the black hole go away. So I propose, from here on, to restrict my discussion to considerations solely within the bounds of Einstein's general theory of relativity.

To begin with, consider the standard picture of a non-rotating black hole. The black hole is characterised by a spherical surface whose radius is proportional to the hole's mass. This surface is called the "absolute event horizon." Its defining property is that signals emitted inside it cannot escape, whereas from any point outside it signals can be emitted that either do escape or fall into the black hole. The radius of the event horizon can be calculated by multiplying twice the mass by the universal gravitational constant and dividing the result by the speed of light squared. Performing this calculation for the sun yields the result that the sun would have to be collapsed into a sphere four miles in diameter if it were to form a black hole. The absolute event horizon would be the surface of this four-mile sphere.

The body whose collapse is responsible for a black hole's existence has fallen deep inside the event horizon. The gravitational field inside the event horizon has become so powerful that even light itself is inevitably dragged inward regardless of the direction in which it is emitted. Outside the event horizon light escapes if it is aimed suitably outward. The closer the emission point is to the event horizon, the more the wave front of the emitted signal is displaced back toward the centre of the black hole. We may intuitively regard this displacement as being caused by the effect of the gravitation on the motion of the light. Owing to the intense gravitational attraction of the black hole, light travels more easily in the direction of the black hole than in the outward direction. Inside the event horizon the inward pull is so strong that outward motion of light has become impossible. This applies not just to light but to any signal originating within the black hole. As for a photon emitted radially outward from the surface of the black hole, on the event horizon, it will mark time, forever hovering in the surface itself at the same distance from the centre of the black hole.

It may seem that this is odd physics indeed, quite unlike the normal situation in relativity theory, where the speed of light has always the same constant value in all directions. But, strange as it may seem, the local physics in the neighbourhood of the absolute event horizon *is* the same as elsewhere. An observer at the event horizon who tries to measure the speed of light must himself be crossing the horizon by falling inwards into the hole. To him the speed of the light hovering on the horizon is indeed the same constant value, in the outward direction.

It would be natural for a reader who is not familiar with general relativity theory to find such a situation confusing. This is partly because so far we have been using a purely spatial description rather than a space-time one — and for many purposes a space-time picture is more illuminating than a spatial one. Strictly speaking, a space-time picture needs to be drawn in four dimensions, but an overall description of the space-time situation can be obtained by suppressing one of the spatial co-ordinates in the space-time diagram and substituting a time co-ordinate. This gives an instantaneous picture of what is going on at all times, and obviates the need for many sequential "snapshots" of a developing situation.

Consider a light flash emitted in all directions from a given point in ordinary space. The wave front of the flash would be a sphere centred on the emitting point and growing larger each moment at the speed of light. A purely spatial representation of the flash would be a sequence of spheres (Figure 1), each sphere larger than the preceding one, marking the position of the light flash's spherical wave front at a given moment in time. A space-time representation of the light flash, however, would be a cone whose vertex represents the time and place at which the light flash is emitted, the cone itself describing the history of the light flash.

By the same token the history of a star's collapse down to a black hole can best be depicted in a space-time representation (Figure 2). The locations of the light cones at various points in space-time show how light signals propagate in the gravitational field. At some points the light cones are drawn as being tipped over, but this is not something that would be noticed by a local observer. Such an observer would follow a path in space-time that proceeds into the interior of the light cone; his speed can never be greater than the locally measured speed of light, and only inside the light cone is this criterion met. But the tipping of the light cones does affect what an observer at large distances can see. Figure 2 shows that material particles and light signals which originate inside the event horizon are inevitably driven further inwards. For a particle or signal to cross the event horizon from the inside to the outside it would have to violate the condition mentioned above; it would have to exceed the local light speed, in violation of relativity. . . .

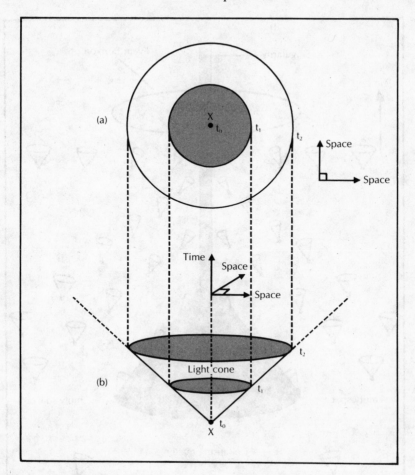

Figure 1. (a) Propagation of light from $X$ in space at times $t_1$ and $t_2$, (b) propagation of light in space-time to give a "light cone."

Although nothing can ever get out of a black hole, things can fall in. Indeed, it is quite possible that stellar astronauts traversing the depths of space in ages to come will run precisely this risk. Not that they will be likely to encounter a black hole by accident — the smallness of black holes compared with the vastness of the universe will see to that. Indeed, they would have to seek out a black hole deliberately if they wished to experience this "ultimate trip," and this might be difficult because black holes are not directly visible.

And what will happen to a hapless astronaut who falls into a black hole? What, indeed, is the fate of the original body which collapsed to produce the black hole? Assuming that the exact spherical symmetry is maintained right down to the centre, the answer pro-

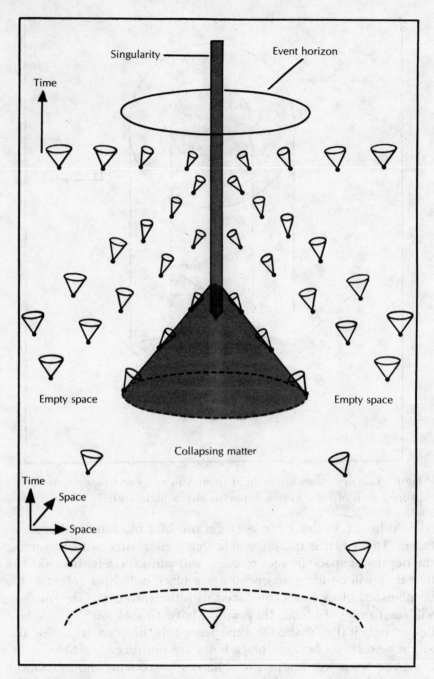

Figure 2. Space-time representation of collapse of a spherically symmetrical body to form a black hole, showing propagation of light inside and outside the event.

vided by general relativity is an alarming one. According to the theory, the curvature of space-time increases without limit as the centre is approached. Not only is the material of the original body squeezed to infinite density at the centre of the black hole — and crushed, effectively, out of existence — but also the vacuum in space-time which is left behind by the body, itself becomes infinitely curved. The effect of this infinite curvature on an observer, were he fool-hardy enough to follow the body inwards, would be that he experiences mounting gravitational tidal forces — tidal forces that mount rapidly to infinity.

The gravitational tidal effect is the most direct physical manifestation of space-time curvature. Einstein pointed out that the gravitational force on a body can be eliminated at any one point simply by choosing a frame of reference that is falling freely. He gave the famous example of a lift that broke its cable and fell toward the earth. Any passenger inside would be falling at the same rate as the lift, so he would feel no net gravitational force relative to the lift, and, indeed, would float free of gravity inside it. Such elimination of the gravitational force by free fall is now a familiar feature of space travel. The tidal effect, however, cannot be so eliminated and is, therefore, an absolute manifestation of the gravitational field. Imagine an observer falling freely in the earth's field. Suppose he is surrounded by a sphere of particles that he initially observed to be at rest with respect to himself. The Newtonian gravitational field of the earth varies as the inverse square of the distance between it and any other body, pulling more strongly on objects closer to its surface than on objects farther away. This non-uniformity of the gravitational field tidally distorts the sphere of particles into an ellipsoid which bulges out in the direction of the earth and in the direction away from it. The earth's ocean tides are a familiar example of an effect of this kind, where in this case the earth is experiencing the tidal effects of the moon and the sun.

Fortunately for us, the tidal effects due to gravity encountered in the solar system are small. Nobody complains that his feet experience more gravitational pull towards the earth than his head. There is a difference, but it is not noticeable in the ordinary way. The space-time curvature responsible for this difference has a radius of about the distance from the earth to the sun. (This is a pure coincidence since the sun is itself irrelevant to this particular tidal effect.) At the surface of a white dwarf, on the other hand, the space-time curvature is considerably larger, the radius of curvature being of the same order as the radius of the sun. This tidal effect would be very notice-able to an astronaut in orbit around the white dwarf. In fact, his head and feet could experience a difference in forces of perhaps one-fifth of the total force the astronaut normally experiences standing on the earth's surface. At the surface of a neutron star, however, the tidal effect is, by ordinary standards, enormous. The radius of space-time curvature there is only about thirty miles. It is clear that no astronaut in a low orbit around a neutron star could possibly survive. For even if he curled himself into a small ball, the gravitational acceleration at various parts of his body would differ by several million times the gravity at the earth's surface.

However, instruments could in principle be built to withstand such high tidal forces — all that would be necessary would be to make them so tiny that the difference between the gravitational attraction on the side nearest the neutron star and the side furthest away is small.

Suppose our astronaut carries such a tiny, rugged instrument as he flies towards a black hole of one solar mass. Long before he reaches the event horizon he will be destroyed by the tidal forces, but his instrument will survive intact as it crosses the event horizon, where it experiences tidal forces about thirty times those at the sur-face of the neutron star. But it survives only for a while, because the tidal forces encountered in a black hole rise to infinity at the centre. As the instrument falls in towards the centre, the mounting tidal forces will rise rapidly, ripping to pieces in turn the material of the instrument, the molecules of which this material is composed, the atoms which constitute these molecules, the atomic nuclei, and finally the fundamental particles which a moment ago had been the building-blocks of these nuclei. And the entire process would not last more than a few thousandths of a second!

So anything falling into a black hole, whether it be a space ship, a hydrogen molecule, an electron, radio-waves or a beam of light can never emerge again. So far as our universe is concerned it disappears completely and forever into nothing. But how can this be? Is it not a basic law of nature that matter or energy can never be completely destroyed but only converted from one form into another? The question is a perfectly reasonable one, but it can be shown by rig-orous argument, based on general relativity, that there must be a region inside a black hole, a region of infinite curvature, called a space-time singularity at which the known laws of physics break down. So there is no known conservation law that can be relied on at the centre of the black hole. There is no reason known why the matter should *not* just be totally destroyed as it reaches the singularity. Eventually, perhaps, laws of nature may be formulated which govern the behaviour of space-time singularities, but no such laws are known at present.

This situation has a more familiar manifestation in a somewhat different context. General relativity, like virtually all viable physical theories, is reversible in time. So corresponding to any solution of the equations in which time runs one way, there must be another in which the time-sense is reversed. This leads us to expect that the above situation could — in principle — exist in a time-reversed form. Initially there would be the space-time singularity. Then matter would appear: elementary particles, light. Only later would these particles collect together into atoms, molecules or stars. In fact, a picture of this kind has been considered for many years by astronomers and cosmologists as a model of the creation of the universe. The initial big bang of the cosmological models is, like the centre of a black hole, also a space-time singularity, where the curvature of space-time becomes infinite. But now, rather than being destroyed, matter is created at the singularity. The cosmological big bang is not precisely the time-reverse of a black hole, however, since the singularity is all-embracing, unlike the relatively localised singularity inside the black hole. The basic difference is one of size, and we may indeed envisage more localised "little bangs," called *white holes*, which are more precisely the time-reverses of black holes. A number of theoreticians have considered such white holes seriously in connection, in particular, with models for quasars. However, I must say that I personally regard the possibility of the existence of white holes with considerable unease. The reason is basically this. Once a black hole is formed there is apparently no means of destroying it. It is created violently, but then settles down and sits around forever — or until the universe re-collapses at the end of time. Now a white hole — the time-reverse of a black hole — would have had to have *been* there since the beginning of time — tamely and invisibly biding its time before making its presence known to us. Then, when its moment arrives, it explodes into ordinary matter. But this moment is of its "own" choosing, governed, apparently, by no definite law. Of course, there is no fundamental reason why this should not occur; the idea simply seems untidy, at variance with thermodynamics and probably also with observation. Nevertheless we *are* still stuck with the big bang and that also seems untidy. But here there appears to be no way out.

But let us now return from the abyss of speculation and pursue the argument concerning black holes. Quite apart from the doubts I have already raised about the validity of the general theory of relativity, there are other questions that need to be settled before one can fully accept the theoretical concept of the black hole as a realistic description of something that actually occurs in nature. In the first place, can we be sure that enough is known about the nature of matter under the extreme conditions required to form a black hole for the predictions to carry conviction? What role does the assumption of exact spherical symmetry play in the discussion? To what extent does the black-hole picture fit in with astronomical observations? Let us consider these questions in turn.

As I have already pointed out, the densities involved in the formation of a black hole need not be excessive. The same applies to space-time curvatures. A black hole with a mass ranging from 10,000 to 100 million solar masses is sometimes considered a possible candidate for what might inhabit the centre of a galaxy, and a collapsing mass equal to 100 million solar masses would reach a black hole situation when its average density was roughly the density of water. The tidal effects at the event horizon are, like the density, proportional to the inverse square of the mass of the hole, so for a body of 100 million solar masses these tidal effects would be somewhat less than those produced at the earth's surface. As an astronaut passed through the absolute event horizon he would notice nothing. He would have no means of telling that an irretrievable situation had developed, because the exact location of the horizon is not something that can be discerned by local measurement. After this he would have but a few minutes to enjoy the experience of life inside a black hole before the tidal effects mounted to infinity. In the case of a black hole of 10 thousand million solar masses he would have about a day.

The question concerning the role played by the assumption of spherical symmetry has to be examined more carefully. If we do not assume spherical symmetry, then we cannot appeal to the exact solutions of Einstein's equations on which we have based the foregoing discussion. Furthermore, even if we assume that initially the deviations from spherical symmetry are slight, we should have every reason to expect that near the central point these asymmetries would be enormously magnified. Might not the different portions of the collapsing body miss one another? Perhaps they could re-emerge after a close encounter and bounce out again. Even if they did not, can we say anything about the final configurations of the gravitational field resulting from the collapse? It is fortunate that, owing to some general theorems that have been proved over the past few years, a remarkably complete picture of asymmetrical collapse has emerged.

Considering the picture in a little detail, suppose that a massive star or a collection of bodies collapses and that deviations from the spherical symmetry are at first comparatively small. We can establish that a point of no return has been passed if a certain criterion is satisfied. That criterion can be stated in several different ways, but the following is the simplest. Imagine that a flash of light is emitted at some instant at some point in space. The flash of light will follow the light cone centred on the point according to our space-time representation (Figure 3). The light rays start out from the point by diverging in all directions. When they pass through matter or through a gravitational field, the matter or the field has a focusing effect on the rays. If enough matter or a sufficiently strong gravitational field is encountered, the amount that the rays diverge can be reduced to such an extent that this divergence is actually reversed,

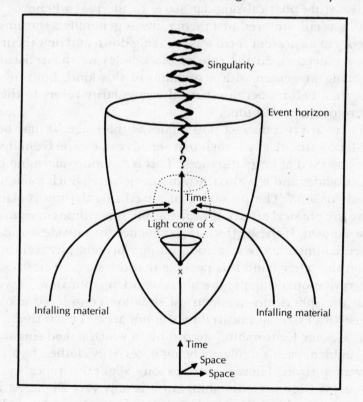

Figure 3. The birth of a black hole.

that is, the rays start to converge. The required criterion for a point of no return is that every light ray from the space-time point encounter enough matter or gravitation for the light cone to be reconverged. It is not hard to show from simple order of magnitude estimates that, for sufficiently large collections of mass, the criterion can indeed be satisfied before densities or curvatures became excessive, and without any assumption of symmetry.

Once this criterion has been satisfied, then according to a precise theorem in general relativity put forward by Stephen Hawking and myself, it follows that there must be a space-time singularity somewhere. The theorem does not say that this singularity is necessarily of the same character as that encountered in the centre of a spherically symmetrical black hole, but it is hard to avoid the inference that tidal effects which approach infinity will occur, producing a region of space-time where infinitely strong gravitational forces literally squeeze matter and photons out of existence.

Physicists are unhappy with a theory that predicts the evolution of such a truly physical singular state. In the past whenever a singularity was encountered in a theory, it was generally a warning that the theory in its present form was breaking down and new theoretical tools were needed. In the case of black holes we theoreticians are again being presented with a situation of this kind, but one more serious than before, because here the singularity refers to the very structure of space and time.

There are two distinct possibilities at this stage. It may be that the resulting singularity is such that signals can escape from it which can be observed at large distances. This is the more alarming of the two possibilities and it is also the more conjectural. Such a singularity is called "naked." The possibility of naked singularities is alarming because the physical effects of near-infinite space-time curvatures are quite unknown. If these effects can influence the outside world, then an essential uncertainty is introduced into present physical theory.

On the other hand it is possible that the singularities resulting from gravitational collapse are always hidden from view, as was the case in the spherically symmetrical situation considered above. In that case this essential uncertainty will not arise. This is the hypothesis of "Cosmic Censorship," according to which naked singularities are forbidden, each singularity being necessarily clothed by an absolute event horizon. There is perhaps some slight theoretical evidence in favour of Cosmic Censorship, but it is only very slight. After all, the consequences of the big bang singularity were decidedly *not* hidden; here was the biggest naked singularity of all time. However, it would be inaccurate to think of the big bang as a violation of Cosmic Censorship. We are concerned here only with singularities which arise in the collapse of perfectly reasonable nonsingular matter. I would personally tend to believe that in situations which do not differ too much initially from that of spherical symmetry, the Cosmic Censorship principle is valid. But in more extreme cases the question is, to my mind, quite open. Perhaps there is even observational evidence against Cosmic Censorship. This is a matter to which I shall return later.

If we assume the Cosmic Censorship hypothesis is true, then once the focusing criterion is satisfied an absolute event horizon must arise. This horizon will have a well-defined cross sectional area which will have a tendency to increase with time (black holes can grow but never shrink) but it seems reasonable to suppose that a black hole, left to itself, will settle down to a stationary state. However, we must be cautious in our use of intuition. It might also seem reasonable to suppose that given the vast range of structures, configurations and complexities of the bodies which could have collapsed to a black hole in the first place, the configuration of the black hole itself could also be complex. Some remarkable work by Werner Israel, Brandon Carter and Stephen Hawking has shown that this is probably not the case. Only a very restricted class of stationary black hole configurations can arise. They are uniquely characterised by the value of the mass, spin and charge of the hole. Einstein's equations for the general theory of relativity have been solved explicitly for this problem by Roy P. Kerr and the solution generalised to include charge by Ezra Newman and his co-workers. The reason the asymmetries present in the collapsing body do not show up in the final state of the black hole is that once the hole is formed the body that produced it has little influence on the hole's subsequent behaviour. The black hole is best thought of as a self-sustaining gravitational field governed by the internal non-linear dynamics of the general theory of relativity. These dynamics allow the asymmetries in the gravitational field of the hole to be carried away in the form of gravitational waves as the hole settles down into a stable configuration.

We have seen that a material object, once swallowed by a black hole, cannot escape. On the other hand, there are mechanisms whereby some of the energy content of the black hole can be extracted. One such mechanism involves the coalescing of two black holes. This process would be accompanied by the copious emission of gravitational waves, whose total energy should be a substantial fraction of the initial rest-mass energy of the black holes. Another mechanism would be to allow a particle to fall into a region close to the event horizon of a rotating black hole. The particle splits into two particles in such a way that one falls into the hole and the other escapes back to infinity with *more* mass energy than the initial particle had. In this way rotational energy of the black hole is transferred to the particle motions outside the hole. In the process the black hole loses mass and spin. In principle, this is an extremely efficient means of converting rest-mass to energy — much more efficient than nuclear fission or fusion! But there is an absolute limit even to this procedure. In the most extreme case the mass of the black hole might conceivably be reduced down to 0.707 of its original value by this sort of general procedure. In any case it is hard to envisage this process being effective in an actual astrophysical situation.

Let us now consider the situation inside the black hole, and the general relativistic implication of the existence of a space-time singularity. Since a "singularity" means a region of breakdown of physical theory, we have the curious situation that, here, general relativity is predicting its own downfall! But perhaps we should not be too surprised at this; after all we are treating general relativity only in its capacity as a classical theory. When the curvature of space-time becomes enormous, quantum effects must eventually play a dominant role. When the radius of space-time curvature becomes as small as, say, $(10^{-13})$ cm (roughly the radius of an elementary particle), then the theory of particle physics as understood at present must break down. If the radius of space-time curvature ever becomes as small as $(10^{-33})$ cm (and the implications of what we have said so far are that it will be that small somewhere inside a black hole — unless theory breaks down before this), then we cannot avoid having to apply quantum mechanics to the structure of space-time itself. At present there is no satisfactory theory for doing this. It should be emphasised that there is no reason to believe that a new theory is needed to deal with situations less extreme than these. Only *at* the singularities would we have trouble. And if Cosmic Censorship holds true, the absolute event horizon would prevent the effects of such new physics from having any influence on the outside world.

A question which often puzzles people in connection with black holes goes roughly as follows: If the absolute event horizon is so effective in shielding the contents of the black hole from outside, how is it that the black hole can still exert a gravitational influence on other bodies? How can it be that the gravitational field of the collapsed body escapes the black hole even though no information or signal can get out? In fact it is not really accurate to say that the gravitational field "escapes." It would be more true to say that the gravitational field is basically that of the body *before* collapse. If subsequently the body is annihilated at the centre of the hole, the exterior gravitational field cannot cut off. It would be *this* which would require information to escape the hole, because the gravitational field would have to "know" when the body producing it had disappeared. So the exterior field does not reflect any change which takes place in the interior. After the body has collapsed in, it is better to think of the black hole as a self-sustaining gravitational field in is own right. It has no further use for the body which originally built it!

Let us return, now, to the observational status of black holes. The most promising place to look for black holes is in binary systems, such as HDE 226868 — Cygnus X-1. As I mentioned earlier there are a number of such systems under consideration, with Cygnus X-1 being presently the most convincing candidate. Apart from the nature of the X-ray emission, the identification of the invisible component as a black hole would come from the estimate of its mass, this mass being detected via its gravitational effect on the visible component of the binary system. When the mass turns out to be considerably too large for a white or black dwarf or a neutron star, then the case for it being a black hole would seem to be a very strong one.

There is another aspect to the role of black holes in observational astronomy. We may compare the situation with that which arose in the case of the neutron stars. For many years astronomers had attempted to detect them by searching for certain effects such as X-ray emission which had been predicted also to occur in connection with neutron stars. However when neutron stars were first detected it was via an effect which was totally unexpected and still not really satisfactorily explained, namely the emission of the rapid and regular sharp pulses of electro-magnetic radiation characteristic of a pulsar. It is quite possible that the detection of other black holes will be by some equally unexpected observational side-effect.

There is no shortage of unexplained phenomena in cosmology today that might conceivably be relevant, from the phenomenal energy output of quasars and radio galaxies, the apparently anomalous red-shifts in the spectra of some quasars and galaxies, the explosions in the centres of some galaxies and the discrepancies in the mass measurements of galaxies, to serious questions that seem to remain even about the spiral-arm structure of normal galaxies. Above all, there is the apparent observation by Joseph Weber of the University of Maryland, of gravitational waves emanating from the centre of our own galaxy. As Professor Martin Rees points out, if these waves are emitted continuously and in all directions uniformly from the galactic centre, the energy they carry would correspond to the loss of mass from the galaxy of many thousands of solar masses per year. This is in gross conflict with other observations.

So far the theory of black holes has not led to a very convincing explanation of this worrying phenomenon, or any of the others, but these are early days yet. In the case of Weber's observations taken at their face value, the best hope would seem to be some mechanism whereby the waves are strongly beamed in the galactic plane. The sun and the solar system are unusually close to the galactic plane — they lie in it to an accuracy of one part in a thousand. With a beaming angle of something like 1/1,000 the observational conflicts would be removed. Attempts have been made to explain this kind of beaming on the basis of a fast rotating black hole at the centre of our galaxy, but so far they are not very convincing. An alternative to beaming would be the possibility that the production of the waves from the galactic centre is a comparatively short-lived phenomenon, lasting for considerably less than a one-hundredth part of the time that our galaxy has been in existence. (This would be a rather uneasy coincidence, since our appearance on this earth presumably need not have occurred just at the time of extreme gravitational activity at the centre of the galaxy.) Another possibility is that there might be some source much nearer to us lying approximately on the straight line containing us and the galactic centre. However, there are objections to this idea too. Of course there is also the possibility that the observations are in some way spurious and that a phenomenon other than gravitational waves is contriving to produce the effects that Weber observes. A number of other groups have now built detectors and it is to be hoped that this uncertainty will be cleared up in a few years.

Let us assume, for the purposes of argument at least, that Weber's results are substantiated; then the question arises as to how nature produces such waves. Although black holes have usually been suggested as the most likely objects which could be responsible, no plausible detailed explanation seems to be forthcoming. I feel that we should not close our minds to the possibility that some other type of gravitational catastrophe might be occurring at the centre of our galaxy. The question of naked singularities has been largely left aside; but as I have said earlier, there is no very convincing theoretical argument in favour of Cosmic Censorship. Could it be that naked singularities *are* in some way responsible? I think that one should not ignore this possibility, particularly since it offers some hope of producing the beaming necessary to fit in with the observations.

Let me present the following very hypothetical picture. Suppose that some form of large rapidly rotating mass at one time collected at the centre of our galaxy and that this mass approached a regime at which the effects of general relativity dominated over all more conventional forces. Suppose also (and this is the big assumption) that such rotating agglomerations have a tendency to settle into a configuration whose exterior field closely approximates a Kerr solution of the Einstein field equations. Recall that black holes do just this — but I am not now thinking of a black hole! If the rotation remains too great, the solution will not in fact describe a black hole. Instead the matter would contract down until a naked singularity is revealed in a ring around the equatorial region. This of course is very speculative but the model possesses one feature which is perhaps suggestive: the naked singularity would be visible only from the equatorial plane! (This follows from some work by Brandon Carter.) Thus any signal originating at or near the singularity would necessarily be beamed very closely in one plane. If we assume a tie-up between this object and the structure of the galaxy then it is not unreasonable to suppose that the beaming plane and the galactic plane coincide. Finally, since ultra-strong gravitational tidal fields would exist near the singularity one would expect that it would be gravitational waves which would be so beamed. Weber's results might conceivably be explained in this way.

It has often been argued that if naked singularities arise then this situation would be disastrous for physics. I do not share such feelings. True we have, as yet, no theory which can cope with space-time singularities. But I am an optimist. I believe that eventually such a theory will be found. In any case we have for some time been confronted by the profound theoretical problems of the big bang singularity, and a theory is needed to cope with this. Would it not be far more exciting if in addition there were other space-time singularities accessible to view *now* which could supply observational means of testing such a new theory? Perhaps then the mysteries of the initial creation could be more readily comprehended.

# STEPHEN W. HAWKING

■

Stephen Hawking (b. 1942), long esteemed by his fellow physicists for his groundbreaking theoretical work in general relativity, achieved more general celebrity in the 1980s when
newspaper and magazine profiles called attention to the fact that
he was managing to conduct his scientific career despite suffering from ALS, a debilitating disease of the central nervous
system that had left him paralyzed since his graduate-student
days. Hawking, who writes and speaks by means of a toggle-
controlled computer and voice synthesizer, published his first
popular book, *A Brief History of Time*, in 1988. It became an
international best-seller. In this chapter from the book he outlines what may be his greatest accomplishment, the theory of
Hawking radiation, which suggests that black holes, thought to
be inescapable, may emit particles and be destined to evaporate.

# *Black Holes Ain't So Black*

BEFORE 1970, MY RESEARCH on general relativity had concentrated
mainly on the question of whether or not there had been a big bang
singularity.* However, one evening in November that year, shortly
after the birth of my daughter, Lucy, I started to think about black
holes as I was getting into bed. My disability makes this rather a slow
process, so I had plenty of time. At that date there was no precise
definition of which points in space-time lay inside a black hole and
which lay outside. I had already discussed with Roger Penrose the
idea of defining a black hole as the set of events from which it was
not possible to escape to a large distance, which is now the generally
accepted definition. It means that the boundary of the black hole,

---

*A "singularity" is a point at which the curvature of space, according to the
general theory of relativity, is infinite. [editor's note]

the event horizon, is formed by the paths in space-time of rays of light that just fail to get away from the black hole, hovering forever just on the edge (Figure 1). It is a bit like running away from the police and just managing to keep one step ahead but not being able to get clear away!

Suddenly I realized that the paths of these light rays could never approach one another. If they did, they must eventually run into one another. It would be like meeting someone else running away from the police in the opposite direction — you would both be caught! (Or, in this case, fall into a black hole.) But if these light rays were swallowed up by the black hole, then they could not have been on the boundary of the black hole. So the paths of light rays in the event horizon had always to be moving parallel to, or away from, each other. Another way of seeing this is that the event horizon, the boundary of the black hole, is like the edge of a shadow — the shadow of impending doom. If you look at the shadow cast by a source at a great distance, such as the sun, you will see that the rays of light in the edge are not approaching each other.

If the rays of light that form the event horizon, the boundary of the black hole, can never approach each other, the area of the event horizon might stay the same or increase with time but it could never decrease — because that would mean that at least some of the rays of light in the boundary would have to be approaching each

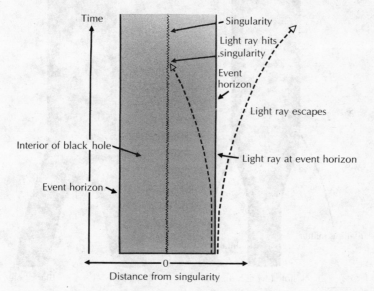

Figure 1

227

other. In fact, the area would increase whenever matter or radiation fell into the black hole (Figure 2). Or if two black holes collided and merged together to form a single black hole, the area of the event horizon of the final black hole would be greater than or equal to the sum of the areas of the event horizons of the original black holes (Figure 3). This nondecreasing property of the event horizon's area placed an important restriction on the possible behavior of black holes. I was so excited with my discovery that I did not get much sleep that night. The next day I rang up Roger Penrose. He agreed with me. I think, in fact, that he had been aware of this property of the area. However, he had been using a slightly different definition of a black hole. He had not realized that the boundaries of the black hole according to the two definitions would be the same, and hence so would their areas, provided the black hole had settled down to a state in which it was not changing with time.

The nondecreasing behavior of a black hole's area was very reminiscent of the behavior of a physical quantity called entropy, which measures the degree of disorder of a system. It is a matter of common experience that disorder will tend to increase if things are left to themselves. (One has only to stop making repairs around the

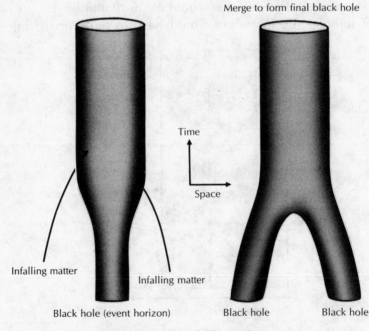

Figure 2 and Figure 3

house to see that!) One can create order out of disorder (for example, one can paint the house), but that requires expenditure of effort or energy and so decreases the amount of ordered energy available.

A precise statement of this idea is known as the second law of thermodynamics. It states that the entropy of an isolated system always increases, and that when two systems are joined together, the entropy of the combined system is greater than the sum of the entropies of the individual systems. For example, consider a system of gas molecules in a box. The molecules can be thought of as little billiard balls continually colliding with each other and bouncing off the walls of the box. The higher the temperature of the gas, the faster the molecules move, and so the more frequently and harder they collide with the walls of the box and the greater the outward pressure they exert on the walls. Suppose that initially the molecules are all confined to the left-hand side of the box by a partition. If the partition is then removed, the molecules will tend to spread out and occupy both halves of the box. At some later time they could, by chance, all be in the right half or back in the left half, but it is overwhelmingly more probable that there will be roughly equal numbers in the two halves. Such a state is less ordered, or more disordered, than the original state in which all the molecules were in one half. One therefore says that the entropy of the gas has gone up. Similarly, suppose one starts with two boxes, one containing oxygen molecules and the other containing nitrogen molecules. If one joins the boxes together and removes the intervening wall, the oxygen and the nitrogen molecules will start to mix. At a later time the most probable state would be a fairly uniform mixture of oxygen and nitrogen molecules throughout the two boxes. This state would be less ordered, and hence have more entropy, than the initial state of two separate boxes.

The second law of thermodynamics has a rather different status than that of other laws of science, such as Newton's law of gravity, for example, because it does not hold always, just in the vast majority of cases. The probability of all the gas molecules in our first box being found in one half of the box at a later time is many millions of millions to one, but it can happen. However, if one has a black hole around, there seems to be a rather easier way of violating the second law: just throw some matter with a lot of entropy, such as a box of gas, down the black hole. The total entropy of matter outside the black hole would go down. One could, of course, still say that the total entropy, including the entropy inside the black hole, has not gone down — but since there is no way to look inside the black hole, we cannot see how much entropy the matter inside it has. It would be nice, then, if there was some feature of the black hole by which observers outside the black hole could tell its entropy, and which would increase whenever matter carrying entropy fell into the black hole. Following the discovery, described above, that the area of the event horizon increased whenever matter fell into a black hole, a research student at Princeton named Jacob Bekenstein suggested that the area of the event horizon was a measure of the entropy of the black hole. As matter carrying entropy fell into a black hole, the area of its event horizon would go up, so that the sum of the entropy of matter outside black holes and the area of the horizons would never go down.

This suggestion seemed to prevent the second law of thermodynamics from being violated in most situations. However, there was one fatal flaw. If a black hole has entropy, then it ought also to have a temperature. But a body with a particular temperature must emit radiation at a certain rate. It is a matter of common experience that if one heats up a poker in a fire it glows red hot and emits radiation, but bodies at lower temperatures emit radiation too; one just does not normally notice it because the amount is fairly small. This radiation is required in order to prevent violation of the second law. So black holes ought to emit radiation. But by their very definition, black holes are objects that are not supposed to emit anything. It therefore seemed that the area of the event horizon of a black hole could not be regarded as its entropy. In 1972 I wrote a paper with Brandon Carter and an American colleague, Jim Bardeen, in which we pointed out that although there were many similarities between entropy and the area of the event horizon, there was this apparently fatal difficulty. I must admit that in writing this paper I was motivated partly by irritation with Bekenstein, who, I felt, had misused my discovery of the increase of the area of the event horizon. However, it turned out in the end that he was basically correct, though in a manner he had certainly not expected.

In September 1973, while I was visiting Moscow, I discussed black holes with two leading Soviet experts, Yakov Zeldovich and Alexander Starobinsky. They convinced me that, according to the quantum mechanical uncertainty principle, rotating black holes should create and emit particles. I believed their arguments on physical grounds, but I did not like the mathematical way in which they calculated the emission. I therefore set about devising a better mathematical treatment, which I described at an informal seminar in Oxford at the end of November 1973. At that time I had not done the calculations to find out how much would actually be emitted. I was expecting to discover just the radiation that Zeldovich and Starobinsky had predicted from rotating black holes. However, when I did the calculation, I found, to my surprise and annoyance, that even nonrotating black holes should apparently create and emit particles at a steady rate. At first I thought that this emission indicated that one of the approximations I had used was not valid. I was afraid that if Bekenstein found out about it, he would use it as a further argument to support his ideas about the entropy of black holes, which I still did not like. However, the more I thought about it, the more it seemed that the approximations really ought to hold. But what finally convinced me that the emission was real was that the spectrum of the emitted particles was exactly that which would be emitted by a hot body, and that the black hole was emitting particles at exactly the correct rate to prevent violations of the second law. Since then the calculations have been repeated in a number of different forms by other people. They all confirm that a black hole ought to emit particles and radiation as if it were a hot body with a temperature that depends only on the black hole's mass: the higher the mass, the lower the temperature.

How is it possible that a black hole appears to emit particles when we know that nothing can escape from within its event horizon? The answer, quantum theory tells us, is that the particles do not come from within the black hole, but from the "empty" space just outside the black hole's event horizon! We can understand this in the following way: What we think of as "empty" space cannot be completely empty because that would mean that all the fields, such as the gravitational and electromagnetic fields, would have to be exactly zero. However, the value of a field and its rate of change with time are like the position and velocity of a particle: the uncertainty principle implies that the more accurately one knows one of these quantities, the less accurately one can know the other. So in empty space the field cannot be fixed at exactly zero, because then it would have both a precise value (zero) and a precise rate of change (also zero). There must be a certain minimum amount of uncertainty, or quantum fluctuations, in the value of the field. One can think of these fluctuations as pairs of particles of light or gravity that appear together at some time, move apart, and then come together again and annihilate each other. These particles are virtual particles like the particles that carry the gravitational force of the sun: unlike real particles, they cannot be observed directly with a particle detector. However, their indirect effects, such as small changes in the energy of electron orbits in atoms, can be measured and agree with the theoretical predictions to a remarkable degree of accuracy. The uncertainty principle also predicts that there will be similar virtual pairs of matter particles, such as electrons or quarks. In this case, however, one member of the pair will be a particle and the other an antiparticle (the antiparticles of light and gravity are the same as the particles).

Because energy cannot be created out of nothing, one of the partners in a particle/antiparticle pair will have positive energy, and the other partner negative energy. The one with negative energy is condemned to be a short-lived virtual particle because real particles always have positive energy in normal situations. It must therefore seek out its partner and annihilate with it. However, a real particle close to a massive body has less energy than if it were far away, because it would take energy to lift it far away against the gravitational attraction of the body. Normally, the energy of the particle is still positive, but the gravitational field inside a black hole is so strong that even a real particle can have negative energy there. It is therefore possible, if a black hole is present, for the virtual particle with negative energy to fall into the black hole and become a real particle or antiparticle. In this case it no longer has to annihilate with its partner. Its forsaken partner may fall into the black hole as well. Or, having positive energy, it might also escape from the vicinity of the black hole as a real particle or antiparticle (Figure 4). To an observer at a distance, it will appear to have been emitted from the black hole. The smaller the black hole, the shorter the distance the particle with negative energy will have to go before it becomes a real particle, and thus the greater the rate of emission, and the apparent temperature, of the black hole.

The positive energy of the outgoing radiation would be balanced by a flow of negative energy particles into the black hole. By Einstein's equation $E = mc^2$ (where $E$ is energy, $m$ is mass, and $c$ is the speed of light), energy is proportional to mass. A flow of negative energy into the black hole therefore reduces its mass. As the black hole loses mass, the area of its event horizon gets smaller, but this decrease in the entropy of the black hole is more than compensated for by the entropy of the emitted radiation, so the second law is never violated.

Moreover, the lower the mass of the black hole, the higher its

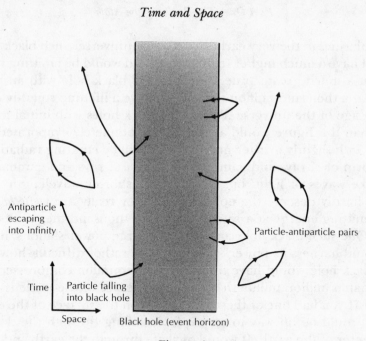

Figure 4

temperature. So as the black hole loses mass, its temperature and rate of emission increase, so it loses mass more quickly. What happens when the mass of the black hole eventually becomes extremely small is not quite clear, but the most reasonable guess is that it would disappear completely in a tremendous final burst of emission, equivalent to the explosion of millions of H-bombs.

A black hole with a mass a few times that of the sun would have a temperature of only one ten millionth of a degree above absolute zero. This is much less than the temperature of the microwave radiation that fills the universe (about 2.7° above absolute zero), so such black holes would emit even less than they absorb. If the universe is destined to go on expanding forever, the temperature of the microwave radiation will eventually decrease to less than that of such a black hole, which will then begin to lose mass. But, even then, its temperature would be so low that it would take about a million, million, million, million, million, million, million, million, million, million, million years (1 with sixty-six zeros after it) to evaporate completely. This is much longer than the age of the universe, which is only about ten or twenty thousand million years (1 with ten zeros after it). On the other hand . . . there might be primordial black holes with a very much smaller mass that were made by the collapse of

233

irregularities in the very early stages of the universe. Such black holes would have a much higher temperature and would be emitting radiation at a much greater rate. A primordial black hole with an initial mass of a thousand million tons would have a lifetime roughly equal to the age of the universe. Primordial black holes with initial masses less than this figure would already have completely evaporated, but those with slightly greater masses would still be emitting radiation in the form of X rays and gamma rays. These X rays and gamma rays are like waves of light, but with a much shorter wavelength. Such holes hardly deserve the epithet *black*: they really are *white hot* and are emitting energy at a rate of about ten thousand megawatts.

One such black hole could run ten large power stations, if only we could harness its power. This would be rather difficult, however: the black hole would have the mass of a mountain compressed into less than a million millionth of an inch, the size of the nucleus of an atom! If you had one of these black holes on the surface of the earth, there would be no way to stop it from falling through the floor to the center of the earth. It would oscillate through the earth and back, until eventually it settled down at the center. So the only place to put such a black hole, in which one might use the energy that it emitted, would be in orbit around the earth — and the only way that one could get it to orbit the earth would be to attract it there by towing a large mass in front of it, rather like a carrot in front of a donkey. This does not sound like a very practical proposition, at least not in the immediate future.

But even if we cannot harness the emission from these primordial black holes, what are our chances of observing them? We could look for the gamma rays that the primordial black holes emit during most of their lifetime. Although the radiation from most would be very weak because they are far away, the total from all of them might be detectable. We do observe such a background of gamma rays: Figure 5 shows how the observed intensity differs at different frequencies (the number of waves per second). However, this background could have been, and probably was, generated by processes other than primordial black holes. The dotted line in Figure 5 shows how the intensity should vary with frequency for gamma rays given off by primordial black holes, if there were on average 300 per cubic light-year. One can therefore say that the observations of the gamma ray background do not provide any *positive* evidence for primordial black holes, but they do tell us that on average there cannot be more than 300 in every cubic light-year in the universe. This limit means

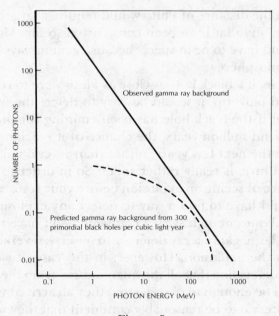

Figure 5

that primordial black holes could make up at most one millionth of the matter in the universe.

With primordial black holes being so scarce, it might seem unlikely that there would be one near enough for us to observe as an individual source of gamma rays. But since gravity would draw primordial black holes toward any matter, they should be much more common in and around galaxies. So although the gamma ray background tells us that there can be no more than 300 primordial black holes per cubic light-year on average, it tells us nothing about how common they might be in our own galaxy. If they were, say, a million times more common than this, then the nearest black hole to us would probably be at a distance of about a thousand million kilometers, or about as far away as Pluto, the farthest known planet. At this distance it would still be very difficult to detect the steady emission of a black hole, even if it was ten thousand megawatts. In order to observe a primordial black hole one would have to detect several gamma ray quanta coming from the same direction within a reasonable space of time, such as a week. Otherwise, they might simply be part of the background. But Planck's quantum principle tells us that each gamma ray quantum has a very high energy, because gamma rays have a very high frequency, so it would not take many quanta to radiate even ten thousand megawatts. And to observe these few coming from the distance of Pluto would require a larger gamma ray detector than any that have been constructed so far. Moreover, the detector would have to be in space, because gamma rays cannot penetrate the atmosphere.

Of course, if a black hole as close as Pluto were to reach the end of its life and blow up, it would be easy to detect the final burst of emission. But if the black hole has been emitting for the last ten or twenty thousand million years, the chance of it reaching the end of its life within the next few years, rather than several million years in the past or future, is really rather small! So in order to have a reasonable chance of seeing an explosion before your research grant ran out, you would have to find a way to detect any explosions within a distance of about one light-year. You would still have the problem of needing a large gamma ray detector to observe several gamma ray quanta from the explosion. However, in this case, it would not be necessary to determine that all the quanta came from the same direction: it would be enough to observe that they all arrived within a very short time interval to be reasonably confident that they were coming from the same burst.

One gamma ray detector that might be capable of spotting primordial black holes is the entire earth's atmosphere. (We are, in any case, unlikely to be able to build a larger detector!) When a high-energy gamma ray quantum hits the atoms in our atmosphere, it creates pairs of electrons and positrons (antielectrons). When these hit other atoms they in turn create more pairs of electrons and positrons, so one gets what is called an electron shower. The result is a form of light called Cerenkov radiation. One can therefore detect gamma ray bursts by looking for flashes of light in the night sky. Of course, there are a number of other phenomena, such as lightning and reflections of sunlight off tumbling satellites and orbiting debris, that can also give flashes in the sky. One could distinguish gamma ray bursts from such effects by observing flashes simultaneously at two or more fairly widely separated locations. A search like this has been carried out by two scientists from Dublin, Neil Porter and Trevor Weekes, using telescopes in Arizona. They found a number of flashes but none that could be definitely ascribed to gamma ray bursts from primordial black holes.

Even if the search for primordial black holes proves negative, as it seems it may, it will still give us important information about the very early stages of the universe. If the early universe had been chaotic or irregular, or if the pressure of matter had been low, one would have expected it to produce many more primordial black holes than the limit already set by our observations of the gamma ray background. Only if the early universe was very smooth and uniform, with a high pressure, can one explain the absence of observable numbers of primordial black holes.

The idea of radiation from black holes was the first example of a prediction that depended in an essential way on both the great theories of this century, general relativity and quantum mechanics. It aroused a lot of opposition initially because it upset the existing viewpoint: "How can a black hole emit anything?" When I first announced the results of my calculations at a conference at the Rutherford-Appleton Laboratory near Oxford, I was greeted with general incredulity. At the end of my talk the chairman of the session, John G. Taylor from Kings College, London, claimed it was all nonsense. He even wrote a paper to that effect. However, in the end most people, including John Taylor, have come to the conclusion that black holes must radiate like hot bodies if our other ideas about general relativity and quantum mechanics are correct. Thus even though we have not yet managed to find a primordial black hole, there is fairly general agreement that if we did, it would have to be emitting a lot of gamma rays and X rays.

The existence of radiation from black holes seems to imply that gravitational collapse is not as final and irreversible as we once thought. If an astronaut falls into a black hole, its mass will increase, but eventually the energy equivalent of that extra mass will be returned to the universe in the form of radiation. Thus, in a sense, the astronaut will be "recycled." It would be a poor sort of immortality, however, because any personal concept of time for the astronaut would almost certainly come to an end as he was torn apart inside the black hole! Even the types of particles that were eventually emitted by the black hole would in general be different from those that made up the astronaut: the only feature of the astronaut that would survive would be his mass or energy. . . .

# PART TWO

## The Wider Universe

The extension of human vision by a factor of ten billion times down into the realm of the atom has been accompanied by a comparable leap out into the realm of the planets, stars, and galaxies. Here, as in the atomic world, we find much that is unfamiliar, yet we also encounter evidence of a curiously intimate relationship between nature and humankind. Comets, invaders from the far reaches of the solar system, have been implicated in the mass dieouts of earthly life that in turn cleared the way for the ascent of new species, our own among them. The explosions of distant stars turn out to have forged the iron atoms with which we build the telescopes we use to study the stars. And our efforts to investigate the grandest questions we know how to ask — questions having to do with the birth and death of the universe — have called upon nuclear and particle physics, in a new alliance of the sciences of the large and the small.

# ANNIE DILLARD

■

"A reader's heart must go out to a young writer with a sense of wonder so fearless and unbridled," Eudora Welty wrote of the American essayist Annie Dillard (b. 1945). Dillard's work combines a naturalist's powers of observation with a poet's command of language. This piece is from her book *Teaching a Stone to Talk*.

# Total Eclipse

## I

IT HAD BEEN LIKE DYING, that sliding down the mountain pass. It had been like the death of someone, irrational, that sliding down the mountain pass and into the region of dread. It was like slipping into fever, or falling down that hole in sleep from which you wake yourself whimpering. We had crossed the mountains that day, and now we were in a strange place — a hotel in central Washington, in a town near Yakima. The eclipse we had traveled here to see would occur early the next morning.

I lay in bed. My husband, Gary, was reading beside me. I lay in bed and looked at the painting on the hotel room wall. It was a print of a detailed and lifelike painting of a smiling clown's head, made out of vegetables. It was a painting of the sort which you do not intend to look at, and which, alas, you never forget. Some tasteless fate presses it upon you; it becomes part of the complex interior junk you carry with you wherever you go. Two years have passed since the total eclipse of which I write. During those years I have forgotten, I assume, a great many things I wanted to remember — but I have not forgotten that clown painting or its lunatic setting in the old hotel.

The clown was bald. Actually, he wore a clown's tight rubber wig, painted white; this stretched over the top of his skull, which was a cabbage. His hair was bunches of baby carrots. Inset in his white clown makeup, and in his cabbage skull, were his small and laughing human eyes. The clown's glance was like the glance of Rembrandt in some of the self-portraits: lively, knowing, deep, and loving. The crinkled shadows around his eyes were string beans. His eyebrows were parsley. Each of his ears was a broad bean. His thin, joyful lips were red chili peppers; between his lips were wet rows of human teeth and a suggestion of a real tongue. The clown print was framed in gilt and glassed.

To put ourselves in the path of the total eclipse, that day we had driven five hours inland from the Washington coast, where we lived. When we tried to cross the Cascades range, an avalanche had blocked the pass.

A slope's worth of snow blocked the road; traffic backed up. Had the avalanche buried any cars that morning? We could not learn. This highway was the only winter road over the mountains. We waited as highway crews bulldozed a passage through the avalanche. With two-by-fours and walls of plyboard, they erected a one-way, roofed tunnel through the avalanche. We drove through the avalanche tunnel, crossed the pass, and descended several thousand feet into central Washington and the broad Yakima valley, about which we knew only that it was orchard country. As we lost altitude, the snows disappeared; our ears popped; the trees changed, and in the trees were strange birds. I watched the landscape innocently, like a fool, like a diver in the rapture of the deep who plays on the bottom while his air runs out.

The hotel lobby was a dark, derelict room, narrow as a corridor, and seemingly without air. We waited on a couch while the manager vanished upstairs to do something unknown to our room. Beside us on an overstuffed chair, absolutely motionless, was a platinum-blond woman in her forties wearing a black silk dress and a strand of pearls. Her long legs were crossed; she supported her head on her fist. At the dim far end of the room, their backs toward us, sat six bald old men in their shirtsleeves, around a loud television. Two of them seemed asleep. They were drunks. "Number six!" cried the man on television, "Number six!"

On the broad lobby desk, lighted and bubbling, was a ten-gallon aquarium containing one large fish; the fish tilted up and down in its water. Against the long opposite wall sang a live canary in its cage. Beneath the cage, among spilled millet seeds on the carpet, were a decorated child's sand bucket and matching sand shovel.

Now the alarm was set for six. I lay awake remembering an article I had read downstairs in the lobby, in an engineering magazine. The article was about gold mining.

In South Africa, in India, and in South Dakota, the gold mines extend so deeply into the earth's crust that they are hot. The rock walls burn the miners' hands. The companies have to air-condition the mines; if the air conditioners break, the miners die. The elevators in the mine shafts run very slowly, down, and up, so the miners' ears will not pop in their skulls. When the miners return to the surface, their faces are deathly pale.

Early the next morning we checked out. It was February 26, 1979, a Monday morning. We would drive out of town, find a hilltop, watch the eclipse, and then drive back over the mountains and home to the coast. How familiar things are here; how adept we are; how smoothly and professionally we check out! I had forgotten the clown's smiling head and the hotel lobby as if they had never existed. Gary put the car in gear and off we went, as off we have gone to a hundred other adventures.

It was before dawn when we found a highway out of town and drove into the unfamiliar countryside. By the growing light we could see a band of cirrostratus clouds in the sky. Later the rising sun would clear these clouds before the eclipse began. We drove at random until we came to a range of unfenced hills. We pulled off the highway, bundled up, and climbed one of these hills.

## II

The hill was five hundred feet high. Long winter-killed grass covered it, as high as our knees. We climbed and rested, sweating in the cold; we passed clumps of bundled people on the hillside who were setting up telescopes and fiddling with cameras. The top of the hill stuck up

243

in the middle of the sky. We tightened our scarves and looked around.

East of us rose another hill like ours. Between the hills, far below, was the highway which threaded south into the valley. This was the Yakima valley; I had never seen it before. It is justly famous for its beauty, like every planted valley. It extended south into the horizon, a distant dream of a valley, a Shangri-la. All its hundreds of low, golden slopes bore orchards. Among the orchards were towns, and roads, and plowed and fallow fields. Through the valley wandered a thin, shining river; from the river extended fine, frozen irrigation ditches. Distance blurred and blued the sight, so that the whole valley looked like a thickness or sediment at the bottom of the sky. Directly behind us was more sky, and empty lowlands blued by distance, and Mount Adams. Mount Adams was an enormous, snow-covered volcanic cone rising flat, like so much scenery.

Now the sun was up. We could not see it; but the sky behind the band of clouds was yellow, and, far down the valley, some hillside orchards had lighted up. More people were parking near the highway and climbing the hills. It was the West. All of us rugged individualists were wearing knit caps and blue nylon parkas. People were climbing the nearby hills and setting up shop in clumps among the dead grasses. It looked as though we had all gathered on hilltops to pray for the world on its last day. It looked as though we had all crawled out of spaceships and were preparing to assault the valley below. It looked as though we were scattered on hilltops at dawn to sacrifice virgins, make rain, set stone stelae in a ring. There was no place out of the wind. The straw grasses banged our legs.

Up in the sky where we stood the air was lusterless yellow. To the west the sky was blue. Now the sun cleared the clouds. We cast rough shadows on the blowing grass; freezing, we waved our arms. Near the sun, the sky was bright and colorless. There was nothing to see.

It began with no ado. It was odd that such a well-advertised public event should have no starting gun, no overture, no introductory speaker. I should have known right then that I was out of my depth. Without pause or preamble, silent as orbits, a piece of the sun went away. We looked at it through welders' goggles. A piece of the sun was missing; in its place we saw empty sky.

I had seen a partial eclipse in 1970. A partial eclipse is very interesting. It bears almost no relation to a total eclipse. Seeing a partial eclipse bears the same relation to seeing a total eclipse as kissing a man does to marrying him, or as flying in an airplane does to falling out of an airplane. Although the one experience precedes the other, it in no way prepares you for it. During a partial eclipse the sky does not darken — not even when 94 percent of the sun is hidden. Nor does the sun, seen colorless through protective devices, seem terribly strange. We have all seen a sliver of light in the sky; we have all seen the crescent moon by day. However, during a partial eclipse the air does indeed get cold, precisely as if someone were standing between you and the fire. And blackbirds do fly back to their roosts. I had seen a partial eclipse before, and here was another.

What you see in an eclipse is entirely different from what you know. It is especially different for those of us whose grasp of astronomy is so frail that, given a flashlight, a grapefruit, two oranges, and fifteen years, we still could not figure out which way to set the clocks for Daylight Saving Time. Usually it is a bit of a trick to keep your knowledge from blinding you. But during an eclipse it is easy. What you see is much more convincing than any wild-eyed theory you may know.

You may read that the moon has something to do with eclipses. I have never seen the moon yet. You do not see the moon. So near the sun, it is as completely invisible as the stars are by day. What you see before your eyes is the sun going through phases. It gets narrower and narrower, as the waning moon does, and, like the ordinary moon, it travels alone in the simple sky. The sky is of course background. It does not appear to eat the sun; it is far behind the sun. The sun simply shaves away; gradually, you see less sun and more sky.

The sky's blue was deepening, but there was no darkness. The sun was a wide crescent, like a segment of tangerine. The wind freshened and blew steadily over the hill. The eastern hill across the highway grew dusky and sharp. The towns and orchards in the valley to the south were dissolving into the blue light. Only the thin river held a trickle of sun.

Now the sky to the west deepened to indigo, a color never seen. A dark sky usually loses color. This was a saturated, deep indigo, up in the air. Stuck up into that unworldly sky was the cone of Mount Adams, and the alpenglow was upon it. The alpenglow is that red light of sunset which holds out on snowy mountaintops long after the valleys and tablelands are dimmed. "Look at Mount Adams," I said, and that was the last sane moment I remember.

I turned back to the sun. It was going. The sun was going, and the world was wrong. The grasses were wrong; they were platinum. Their every detail of stem, head, and blade shone lightless and artificially distinct as an art photographer's platinum print. This color has never been seen on earth. The hues were metallic; their finish was matte. The hillside was a nineteenth-century tinted photograph from which the tints had faded. All the people you see in the photograph, distinct and detailed as their faces look, are now dead. The sky was navy blue. My hands were silver. All the distant hills' grasses were finespun metal which the wind laid down. I was watching a faded color print of a movie filmed in the Middle Ages; I was standing in it, by some mistake. I was standing in a movie of hillside grasses filmed in the Middle Ages. I missed my own century, the people I knew, and the real light of day.

I looked at Gary. He was in the film. Everything was lost. He was a platinum print, a dead artist's version of life. I saw on his skull the darkness of night mixed with the colors of day. My mind was going out; my eyes were receding the way galaxies recede to the rim of space. Gary was light-years away, gesturing inside a circle of darkness, down the wrong end of a telescope. He smiled as if he saw me; the stringy crinkles around his eyes moved. The sight of him, familiar and wrong, was something I was remembering from centuries hence, from the other side of death: yes, *that* is the way he used to look, when we were living. When it was our generation's turn to be alive. I could not hear him; the wind was too loud. Behind him the sun was going. We had all started down a chute of time. At first it was pleasant; now there was no stopping it. Gary was chuting away across space, moving and talking and catching my eye, chuting down the long corridor of separation. The skin on his face moved like thin bronze plating that would peel.

The grass at our feet was wild barley. It was the wild einkorn wheat which grew on the hilly flanks of the Zagros Mountains, above the Euphrates valley, above the valley of the river we called *River*. We harvested the grass with stone sickles, I remember. We found the grasses on the hillsides; we built our shelter beside them and cut them down. That is how he used to look then, that one, moving and

246

living and catching my eye, with the sky so dark behind him, and the wind blowing. God save our life.

From all the hills came screams. A piece of sky beside the crescent sun was detaching. It was a loosened circle of evening sky, suddenly lighted from the back. It was an abrupt black body out of nowhere; it was a flat disk; it was almost over the sun. That is when there were screams. At once this disk of sky slid over the sun like a lid. The sky snapped over the sun like a lens cover. The hatch in the brain slammed. Abruptly it was dark night, on the land and in the sky. In the night sky was a tiny ring of light. The hole where the sun belongs is very small. A thin ring of light marked its place. There was no sound. The eyes dried, the arteries drained, the lungs hushed. There was no world. We were the world's dead people rotating and orbiting around and around, embedded in the planet's crust, while the earth rolled down. Our minds were light-years distant, forgetful of almost everything. Only an extraordinary act of will could recall to us our former, living selves and our contexts in matter and time. We had, it seems, loved the planet and loved our lives, but could no longer remember the way of them. We got the light wrong. In the sky was something that should not be there. In the black sky was a ring of light. It was a thin ring, an old, thin silver wedding band, an old, worn ring. It was an old wedding band in the sky, or a morsel of bone. There were stars. It was all over.

### III

It is now that the temptation is strongest to leave these regions. We have seen enough; let's go. Why burn our hands any more than we have to? But two years have passed; the price of gold has risen. I return to the same buried alluvial beds and pick through the strata again.

I saw, early in the morning, the sun diminish against a backdrop of sky. I saw a circular piece of that sky appear, suddenly detached, blackened, and backlighted; from nowhere it came and overlapped the sun. It did not look like the moon. It was enormous and black. If I had not read that it was the moon, I could have seen the sight a hundred times and never thought of the moon once. (If, however, I had not read that it was the moon — if, like most of the world's people throughout time, I had simply glanced up and seen this thing — then I doubtless would not have speculated much, but would have, like Emperor Louis of Bavaria in 840, simply died of fright on the spot.) It did not look like a dragon, although it looked more like a dragon than the moon. It looked like a lens cover, or the lid of a pot. It materialized out of thin air — black, and flat, and sliding, outlined in flame.

Seeing this black body was like seeing a mushroom cloud. The heart screeched. The meaning of the sight overwhelmed its fascination. It obliterated meaning itself. If you were to glance out one day and see a row of mushroom clouds rising on the horizon, you would know at once that what you were seeing, remarkable as it was, was intrinsically not worth remarking. No use running to tell anyone. Significant as it was, it did not matter a whit. For what is significance? It is significance for people. No people, no significance. This is all I have to tell you.

In the deeps are the violence and terror of which psychology has warned us. But if you ride these monsters deeper down, if you drop with them farther over the world's rim, you find what our sciences cannot locate or name, the substrate, the ocean or matrix or ether which buoys the rest, which gives goodness its power for good, and evil its power for evil, the unified field: our complex and inexplicable caring for each other, and for our life together here. This is given. It is not learned.

The world which lay under darkness and stillness following the closing of the lid was not the world we know. The event was over. Its devastation lay round about us. The clamoring mind and heart stilled, almost indifferent, certainly disembodied, frail, and exhausted. The hills were hushed, obliterated. Up in the sky, like a crater from some distant cataclysm, was a hollow ring.

You have seen photographs of the sun taken during a total eclipse. The corona fills the print. All of those photographs were taken through telescopes. The lenses of telescopes and cameras can no more cover the breadth and scale of the visual array than language can cover the breadth and simultaneity of internal experience. Lenses enlarge the sight, omit its context, and make of it a pretty and sensible picture, like something on a Christmas card. I assure you, if you send any shepherds a Christmas card on which is printed a three-by-three photograph of the angel of the Lord, the glory of the Lord, and a multitude of the heavenly host, they will not be sore afraid. More fearsome things can come in envelopes. More moving photographs than those of the sun's corona can appear in magazines. But I pray you will never see anything more awful in the sky.

You see the wide world swaddled in darkness; you see a vast breadth of hilly land, and an enormous, distant, blackened valley; you see towns' lights, a river's path, and blurred portions of your hat and scarf; you see your husband's face looking like an early black-and-white film; and you see a sprawl of black sky and blue sky together, with unfamiliar stars in it, some barely visible bands of cloud, and over there, a small white ring. The ring is as small as one goose in a flock of migrating geese — if you happen to notice a flock of migrating geese. It is one 360th part of the visible sky. The sun we see is less than half the diameter of a dime held at arm's length.

The Crab Nebula, in the constellation Taurus, looks, through binoculars, like a smoke ring. It is a star in the process of exploding. Light from its explosion first reached the earth in 1054; it was a supernova then, and so bright it shone in the daytime. Now it is not so bright, but it is still exploding. It expands at the rate of seventy million miles a day. It is interesting to look through binoculars at something expanding seventy million miles a day. It does not budge. Its apparent size does not increase. Photographs of the Crab Nebula taken fifteen years ago seem identical to photographs of it taken yesterday. Some lichens are similar. Botanists have measured some ordinary lichens twice, at fifty-year intervals, without detecting any growth at all. And yet their cells divide; they live.

The small ring of light was like these things — like a ridiculous lichen up in the sky, like a perfectly still explosion 4,200 light-years away: it was interesting, and lovely, and in witless motion, and it had nothing to do with anything.

It had nothing to do with anything. The sun was too small, and too cold, and too far away, to keep the world alive. The white ring was not enough. It was feeble and worthless. It was as useless as a memory; it was as off kilter and hollow and wretched as a memory.

When you try your hardest to recall someone's face, or the look of a place, you see in your mind's eye some vague and terrible sight such as this. It is dark; it is insubstantial; it is all wrong.

The white ring and the saturated darkness made the earth and the sky look as they must look in the memories of the careless dead. What I saw, what I seemed to be standing in, was all the wrecked light that the memories of the dead could shed upon the living world. We had all died in our boots on the hilltops of Yakima, and were alone in eternity. Empty space stoppered our eyes and mouths; we cared for nothing. We remembered our living days wrong. With great effort we had remembered some sort of circular light in the sky — but only the outline. Oh, and then the orchard trees withered, the ground froze, the glaciers slid down the valleys and overlapped the towns. If there had ever been people on earth, nobody knew it. The dead had forgotten those they had loved. The dead were parted one from the other and could no longer remember the faces and lands they had loved in the light. They seemed to stand on darkened hilltops, looking down.

## IV

We teach our children one thing only, as we were taught: to wake up. We teach our children to look alive there, to join by words and activities the life of human culture on the planet's crust. As adults we are almost all adept at waking up. We have so mastered the transition we have forgotten we ever learned it. Yet it is a transition we make a hundred times a day, as, like so many will-less dolphins, we plunge and surface, lapse and emerge. We live half our waking lives and all of our sleeping lives in some private, useless, and insensible waters we never mention or recall. Useless, I say. Valueless, I might add — until someone hauls their wealth up to the surface and into the wide-awake city, in a form that people can use.

I do not know how we got to the restaurant. Like Roethke, "I take my waking slow." Gradually I seemed more or less alive, and already forgetful. It was now almost nine in the morning. It was the day of a solar eclipse in central Washington, and a fine adventure for everyone. The sky was clear; there was a fresh breeze out of the north.

The restaurant was a roadside place with tables and booths. The other eclipse-watchers were there. From our booth we could see their cars' California license plates, their University of Washington parking stickers. Inside the restaurant we were all eating eggs or waffles; people were fairly shouting and exchanging enthusiasms, like fans after a World Series game. Did you see . . . ? Did you see . . . ? Then somebody said something which knocked me for a loop.

A college student, a boy in a blue parka who carried a Hasselblad, said to us, "Did you see that little white ring? It looked like a Life Saver. It looked like a Life Saver up in the sky."

And so it did. The boy spoke well. He was a walking alarm clock. I myself had at that time no access to such a word. He could write a sentence, and I could not. I grabbed that Life Saver and rode it to the surface. And I had to laugh. I had been dumbstruck on the Euphrates River, I had been dead and gone and grieving, all over the sight of something which, if you could claw your way up to that level, you would grant looked very much like a Life Saver. It was good to be back among people so clever; it was good to have all the world's words at the mind's disposal, so the mind could begin its task. All those things for which we have no words are lost. The mind — the culture — has two little tools, grammar and lexicon: a decorated sand bucket and a matching shovel. With these we bluster about the continents and do all the world's work. With these we try to save our very lives.

There are a few more things to tell from this level, the level of the restaurant. One is the old joke about breakfast. "It can never be satisfied, the mind, never." Wallace Stevens wrote that, and in the long run he was right. The mind wants to live forever, or to learn a very good reason why not. The mind wants the world to return its love, or its awareness; the mind wants to know all the world, and all eternity, and God. The mind's sidekick, however, will settle for two eggs over easy.

The dear, stupid body is as easily satisfied as a spaniel. And, incredibly, the simple spaniel can lure the brawling mind to its dish. It is everlastingly funny that the proud, metaphysically ambitious, clamoring mind will hush if you give it an egg.

Further: while the mind reels in deep space, while the mind grieves or fears or exults, the workaday senses, in ignorance or idiocy, like so many computer terminals printing out market prices while the world blows up, still transcribe their little data and transmit them to the warehouse in the skull. Later, under the tranquilizing influence of fried eggs, the mind can sort through this data. The restaurant was a halfway house, a decompression chamber. There I remembered a few things more.

The deepest, and most terrifying, was this: I have said that I heard screams. (I have since read that screaming, with hysteria, is a common reaction even to expected total eclipses.) People on all the hillsides, including, I think, myself, screamed when the black body of the moon detached from the sky and rolled over the sun. But something else was happening at that same instant, and it was this, I believe, which made us scream.

The second before the sun went out we saw a wall of dark shadow come speeding at us. We no sooner saw it than it was upon us, like thunder. It roared up the valley. It slammed our hill and knocked us out. It was the monstrous swift shadow cone of the moon. I have since read that this wave of shadow moves 1,800 miles an hour. Language can give no sense of this sort of speed — 1,800 miles an hour. It was 195 miles wide. No end was in sight — you saw only the edge. It rolled at you across the land at 1,800 miles an hour, hauling darkness like plague behind it. Seeing it, and knowing it was coming straight for you, was like feeling a slug of anesthetic shoot up your arm. If you think very fast, you may have time to think, "Soon it will hit my brain." You can feel the deadness race up your arm; you can feel the appalling, inhuman speed of your own blood. We saw the wall of shadow coming, and screamed before it hit.

This was the universe about which we have read so much and never before felt: the universe as a clockwork of loose spheres flung at stupefying, unauthorized speeds. How could anything moving so fast not crash, not veer from its orbit amok like a car out of control on a turn?

Less than two minutes later, when the sun emerged, the trailing edge of the shadow cone sped away. It coursed down our hill and raced eastward over the plain, faster than the eye could believe; it swept over the plain and dropped over the planet's rim in a twinkling. It had clobbered us, and now it roared away. We blinked in the light. It was as though an enormous, loping god in the sky had reached down and slapped the earth's face.

Something else, something more ordinary, came back to me along about the third cup of coffee. During the moments of totality, it was so dark that drivers on the highway below turned on their cars' headlights. We could see the highway's route as a strand of lights. It was bumper-to-bumper down there. It was eight-fifteen in the morning, Monday morning, and people were driving into Yakima to work. That it was as dark as night, and eerie as hell, an hour after dawn, apparently meant that in order to *see* to drive to work, people had to use their headlights. Four or five cars pulled off the road. The rest, in a line at least five miles long, drove to town. The highway ran between hills; the people could not have seen any of the eclipsed sun at all. Yakima will have another total eclipse in 2086. Perhaps, in 2086, businesses will give their employees an hour off.

From the restaurant we drove back to the coast. The highway crossing the Cascades range was open. We drove over the mountain like old pros. We joined our places on the planet's thin crust; it held. For the time being, we were home free.

Early that morning at six, when we had checked out, the six bald men were sitting on folding chairs in the dim hotel lobby. The television was on. Most of them were awake. You might drown in your own spittle, God knows, at any time; you might wake up dead in a small hotel, a cabbage head watching TV while snows pile up in the passes, watching TV while the chili peppers smile and the moon passes over the sun and nothing changes and nothing is learned because you have lost your bucket and shovel and no longer care. What if you regain the surface and open your sack and find, instead of treasure, a beast which jumps at you? Or you may not come back at all. The winches may jam, the scaffolding buckle, the air conditioning collapse. You may glance up one day and see by your headlamp the canary keeled over in its cage. You may reach into a cranny for pearls and touch a moray eel. You yank on your rope; it is too late.

Apparently people share a sense of these hazards, for when the total eclipse ended, an odd thing happened.

When the sun appeared as a blinding bead on the ring's side, the eclipse was over. The black lens cover appeared again, backlighted, and slid away. At once the yellow light made the sky blue again; the black lid dissolved and vanished. The real world began there. I remember now: we all hurried away. We were born and bored at a stroke. We rushed down the hill. We found our car; we saw the other people streaming down the hillsides; we joined the highway traffic and drove away.

We never looked back. It was a general vamoose, and an odd one, for when we left the hill, the sun was still partially eclipsed — a sight rare enough, and one which, in itself, we would probably have driven five hours to see. But enough is enough. One turns at last even from glory itself with a sigh of relief. From the depths of mystery, and even from the heights of splendor, we bounce back and hurry for the latitudes of home.

# CARL SAGAN AND ANN DRUYAN

∎

*Comet*, by the astronomer Carl Sagan (b. 1934) and his wife, the novelist Ann Druyan (b. 1949), was the most elegant of the many books published to celebrate the 1986 return of Halley's comet. In this chapter Sagan and Druyan describe the passage of a comet through the inner solar system, from the perspective of the comet itself.

# *Astride the Comet*

THESE ARE THE SNOWS of yesteryear, the pristine remnants of the origin of the solar system, waiting frozen in the interstellar dark. Out here trillions of orbiting snowbanks and icebergs are stored, gently suspended about the Sun. They cruise no faster than a small propeller-driven aircraft would, buzzing through the blue skies of far-off Earth. The slowness of their motion just balances the gravity of the distant Sun, and, poised between feeble contending forces, they take millions of years to complete one orbit around that yellow point of light. Out here you are a third of the way to the nearest star. Or rather, to the next nearest star: In the depth and utter blackness of the dark sky around you, it is entirely clear that the Sun is one of the stars. It is not even the brightest star in the sky. Sirius is brighter, and Canopus. If there are planets circling the star called the Sun, there is no hint of them from this remote vantage point.

These trillions of floating icebergs fill an immense volume of space; the nearest one is three billion kilometers away from you, about the distance of the Earth from Uranus. There are many icebergs, but the space they fill, a thick shell surrounding the Sun, is incomprehensibly vast. Most of them have been out here since the solar system began, quarantined from whatever mischief may be going on down there, in that alien and hostile region bordering the Sun.

Beyond the occasional soft ping of a cosmic ray from some collapsed star at the other end of the Milky Way, hardly anything ever happens here. It is very peaceful. But something *has* happened, a gravitational intrusion, not by the Sun or its possible planets, but by another star. It was slow in coming, and at its closest it was never very near. You can see it over there, glowing faintly red, much dimmer than the Sun. This cloud of icebergs has been carried with the Sun on its motions through the Milky Way Galaxy. But other stars have their own characteristic motions, and sometimes by accident approach us. So on occasion, as now, there is a little gravitational rumbling, and the cloud trembles.

Since your iceberg is bound so weakly to the Sun, even a little push or tug is enough to throw it onto some new trajectory. The neighboring icebergs — much too small and distant for you to see directly — have been similarly affected, and are now hurrying off in many directions. Some have been shaken loose from the gravitational shackles that had bound them to the Sun, and are now liberated from their ancient servitude, embarking on odysseys into the vast spaces between the stars. But for your iceberg there is a different destiny working itself out: You have been tugged in such a way that you are now falling, slowly at first, but with gradually increasing speed — down, down, down toward the point of light about which this vast collection of little worlds slowly revolves.

Imagine that you are as patient and long-lived as the iceberg on which you are standing, that you have adequate life expectancy and life-support equipment for a journey of a few million years. You are falling toward the bright yellow star. Your worldlet and its brethren have been given a name. Comets, they are called. Your comet is an emissary from the kingdom of ice to the infernal realm near the Sun.

Out here a comet is only an iceberg. Later on, the iceberg will be just one part of the comet, called the nucleus. A typical cometary nucleus is a few kilometers across. Its surface area is the size of a small city. If you were standing on it you would see the smoothly curving contours of gracefully sculpted hillocks built of dark, reddish-brown ice. There is no air on this small world, nothing liquid, and — apart from you yourself — nothing alive, at least so far as you can see. You can, over the following millions of years, explore every corner, every mountain, every crevice. With the skies perfectly clear, and with no particularly urgent tasks before you, you can also spend a little time studying the magnificent array of unwinking bright stars that surrounds you.

Your footprints are deep, because the snow beneath your feet is weak. In a few places there are patches of ground so fragile that were you to walk upon them imprudently you would fall through — as in the legendary quicksand of Earth — onto deep shadowed ice, perhaps meters below. Your fall would be slow, though, almost languorous, because the acceleration caused by gravity, the downward force you feel here on the comet, is only a few thousandths of a percent of the familiar 1 g of Earth.

On more solid ground, the low gravity might tempt you into unprecedented athletic feats. But you must be careful. If you so much as stride purposefully, you walk off the comet altogether. With only a little effort, from a standing position you leap thirty kilometers into space, taking almost a week to reach the peak of your trajectory. There, gently tumbling, you have a comprehensive view of the comet slowly rotating beneath you, its axis by accident almost pointing toward the Sun. You can make out its lumpy shape; the comet is far from a perfect sphere. Perhaps you worry that you have jumped too high, that you will not fall back to the comet, that you will drift alone through space forever. But no, you see that your outbound velocity is gradually diminishing, and eventually, ten or twelve days after making this modest exertion, you tumble lightly back onto the somber snows. On this world you are dangerously strong.

Since it is hard to take a step without launching yourself on a small parabolic trajectory, team sports would be played in agonizing slow motion, the cluster of players rising and spinning in the space surrounding the comet like a swarm of gnats sizing up a grapefruit. A game of baseball would take years to complete — which is just as well, since you have a million years or so to idle away. But the ground rules would be unorthodox.

You pack the snow into an odd dark snowball and easily fling it off the comet, never to return again. With a flick of the wrist, without even engaging your arm in the throwing motion, you have launched a new comet on its own long, falling trajectory into the inner solar system. High above the equator of your comet, you can lightly pitch a snowball so that it hovers forever above the same point on the surface. You can make arrays of objects stationary in space, vast, apparently motionless three-dimensional assemblages, poised above the cometary surface.

As the millennia pass, you cannot help but notice that the yellow star is gradually growing more intense, until it has become by far the brightest star in the sky. The early phase of your voyage has been tedious, even if you are endowed with heroic patience, and in several million years hardly anything has happened. But you can at least see your surroundings more clearly now. The icy ground beneath you has hardly changed at all. The journey has been so long that you have been able to detect variations in the positions and, you think, even the brightness of many nearby stars. Your world is moving faster now, but otherwise everything is still, silent, cold, dark, changeless.

The comet eventually begins crossing the orbits of other kinds of objects, much larger bodies that are also bound in gravitational thrall to that beckoning point of light. As you pass close by them, you career perceptibly. Their gravities are so large that they retain massive atmospheres. Your comet, by contrast, is so insubstantial that any puff of gas released escapes almost instantly to space. Accompanying the giant gas planets with their multicolored clouds is a retinue of smaller, airless worlds, some of them made of ice — much more kin to the comets than the huge ball of hydrogen that fills your sky.

You can feel the warmth of the Sun increasing. The comet feels it too. Little patches of snow are becoming agitated, frothy, unstable. Grains of dust are being levitated over the patches. Considering the feeble gravity, it is no surprise that even gentle puffs of gas send grains of ice and dust swirling skyward. A powerful jet gushes up from the ground and a fountain of fine particles is launched far above you. The ice crystals sparkle prettily in the sunlight. After a while the ground becomes covered with a light snow. As you plunge onward, closer to the Sun, its disk now easily visible, such blowoffs become more frequent. While on one of your excursions aloft, you chance to see an active jet, a geyser pouring out of the ground. You give it wide berth. But it reminds you of the instability — the literal volatility — of this tiny world.

Far out in space, the vanguards and outriders of the columns of crystals are being blown back by some invisible influence. Eventually the cometary nucleus on which you have been riding is enveloped in a cloud of dust particles, ice crystals and gas, and the material being blown back behind you slowly forms an immense but graceful tail. If you stand on solid ground, far from the kinds of unstable ices that produce the geysers, you can still see a fairly clear sky, and track your motion by the stars. When the big jets go off, you can feel the ground shift. Here and there the ice has sheared or cracked or fallen, revealing intricately stratified layers of various colors and darkness — a historical record of the building of the comet from interstellar debris billions of years ago. By looking at the stars, you can tell that your worldlet is darting a little, rebounding in the opposite direction every time a new gusher erupts. The fountains of fine particles cast diffuse shadows on the ground, and there are now a sufficient number of them — most still of modest dimensions — that the field of darkish ice has taken on a mottled, dappled appearance.

The fine, icy grains evaporate in a few moments when heated by the increasingly fearsome sunlight, and only the dark grains that they contain will persist as solids. The substance of the comet is being converted into gas before your eyes. And the gas, illuminated by sunlight, is glowing eerily. You realize that there is now not just one tail, but several. There are straight blue tails of gas, and curving yellow tails of dust. No matter in what direction the jet happens to gush at the beginning, the unseen hand carries it away from the Sun. As the jets turn on and off, and the streamers above you curve because of the rotation of the comet, a rococo skywriting takes form. But everything aloft is relentlessly redirected by the invisible forces, the pressure of light and the wind from the Sun. The solar wind seems intermittent. So the gradually burgeoning tails form, merge, separate, and dissipate, and knots of higher brightness abruptly accelerate and then decelerate leeward from the Sun. And windward, sheets of gas and fine particles form complex and exquisite veils that change their aspect in the twinkling of an eye. This is a kind of polar fairyland, and its beauty momentarily distracts you from how dangerous it has become.

Because of the evaporation of so much ice, the ground near the exhausted gushers is friable, delicate, fragile, often no more than a matrix of fine particles stuck together billions of years ago. Soon hills of snow that had stolidly resisted the importuning sunlight for eons show signs of stirring. There is an internal motion. The ground buckles. Tentative puffs of gas are released to space, then many geysers simultaneously erupt, and you know that nowhere on the entire surface of this cometary nucleus is there a safe refuge. The comet has awakened from a four-billion-year-long trance into a wild and manic frenzy.

Later, after you pass the Sun and retreat into the interstellar night, the comet will lose its tail, and settle down. Its orbit will one day carry it back again to the inner solar system. Perhaps in some future pass by the Sun, millions of years hence, the surface will be safer because all the outer layers of ice will already have been vaporized by the heat. Only dusty and rocky stuff will be left. After many passages, a comet becomes less active, produces fewer jets, generates a less spectacular tail. As they get older, they settle down. But you are aboard a new comet, and swirling geysers are spraying the skies with dust. Maiden voyages are always the most dangerous.

You are venturing still closer to the Sun, and although the sky above you is overcast, the temperature is rising. But the surrounding haze of fine, bright particles that diffuses the sunlight also reflects it back to space, and evaporating the ice uses energy that otherwise would go into heat. If not for this protection the cometary nucleus and you yourself might become dangerously hot.

You are rounding the Sun now, racing, hurtling through this treacherous regime. You have never moved this fast before. The ground is creaking and straining, new fountains are violently erupting. You take refuge at the shadowed base of a hillock of ice whose sunward side is crumbling, evaporating, and shooting pieces of itself out into space. But eventually the activity subsides, and the skies partially clear. Formerly the Sun was ahead of you; now it is behind, on the other side of the nucleus.

Through a break in the surrounding nebulosity you realize that you are passing close to a small, blue world with white clouds and a single battered moon. It is the Earth in an unknown epoch. There may be beings there who will look up and see this apparition in their skies, who will note the great blue and yellow tails streaming away from the Sun, the complex pinwheel-shaped patterns of fine-grained material being jetted off into space, and they will wonder what it means. Some of them might even wonder what it is.

It is surprising how closely you are passing, and it occurs to you that sooner or later some comet is going to run smack into this little planet. The Earth would survive such an impact, of course — although there would doubtless be minor attendant changes, some species of life having failed, perhaps, and others newly promoted. But the comet would not survive. It would fall deeper and deeper into that atmosphere, large fragments, whole hills of the nucleus separating off, flames licking through the crevices into its hidden interior. Perhaps enough of the comet would survive to make a huge explosion, generating a large hole in the ground down there, and spraying up a cloud of surface dust. But of the comet itself, all the ices would have vaporized. All that would be left would be a sprinkling of fine, dark grains scattered like birdseed or buckshot upon this alien land.

But, you reassure yourself, running into a world is an unlikely event. On this swing past the Sun, at any rate, you will not collide with anything larger than occasional motes of interplanetary dust, the remnants of past comets that have spent their substance dashing through the realm of fire. With a last glance at the blue world, you silently wish its inhabitants well. Perhaps they will consider their skies comparatively drab and cheerless once your comet has departed. For yourself, you are relieved to be on the return trajectory, out of the deadly heat and light, and heading back to the placid cold and dark, where, except for an occasional unlucky jostling by a passing star, comets can live forever.

The long and lovely tails now precede you on your journey; the wind from the Sun is at your back.

# RICHARD MULLER

■

Richard Muller (b. 1944) is an experimental physicist who serves on the faculties of the University of California, Berkeley, and Lawrence Berkeley Laboratory. Like his mentor Luis Alvarez, he has taken an interest in a great variety of questions, from particle physics to geology to astronomy. The following passages comprise the first and final chapters of Muller's book *Nemesis*, which discusses the hypothesis that mass extinctions of living species on Earth have been triggered by the periodic approach of a dark star to the sun.

# *Cosmic Terrorist*

LUIS ALVAREZ WALKED into my office looking as though he was ready for a fight. "Rich, I just got a crazy paper from Raup and Sepkoski. They say that great catastrophes occur on the Earth *every* 26 million years, like clockwork. It's ridiculous."

I recognized the names of the two respected paleontologists. Their claim *did* sound absurd. It was either that, or revolutionary, and one recent revolution had been enough. Four years earlier, in 1979, Alvarez had discovered what had killed the dinosaurs. Working with his son Walter, a geologist, and Frank Asaro and Helen Michel, two nuclear chemists, he had shown that the extinction had been triggered 65 million years ago by an asteroid crashing into the Earth. Many paleontologists had initially paid no attention to this work, and one had publicly dismissed Alvarez as a "nut," regardless of his Nobel Prize in physics. Now, it seemed, the nuts were sending their theories to Alvarez.

"I've written them a letter pointing out their mistakes," Alvarez continued. "Would you look it over before I mail it?"

It sounded like a modest request, but I knew better. Alvarez expected a lot. He wanted me to study the "crazy paper," understand it in detail, and then do the same with his letter. He wanted each of his calculations redone from scratch. It would be a time-consuming and tedious task, but I couldn't turn him down. He and I depended on each other for this kind of work. We knew we could trust each other to do a thorough job. Moreover, we had enough mutual respect so that we didn't mind looking foolish to each other, although neither of us liked looking foolish to the outside world. So I reluctantly accepted the task, as I had many times before.

The Alvarez theory had slowly been gaining acceptance in the scientific world. The astronomers had been the most receptive, perhaps because their photographs often showed large asteroids and comets floating around in space in orbits that crossed the path of the Earth. They knew that disastrous impacts must have taken place frequently in the past. Many geologists had likewise been won over. But a majority of paleontologists still seemed opposed to the theory, which was disruptive to their standard models of evolution. Alvarez took pride in the fact that some of the most respected paleontologists nevertheless liked his theory, including Stephen Jay Gould, Dale Russell, David Raup, and J. John Sepkoski.

I began my assignment by reading the paper by Raup and Sepkoski. They had collected a vast amount of data on family extinctions in the oceans, far more than had previously been assembled. That fact disturbed me; I hate to dismiss the conclusions of experts, especially conclusions based on such minute study. Their analysis showed that there were intense periods of extinctions every 26 million years. It wasn't surprising that there should be extinctions this often, but it *was* surprising that they should be so regularly spaced. Alvarez's work had shown that at least two of these extinctions were caused by asteroid impacts, the one that killed the dinosaurs at the end of the Cretaceous period, 65 million years ago, and the one that killed many land mammals at the end of the Eocene, 35–39 million years ago. (The age was uncertain because of the difficulty of dating old rocks.)

Astrophysics was a field I thought I knew; my work in it had earned me a professorship in physics at Berkeley and three prestigious national awards. But the paper beggared my understanding. I found it incredible that an asteroid would hit precisely every 26 million years. In the vastness of space, even the Earth is a very small target. An asteroid passing close to the sun has only slightly better than one chance in a billion of hitting our planet. The impacts that do occur should be randomly spaced, not evenly strung out in time.

What could make them hit on a regular schedule? Perhaps some cosmic terrorist was taking aim with an asteroid gun. Ludicrous results require ludicrous theories.

I hurried to the end of their paper, like a reader cheating on a mystery novel, to see how Raup and Sepkoski would explain the periodicity. I was disappointed to find that they had no theory, only facts. Physicists have a wry saying: "If it happens, then it must be possible." Many discoveries had been missed because scientists ignored data that didn't fit into their established mode of thinking, their paradigm, and I didn't want to fall into that trap. Maybe it would be best to review their data, I thought, and try to judge them independently of theory. On a chart, they had plotted the varying extinction rate for the last 250 million years. The big peaks in the rate were spaced 26 million years apart.

Next I turned to Alvarez's letter. He thought there were several mistakes in the way that Raup and Sepkoski had analyzed their data. Several of the apparent peaks, he argued, should be removed from the analysis because of their low statistical certainty. Likewise, both the Cretaceous and Eocene extinctions should not be considered as part of a periodic pattern, since they were due to asteroid impacts and therefore must be random in time. This had been as obvious to Alvarez as to me. With these extinctions removed, the remaining ones were so widely separated that it looked like all evidence for periodicity had vanished.

Alvarez's approach was convincing, but was it right? It was my job to be the devil's advocate, to defend the conclusions of Raup and Sepkoski. I went back to their paper and looked at the chart again. I mustn't be too skeptical, I thought. I replotted their data, substituting the conventions of physicists for those of paleontologists. I gave each extinction an uncertainty in age as well as in intensity. The new chart looked more impressive than I had expected. It was a rough version of the one shown on page 264.

I had placed the arrows at the regular 26-million-year intervals. Eight of them pointed right at extinction peaks; only two missed. The peaks certainly seemed to be evenly spaced.

*Maybe they were right.* I realized I had better reexamine Alvarez's case, and see if *it* was flawed. This job was turning out to be more fun than I had expected.

On my second reading of Alvarez's letter, I found it particularly dubious that the Cretaceous and Eocene extinctions should be excluded. How do we know that asteroids do not hit the Earth

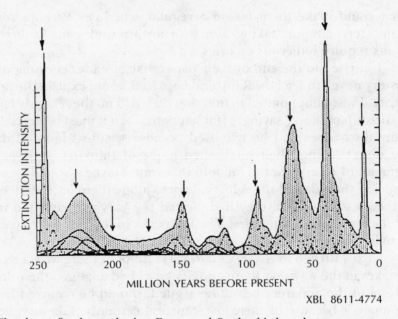

XBL 8611-4774

The data of paleontologists Raup and Sepkoski that show great mass extinctions occurring every 26 million years, as replotted by the author. This plot inspired the Nemesis theory.

periodically? I asked. Maybe our failure to arrive at a theory just meant that we hadn't been clever enough. Not finding something is not the same as proving it is not there. I decided to reserve judgment.

A few minutes later Alvarez stopped by to see if I had finished, and I told him that I had found a mistake in his logic. It had been improper to exclude the Cretaceous and Eocene mass extinctions, I said. I presented my case like a lawyer, interested in proving my client innocent, even though I wasn't totally convinced myself.

Alvarez rejoined strongly, like a lawyer himself. "To keep those extinctions in the analysis would be cheating," he said. His belligerent offense threw me momentarily off balance. "You're taking a no-think approach," he continued. "A scientist is not allowed to ignore something he knows to be true, and we know those events were due to asteroid impacts."

I knew Alvarez far too well to acquiesce in his onslaught. My approach was not no-think, I said. It was proper to ignore certain "prior knowledge" in testing a hypothesis. He had no right to assume that the Cretaceous and Eocene extinctions could not be a part of a larger periodic pattern. Maybe if we were clever enough to find the right explanation, we would see that asteroid impacts can be periodic.

Alvarez repeated his previous argument, with a little more emphasis on the phrase "no-think." His body language seemed to say, "Why doesn't Rich understand me? How can he be so dumb?" I repeated my old arguments. We were talking right past each other. He knew he was right. I knew I was right. We weren't getting anywhere. This was not a question of politics or religion or opinion. It was a question of data analysis, something all physicists should be able to agree on. Certainly Alvarez and I should be able to agree, after nearly two decades of working together.

I tried again. "Suppose someday we found a way to make an asteroid hit the Earth every 26 million years. Then wouldn't you have to admit that you were wrong, and that all the data should have been used?"

"What is your model?" he demanded. I thought he was evading my question.

"It doesn't matter! It's the possibility of such a model that makes your logic wrong, not the existence of any particular model."

There was a slight quiver in Alvarez's voice. He, too, seemed to be getting angry. "Look, Rich," he retorted, "I've been in the data-analysis business a long time, and most people consider me quite an expert. You just can't take a no-think approach and ignore something you know."

He was claiming authority! Scientists aren't allowed to do that. Hold your temper, Rich, I said to myself. Don't show him you're getting annoyed.

"The burden of proof is on you," I continued, in an artificially calm voice. "I don't have to come up with a model. Unless you can demonstrate that no such models are possible, your logic is wrong."

"How could asteroids hit the Earth periodically? What is your model?" he demanded again. My frustration brought me close to the breaking point. Why couldn't Alvarez understand what I was saying? He was my scientific hero. How could *he* be so stupid?

Damn it! I thought. If I have to, I'll win this argument on *his* terms. I'll invent a model. Now my adrenaline was flowing. After another moment's thought, I said: "Suppose there is a companion star that orbits the sun. Every 26 million years it comes close to the Earth and does something, I'm not sure what, but it makes asteroids hit the Earth. Maybe it brings the asteroids with it."

I was surprised by Alvarez's thoughtful silence. He seemed to be taking the idea seriously and mentally checking to see if there was anything wrong with it. His anger had disappeared.

Finally he said, "You surprised me, Rich. I was sure you would come up with a model that brought in dust or rocks from *outside* the solar system, and then I was going to hit you with a fact I bet you didn't know, that the iridium layer associated with the disappearance of the dinosaurs came from *within* our own solar system. The rhenium-187/rhenium-185 ratio in the boundary clay is the same as in the Earth's crust. I figured that you didn't know this. But your companion star was presumably born along with the sun, and so it would have the same isotope ratios as the sun. The argument I was holding in reserve is no good. Nice going."

Alvarez paused. He had been trying to think a step ahead of me, anticipating my moves, like a chess master. He had guessed what my criticism would be and had his answer ready — but I had made a different move. He seemed pleased that his former student could surprise him. He finally said, "I think that your orbit would be too big. The companion would be pulled away by the gravity of other nearby stars."

I hadn't expected the argument to cool down so suddenly. We were back to discussing physics, not authority or logic. I hadn't meant my model to be taken that seriously, although I had felt that my point would be made if the model could withstand assault for at least a few minutes. He was right that I was ignorant of the rhenium discovery. Alvarez's son Walter, a geologist, had found a clay layer that had been deposited in the oceans precisely at the time the dinosaurs were destroyed. This clay layer, the elder Alvarez hypothesized, had been created by the impact of an extraterrestrial body (such as a comet or an asteroid) on the Earth. Rhenium comes in several forms — among others, rhenium-185, which is stable, and rhenium-187, which is radioactive and disappears with a half-life of 40 billion years. In the 4.5 billion years since the formation of the solar system, approximately 8% of the rhenium-187 should have disintegrated. And, in fact, roughly that amount had. Unless the rhenium in the clay had been produced at the same time as the rhenium in the Earth (i.e., at the formation of the solar system), the ratios were very unlikely to be so nearly identical. In other words, the extraterrestrial body would appear to have been born at the same time as the sun.

Now I took the initiative. "Let's see if you are right that the star would be pulled away from the sun. We can calculate how big the orbit would be." I wrote Kepler's laws of gravitational motion on the blackboard. The major diameter of an elliptical orbit is the period of the orbit, in this case 26 million years, raised to the 2/3 power, and multiplied by 2. My Hewlett-Packard 11C pocket calculator quickly yielded the answer: 176,000 astronomical units, i.e., 176,000 times as far as the distance from the Earth to the sun, about 2.8 light-years. (A light-year is the distance that light travels in one year.) That put the companion star close enough to the sun so it would not get pulled away by other stars. Alvarez nodded. The theory had survived five minutes, so far.

"It looks good to me. I won't mail my letter." Alvarez's turna-round was as abrupt as his argument had been fierce. He had switched sides so quickly that I couldn't tell whether I had won the argument or not. It was my turn to say something nice to him, but he spoke first. "Let's call Raup and Sepkoski and tell them that you found a model that explains their data."

So was born the Nemesis hypothesis, though I had no idea at the time where this would lead me. . . .

# I Think I See It

"WHICH ONE IS Nemesis, Daddy?" My seven-year-old daughter was lying in her sleeping bag staring at the stars sparkling in the sky above. We were in the midst of a four-day backpack trip in the Sierra Nevada, just outside Yosemite. "I don't know, Betsy. We haven't found it yet. But we're pretty sure it's not one of the brighter stars." Betsy was undaunted. Suddenly she said, "I think I see it!"

I wish it were that easy. Eighteen months earlier, I had been overly optimistic and had predicted that it would take three months to find the star. "It is a scientist's *duty* to be optimistic," Edwin Land was fond of saying. The search is taking years, not months, but our enthusiasm and optimism haven't waned. We have already ruled out all of the naked-eye stars, Betsy notwithstanding.

Nemesis, if we are right, is currently as lost as a needle in a haystack among a million brighter stars. But when you find a needle, you can tell that it ain't hay. If we knew which one it was, we could see it through binoculars. With a small telescope its distance from the sun and its orbit could easily be measured, once we knew which one it was. Some skeptics say the Nemesis theory is pure speculation. They won't pay attention until we have found the star. I admit that I can't prove for sure that Nemesis is out there, but I think the odds make it a very good bet, good enough to bet several years of my career. First we will finish our star survey of the Northern and Southern Hemispheres. If the mass of Nemesis is greater than about 1/20 of the sun's mass, we will find it. Soon the Hipparcos satellite will be launched into space with the capability of making a sweeping survey of nearby stars and looking for smaller stars. If Nemesis is not found by this satellite, it will be time to look for another theory. But I am hoping that we will not need the help of Hipparcos.

What *do* we know for sure? We know that an extraterrestrial object, either a comet or an asteroid, hit the Earth 65 million years ago and brought to an end the great Cretaceous period of the dinosaurs. I think this conclusion is as firmly established as any theory can be after half a decade. I can say this freely because my contribution to this early work was negligible. I have watched other great discoveries go from controversy to acceptance, and other great discoveries go from controversy to retraction. This one will not go away. In the words of a court of law, I believe that the case has been proven beyond a reasonable doubt.

I also believe that the Earth is subjected to periodic storms of comets or asteroids. The important discovery of periodic mass extinctions by Dave Raup and Jack Sepkoski lies on firm and careful analysis of the data. I would not claim, however, that this conclusion is as solidly established as the Alvarez impact discovery. It is conceivable that we were just very unlucky, and Nature happened to cause mass extinctions in a way that just *looked* periodic. But I don't think so. The periodic extinctions, and the periodic cratering that goes along with them, are firmly established, to my mind. To prove a case in a civil court it is unnecessary to prove it beyond a reasonable doubt; it is only necessary to show that the preponderance of the evidence supports the case. I believe that this is true for the periodic comet storms.

Nemesis is in a different category. It is a beautiful and simple solution to the mystery of the periodic mass extinctions. I also believe that it is the only theory suggested so far that is consistent with everything we know about physics, astronomy, geology, and paleontology. But the evidence for Nemesis is circumstantial, and it hasn't yet convinced the bulk of the scientific community. It is an elegant theory, a marvelous prediction, that needs verification. We had looked at rock under our feet and predicted a star. We need direct evidence, a smoking gun, a body, the star. When we find it, hundreds of astronomers around the world could verify within a week that the star we have found is part of the solar system, orbiting the sun. (More correctly, the sun and Nemesis would both be orbiting their common center of mass.) It would immediately put an end to most of the controversy that has surrounded the mass-extinction work, because unlike many prior discoveries, it is easily checked. A piece of hay will turn out to be a needle.

The discovery of Nemesis would fill in the last piece of the jigsaw puzzle. Or would it? We've been surprised before. Every time it appeared that the puzzle was complete we were suddenly led in a wonderful new direction. Walter Alvarez had become interested in "an inconspicuous layer of clay in the Apennines." Luis Alvarez had suggested that trace analysis of iridium could be used to measure sedimentation rate, but to everyone's surprise it demonstrated that an extraterrestrial impact had taken place. The clay layer was found worldwide, and analysis showed it was about 10% asteroid or comet material, the rest coming from the vaporized rock thrown up by the impact. Five mass extinctions are now known to have iridium signals. There is the original one at the end of the Cretaceous period, 65 million years ago, and another at the end of the Eocene, 35 to 39 million years ago. More recently, Digby McLaren, a geologist at the University of Ottawa, Carl Orth of the U.S. Geological Survey, and their collaborators at Los Alamos found an iridium layer at the Frasnian-Famennian boundary, 367 million years ago, at the time of a major global extinction in the late Devonian period, and a Polish team has found iridium at the Callovian-Oxfordian boundary, 163 million years old, at locations in Poland and Spain. We found iridium 90 million years old at the Bonarelli; the layer was found independently by Orth. Of course none of the new discoveries has yet been subjected to the scrutiny that has been given to the Cretaceous boundary, so there may be even more surprises awaiting us.

The studies of the effects on climate of dust thrown into the air by the impacts led to the discovery of the nuclear winter. The loop connecting mass extinctions with nuclear winter was closed when Edward Anders found soot in the boundary clay layer, suggesting that vast fire storms had been set by the impact. As Walt once expressed it, "Only gradually have we come to appreciate the appalling violence of the geologic event that produced that thin layer of clay." The attempts of Raup and Sepkoski to show that mass extinctions occur frequently led them to a surprising, and at first totally inexplicable, conclusion: that mass extinctions take place on a nearly regular 26-to-30-million-year schedule. Based on this discovery, Marc Davis, Piet Hut, and I proposed Nemesis, a companion star to the sun that triggers comet storms, a theory simultaneously proposed by David Whitmire and Albert Jackson. Our theory immediately led Walt and me to the discovery that impacts on the Earth follow the same schedule as the mass extinctions, a correlation found independently by Michael Rampino and Richard Stothers. The concept of storms of comets proved to be more general, and testable, than the Nemesis theory itself, and led to the discovery that the mass extinctions were punctuated during the several million years of their duration. A belief in comet storms led Donald Morris and me to a method of explaining some of the geomagnetic reversals.

What else could be out there, having left behind subtle clues, to challenge our observational abilities and intelligence? We know what we are looking for, but in the past such knowledge has not been

If the standards are bad, the results will be bad. Your clock depends on another clock, which, in turn, probably depends on the Arlington radio time signals, and they depend on astronomical measures of star transits, which depend on the rotation of the Earth. All measures, in the last analysis, are clearly relative.

When we measure the distances of the stars and the dimensions of our Galaxy, and when we measure the rotation period of a planet, or the life-history of a star, we must have fundamental units in terms of which we can conveniently express the measures. As a matter of convention we have, for practical purposes, adopted the rotation period of the Earth as the unit of time; that is, we express time in fractions or multiples of a day. We express the measurements of the three space coordinates, length, breadth, and thickness, in terms of a standard length; the international meter that is carefully preserved in Paris. Clocks and watches are thus based on the time of the rotation of the Earth, and foot rules and the chains of surveyors are based on copies of the international meter stick.

But we cannot always use our rulers and our clocks directly. Your wrist watch would not help much in measuring geological ages; nor would a yardstick give you the altitude of the mountains on the Moon. We must change our methods as we get out into space. I shall use most of my time tonight in explaining the astronomical methods of measuring distance. They are very simple in principle, as you will see; and if they are complex in practice, it is because we are very ambitious for accuracy and distance.

Let us suppose you desire to measure the height of the room in which you are now sitting. Probably you would use a yardstick, applying it directly to the wall. That is all right for the room, but it will not do for the stars. If you want to measure the height of a tall tree, you can either cut it down and measure its length, or climb it, with a tape measure, using the same direct method of applying the standard to the object to be measured. But you cannot cut down a star, of course, or climb it.

In the case of the tree, however, you might also back off a little way, measure as a base line your distance from the foot of the tree, and taking an engineer's surveying telescope measure the altitude, that is, measure the angle subtended by the length of the tree. Then, with the base line and the angle measured, you would have the necessary material for solving a simple triangle, and thereby computing the height of the tree. Moreover, you would introduce a new procedure, the surveyor's method of measuring distance. Using this same method, but measuring two angles instead of one, you could find the distance across a river at any point; and you could also measure the distance across space to the Moon, for the surveyor's trick of measuring triangles is the most direct way astronomers have for getting the distance to the Moon, to the other planets, and the Sun, and even to the nearer stars.

To obtain distances across a river, or across interplanetary space, an accurately measured base line is very essential. If the base line is too short, compared with the distance that is to be measured, accuracy is impossible, because the directions from the ends of the measured line to the distant object are then too nearly alike. A base line a hundred yards long is quite satisfactory to the surveyor who is measuring the width of a river, but it is hopelessly short for astronomical surveying.

To measure the distance to the Moon it is sufficient to make simultaneous observations from one observatory at Chicago and another one at Boston. Knowing the distance between the two cities, the Moon's distance in space can be computed from the angular measurements. A practical variation on this method would be to make both measures from the Chicago observatory at an interval of two hours, for during that interval of time the rotation of the Earth will have moved the Chicago observatory more than a thousand miles eastward, thus providing an ample base line for the measurement of the distance to the Moon.

There are, of course, in practice, many technical difficulties in measuring the distances to celestial bodies — difficulties arising from the effects of our own atmosphere on the apparent positions of stars and planets in the sky, and difficulties arising from the motions of the astronomical bodies in their complicated orbits, and also other troubles; but the essence of this method of measuring distance is as I have described — the determination of angular positions in the sky with reference to some fixed (or relatively fixed) objects, like the remote stars, and the determination of the base line along which the measures are made.

The yearly motion of the Earth in its orbit around the Sun is continually shifting our observing stations with respect to the planets and the stars. In fact, it is the annual orbital motion of the Earth that provides a sufficiently long base line to enable the astronomer to measure the distances to the nearest stars. In six months from now the Earth will be on the other side of the Sun, nearly two hundred million miles from its present position. Observations of the position in the sky of a nearby star made six months hence and compared with observations made tonight would show a small displacement with respect to the fainter and much more remote comparison stars. The nearer the star, the greater that relative shift will be. The amount of the displacement is the measure of distance; we call it the parallax of a star.

You can easily grasp this idea of measuring parallaxes and getting the distances of the stars by a little experiment performed as you sit reading my remarks. Close one eye, hold a pencil or one finger up in front of your face and keep your gaze fixed upon it. Now, keeping your shoulders still, rotate your head round and round. As your head moves to the left and right, your finger appears to be shifting to the right and left with respect to the background of wall or window, or whatever object you are facing. When you hold your finger near your eye, the displacement is large — the parallax is large, I might say — and if the finger is held as far as possible from your face, the shift is much smaller. The smaller the shift, that is, the smaller the parallax of your finger, the more remote it is.

In this arm chair illustration, your open eye is the observer on the Earth with his telescope, seeking to measure the universe. The wagging back and forth of your head is the yearly trip of the Earth around the Sun. Your finger is the nearby star, and the background represents the faint and very distant stars that happen to be in the same field of vision as the nearby object you are measuring.

If the finger is at a great distance from the angle-measuring eye, the displacement will be too small, and the surveyor's method will fail to give accurate results. Only nearby stars and fingers can be measured in this way. We need a longer base line for more remote stars. We cannot enlarge the orbit of the Earth so as to give us the longer base, but if we are patient we can take advantage of the fact that the Sun, Earth, and other planets are all travelling together through space at high speed, and thereby they are producing for us a base line that is growing continually longer. Astronomers make use of this general drift of the solar system to get the distance of many more stars than are available by the ordinary methods; but the modified procedure has various limitations that I will not trouble to describe.

Before leaving our simple illustration of measuring space, let us apply it also to an additional example. If someone else holds a pencil or finger while you, with one eye shut and wagging your head, drift slowly across the room, you will picture for yourself the combination of the two methods of observing parallax. The revolution of the

Earth around the Sun in a year provides the annual base line, and the drift of the planetary system affords a slowly growing base line for measuring those more remote.

The displacements of even the nearest stars are extremely small when compared with the shift you have just observed for your finger projected against the wall. The total yearly displacement for the nearest star, Alpha Centauri, is but one and a half seconds of arc, an angle which corresponds to a shift of one inch when seen from a distance of about two miles. Stars one hundred times as remote can be measured by this surveyor's method, but that is about the limit.

The orbit of the Earth seen from such a distance, one hundred times that of Alpha Centauri, would have a diameter but a little greater than that of a cent seen from the distance separating Boston and New York. This indicates to what a high degree of perfection the astronomical instruments have been carried in recent years. The work on star parallaxes by these triangulation methods is done only with large telescopes, and with highly specialized photographic apparatus.

Relatively few of the stars are within the distance of four or five hundred light years that can be surveyed by the methods I have just described. So long as we rely on the surveyor's method of measuring the distances to stars, we have a very restricted view of the depths of the universe; even when we go much farther in our methodology and in our measurement, we do not know too much about its dimensions and structure.

A revolution in our methods of sounding space began a few years ago when astronomers discovered ways of determining the actual light output, or candle power, of stars. I think it will be clear to all of you that if we know the candle power of a street lamp or of a star, and measure, with some reliable apparatus, how much of the light from the lamp or star comes to us, then we can determine the distance. The nearer we are to the object of known candle power, the brighter it will seem to be; the more distant it is in space, or in the street, the feebler it will appear. That is, for an object of known candle power, the appearance, or, as we say in astronomy, the *apparent magnitude,* is a direct indication of the distance.

For all the stars within a few hundred light years, of which we have measured the distances directly, we can compute the candle power as soon as we have measured how bright they appear to be. These nearby stars give us standards of star luminosity. Close examination of the motions, light variations, and spectral peculiarities have shown that certain characteristics often indicate the true luminosities of stars. For instance, some nearby stars with identical spectra are found also to have identical candle power. If, then, we find other stars in space, beyond the range of the surveyor's method, which have exactly this same type of spectrum, we are fairly safe in concluding that their candle powers also are the same. They appear fainter because they are more distant, but the real brightnesses are equivalent. Granting this assumption, we need only measure the apparent magnitudes of these farther stars and we readily find their distances.

Astronomers have discovered several ways of estimating candle power from the spectrum, and all the various spectroscopic methods are obviously very powerful. They permit us to estimate the distances of objects, no matter how remote, providing only that the light is strong enough for spectrum analysis. We thus transcend the range of the surveyor, and open up vast regions of space.

Another peculiarity of stars, the variation in brightness of that class known as Cepheid variables, has been valuable in the recent measurements of remote parts of the stellar universe. It has been found, mainly through the studies carried on at the Harvard and Mount Wilson Observatories, that the period of variation of these stars is an index of candle power, and therefore an index of distance, as soon as the readily obtained apparent magnitudes have also been determined.

If the time from one maximum brightness of a Cepheid variable to the next maximum brightness is but ten or twelve hours, we know that the variable is about one hundred times as bright as the Sun; it is a giant star. If the interval from one maximum to the next is five days, the variable is, on the average, about one thousand times as bright as the Sun; it is then a super-giant star, and being so bright can be seen from enormous distances. In fact the Cepheid variable stars, especially those of long period and great brightness, are among the remotest objects known to us. Fortunately such stars are frequent in the globular clusters. They are also extremely numerous in the great star clouds of Magellan, and they occur in the remote spiral nebulae, which now seem to be other galaxies of stars. Wherever these Cepheid variables appear, we need only measure the period of their light variations, and their apparent brightness, and we can then derive their distances from the Earth. If they are members of some remote cluster or star cloud, we get at the same time the distance of all their thousands of fellow members in the clouds or clusters.

The stars in the sky are so hopelessly numerous that we can never expect to find the distances of even one-thousandth part of them. The best we can do is to get the distances of special types of stars, or of all the stars in sample regions, or of objects in peculiar situations in the sky. Putting together the material which we thus collect, we hope to piece together slowly and uncertainly a rough picture of the form and dimensions of the stellar universe. It will be difficult to get the exact boundaries; and behind the dark cosmic nebulae, through which starlight does not pass, there will always remain extensive unexplored regions. The dark nebulae that hide these mysteries of the Milky Way will not dissolve or move out of the way in our time. The lives of astronomers, or even of civilizations, are far too short compared with the cosmic processes. We cannot wait to get all the material about the distribution of stars. We build and interpret with what we have.

From the scanty material now on hand, we have in recent years drawn up the familiar picture of a flattened Galaxy composed of stars, nebulae, and star clouds. The whole is disc-shaped like a watch. The ordinary surveys, however, merely permit the exploration of the neighborhood of the Sun. It is only since the new methods were introduced and developed — those methods involving candle power, the spectra of stars, and the Cepheid variables — that we have been able to explore deeply into the watch-shaped Galaxy. And then we find that the Sun and the planets are far from the center. It probably takes light more than fifty thousand years to travel from the center of the Galaxy to the region out toward one edge, where the Sun and our naked-eye stars are found. . . .

# BART J. BOK

∎

The American astronomer Bart J. Bok (1906–1983), a warm and witty man who brought great enthusiasm to his research, was devoted to trying to understand the Milky Way, the vast galaxy of stars to which the sun belongs. His accounts of the anatomy of the galaxy continue to inspire astronomers today.

# *The Milky Way Galaxy*

ON A CLEAR AND MOONLESS NIGHT free of the lights of civilization the most arresting thing in the sky is the luminous band of the Milky Way. Even without a telescope there is much one can say about it. One can see, for example, that stars become more numerous as one directs one's gaze across the sky and into the band. The fact that the band itself is made up mostly of stars then seems less astonishing. One can see that the band follows a great circle that bisects the celestial sphere. Hence the earth is embedded in the central plane of the band. One can see, moreover, that the band is widest and brightest in the direction of the constellation Sagittarius. Surely this is the direction to the center of the system. The system is the Milky Way galaxy.

A telescope will reveal that certain starlike points in the Milky Way are actually great aggregations of stars. They are plainly distant objects; they are known as globular clusters. Between 1918 and 1921 Harlow Shapley employed the telescopes of the Mount Wilson Observatory to demonstrate that the clusters, like the stars in the band, are commonest in Sagittarius. In an area of Sagittarius that makes up only 2 percent of the sky Shapley plotted a third of all the globular clusters then known. Evidently, therefore, the solar system is far from the center of the galaxy; the distance from the sun to the center is now estimated to be 8,500 parsecs. (A parsec is 3.26 light-years.) It has since been established that the solar system, along with the rest of the mass in the central plane of the galaxy, revolves about the center. The sun revolves at a rate now taken to be some 230 kilometers per second. It thus completes a revolution every 200 million years.

In 1930 Robert J. Trumpler of the Lick Observatory showed that an interstellar medium of gas and dust dims the light of the stars, particularly the stars in the central plane of the galaxy. In 1951 William W. Morgan of the Yerkes Observatory and his students Donald E. Osterbrock and Stewart L. Sharpless found evidence that the band is an edge-on view from the earth of the galaxy's spiral structure. Evidence for spiral features soon followed at radio wavelengths. In the 1960's Chia Chiao Lin of the Massachusetts Institute of Technology and Frank H. Shu, then at the Harvard College Observatory, suggested that waves of increased density in the interstellar medium precipitate the spiral features and are responsible for the formation of stars. Meanwhile the advent of infrared astronomy provided a means for exploring the interior of dark interstellar clouds. It now appears that the clouds are where most of the galaxy's new stars form. By 1970 radio astronomy had begun to reveal the composition of the dark clouds. They are made up of dust and hydrogen, with an admixture of some surprisingly complex molecules, many of them organic.

For many years I have been a night watchman of the Milky Way galaxy. I remember the mid-1970's as a time when I and my fellow watchers were notably self-assured. The broad outlines of the galaxy seemed reasonably well established. The galaxy had two main components: a dense central bulge of stars with a boundary between 4,000 and 5,000 parsecs from the center, and a flat, much thinner disk of stars and interstellar gas and dust whose inner margin abutted the bulge and whose outer margin lay some 15,000 parsecs from the center. The bulge is in Sagittarius; the disk is flung across the sky.

The combined mass of the disk and the bulge was then calculated to be well under 200 billion times the mass of the sun. The disk and the bulge were surrounded, however, by a "halo" of matter on each side of the galaxy's central plane. In overall form the halo is a slightly flattened sphere. In the plane of the galaxy it has a radius on the order of 20,000 parsecs. It is notable for old stars, and also for a scattering of perhaps 100 globular clusters. (Another 100 globular clusters lie in or near the galactic disk.) The halo might add at most 100 billion solar masses to the total mass of the galaxy.

Much work was still to be done. We felt confident nonetheless that the basic findings would stand, and that astronomers would be undistracted as they looked into such matters as how stars form and how the Milky Way's spiral features evolve. We did not suspect it would soon be necessary to revise the radius of the Milky Way upward by a factor of three or more and to increase its mass by as much as a factor of 10. The revisions are emblematic of a number of recent upheavals. Here I shall take up several aspects of the current effort to understand the Milky Way. I include most of them because they are areas of notable ferment where progress in understanding may be imminent. I include one, in contrast, because I fear it is reaching a dead end.

Suggestions that the Milky Way galaxy is unexpectedly large and massive were offered as early as 1974. On the theoretical side they came principally from Donald Lynden-Bell of the University of Cambridge and from Jeremiah P. Ostriker, P.J.E. Peebles and Amos Yahil at Princeton University. The basic argument was that the dynamical stability and permanence of the galaxy cannot be guaranteed unless the galactic disk is surrounded, and thereby stabilized gravitationally in spite of its thinness and its delicate spiral structure, by an extended and massive halo.

J. Einasto and his associates at the Tartu Observatory in Estonia came forward with a different chain of reasoning. For several years investigators looking into the dynamics of the Milky Way had argued that the velocity of the sun with respect to the globular clusters in the halo of the galaxy is only some 180 kilometers per second. Since the clusters are scattered throughout a large spherical volume, it seems plain that on the whole they do not participate in whatever rotation the galactic disk may have. The distribution of the clusters thus should be more or less stationary with respect to the center of the galaxy. Specifically, it is thought the rotational velocity of the system of globular clusters about the center of the Milky Way can be no more than 50 kilometers per second. Hence the rotational velocity of the sun about the center of the galaxy should be no more than 230 kilometers per second, and certainly no more than 250.

Meanwhile other investigators had determined the velocity of the sun with respect to the average motion of nearby galaxies, namely those whose distance from the sun is no greater than a million parsecs. Their result for the rotational velocity of the sun about the center of the Milky Way was 300 kilometers per second. The difference between the two results could best be interpreted as indicating that the center of the Milky Way galaxy is moving at a velocity of 50 to 80 kilometers per second with respect to the nearby galaxies. That velocity seemed surprisingly large.

Einasto argued that the result might nonetheless be correct and that the unsuspected massiveness of the Milky Way system might be the cause of it. To test his hypothesis he examined the motion of the Milky Way within the group of galaxies of which it is a member. Einasto proposed that there are subgroups within the group and that in one of the subgroups the Milky Way is gravitationally dominant over several other aggregations of stars, such as the two small nearby galaxies called the Large and Small Clouds of Magellan and a number of dwarf spheroidal galaxies, of which seven are now known. One of the dwarf spheroidal galaxies lies some 150,000 parsecs from the center of the Milky Way. The velocity of the sun with respect to the average motion of these galactic companions proved to be almost 300 kilometers per second. It too was surprisingly high. Einasto interpreted the velocity as an effect of the great mass of our galaxy on its nearest neighbor galaxies.

Einasto was therefore strengthened in his suspicion that the Milky Way is more extended and more massive than had been supposed. In 1976 he offered a model of the Milky Way system in which the mass of the central bulge, the disk and an extended halo is 900 billion solar masses. Even that mass is insufficient to account for the great velocities observed among the galaxy and its companions. Einasto therefore proposed that the bulge, the disk and the halo are embedded in a still larger but nonetheless unseen component of the galaxy, the corona (as he called it), which extends out to at least 100,000 parsecs from the center and has a mass of 1.2 trillion solar masses. The total mass of the Milky Way would then be 2.1 trillion solar masses, or at least seven times the value accepted in 1975.

Supporting evidence was forthcoming from several investigations. First, Vera C. Rubin, W. Kent Ford, Jr., and Norbert Thonnard of the Department of Terrestrial Magnetism of the Carnegie Institution of Washington examined the Doppler shifts of spectral lines in the light emitted by matter in the outer part of 17 galaxies. Each such shift is a displacement of a spectral line to a wavelength different from the one it would have if the source of the radiation were motionless with respect to the instrument that receives it. The investigators concluded that the outlying matter in each galaxy circles the center of the galaxy just as fast as the matter nearer the center.

To put it another way, a rotation curve — a graph of rotational velocity as a function of distance from the center of the galaxy — was essentially a horizontal line for the outer part of each galaxy they studied.

The discovery of such flat rotation curves is remarkable. After all, the distribution of brightness in the optical image of a typical spiral galaxy leads one to infer that the galaxy's visible matter is concentrated toward the center and gets sparse at the periphery. From this concentration one can deduce that the outermost visible matter ought to be moving in response to forces analogous to those acting on the outermost planets of the solar system. The peripheral matter ought to be rotating about the center of the galaxy at a lower circular velocity (expressed in kilometers per second) than the matter nearer the center. Evidently it does not. The rotation curves calculated by Rubin, Ford and Thonnard thus imply the presence of unseen matter in large quantity beyond the apparent periphery of each of the galaxies.

Studies similar to those of Rubin, Ford and Thonnard bear directly on the distribution of mass in the Milky Way. In one of them F.D.A. Hartwick of the University of Victoria and Wallace L. W. Sargent of the California Institute of Technology calculated the velocities of globular clusters at distances greater than 20,000 parsecs from the center of the galaxy. In other studies James E. Gunn, Gillian R. Knapp and Scott D. Tremaine employed data gathered at the Owens Valley Radio Observatory to determine the velocities of clouds of interstellar hydrogen atoms. Maurice P. Fitzgerald of the University of Waterloo, together with Peter D. Jackson of the University of Maryland and Anthony Moffat of the University of Montreal, determined the velocities of stars and star clusters as much as 17,000 parsecs from the center of the galaxy. William L. H. Shuter of the University of British Columbia determined the velocities of clouds of hydrogen and carbon monoxide that lie in the galactic anticenter: the direction opposite to the line of sight from the earth to the center of the galaxy. Recently Leo Blitz and his colleagues at the University of California at Berkeley have measured the Doppler shifts in both the optical and the radio parts of the electromagnetic spectrum for lines from 184 nebulas and large clouds of interstellar hydrogen and carbon monoxide in the anticenter. All the results are consistent with a rotation curve that does not fall.

From the evidence available today it seems fair to conclude that the rotation curve for the Milky Way attains a value of 230 kilometers per second at 8,500 parsecs, the distance from the galactic center that marks the position of the sun. From there the rotational velocity continues to increase. It reaches 300 kilometers per second at a distance of 20,000 parsecs.

Several objects more distant than 20,000 parsecs are visible; a list of them was published by the International Astronomical Union in 1979. Four globular clusters lie between 20,000 and 40,000 parsecs from the galactic center; the Large Cloud of Magellan and two globular clusters lie between 40,000 and 60,000 parsecs; two dwarf spheroidal galaxies and the Small Cloud of Magellan lie between 60,000 and 80,000 parsecs, and one dwarf spheroidal galaxy and three globular clusters lie between 80,000 and 100,000 parsecs. Four more dwarf spheroidal galaxies and two globular clusters lie between 100,000 and 220,000 parsecs, but their claims to membership in the corona of the Milky Way are more dubious.

Plainly our galaxy is far more extended and of much greater mass than was hitherto thought; the Milky Way has been elevated to the rank of a major spiral galaxy. Taken together, however, the visible constituents of the corona have only a tiny fraction of the hundreds of billions of solar masses that ought to be there. Apparently the Milky Way shares with the galaxies studied by Rubin, Ford and Thonnard the property that much of its outlying matter is dark. Indeed, John N. Bahcall and Raymond M. Soneira of the Institute for Advanced Study infer from the invisibility of the hypothetical matter that if there are stars in the corona that are not bound into clusters or into dwarf galaxies, their intrinsic brightness should be less than a thousandth that of the sun.

What, then, is the unseen mass? Three facts are worth noting. First, dwarf galaxies and globular clusters consist mainly of old stars. Second, old stars are not highly luminous. Third, no one has detected from the corona the spectral lines that characterize clouds of gaseous matter such as hydrogen and carbon monoxide in more central parts of the galaxy. At present, therefore, the best suggestion is that the corona of the Milky Way is composed mainly of old, burned-out stars. On the other hand, the unseen mass of the galaxy's corona may not fit any of the categories based on what can be seen in more accessible regions. We do not know yet what is out there.

It may come as a surprise that the center of the Milky Way is no less mysterious than the galactic corona. In fact it is only slightly more visible to observers on the earth. Twenty-five years ago, when observations could be made only at visible and radio wavelengths, three kinds of object were known to be plentiful in the galactic center. Each of them is old. The first is globular clusters. The second is RR Lyrae variables. These are old stars that alternately brighten and dim with a period on the order of a day. The third is planetary nebulas. Most of them are the old, collapsed stars called white dwarfs, each one surrounded by a cloud of gas that is thought to be the shed atmosphere of the star. The galactic center itself was hidden from optical view by layers and shells of dust that are estimated to dim the light from the center by as much as 30 astronomical magnitudes, or a factor of $10^{12}$. The dust is particularly effective at blotting out the blue light to which the photographic emulsions employed in earlier days of astronomical photography were often made most sensitive. Some of the dust lies in a lane at the margin of the central bulge, a few thousand parsecs from the solar system. Some of it lies only 300 parsecs from the solar system, in the constellation Ophiuchus, along the line of sight from the earth to the center. The only thing then known about the center was that it harbored a strong radio source. Electromagnetic radiation in the radio part of the spectrum is not absorbed by the dust.

Some 15 years ago the prospects began to brighten. The detection of radio emissions at a wavelength of 21 centimeters served to delineate the sources of such radiation: clouds of neutral (un-ionized) hydrogen atoms. The detection of emissions at other wavelengths revealed the presence of dark interstellar clouds consisting mostly of molecules. Above all, the advent of infrared astronomy opened a window to the center of the galaxy at wavelengths of from a micrometer to a millimeter. Infrared radiation constitutes a second band of wavelengths that the interstellar dust does not heavily absorb. One of the most useful new techniques is the detection of the infrared radiation emitted at a wavelength of 12.8 micrometers by ionized neon atoms. Because neon is a by-product of energetic events such as stellar explosions it becomes a ubiquitous constituent of interstellar clouds. The neon therefore serves as a tracer.

The observations at radio and infrared wavelengths yield four kinds of data. First, they reveal the presence of local maximums in the rain of radiation from certain directions in the sky. Second, the Doppler shift of spectral lines in the radiation reveals the radial velocity of such a source: the speed of the source either toward the solar system or away from it. Third, the broadening of a spectral line can suggest that the source is expanding or contracting. Fourth, the relative intensities of certain lines in the spectrum suggest the temperature of the source. Even when all possible information has been extracted from the radiation, however, much remains uncertain. It cannot be ascertained, for example, whether a source in approximately the direction of the center of the galaxy lies in front of the center or behind it, or whether the radial velocity of a source actually means it is rotating about the center.

Let me summarize the present state of knowledge about the central bulge of the Milky Way. In its overall shape the central bulge is a slightly flattened sphere. Its outer boundary, which lies 5,000 parsecs from the center, is marked by a ring of what are now called giant molecular complexes. They are large, dark, clumpy interstellar clouds made up mostly of hydrogen molecules. I shall be returning to them. The bulge itself consists in general of a dense clustering of old stars in a rather thin matrix of interstellar gas and dust. The stars are known from their infrared radiation, which can be distinguished from that of gas or dust. One way to account for the relative scarcity of gas and dust in the bulge is to assume that much of it placidly condensed long ago to form the stars of the bulge. On the other hand, several interstellar features in the bulge suggest that the center of the galaxy has had a complex and violent history.

The outermost detectable feature of the bulge is a ring of neutral hydrogen at a distance of 3,000 parsecs from the center. The ring was discovered in 1964 by Jan H. Oort and G. W. Rougoor of the Leiden Observatory. The Doppler shifts of the radiation it emits show it is rotating and, more important, expanding, with velocities away from the center that range from 50 to 135 kilometers per second. Perhaps the ring is a new spiral arm unfurling. One is equally tempted, however, to speculate that the center of the galaxy expelled a kind of smoke ring some 30 million years ago. It is as if there had been a titanic explosion there. Perhaps the explosion swept away much of the gas and dust in the bulge.

The next feature inward lies at a distance of about 1,500 parsecs from the center. It is construed by its discoverers, Butler Burton of the University of Minnesota and Harvey S. Liszt of the National Radio Astronomy Observatory, to be a disk of both atomic and molecular hydrogen. It too is both rotating and expanding. Surprisingly, the best way to interpret the data is to assume that the disk is tilted at an angle of 15 to 20 degrees to the plane of the galaxy.

One might hope that the composition of the central bulge would be more or less homogeneous from the Burton-Liszt ring to the center, if only to simplify the task facing those who attempt to understand the structure of the bulge. More surprises, however, await. Another smoke ring is evident some 300 parsecs from the center. This one too is a mixture of molecular complexes, dust clouds and regions of atomic and molecular hydrogen. The atomic hydrogen is ionized in places, which means it is quite hot: well over 10,000 degrees Kelvin. Associated with these hot spots are clusters of newly formed blue-white supergiant stars. Why should these realms of high temperature and star formation lie precisely in this ring? One is particularly puzzled because a cooler ring of only mildly ionized atoms at a temperature of 5,000 degrees lies a mere 10 parsecs from the center. The 10-parsec ring is rather dense, and it is rotating.

The central three parsecs of the galaxy evidently includes several million stars; they give the center the densest packing of stars in the galaxy. The core region also includes a number of compact clouds of ionized gas; a group of workers led by John H. Lacy and Charles H. Townes of the University of California at Berkeley has detected 14 of them. A typical cloud has about the mass of the sun and a diameter of a fraction of a parsec, and it is speeding around the center: it completes an orbit in some 10,000 years (compared with the sun's orbital period of some 200 million years). Luis Rodriguez and Eric J. Chaisson of the Harvard College Observatory have shown that the velocity of the ionized gas increases with proximity to the center. All of this suggests the clouds are satellites of a supermassive innermost object.

Whatever it is at the very center appears as a bright infrared source in maps made by Eric E. Becklin and Gerry Neugebauer of the California Institute of Technology. According to Bruce Balick of the University of Washington and Robert L. Brown of the National Radio Astronomy Observatory, who have studied the radio emissions of the central object, it has a diameter no greater than 10 times the distance from the earth to the sun. Its mass may be as great as 50 million solar masses. The most likely conjecture is that the very center of the Milky Way harbors a black hole created by the infalling of hundreds of thousands of stars. The center would then be in essence a stellar graveyard.

When the periphery and the center of the Milky Way have been considered, there remains a middle region, which has the solar system in its midst. It is the part of the galaxy in which spiral structure prevails.

The tracing of the spiral arms of the Milky Way began in earnest three decades ago, when Morgan, Osterbrock and Sharpless distinguished three spiral-arm segments. To trace them they plotted the positions of the blue-white supergiant stars classified on the basis of their pattern of spectral lines as *O* and *B* stars, together with the bright clouds of ionized hydrogen atoms that often surround such stars. Their diagrams show an Orion arm, of which the sun is a member; a Perseus arm, 2,000 parsecs farther from the center of the galaxy, and a Sagittarius arm, 2,000 parsecs closer to the center of the galaxy. The naming of the arms reflects the general practice in astronomy of employing the constellations to signify directions in the sky. To the three arms distinguished by Morgan and his colleagues has since been added the Carina arm, which may be a continuation of the Sagittarius arm; the concatenation is called the Sagittarius-Carina arm. The two segments that compose it meet at the Great Nebula in Carina, which enmeshes a large number of *O* and *B* stars. There is evidence for other segments as well.

Recent work by Roberta M. Humphreys of the University of Minnesota confirms that *O* and *B* stars are abundant in the principal arms that were recognized three decades ago. This is cheerful news. It means the spiral arms are indeed delineated by very hot, blue-white supergiant stars, by the clusters made up of such stars and by the bright clouds of gas in which such stars and clusters are found. *O* and *B* stars are quite young; the ones seen now in the spiral-arm segments of the Milky Way were formed no more than 10 million years ago. There is no question the spiral arms are regions of star formation.

The inner margins of most spiral arms are marked, it seems, by dark nebulas: cold clouds of atoms, molecules and dust. According to the density-wave theory of Lin and Shu, the dark nebulas signal the compression of the interstellar matter by a wave of pressure that advances through the galactic disk. The compression precipitates the condensation of the matter into the stars of the spiral segment. The young stars might be expected to lie behind the advancing density wave and dark nebulas to lie inside it. According to the theory, however, the wave moves at only two-thirds the speed of the galactic disk's rotation. Hence the wave is overtaken by the stars that formed inside it, and the loci of dark nebulas come to lie at the trailing edge of the stars.

Certain of Humphreys' results, it should be said, are less than cheerful. Some of the stars called Cepheid variables are as young as *O* and *B* stars; in particular the Cepheid variables that alternately dim and brighten with a period longer than 15 days have ages of no more than 10 million years. Moreover, they too are large and bright, and so they can be seen at great distances. Since they are young, they should not have moved far from their birthplace: presumably a dark nebula at the inner edge of a spiral arm. All things considered, the long-period Cepheid variables should be excellent delineators of spiral structure. They are not. The Cepheid variables plotted by Humphreys form an essentially random distribution of points in the galactic plane.

One thing must always be kept in mind by those who search for spiral structure in optical observations of the Milky Way. The usual method of calculating the distance to a star is to compare its observed luminosity with what its intrinsic luminosity is taken to be. The intrinsic luminosity is assigned on the basis of the pattern of lines in the spectrum of the star's radiation. The diminution of the brightness is a correlate of the distance. Even with the best techniques, however, the distance of an *O* star or a *B* star is uncertain by plus or minus 10 percent. For example, a star that is calculated to be 2,700 parsecs from the sun — say a star in the Carina nebula — could be as little as 2,400 or as much as 3,000 parsecs away. The result is a purely observational blurring of several hundred parsecs in the plotting of what might actually be a spiral feature.

With the most modern telescopes, spectrographs and photographic equipment, *O* and *B* stars can be detected, and their spectra can be examined, at distances calculated to be as great as 8,000 parsecs from the sun. At that distance, however, the uncertainty in the calculation is plus or minus 800 parsecs. In such a case a spiral feature may well be unrecognizable. There seems little prospect of tracing the spiral structure of the Milky Way by optical means to distances beyond 8,000 parsecs.

How do the efforts to trace spiral structure fare at radio wavelengths? Again the prospects are far from encouraging. By the early 1950's, when Morgan, Osterbrock and Sharpless presented the key optical features of the spiral structure, Harold I. Ewen and Edward

M. Purcell of Harvard University had detected radiation from interstellar clouds of neutral atomic hydrogen at a wavelength of 21 centimeters. Within a few years the first radio maps of the galaxy were
available. They suggested that at least in the outer parts of the galactic disk the clouds of neutral atomic hydrogen are arrayed in nearly
circular spiral features.

Some details of the way in which such data are amassed and
interpreted are worth examining. When a radio telescope detects
radiation at and near a wavelength of 21 centimeters from a small
area of the sky, it is actually receiving radiation from a number of
clouds of neutral atomic hydrogen along a single line of sight. Each
cloud has its own velocity of approach or recession with respect to
the telescope, and hence its radiation has a distinctive Doppler shift.
As a result the telescope receives from a single direction a profile of
peaks and valleys in a graph of intensity v. wavelength. The telltale
21-centimeter line is actually a set of closely spaced lines with various
intensities, various degrees of broadening and various Doppler shifts.
Since neutral atomic hydrogen is assumed to have its greatest density
at the trailing edge of spiral features, it seems reasonable to assume
that a peak of great strength in the profile should lie at the wavelength corresponding to the velocity of approach or recession of hydrogen where the line of sight crosses a spiral feature. With the aid
of a rotation curve for the Milky Way one should then be able to
determine the distance to the feature.

This chain of reasoning is simple and lovely, but it does not
stand up under scrutiny. The first complication is that the clouds of
neutral atomic hydrogen have motions of their own, quite apart
from the motion of the spiral features. These independent motions
can easily alter the total velocity of a cloud by as much as six kilometers per second. An additional complication is that the matter of
the galactic disk apparently exhibits its large-scale streamings.
Burton and his associates at the National Radio Astronomy Observatory have demonstrated that even slight motions of this kind can
give rise to peaks of intensity in the 21-centimeter radiation for directions in which the line of sight does not cross a spiral feature. Further
still, there are directions in the sky for which the velocity of approach
or recession of a cloud of gas may change only slowly with distance.
The 21-centimeter profile may then show a peak caused by contributions from the width of a single great expanse of gas of uniform
density.

Several other approaches can be made to the study of spiral

structure. One is to observe the spectral lines emitted by molecules of carbon monoxide at radio wavelengths close to 2.6 millimeters. The carbon monoxide lines give access to a class of clouds that are cooler than the ones composed of neutral atomic hydrogen. The cooler clouds are composed mostly of molecular hydrogen, with some admixture of carbon monoxide and other substances. (The general principle is that increasing temperature leads first to the dissociation of molecules into neutral atoms and then to the ionization of the atoms.) William Herbst of Wesleyan University has shown that the presence of carbon monoxide in an interstellar cloud is correlated with the presence of dust. The carbon monoxide lines therefore aid in the detection of distant dust clouds and dust complexes that might otherwise escape notice.

A low-resolution survey of carbon monoxide clouds by Richard S. Cohen and Patrick Thaddeus of the Goddard Institute for Space Studies and Thomas M. Dame of Columbia University shows evidence of the existence of spiral features. Indeed, in some instances the carbon monoxide clouds seem to fill gaps between clouds of atomic hydrogen in some of the recognized spiral features. Still, I doubt the spiral structure of the Milky Way would have been discovered if the only data available had been the radio observations at a wavelength of 2.6 millimeters.

I thoroughly dislike having to be pessimistic about the prospects for mapping the spiral structure of the Milky Way beyond a distance of about 8,000 parsecs from the sun, particularly since the structure seems firmly established up to that distance, but I see no chance for improvement in the next decade or two. This is not to say there are now no possibilities whatsoever for the study of spiral structure. Among optical telescopes the great reflectors already in place and the Space Telescope, an instrument that will orbit the earth, promise to reveal the fine details of spiral structure and the motions of small-scale features. Among radio telescopes the Westerbork Array in the Netherlands and the Very Large Array in New Mexico, which synthesize images from a set of detectors, will advance the investigation of the cold, dark clouds in spiral features and no doubt will suggest much about the causes of spiral structure. These promising developments, however, apply mainly to spiral galaxies other than our own. The spiral galaxies Messier 31, 33, 51, 81 and 101 are a few fairly nearby examples.

One realm of research that is flourishing in both galactic and extragalactic astronomy is the examination of globular clusters. The clusters are important in part because they seem to be the oldest objects in the Milky Way. They therefore hint at the birth and evolution of the galaxy and indeed of the early universe. The simplest hypothesis about the clusters is that they all formed within a short time (say a billion years) of the big bang: the instant when all the matter in the present universe emerged explosively, it is thought, from a single point. The clusters would then have been among the first objects to condense as the galaxies took shape, each cluster evolving from a large blob of gas. It counts in favor of this hypothesis that half of the roughly 200 globular clusters in the Milky Way lie scattered throughout the almost spherical volume of the galactic halo. Presumably they formed there well before the galactic disk took shape. The orbit of such a cluster is typically a rather eccentric ellipse and not the more nearly circular orbit characteristic of matter in the disk. The major axis of the orbit is sometimes several tens of thousands of parsecs, and once every billion years the orbit sends the cluster rushing through the thickness of the disk, which is only a few hundred parsecs.

On the simplest hypothesis the stars of the globular clusters would have formed at a time when the available matter in the galaxy was mainly hydrogen and helium, the two chemical elements presumed to have been created in the immediate aftermath of the big bang. In contrast, a later-born star would condense from interstellar gas part of which had been cycled through the interior of the first-born stars. Some of the gas, for example, would have been propelled into space by supernova explosions. It would be matter in which heavier atoms had been created by thermonuclear fusion. The later-born star would therefore have higher concentrations of the heavy chemical elements; in the shorthand of astrophysicists all such elements are known as metals.

It is the concentration of metals in the various globular clusters that imperils the simplest hypothesis. To be sure, the ratio of metals to hydrogen and helium is. 100 times greater in the sun than it is in the stars of metal-poor globular clusters such as M3 (for Messier 3). Moreover, the metal-poor clusters tend to be the ones that are outermost in the galaxy. An age of 15 billion years is now assigned to them. On the other hand, the globular cluster 47 Tucanae is relatively metal-rich. Its age is thought to be 10 billion years, which is twice the age of the sun. Omega Centauri, the most impressive globular cluster in the Milky Way, shows a range of metal concentrations. Evidently it is an idiosyncratic agglomeration of stars that were born at different times. The spread of ages for the globular clusters conflicts with current models of how the galaxy evolved. No model allows as much as several billion years for the galactic disk to have condensed.

With respect to the question of age it is notable that the maximum age of the universe can be inferred from the velocities at which galaxies are receding from one another. The current value for the maximum is close to 15 billion years. Recent investigations of the recession rate have tended to reduce the maximum age. If the trend continues, the age could conceivably fall to as little as 10 billion years. Then one would have to explain the finding that certain globular clusters seem to exceed the age assigned to the universe. It is clearly important that trustworthy values be calculated for the ages of the globular clusters. It would be particularly useful if such ages were available by the end of this decade, because by then the Space Telescope should have yielded far more reliable values for the recession velocities and the distances of the galaxies. It will thus have yielded a far more reliable value for the maximum age of the universe.

The internal dynamics of the globular clusters are well described by three characteristic times. The first is the crossing time: the time it takes a star to move across the cluster under the gravitational attraction of the cluster as a whole. The second is the relaxation time: the time in which the star settles down and becomes a stable member of the cluster under the influence of its gravitational interactions with nearby stars. The third is the evolution time: the time in which a stable cluster changes its form and its stellar composition significantly. For a globular cluster rich in stars the crossing time is much shorter than the relaxation time, which in turn is much shorter than the evolution time.

The work of Ivan R. King of the University of California at Berkeley has suggested some further quantifications. When a star-rich globular cluster has been in existence for roughly 50 crossing times, its stars will have settled down; the statistics of their velocities will be much like those for the velocities of the molecules in a cloud of gas. This state of equilibrium might persist indefinitely if the stars never changed internally and if no stars left the cluster.

As early as the 1930's, however, Lyman Spitzer, Jr., of Princeton and Victor A. Ambartsumian of the Byurakan Astrophysical Observatory in Armenia had concluded that the stars of lowest mass in a cluster were most likely to have the greatest velocity, and that in many cases the stars with the greatest velocity would escape from the cluster. Because of their great velocity the escaping stars would take with them more than the average share per star of the total energy of the cluster. As a result the cluster would contract. Over a period of time on the order of the evolution time the cluster would lose appreciably in average energy per star, and the most massive stars would settle close to the center.

Internal changes among the stars that remain can only quicken this trend. To begin with, the loss of mass by stars is now recognized to be an astrophysical commonplace. Early in its evolution a typical star is embedded in a shell of gas that is gradually expelled by the outward-directed pressure of the star's electromagnetic radiation. (Each photon, or quantum of the radiation, has momentum, and so it pushes whatever absorbs it.) Later the star begins to shed a gentle stellar wind of particles. Still later certain kinds of star explode. In a nova the exploding star ejects into space the equivalent of the mass of a planet such as Jupiter. A supernova is an explosion in which an entire star is destroyed. Because mass is lost through mechanisms such as these the stars in a globular cluster move systematically to classes with lower mass (and also fainter intrinsic brightness). Over a period of time on the order of the relaxation time they acquire the higher velocity of a less massive star. Hence their chance for escape from the cluster improves.

Throughout the hundreds of billions of years a cluster may spend in the thin outer halo (or even the corona) of the Milky Way the evolution of the cluster is affected only by the internal events I have just described. There comes a time, however, when the cluster traverses the galactic disk. At such a time the cluster is at risk not so much of collision as of gravitational interaction with the matter in and near the galactic plane. The force of the interaction may tear the cluster apart.

As the orbit of a cluster brings it close to the center of the galaxy a further threat arises. The threat exists because the central mass of the galaxy exerts a greater attractive force on the side of the cluster that passes nearest the center than it does on the side farthest away. The difference between the inner and the outer force can deform the cluster. Indeed, for any globular cluster there is a critical radius: a distance from the galactic center within which some of the stars in the cluster will be sheared off as the cluster makes its closest approach to the center. The critical distance is called the tidal radius because the mechanism that deforms the cluster is similar to the one that raises tides in the oceans. It is not surprising that the globular clusters inside the galaxy's central bulge are smaller on the average than the ones in either the galactic halo or the galactic corona.

One final factor may affect the globular clusters, but I mention it with some misgivings. As recently as 15 years ago few astronomers would have thought the formation of double star systems would be important in star clusters. Then Sebastian von Hoerner of the National Radio Astronomy Observatory made the first computer simulations of open clusters: aggregations of stars that differ from globular clusters in that the stars are young, fewer and more widely dispersed. Von Hoerner based the simulations on equations that represent the gravitational interactions among small numbers of stars. The simulations consistently showed that the stars in a cluster tend to unite by pairs into binary systems.

In more recent computer simulations incorporating first 100 and later 500 and even 1,000 interacting stars, the same effect was noted. In 1975 Douglas C. Heggie of the University of Edinburgh showed that a newly formed binary system is a formidable sink of the energy available in the cluster. In essence the formation of each binary system takes kinetic energy from the motion of the stars and converts it into energy that binds the two stars into orbit, each one around the other. The pair accordingly sink toward the center of the cluster. Still more recently, Spitzer and his associates at Princeton have devised models suggesting that in the crowded central region of a globular cluster, where the most massive stars are huddled, each star is likely to capture one or even two or more companions. As a result the central region may become even more densely packed as the globular cluster evolves.

The reason for my misgivings in the face of all this circumstantial evidence is simple. Massive binary star systems do indeed prevail in open clusters. Mizar is a prime example. It is a star in the Ursa Major open cluster. More recognizably, it is the star at the bend of the handle of the Big Dipper. Mizar has a companion called Alcor, which is faintly visible to someone with excellent eyesight. Evidently, then, Mizar is a binary system. Actually both Mizar and Alcor have companion stars. Each one is a binary system. That makes it all the more exasperating that the search for instances of massive binary star systems near the center of globular clusters has not been equally successful. In fact, it has had no successes at all.

Something else has been found, however. In 1976 Jonathan E. Grindlay and Herbert Gursky of the Center for Astrophysics of the

Harvard College Observatory and the Smithsonian Astrophysical Observatory reported that an X-ray burst of terrific strength had reached the solar system from a direction close to that of the center of the globular cluster NGC 6624. The burst lasted for only eight to 10 seconds. Similar bursts have since been detected repeatedly from six globular clusters. According to one hypothesis, there is a black hole with a mass not less than 100 solar masses at the center of each such cluster. The bursts would then result from the intermittent collision of interstellar gas with a hot accretion disk of gas that surrounds the black hole. It comes to mind that the hypothetical black hole might be the ultimate consequence of the gradual collapse of stars toward the center of the cluster, a collapse to which the formation of binary star systems may have made a contribution, and perhaps even given the final pushes.

It was apparent even 35 years ago that the interstellar medium in the disk of the Milky Way includes clouds of gas and dust in which newborn stars are condensing. After all, some of the most striking objects in the galaxy are emission nebulas: bright clouds of ionized hydrogen atoms that are energized by the groups of giant stars inside them. The stars emit radiation at such a prodigious rate that they cannot have been doing so for more than a few tens of millions of years. The stars are therefore quite young. With only optical observations of the clouds, however, it was difficult to advance our understanding of how the stars actually formed.

All of this has now changed. Radio astronomy has revealed the presence in the clouds of more than 50 kinds of interstellar molecules, from hydrogen molecules, the lightest and by a factor of 1,000 the commonest, to a nine-carbon chain, the heaviest. Each species of molecule emits electromagnetic radiation at characteristic radio wavelengths. Moreover, infrared astronomy is now equipped to reveal the incipient stars themselves within their dense, obscuring clouds.

In almost all the processes in which a new star is thought to form, the initial step is the development of a concentration of matter — I shall call it a nodule — inside a cloud of interstellar atoms and molecules (for the most part hydrogen molecules), with a small admixture of dust. Some of the clouds are actually enormous clumpy distributions of matter. A giant molecular complex, for example, can have a mass of several hundred thousand solar masses. Other clouds are much smaller. Some of the clouds called globules have masses of only 20 solar masses. What all the nodules have in common is that they tend to collapse, mostly under the influence of their own gravitation, with perhaps occasional pressure from outside. An outside pressure that may be quite important is the one exerted on a cloud and the incipient nodules inside it when a density wave passes through and a spiral arm begins to form. The wave may well accelerate each of the processes I shall now describe.

The conditions in the roughly 4,000 giant molecular complexes that lie within 13,000 parsecs of the center of the galaxy seem ready-made for the formation of new stars. For one thing, each complex has plenty of matter. The mass of a typical complex is as great as several hundred thousand solar masses. Its diameter is about 50 parsecs. It is the most massive object in the galaxy (unless the galactic center has a supermassive black hole). The mass of a giant molecular complex is thought to be almost entirely hydrogen molecules; at a temperature of 20 degrees K., the cloud is too cold for the molecules to dissociate into atoms. The cloud is also too cold in most places for the hydrogen molecules to emit detectable amounts of radiation at their characteristic wavelengths. The presence of molecular hydrogen must therefore be inferred. The cloud is detected best by the emission (at 2.6 millimeters) of the next-commonest molecule, carbon monoxide, and by the emissions of still less common molecules such as formaldehyde.

One other component of a giant molecular complex is important for star formation. That component is the dust. The dust particles are sites on which the surrounding gas can collect. The dust also shields the incipient stars from ultraviolet radiation, which would disrupt the condensation. It is thought that for every 100 to 200 grams of molecular hydrogen in the giant molecular complex there is a gram of dust.

Two nearby complexes of carbon monoxide and dust (and presumably hydrogen in abundance) have been particularly well studied. They are the Orion complex (which is centered on the bright region of ionized hydrogen and newborn giant stars called the Great Nebula in Orion) and the Ophiuchus complex (which blocks the light from the center of the galaxy). In the Orion complex Becklin, Neugebauer and their associates at Cal Tech, together with Frank J. Low of the University of Arizona, have found evidence at infrared wavelengths for the presence of condensed objects that seem to be intrinsically very red. Each one may be a newborn star still embedded in a thick cocoon of dust that the star's ultraviolet radiation has not yet blown away completely. The radiation heats the dust, which then radiates in the infrared. The star may have formed in the first place when the cooling of a small part of the complex reduced the pressure that results from the heat of a gas. The cooled region would thus have begun to collapse under the influence of its own gravitation.

In the Ophiuchus complex a group of 30 stars has been found by Gary L. Grasdalen of the University of Wyoming, Stephen E. and Karen M. Strom of the University of Arizona and Frederick J. Vrba of the U.S. Naval Observatory. The stars were not detected earlier because they are hidden by at least 30 magnitudes of optical absorption; it took an infrared search to find them. More recent observations by Charles J. Lada and Bruce A. Wilking of the University of Arizona in an extremely dense dust cloud near the star Rho Ophiuchi have revealed the presence of 20 similar stars whose optical radiation is dimmed by as much as 100 magnitudes. The stars lie only half a parsec from the ones discovered earlier.

The evidence to date suggests that giant molecular complexes give rise spontaneously to stars that have masses no greater than a few times the mass of the sun. In particular it seems they give rise to stars of the spectral classes *B, A, F* and *G*. (The sun is a star of the class *G*.) George H. Herbig and his associates at the Lick Observatory have found near the borders of the well-studied giant molecular complexes some groups of small, rather dim and nebulous stars that are said to be young. Almost all of them have a brightness that varies irregularly. They are called T Tauri stars. Perhaps they are products of the interrupted condensation of nodules. They may have wandered out of the complex. Often they are seen in places where the ultraviolet radiation of newborn massive stars has blown away the dark nebula's gas and dust.

Star formation of another kind is typical, it appears, in places where an emission nebula with supergiant *O* and *B* stars embedded in it lies next to a giant molecular complex. The clearest exposition of what happens in such cases has been given by Bruce G. Elmegreen of Columbia and by Lada. According to Elmegreen and Lada, the *O* and *B* stars emit ultraviolet radiation whose pressure piles up cold gas and dust at the outer edge of the complex. The result is the condensation of protostars there. In places such as the Orion nebula the process appears to be advancing sequentially. In the Orion nebula a group of *O* and *B* supergiants is fading after a lifetime of a few tens of millions of years. The radiation from these stars has triggered the formation of a younger generation of *O* and *B* supergiants, and the younger stars in turn are now emitting radiation that eats its way slowly but persistently into the giant molecular complex, where a third set of *O* and *B* supergiants will presumably form. Why the process should give rise to *O* and *B* giants and supergiants rather than the smaller *B, A, F* and *G* stars that condense spontaneously in the complex is not yet understood.

I turn now to the class of dark clouds known as globules. Roughly 200 of these objects have been found within 500 parsecs of the sun. They have remarkably similar properties. Each one is dark and distinct and on a photographic plate is almost circular. No doubt they are nearly spherical. Their radius varies from .2 to .6 parsec, their mass from 20 to 200 solar masses and their internal temperature from five to 15 degrees K. They are impenetrable to visible light. On the other hand, some images recorded at near-infrared wavelengths, either photographically or by electronic imaging techniques, show the stars behind the globule. The dimming of the infrared radiation from such stars allows an estimate of the globule's content of dust.

Radio observations at 2.6 millimeters show that the globules are rich in carbon monoxide. Other molecules, notably formaldehyde and ammonia, have now been found as well. Evidently, then, a globule is a small and often isolated spherule of darkness quite similar in composition to a giant molecular complex. Presumably the globule is composed predominantly of molecular hydrogen too cold to emit detectable radiation.

The radio data also show that several of the globules are collapsing under the influence of self-gravitation. The rate of collapse is roughly half a kilometer per second, which corresponds to half a parsec per million years. Since the radius of a typical globule is half a parsec, a million years is roughly the time it takes for the collapse to be completed. The collapse again is crucial to models of star formation. In almost every model the center of the globule collapses faster than the periphery, and so a nodule forms. The collapse converts into kinetic energy the gravitational potential of the infalling matter. Eventually the energy at the center raises the temperature of the matter enough for thermonuclear fusion to begin. This signals the birth of a star. If the star is large, it emits enough radiation to blow away the gas and dust that surround it.

In short, a globule ought to give rise to a single star in about a million years. Considering the time scale of the process and estimating the number of globules in the Milky Way, one concludes that the globules might account for the formation of 25,000 stars per million years, or about a sixth of the overall rate at which the stars in the galaxy form. (The formation of stars in giant molecular complexes is perhaps a more fecund process.) In 1977, however, Richard Schwartz of the University of Missouri observed what appears to be a pair of incipient, nebulous stars that are being expelled from a very dark globule some 300 parsecs from the solar system in the part of the southern sky marked by the constellation Vela. The creation of two stars rather than one is surprising, and so is their expulsion from the globule. One wonders where the energy to expel them came from. The two stars are connected by a luminous strand that might be described somewhat fancifully as a stellar umbilical cord.

I should like to give one final example to suggest how much remains to be learned. Investigators attempting to model the processes by which stars form have tacitly assumed that the protection of a cloud of dust is required. Inside such a cloud the temperature is low and the accretion of matter is undisturbed. In particular the dust shields the interior of the cloud from disruption by ultraviolet radiation from outside sources. Consider, however, the Clouds of Magellan, which lie in the galactic corona but defy the generalization that all the content of the corona is old.

In the Large Cloud of Magellan there is a grouping of roughly 50 luminous *O* and *B* stars known as Shapley's Constellation I. It is likely that their age is no greater than 20 million years. Their velocities are on the order of only 10 kilometers per second. Hence each star has moved no more than 200 parsecs from where it was born. Within a radius of from 200 to 300 parsecs of where the stars are now, measurements at 21 centimeters have revealed five million solar masses of neutral atomic hydrogen and measurements at optical wavelengths have revealed 60,000 solar masses of ionized hydrogen. But the sky in that area is transparent. It probably harbors no giant molecular complex; it has little or no molecular hydrogen and it has no cosmic dust. How did the stars form there?

## ALLAN SANDAGE

∎

Cosmology, the study of the universe as a whole, was once the province of philosophers and theologians. Today it has become a science, in which theories can be tested against the evidence of observation. The American astronomer Allan Sandage (b. 1926), one of the leaders in this development, here assesses the status of observational cosmology in the late 1980s.

# Cosmology: The Quest to Understand the Creation and Expansion of the Universe

NEW YEAR'S DAY 1925; WASHINGTON, D.C.: The thirty-third meeting of the American Astronomical Society had been in progress since December 30. In a late session on the last day a paper was read *in absentia,* submitted by a young astronomer from Pasadena in remote southern California. Working with the Mount Wilson 100-inch reflector, then the world's largest telescope, Edwin Hubble, age thirty-two, had discovered variable stars of a certain type — the *cepheids* — in a vast spiral-shaped cloud then called the Great Nebula in Andromeda (or M31). These stars serve astronomers as standard beacons in the universe, enabling them to measure cosmic distances. The cepheids had appeared at much fainter brightness levels on Hubble's photographs of M31 than in the Magellanic Clouds, the satellites to the Milky Way. Thus, Hubble was able to prove beyond doubt that the Andromeda nebula was at a distance far beyond the limits of the Milky Way — it was a separate galaxy.

Hubble's paper, published early in 1925, marked the closing of a long debate on the nature of the white nebulae. Since the time of the great astronomer William Herschel in the eighteenth century, observers had seen these objects in the sky in regions away from the band of the Milky Way, but there was no agreement about their distance. They could either be relatively close, and would then be part of the Milky Way system, or they could be remote independent systems — *island universes,* similar to the Milky Way but isolated. Clearly, this was a question of fundamental importance to understanding the large-scale structure of the universe.

The island universe hypothesis was an old idea, begun in speculation as early as the 1700's. Emanuel Swedenborg (1688–1772), Thomas Wright (1711–1786), Immanuel Kant (1724–1804), among others, had discussed the hypothesis. Theirs were the first ideas of a new *cosmology* (or theory of the universe) of a vast dimension, but the subject remained speculative for lack of new observational data.

The situation changed decisively in the final decade of the last century. Large telescopes began to be built, primarily in the western United States. Major observatories were founded at which the problem of the nebulae began to be studied, using the new powerful methods of observational astronomy. The Lick Observatory in California, Lowell Observatory in Arizona, Yerkes Observatory in Wisconsin, and Mount Wilson Observatory became the centers for this study.

Hubble's discovery of the M31 cepheid variables had not been made in a vacuum. The island universe hypothesis *had* begun to be heavily discussed again in the early years of this century on the basis of a variety of observations made with the new telescopes. A number of astronomers in the United States and Europe argued for the island universe hypothesis, but others were not convinced. Principal among the opposition was Harlow Shapley, who was the influential director of the Harvard Observatory in 1924 and who had been a former colleague of Hubble at Mount Wilson.

Hubble's observations proved to be conclusive, showing to both sides that the Andromeda nebula was distant. By inference it was clear that other nebulae similar to M31 were also very likely to be at remote distances. It is fair to say that with Hubble's demonstration, modern cosmology *as a science* began the accelerated development that has led to our present ideas of the creation of the universe itself.

\* \* \*

The tapestry of our modern theory of the origin and large-scale properties of the cosmos is breathtaking. Its scope is awesome. Based on astronomical observations, on experiments in particle physics, and on a theoretical understanding of the forces that govern phenomena on both the large scale (the realm of the galaxies) and on the small scale (the realm of the atom), it has a strong ring of truth. Besides, it has that exquisite beauty which is essential for truth. But further, the theory is to a large extent testable and, as we shall see, its central predictions have, rather surprisingly, already been verified.

All cultures have produced their own peculiar cosmology; ours is not unique in the attempt. Cosmology is the quest for a cogent world view, without which society seldom functions well. Examples of ancient cosmologies that shaped the societies that made them are the Taoism and Buddhism of the Chinese, the Egyptian view of life after death, the Greek universe governed by gods of nature, the Norse legends of Valhalla and its citizens, and the medieval reliance on astrology based on beliefs about the influence of the stars.

What is cosmology? Pushed to its speculative limit, the widest definition would be "a system of beliefs leading to an explanation of the mystery of existence." But this would put much of the practice beyond the scientific method, which insists on prediction and testability. A narrower and more useful dictionary definition is: "Cosmology is that branch of natural philosophy that deals with the character of the universe as a cosmos — that branch that deals with *processes* of nature and the relation of its parts." We shall later claim that, still keeping within the scientific method, we can broaden the definition slightly by recognizing the recent advances in high-energy particle physics and state that "Cosmology also deals with the ontology of things — that is, inquires into *being* — in this case, the existence or being of the cosmos and its contents." Clearly, this then becomes nothing less than an inquiry into the question of creation. It can say nothing about the *reason* for the event but rather provides a description of the consequent *processes* that began just after the instant of creation.

Early on, astronomy naturally provided the way to approach cosmology, both in ancient cultures and in our own. The objects of astronomical study are not of this world. We look out as we look up — an idea as old as the Chinese. As we look out, we begin to sample a wider world beyond our neighborhood. Each cosmology has been born from this realization.

\* \* \*

Following Hubble's discovery, study of the galaxies took on a wholly new significance. One key question was: Are galaxies the structures with which to map the basic character of the universe, or are they only an intermediate component, leading to a larger hierarchy as we sample larger and larger parts of the universe? If galaxies are only an intermediate structure, then as we look to larger and larger distances, they should decrease in importance relative to the larger structures they compose. Hubble set out to test this idea.

In one of his least-known but most important observational programs, Hubble began to survey many (about a thousand) areas of the sky with the Mount Wilson telescopes in the middle 1920's. The method was to count the number of galaxies at progressive levels of faintness. In general, the fainter a galaxy, the farther away it is. If galaxies are spread uniformly in space in all directions and at all distances, then the number of them seen at fainter intensity levels should increase in proportion to the volume surveyed. This would not be true if galaxies are merely a local phenomenon relative to a much larger organization.

Hubble announced the results of his survey over a seven-year period from 1931 to 1938 in a series of reports, supplemented by a major count survey performed by Nicholas Mayall at the Lick Observatory in 1934. The result was that the number of galaxies does, in fact, increase with increasing faintness in the way expected if they are distributed homogeneously in space at all distances. From this, Hubble concluded in 1930, 1934, and 1936 that galaxies can give us a fair sampling of the universe at large.

To be sure, it has subsequently been found that galaxies are organized into clusters and often lie on sheets resembling bubbles that encircle voids. Nevertheless, Hubble's discovery still stands when averages are taken over many of the bubbles, voids, and sheets — that is, over an appreciable fraction of the sky: The universe is homogeneous in the large.

It might have been thought that here at last was the solution to the cosmological problem. The basic building blocks of the universe had been found. They were distributed homogeneously on the largest scales that were measured. They show no sign of an edge to their distribution, to the limit of the largest telescopes.

True enough — but there was one other (crucial) discovery yet to be understood. Astronomers had found — in the second decade of the twentieth century — that the lines in the spectra of all but the nearest galaxies showed a *red shift*. . . . A shift of the colors in the spectrum toward the red means the galaxy is moving away from us. The systematic red shift of the galaxies is interpreted to mean that the universe is in general expansion. It surely is one of the capital discoveries in all of science.

The time was 1922. Einstein's theory of general relativity, describing the relationships between space, time, and gravity, had been published in 1916. Scientists generally assumed that space was "stable" or static — it did not move. Newton, two hundred years before, had nevertheless puzzled over how this could be. Because Newton's law of gravity requires that masses attract each other, all objects in a static universe would be expected to fall together to form a massive "central lump." In the solar system the reason this does not happen is understood. The planets move. The force of gravity inward is counterbalanced by the outward centrifugal force caused by their circular motion around the Sun. Newton wondered if there might, in fact, be a large-scale motion of the universe outward to counter his inward force, but the wonder remained just that.

Then, in 1922, the Russian mathematician Alexander Friedmann found a solution to Einstein's complicated set of relativity equations that suggested a general expansion of space, carrying test particles with it as if they had radial velocities outward. When Friedmann published the solution to Einstein's equations that predicted the expanding universe, Einstein was not impressed and published a two-sentence rebuttal of Friedmann's paper the same year. He then published a further two-sentence paper in 1923 withdrawing his earlier one, admitting *his* mistake, and agreeing with Friedmann. It is the Friedmann equations that form the loom upon which the current discipline of theoretical cosmology is woven.

The next person to generalize the theory and take the expansion prediction seriously was a Belgian theologian, the Abbé Georges-Henri Lemaître, who believed that here was the *scientific prediction of the creation event*. Everyone now agrees, but his 1927 paper with its famous "coasting" solution for the motion of the universe was buried in a relatively obscure journal. It was found in 1930 by George McVittie, then a research student working with the great British astronomer Arthur Eddington, who thereafter wrote extensively on the expanding universe. The idea began to spread.

It is not clear how far any of the theoretical framework had penetrated into the Mount Wilson and the Lick observatories in

1927. These were the two largest observatories in the world at the time, and one would have thought that any observational test of such theories would have to be made at one of them. But this turned out not to be the case, at least at first.

At Percival Lowell's private observatory in Flagstaff, Arizona, Vesto Slipher, one of a pair of famous astronomer brothers, had begun in about 1915 to obtain spectra of spiral nebulae. In early 1921, *The New York Times* (January 19) reported Slipher's discovery of what was then considered an enormous red shift for a nebula known as NGC 584. Translating the shift into motion, the discovery indicated that the nebula was moving away at a velocity of about two thousand kilometers per second! Slipher had perfected his own design of an instrument for taking spectra of very faint objects, and took spectra for a total of about thirty galaxies between 1915 and 1924. He discovered that most had shifts of their spectral lines toward the red. Was this the expansion of space predicted by Friedmann and Lemaître?

Slipher never published his results on NGC 584 or others on his list. He did send the velocities to Gustav Strömberg at Mount Wilson and to Eddington in England. Strömberg made an analysis of them in a 1925 publication, concluding that here was a strange phenomenon indeed, because no other astronomical objects had such large apparent motions, most of which were outward.

In the mid-1920's the mathematician and theoretical physicist Harold Percy Robertson had become interested in the Friedmann expanding spaces. In a highly theoretical paper published in 1928 he combined Slipher's red shifts with brightness measures of the same galaxies published by Hubble in 1926. In a small paragraph in his nearly forgotten paper one reads words paraphrased as: *It appears that a relation between red shift and distance can be established from these data giving a rough verification of Friedmann's prediction.* Robertson, who soon become famous in cosmology for his proof of the mathematical equation for the geometry of a homogeneous space, recounts this incident in a single sentence in his 1963 article on cosmology in a new edition of the *Encyclopædia Britannica.* He told the story privately in 1960, but he made no other point of it and history seems to have passed him by in this central discovery.

Hubble had also been studying the galaxy data in the late 1920's, trying like many others to interpret them. Slipher's largest velocities were about two thousand kilometers per second. In 1925, Lundmark had published an analysis of the existing data where he fitted a mathematical series to the correlation of red shift and distance. Besides the linear term, which later was to become so famous, he added a square term in the distance. He found a negative coefficient for it, concluding that no red shifts should exist larger than about three thousand kilometers per second. If true, clearly this could not be the Friedmann phenomenon, which required larger and larger red shifts as you went to greater and greater distances.

Hubble knew of Lundmark's work. It was crucial to determine the reality of the squared term; what was needed were more red-shift data. Hubble suggested a plan to Milton Humason, a self-educated man at Mount Wilson who progressed between 1918 and 1963 from mule driver to janitor to night assistant to observer to staff astronomer to secretary of the Mount Wilson and Palomar observatories. Humason was to observe the spectra of fainter galaxies that Hubble thought would show larger red shifts if this was the Friedmann effect. In 1929, Humason obtained spectra with the Mount Wilson equipment, exposing photographic plates on the galaxy NGC 7619 for thirty-six hours with a confirming spectrum of forty hours. He obtained a red shift of four thousand kilometers per second. Hubble and Humason then published back-to-back announcements in volume 15 (1929) of the *Proceedings of the National Academy of Sciences*. Humason gave the spectral data. Hubble combined all available data and announced that they showed a linear relationship between the speed of the galaxies and their distance. The farther the galaxy, the faster it was receding from us. Here was the pattern of the expanding universe. This has turned out to be the most important discovery in astronomy, certainly in the twentieth century, and debatedly since Copernicus and Kepler. It has changed our conception of cosmology and has opened the door into the modern, scientific theory of creation.

Hubble and Humason began thereafter a campaign of observation with the Mount Wilson 100-inch reflector, proving that the relation was general and that it applied everywhere and at all distances. By 1950 the relation had been eventually extended to red shifts, expressed as a velocity of 60,000 kilometers per second. This *was* clearly the Friedmann-Lemaître effect. Important papers were published by Hubble and Humason in 1930, 1931, 1934, and 1936. The work was begun again by Humason using the 200-inch telescope at Palomar in 1949 and has continued, in its several aspects, to this day both at Palomar and now principally at Lick.

\* \* \*

All the work from Hubble's day onward has shown that the relation is *linear* — that is, the red shifts increase in direct proportion to a galaxy's distance from us. This form of the relation is so crucial that much of the early work at Palomar with the newly completed 200-inch reflector was spent proving this. A linear relation is the only mathematical form of such a law where every observer, on every galaxy, sees the same type of velocity-distance relation. Each observer appears (to himself) to be the center of the expansion. But in reality there is no center — every place in the expanding universe is the center.

To understand this, consider Eddington's famous example of the surface of a balloon painted with dots. The curved surface of this balloon is an analogy (with one fewer dimension) for the curved space of the universe in Einstein's general theory of relativity. . . . On the curved surface of a balloon, as on the surface of the Earth, there is no center or edge. Any point on the surface has as much right to consider itself the center *of that surface* as any other point. (Of course, the whole balloon has a center, but the *surface* does not!)

Now we add some additional air to the balloon and watch it expand. Place yourself on any dot and look at other dots in any direction. All the dots move away from all others. A linear relation (which is what our balloon obeys) is the only velocity law where all points are equivalent.

The second crucial thing to realize is that a linear relation is the only law that permits all points to have had a common origin at some time in the past. Let the air out of a balloon and let its surface become smaller and smaller until it has zero area (in the limit of the imaginary balloon). When this happens it is clear to see that all points are together at a common point. Said differently, consider two galaxies seen from our vantage point and say one is twice as far away as the other. The linear law says that the velocity of the distant one is twice as great as that of the nearer galaxy. In your mind's eye, now reverse the velocities so that the galaxies approach us. The near one reaches us at a particular time. The farther one reaches us at the *same* time. It has twice as far to go but it is moving twice as fast. The same will be true for all galaxies in the space. Hence, if the velocity law is linear, then there could be a creation event in the past, when all matter was together and where all parts of space coincided in a sort of cosmic egg. Lemaître called it the *primeval atom* in his famous book by that title — a book that became a primer for students to learn something of the mysterious expansion.

\* \* \*

All this concerning the expansion is, of course, well known to those who follow astronomical developments. The general idea has even become part of common culture, worldwide; everyone has heard of the expanding universe. It is the centerpiece of the modern cosmological world view. But is it true?

Hubble was not sure. In all his writing he was cautious and never committed himself to the view that red shift means expansion and hence that it is a clear signal for a creation event. There were two reasons. The first resulted from a highly technical analysis he made with some data that turned out to be incorrect.

The more serious problem concerns the time for the creation event, calculated by the argument we discussed above. By reversing the velocities of the expanding galaxies, we can calculate when all galaxies must have been together by dividing their distances by their velocities. When Robertson in 1928 and Hubble in 1929 did that, they obtained a very short time for the expansion age of the universe — less than 2 billion years. But 2 billion years was less than the age of the crust of the Earth, already in the 1930's put at about 3.5 billion years by the geologist Arthur Holmes and others. How could the whole universe be younger than the Earth?

This time-scale problem has persisted until recently. The difficulty rests with finding the value of the distances to galaxies by which to divide the measured velocities to calculate when the clock began. . . . Distances to galaxies have been enormously hard to measure. It has turned out, after thirty years of effort — primarily at Palomar — that we can identify a small class of distance indicators in addition to the cepheid variables to be used in the measurements. These indicators are brighter than cepheids, permitting their discovery in galaxies at much larger distances than the cepheids. The work has been to find ways of calibrating the absolute brightness of such indicators, and then to apply the calibration to actual galaxies of known velocity. In this way, the crucial ratio of distance divided by velocity — which is often called the Hubble time — can be calculated.

Early in the Palomar work in the middle 1950's it was known that Hubble's scale of distance was too small — but by how much? In 1950 the correction to the scale was put by Walter Baade to be about two times; Hubble's distances should be multiplied by two. This would increase the age from 2 billion to 4 billion years — not enough, because the age of the Sun had been found in the early

1950's to be about 5 billion years. As more and more was learned about the distance indicators from Mount Wilson and from Palomar from 1950 to the early 1970's, it became clear that the correction factor was much more than two. By 1956 it was thought to be at least 3, giving an age of at least 6 billion.

Without going into the technical details, it suffices here to note that astronomers now working on the distance scale problem agree that the correction to Hubble's 1936 distance scale is at least a factor of six and could be as large as twelve — the preferred value at present. This puts the Hubble time between about 12 billion and 24 billion years.

This is not exactly the age of the universe, because the expansion may have slowed down with time — pulling in on itself due to the self-gravity of the entire system. In the most popular model, where space-time is precisely "flat" in the Euclidean geometrical sense, this slowing will cause the true age of the universe to be two thirds of the Hubble time. If so, then the range given by the correction factor between six and twelve gives a universal age of between 8 billion and 16 billion years.

How then can we tell if the expansion of the universe implied by these measurements is real? The crucial test is to compare this age with the age obtained by other methods.

By the 1950's, astronomers had learned how to age-date the stars. This came from an understanding that the source of starlight was the nuclear conversion of hydrogen into helium deep in the interiors of the stars, where the temperatures are high enough for the nuclear reactions to work. Knowing how much hydrogen there is to convert to helium, and knowing the rate at which any given star is radiating away this released nuclear energy, tells how long the star can go on burning its fuel. In the 1950's, astronomers learned how to identify a phase in a star's life when it had converted about 10 percent of its hydrogen to helium. Knowing how much energy had been released by the conversion of that amount of matter to helium tells us when that particular star began to shine. The game then became one of finding the oldest stars in our Galaxy. By age-dating those stars, we can straightaway establish that our Galaxy is at least that old.

The theory of stellar evolution as developed between 1935 and about 1965 has been so convincingly borne out by a number of different tests that there is no question of its correctness in general outline. It then became a matter of improving the calculations to obtain a *precise* measurement of stellar ages. This task is complicated by the requirement that we must know the abundance of certain critical elements such as oxygen, nitrogen, and carbon in a star before we get the correct answer for the nuclear burning rates. The best estimates in 1987 for the ages of its oldest stars put the age of our Galaxy at greater than 12 billion years but less than about 18 billion, with the best guess at close to 15 billion.

The good agreement with the range of possible Hubble ages and the corresponding age of the universe (depending on the slow-down rate of the expansion) is remarkable. If the observed red shifts do not mean expansion, then no such agreement would be expected. The agreement, which most astronomers believe is real, removes Hubble's second objection: The expansion does appear to be real. Of course, there are still the small remaining details of the time-scale problem. Do we have the crucial factor of two thirds, or not? Is the correction factor six or twelve? Are the oldest stars in the Galaxy 12 billion or 18 billion years old? These are important questions because their answers determine the eschatology of the cosmos — the *ultimate fate* of the universe. We know the gravity of all the matter in the universe is slowing down the expansion. Is there enough matter to stop the expansion completely sometime in the future, whereupon the collapse of the entire system of galaxies should begin? Or does the expansion have enough energy to continue forever? The time-scale data are simply not good enough yet to decide this further question.

Where then does this leave the search for a coherent cosmology? We have emphasized that the expansion appears to be real. It may require a creation event. (The velocity-distance relation is linear.) Is there other evidence for such an event? The conclusion reached by taking the Friedmann model seriously is that the universe is a most remarkable machine. If it is expanding, it must once have been denser and hotter than it is now. By considering the physical conditions at the earliest times that are still describable by the Friedmann equations, the cosmological problem becomes much richer than merely a description of the organization of the present universe.

George Gamow and his students Ralph Alpher, Robert Herman, and James Wightman Follin took the Friedmann model quite seriously in the 1950's and asked what would happen if they were to run the equations backward toward the high-density and -temperature era, assuming, in fact, a creation event. In a classic series of papers in the 1940's and in an important book by Gamow in 1952 called *The Creation of the Universe,* these men set down the foundations of what is now called the *new cosmology.* Their aim was to describe the creation of the stable matter we have in the universe today, out of the "cauldron" of energy at the beginning. For example, about 0.01 seconds after the creation event itself, free neutrons and protons had cooled to temperatures and expanded to densities that were low enough for them to react by standard, well-understood reactions of nuclear physics. Gamow and his students showed that they would then form the stable nuclei of heavy hydrogen (deuterium), two isotopes of helium with mass 3 and 4, and lithium 7 (with 3 protons and 4 neutrons). The universe, continuing to expand, cooled until at about one thousand years after the event, it was cold enough for the free electrons to combine with the free protons to form neutral hydrogen atoms. Before this time the free *photons,* the packets of electromagnetic energy, were being absorbed by the electrons. This kept the electrons from forming atoms and kept the photons from traveling very far. After the electrons combined with protons into atoms, the photons became independent particles and continued to cool. The universe became transparent to them and they could now essentially survive forever. Alpher and Herman could predict the temperature that these photons should have today, knowing how high the temperature and density had to be at the time of the formation of the two kinds of helium and the lithium. Their predicted temperature for the leftover radiation was five degrees above absolute zero — very cold and difficult to observe. The prediction, made in the early 1940's, was lost in the pages of the physics journal called the *Physical Review* but not forgotten by them or by Follin, the fourth member of the team. In the 1950's, Follin had begun to think about how to detect this leftover radiation using the new techniques of rocketry, which were beginning to be applied to astronomy. Although no public record exists of experiments considered by Follin or his students, it is not true, as you sometimes read in popular books, that Gamow's team had not begun to think of how their prediction could be tested, or that they did not believe in them.

The radiation left over from the hot epoch of the universe was indeed discovered in 1963, almost by accident, by the Bell Telephone astronomers Arno Penzias and Robert Wilson, who eventually received the Nobel Prize in physics for this work. Their serendipitous discovery came just before that of a Princeton team which had mounted a specific experiment to find such radiation, following a new and independent prediction of their own. Now confirmed by many experimenters, both from the ground and using rockets, the *cosmic background* radiation (as it has come to be called) is at roughly three degrees above absolute zero. Flooding in on us from everywhere in the universe, it is our most direct confirmation of the hot, dense early period in the history of the universe predicted by the new cosmology.

There has also been confirmation of the calculations by the Gamow team of the abundances of the light elements produced during the early phases of the universe. These calculations had been made more precise in the 1960's by Robert Wagoner and have now become a cottage industry among the physicists. Recently, new techniques in astronomical spectroscopy have allowed astronomers to measure the abundances of helium and lithium with much greater precision. Their work has shown spectacular agreement with the predictions, providing yet another strong verification of the hot early universe.

But the new cosmology has been even bolder in an attempt to understand even earlier times. . . . Physicists are now trying to calculate the properties of the hot early universe back to such high energies that the times are said to be only $-10^{-24}$ seconds after the creation event — which astronomers like to call the Big Bang. The reasons for believing that these ideas are more than speculation are that particle physics has witnessed an internal revolution of its own in our understanding of the four fundamental forces of nature.

The basic theory of these forces and how they can be unified . . . is so elegant that many believe it must be true. Furthermore, a number of its predictions have already been confirmed in our largest particle accelerators.

The logical extension of the theory to even higher energies predicts a unification of the strong nuclear force and the now unified "electroweak" force at energies far beyond what can be produced in particle accelerators at $10^{18}$ MeV. However, if the new cosmology is correct, the early universe had such energies some $10^{-36}$ seconds after the Big Bang.

Fantasy? Perhaps. But remember that the same type of people that gave us the 3° radiation, the nucleosynthesis of $^3$He, $^4$He, and $^7$Li, and the W and heavy bosons have given us this extension of the theory. What are the consequences?

A direct prediction of these ideas is that the universe must today be very close to flat in shape — as the specialists say, it must have zero space-time curvature. In that case, according to Einstein's general theory, the density of matter must be high enough to just create an exact Euclidean geometry. This leads to a very definite prediction for the matter density of the present universe. If we could spread out all the matter in the cosmos evenly, its average density would be near $10^{-30}$ grams per cubic centimeter. The flat geometry also requires that the age of the universe be precisely two thirds of the Hubble distance-velocity ratio. These two predictions return the ball once again to the astronomers' court. Ideally, both the mean density of matter in the universe and, as we saw, its time scales should be observable at the telescope.

The time-scale test almost works, as we saw. It does so precisely if the observed Hubble time is 24 billion years and the age of the galaxies is 16 billion, because $2/3 \times 24 = 16$. This is why the time-scale test is now being looked at so carefully at many observatories.

The mean density test is more difficult. We see only the *visible* matter in the universe (using a broad definition of *visible* to include radio-waves and other forms of radiation). We cannot feel (via gravity) any dark matter that is spread uniformly in space, because there are no unbalanced forces from such a distribution. Counting up the visible mass that produces the visible radiation gives, at most, only one one-hundredth the needed density to make a flat universe. Therefore, for the new cosmology to be correct requires at least one hundred times more mass in the universe than we can see.

Some consider the internal particle physics theories to be so persuasive as to believe in this amount of dark matter. Some find the conclusion on dark matter to be so outrageous as to disbelieve the particle physics theories. If dark matter exists, it would be crucial to detect it in some way. We cannot rely on the time-scale test alone, since we do not know if we shall ever be able to determine the age of the Galaxy accurately enough, even if the problem of the distances to galaxies is solved to great precision. A great effort is, therefore, presently being put toward the direct detection of the dark matter in the universe, in an attempt to prove or disprove its existence. On that may rest the fate of much of the present superstructure of cosmological theory.

Nevertheless, the scientific cosmology built from the foundations of observations of the galaxies, almost entirely in this century, is surely essentially correct. With it has come the realization of the vastness of the universe, an elaboration of its expansion, and an idea of its creation, mysterious as that event remains.

# EDWIN HUBBLE

■

Edwin Hubble (1889–1953) was a Rhodes scholar, boxer, and attorney who switched to astronomy when he found that his earlier endeavors left him unsatisfied. "Astronomy mattered," he said. Working at Mount Wilson Observatory in California, Hubble established that the hazy patches of light called spiral nebulae are actually galaxies of stars, and that the galaxies are moving apart from one another — evidence that the universe is expanding. This excerpt is from his book *The Realm of the Nebulae*, published in 1936.

# The Exploration of Space

THE EXPLORATION OF SPACE has penetrated only recently into the realm of the nebulae.* The advance into regions hitherto unknown has been made during the last dozen years with the aid of great telescopes. The observable region of the universe is now defined and a preliminary reconnaissance has been completed. . . .

The earth we inhabit is a member of the solar system — a minor satellite of the sun. The sun is a star among the many millions which form the stellar system. The stellar system is a swarm of stars isolated in space. It drifts through the universe as a swarm of bees drifts through the summer air. From our position somewhere within the system, we look out through the swarm of stars, past the borders, into the universe beyond.

The universe is empty, for the most part, but here and there, separated by immense intervals, we find other stellar systems, comparable with our own. They are so remote that, except in the nearest systems, we do not see the individual stars of which they are com-

---

*By "nebulae" Hubble meant what we today call galaxies — vast systems of billions of stars. [editor's note]

posed. These huge stellar systems appear as dim patches of light. Long ago they were named "nebulae" or "clouds" — mysterious bodies whose nature was a favorite subject for speculation.

But now, thanks to great telescopes, we know something of their nature, something of their real size and brightness, and their mere appearance indicates the general order of their distances. They are scattered through space as far as telescopes can penetrate. We see a few that appear large and bright. These are the nearer nebulae. Then we find them smaller and fainter, in constantly increasing numbers, and we know that we are reaching out into space, farther and ever farther, until, with the faintest nebulae that can be detected with the greatest telescope, we arrive at the frontiers of the known universe.

This last horizon defines the observable region of space. It is a vast sphere, perhaps a thousand million light-years in diameter. Throughout the sphere are scattered a hundred million nebulae. . . . The nebulae are distributed singly, in groups, and occasionally in great clusters, but when large volumes of space are compared, the tendency to cluster averages out. To the very limits of the telescope, the large-scale distribution of nebulae is approximately uniform.

One other general characteristic of the observable region has been found. Light which reaches us from the nebulae is reddened in proportion to the distance it has traveled. This phenomenon is known as the velocity-distance relation, for it is often interpreted, in theory, as evidence that the nebulae are all rushing away from our stellar system, with velocities that increase directly with distances.

This sketch roughly indicates the current conception of the realm of the nebulae. It is the culmination of a line of research that began long ago. The history of astronomy is a history of receding horizons. Knowledge has spread in successive waves, each wave representing the exploitation of some new clew to the interpretation of observational data.

The exploration of space presents three such phases. At first the explorations were confined to the realm of the planets, then they spread through the realm of the stars, and finally they penetrated into the realm of the nebulae.

The successive phases were separated by long intervals of time. Although the distance of the moon was well known to the Greeks, the order of the distance of the sun and the scale of planetary distances was not established until the latter part of the seventeenth century. Distances of stars were first determined almost exactly a century ago, and distances of nebulae, in our own generation. The distances were the essential data. Until they were found, no progress was possible.

The early explorations halted at the edge of the solar system, facing a great void that stretched away to the nearer stars. The stars were unknown quantities. They might be little bodies, relatively near, or they might be gigantic bodies, vastly remote. Only when the gap was bridged, only when the distances of a small, sample collection of stars had been actually measured, was the nature determined of the inhabitants of the realm beyond the solar system. Then the explorations, operating from an established base among the now familiar stars, swept rapidly through the whole of the stellar system.

Again there was a halt, in the face of an even greater void, but again, when instruments and technique had sufficiently developed, the gap was bridged by the determination of the distances of a few of the nearer nebulae. Once more, with the nature of the inhabitants known, the explorations swept even more rapidly through the realm of the nebulae and halted only at the limits of the greatest telescope.

This is the story of the explorations. They were made with measuring rods, and they enlarged the body of factual knowledge. They were always preceded by speculations. Speculations once ranged through the entire field, but they have been pushed steadily back by the explorations until now they lay undisputed claim only to the territory beyond the telescopes, to the dark unexplored regions of the universe at large.

The speculations took many forms and most of them have long since been forgotten. The few that survived the test of the measuring rod were based on the principle of the uniformity of nature — the assumption that any large sample of the universe is much like any other. The principle was applied to stars long before distances were determined. Since the stars were too far away for the measuring instruments, they must necessarily be very bright. The brightest object known was the sun. Therefore, the stars were assumed to be like the sun, and distances could be estimated from their apparent faintness. In this way, the conception of a stellar system, isolated in space, was formulated as early as 1750. The author was Thomas Wright (1711–86), an English instrument maker and private tutor.

But Wright's speculations went beyond the Milky Way. A single stellar system, isolated in the universe, did not satisfy his philosoph-

337

ical mind. He imagined other, similar systems and, as visible evidence of their existence, referred to the mysterious clouds called "nebulae."

Five years later, Immanuel Kant (1724–1804) developed Wright's conception in a form that endured, essentially unchanged, for the following century and a half. Some of Kant's remarks concerning the theory furnish an excellent example of reasonable speculation based on the principle of uniformity. . . .

The theory, which came to be called the theory of island universes, found a permanent place in the body of philosophical speculation. . . . The astronomers themselves took little part in the discussions: they studied the nebulae. Toward the end of the nineteenth century, however, the accumulation of observational data brought into prominence the problem of the status of the nebulae and, with it, the theory of island universes as a possible solution.

A few nebulae had been known to the naked-eye observers and, with the development of telescopes, the numbers grew, slowly at first, then more and more rapidly. At the time Sir William Herschel (1738–1822), the first outstanding leader in nebular research, began his surveys, the most extensive published lists were those by Messier, the last of which (1784) contained 103 of the most conspicuous nebulae and clusters. These objects are still known by the Messier numbers — for example, the great spiral in Andromeda is M31. Sir William Herschel catalogued 2,500 objects, and his son, Sir John (1792–1871), transporting the telescopes to the southern hemisphere (near Capetown in South Africa), added many more. (Sir John Herschel's general catalogue, representing the first systematic survey of the entire sky to a fairly uniform limit of apparent faintness, was published in 1864, and contained about 4,630 nebulae and clusters observed by his father and himself, together with about 450 discovered by others. The catalogue was replaced by Dreyer's *New General Catalogue* in 1890.) Positions of about 20,000 nebulae are now available, and perhaps ten times that number have been identified on photographic plates. The mere size of catalogues has long since ceased to be important. Now the desirable data are the numbers of nebulae brighter than successive limits of apparent faintness, in sample areas widely distributed over the sky.

Galileo, with his first telescopes, resolved a typical "cloud" — Praesepe — into a cluster of stars. With larger telescopes and continued study, many of the more conspicuous nebulae met the same fate. Sir William Herschel concluded that all nebulae could be

338

resolved into star-clusters, if only sufficient telescopic power were available. In his later days, however, he revised his position and admitted the existence, in certain cases, of a luminous "fluid" which was inherently unresolvable. Ingenious attempts were made to explain away these exceptional cases until Sir William Huggins (1824–1910), equipped with a spectrograph, fully demonstrated in 1864 that some of the nebulae were masses of luminous gas.

Huggins' results clearly indicated that nebulae were not all members of a single, homogeneous group and that some kind of classification would be necessary before they could be reduced to order. The nebulae actually resolved into stars — the star-clusters — were weeded out of the lists to form a separate department of research. They were recognized as component parts of the galactic system, and thus had no bearing on the theory of island universes.

Among unresolved nebulae, two entirely different types were eventually differentiated. One type consisted of the relatively few nebulae definitely known to be unresolvable — clouds of dust and gas mingled among, and intimately associated with, the stars in the galactic system. They were usually found within the belt of the Milky Way and were obviously, like the star-clusters, members of the galactic system. For this reason, they have since been called "galactic" nebulae. They are further subdivided into two groups, "planetary" nebulae and "diffuse" nebulae, frequently shortened to "planetaries" and "nebulosities."

The other type consisted of the great numbers of small, symmetrical objects found everywhere in the sky except in the Milky Way. A spiral structure was found in most, although not in all, of the conspicuous objects. They had many features in common and appeared to form a single family. They were given various names but, to anticipate, they are now known as "extragalactic" nebulae and will be called simply "nebulae." (The term "external galaxies," revived by Shapley, is also widely used, as is a third term, "anagalactic" nebulae, introduced by Lundmark.)

The status of the nebulae, as the group is now defined, was undetermined because the distances were wholly unknown. They were definitely beyond the limits of direct measurement, and the scanty, indirect evidence bearing on the problem could be interpreted in various ways. The nebulae might be relatively nearby objects and hence members of the stellar system, or they might be very remote and hence inhabitants of outer space. At this point, the development of nebular research came into immediate contact with the philosophical theory of island universes. The theory represented, in principle, one of the alternative solutions of the problem of nebular distances. The question of distances was frequently put in the form: Are nebulae island universes?

The situation developed during the years between 1885 and 1914: from the appearance of the bright nova in the spiral M31, which stimulated a new interest in the question of distances, to the publication of Slipher's first extensive list of radial velocities of nebulae, which furnished data of a new kind and encouraged serious attempts to find a solution of the problem.

The solution came ten years later, largely with the help of a great telescope, the 100-inch reflector, that had been completed in the interim. Several of the most conspicuous nebulae were found to be far beyond the limits of the galactic system — they were independent, stellar systems in extragalactic space. Further investigations demonstrated that the other, fainter nebulae were similar systems at greater distances, and the theory of island universes was confirmed.

The 100-inch reflector partially resolved a few of the nearest, neighboring nebulae into swarms of stars. Among these stars various types were recognized which were well known among the brighter stars in the galactic system. The intrinsic luminosities (candle powers) were known, accurately in some cases, approximately in others. Therefore, the apparent faintness of the stars in the nebulae indicated the distances of the nebulae.

The most reliable results were furnished by Cepheid variables, but other types of stars furnished estimates of orders of distance, which were consistent with the Cepheids. Even the brightest stars, whose intrinsic luminosities appear to be nearly constant in certain types of nebulae, have been used as statistical criteria to estimate mean distances for groups of systems.

The nebulae whose distances were known from the stars involved furnished a sample collection from which new criteria, derived from the nebulae and not from their contents, were formulated. It is now known that the nebulae are all of the same order of intrinsic luminosity. Some are brighter than others, but at least half of them are within the narrow range from one half to twice the mean value, which is 85 million times the luminosity of the sun. Thus, for statistical purposes, the apparent faintness of the nebulae indicates their distances.

With the nature of the nebulae known and the scale of nebular distances established, the investigations proceeded along two lines. In the first place the general features of the individual nebulae were studied; in the second, the characteristics of the observable region as a whole were investigated.

The detailed classification of nebular forms has led to an ordered sequence ranging from globular nebulae, through flattening, ellipsoidal figures, to a series of unwinding spirals. The fundamental pattern of rotational symmetry changes smoothly through the sequence in a manner that suggests increasing speed of rotation. Many features are found which vary systematically along the sequence, and the early impression that the nebulae were members of a single family appears to be confirmed. The luminosities remain fairly constant through the sequence (mean value, $8.5 \times 10^7$ suns, as previously mentioned), but the diameters steadily increase from about 1,800 light-years for the globular nebulae to about 10,000 light-years for the most open spirals. (The numerical values refer to the main bodies, which, as will be explained later, represent the more conspicuous portions of the nebulae.) The masses are uncertain, the estimates ranging from $2 \times 10^9$ to $2 \times 10^{11}$ times the mass of the sun.

Investigations of the observable region as a whole have led to two results of major importance. One is the homogeneity of the region — the uniformity of the large-scale distribution of nebulae. The other is the velocity-distance relation.

The small-scale distribution of nebulae is very irregular. Nebulae are found singly, in pairs, in groups of various sizes, and in clusters. The galactic system is the chief component of a triple nebula in which the Magellanic Clouds are the other members. The triple system, together with a few additional nebulae, forms a typical, small group that is isolated in the general field of nebulae. The members of this local group furnished the first distances, and the Cepheid criterion of distance is still confined to the group.

When large regions of the sky, or large volumes of space, are compared, the irregularities average out and the large-scale distribution is sensibly uniform. The distribution over the sky is derived by comparing the numbers of nebulae brighter than a specified limit of apparent faintness, in sample areas scattered at regular intervals.

The true distribution is confused by local obscuration. No nebulae are seen within the Milky Way, and very few along the borders. Moreover, the apparent distribution thins out, slightly but systematically, from the poles to the borders of the Milky Way. The explanation is found in the great clouds of dust and gas which are scattered throughout the stellar system, largely in the galactic plane. These clouds hide the more distant stars and nebulae. Moreover, the sun is embedded in a tenuous medium which behaves like a uniform layer extending more or less indefinitely along the galactic plane. Light from nebulae near the galactic poles is reduced about one fourth by the obscuring layer, but in the lower latitudes, where the light-paths through the medium are longer, the absorption is correspondingly greater. It is only when these various effects of galactic obscuration are evaluated and removed, that the nebular distribution over the sky is revealed as uniform, or isotropic (the same in all directions).

The distribution in depth is found by comparing the numbers of nebulae brighter than successive limits of apparent faintness, that is to say, the numbers within successive limits of distance. The comparison is effectively between numbers of nebulae and the volumes of space which they occupy. Since the numbers increase directly with the volumes (certainly as far as the surveys have been carried, probably as far as telescopes will reach), the distribution of the nebulae must be uniform. In this problem, also, certain corrections must be applied to the apparent distribution in order to derive the true distribution. These corrections are indicated by the velocity-distance relation, and their observed values contribute to the interpretation of that strange phenomenon.

Thus the observable region is not only isotropic but homogeneous as well — it is much the same everywhere and in all directions. The nebulae are scattered at average intervals of the order of two million light-years or perhaps two hundred times the mean diameters. The pattern might be represented by tennis balls fifty feet apart.

The order of the mean density of matter in space can also be roughly estimated if the (unknown) material between the nebulae is ignored. If the nebular material were spread evenly through the observable region, the smoothed-out density would be of the general order of $10^{-29}$ or $10^{-28}$ grams per cubic centimeter — about one grain of sand per volume of space equal to the size of the earth.

The size of the observable region is a matter of definition. The dwarf nebulae can be detected only to moderate distances, while giants can be recorded far out in space. There is no way of distinguishing the two classes, and thus the limits of the telescope are most conveniently defined by average nebulae. The faintest nebulae that have been identified with the 100-inch reflector are at an average distance of the order of 500 million light-years, and to this limit about 100 million nebulae would be observable except for the effects of galactic obscuration. Near the galactic pole, where the obscuration is least, the longest exposures record as many nebulae as stars.

The foregoing sketch of the observable region has been based almost entirely upon results derived from direct photographs. The region is homogeneous and the general order of the mean density is known. The next — and last — property to be discussed, the velocity-distance relation, emerged from the study of spectrograms.

When a ray of light passes through a glass prism (or other suitable device) the various colors of which the light is composed are spread out in an ordered sequence called a spectrum. The rainbow is, of course, a familiar example. The sequence never varies. The spectrum may be long or short, depending on the apparatus employed, but the order of the colors remains unchanged. Position in the spectrum is measured roughly by colors, and more precisely by wave-lengths, for each color represents light of a particular wave-length. From the short waves of the violet, they steadily lengthen to the long waves of the red.

The spectrum of a light source shows the particular colors or wave-lengths which are radiated, together with their relative abundance (or intensity), and thus gives information concerning the nature and the physical condition of the light source. An incandescent solid radiates all colors, and the spectrum is *continuous* from violet to red (and beyond in either direction). An incandescent gas radiates only a few isolated colors and the pattern, called an *emission* spectrum, is characteristic for any particular gas.

A third type, called an *absorption* spectrum and of special interest for astronomical research, is produced when an incandescent solid (or equivalent source), giving a continuous spectrum, is surrounded by a cooler gas. The gas absorbs from the continuous spectrum just those colors which the gas would radiate if it were itself incandescent. The result is a spectrum with a continuous background interrupted by dark spaces called absorption lines. The pattern of dark absorption lines indicates the particular gas or gases that are responsible for the absorption.

The sun and the stars give absorption spectra and many of the known elements have been identified in their atmospheres.

343

Hydrogen, iron, and calcium produce very strong lines in the solar spectrum, the most conspicuous being a pair of calcium lines in the violet, known as H and K.

The nebulae in general show absorption spectra similar to the solar spectrum, as would be expected for systems of stars among which the solar type predominated. The spectra are necessarily short — the light is too faint to be spread over long spectra — but the H and K lines of calcium are readily identified and, in addition, the G-band of iron and a few hydrogen lines can generally be distinguished.

Nebular spectra are peculiar in that the lines are not in the usual positions found in nearby light sources. They are displaced toward the red of their normal position, as indicated by suitable comparison spectra. The displacements, called red-shifts, increase, on the average, with the apparent faintness of the nebula that is observed. Since apparent faintness measures distance, it follows that red-shifts increase with distance. Detailed investigation shows that the relation is linear.

Small microscopic shifts, either to the red or to the violet, have long been known in the spectra of astronomical bodies other than nebulae. These displacements are confidently interpreted as the results of motion in the line of sight — radial velocities of recession (red-shifts) or of approach (violet-shifts). The same interpretation is frequently applied to the red-shifts in nebular spectra and has led to the term "velocity-distance relation" for the observed relation between red-shifts and apparent faintness. On this assumption, the nebulae are supposed to be rushing away from our region of space, with velocities that increase directly with distance.

Although no other plausible explanation of red-shifts has been found, the interpretation as velocity-shifts may be considered as a theory still to be tested by actual observations. Critical tests can probably be made with existing instruments. Rapidly receding light sources should appear fainter than stationary sources at the same distances, and near the limits of telescopes the "apparent" velocities are so great that the effects should be appreciable.

A completely satisfactory interpretation of red-shifts is a question of great importance, for the velocity-distance relation is a property of the observable region as a whole. The only other property that is known is the uniform distribution of nebulae. Now the observable region is our sample of the universe. If the sample is fair, its observed

344

characteristics will determine the physical nature of the universe as a whole.

And the sample may be fair. As long as explorations were confined to the stellar system, the possibility did not exist. The system was known to be isolated. Beyond lay a region, unknown, but necessarily different from the star-strewn space within the system. We now observe that region — a vast sphere, through which comparable stellar systems are uniformly distributed. There is no evidence of a thinning-out, no trace of a physical boundary. There is not the slightest suggestion of a supersystem of nebulae isolated in a larger world. Thus, for purposes of speculation, we may apply the principle of uniformity, and suppose that any other equal portion of the universe, selected at random, is much the same as the observable region. We may assume that the realm of the nebulae is the universe and that the observable region is a fair sample.

The conclusion, in a sense, summarizes the results of empirical investigations and offers a promising point of departure for the realm of speculation. That realm, dominated by cosmological theory, will not be entered in the present summary. The discussions will be largely restricted to the empirical data — to reports of the actual explorations — and their immediate interpretations.

Yet observation and theory are woven together, and it is futile to attempt their complete separation. Observations always involve theory. Pure theory may be found in mathematics but seldom in science. Mathematics, it has been said, deals with possible worlds — logically consistent systems. Science attempts to discover the actual world we inhabit. So in cosmology, theory presents an infinite array of possible universes, and observation is eliminating them, class by class, until now the different types among which our particular universe must be included have become increasingly comprehensible.

The reconnaissance of the observable region has contributed very materially to this process of elimination. It has described a large sample of the universe, and the sample may be fair. To this extent the study of the structure of the universe may be said to have entered the field of empirical investigation.

# ARTHUR STANLEY EDDINGTON

■

The English astrophysicist Arthur Stanley Eddington (1882–1944)
was one of the first to appreciate the significance to cosmology
of Einstein's general theory of relativity. He wrote the following
nontechnical explanation of Einstein's curved space for his book
*The Expanding Universe*, published in 1933.

# *Spherical Space*

WHEN A PHYSICIST refers to curvature of space he at once falls under
suspicion of talking metaphysics. Yet space is a prominent feature of
the physical world; and measurement of space — lengths, distances,
volumes — is part of the normal occupation of a physicist. Indeed it
is rare to find any quantitative physical observation which does not
ultimately reduce to measuring distances. Is it surprising that the
precise investigation of physical space should have brought to light
a new property which our crude sensory perception of space has
passed over?

Space-curvature is a purely physical characteristic which we may
find in a region by suitable experiments and measurements, just as
we may find a magnetic field. In curved space the measured distances
and angles fit together in a way different from that with which we
are familiar in the geometry of flat space; for example, the three
angles of a triangle do not add up to two right angles. It seems rather
hard on the physicist, who conscientiously measures the three angles
of a triangle, that he should be told that if the sum comes to two
right angles his work is sound physics, but if it differs to the slightest
extent he is straying into metaphysical quagmires.

In using the name "curvature" for this characteristic of space,
there is no metaphysical implication. The nomenclature is that of the
pure geometers who had already imagined and described spaces with
this characteristic before its actual physical occurrence was suspected.

346

Primarily, then, curvature is to be regarded as the technical name for a property discovered observationally. It may be asked, How closely does "curvature" as a technical scientific term correspond to the familiar meaning of the word? I think the correspondence is about as close as in the case of other familiar words, such as Work, Energy, Probability, which have acquired a specialised meaning in science.

We are familiar with curvature of *surfaces*; it is a property which we can impart by bending and deforming a flat surface. If we imagine an analogous property to be imparted to *space* (three-dimensional) by bending and deforming it, we have to picture an extra dimension or direction in which the space is bent. There is, however, no suggestion that the extra dimension is anything but a fictitious construction, useful for representing the property pictorially, and thereby showing its mathematical analogy with the property found in surfaces. The relation of the picture to the reality may perhaps best be stated as follows. In nature we come across curved surfaces and curved spaces, i.e. surfaces and spaces exhibiting the observational property which has been technically called "curvature." In the case of a surface we can ourselves remove this property by bending and deforming it; we can therefore conveniently describe the property by the operation (bending or curving) which we should have to perform in order to remove it. In the case of a space we cannot ourselves remove the property; we cannot alter space artificially as we alter surfaces. Nevertheless we may conveniently describe the property by the imaginary operation of bending or curving, which would remove it if it could be performed; and in order to use this mode of description a fictitious dimension is introduced which would make the operation possible.

Thus if we are not content to accept curvature as a technical physical characteristic but ask for a picture giving fuller insight, we have to picture more than three dimensions. Indeed it is only in simple and symmetrical conditions that a fourth dimension suffices; and the general picture requires six dimensions (or, when we extend the same ideas from space to space-time, ten dimensions are needed). That is a severe stretch on our powers of conception. But I would say to the reader, do not trouble your head about this picture unduly; it is a stand-by for very occasional use. Normally, when reference is made to space-curvature, picture it as you picture a magnetic field. Probably you do *not* picture a magnetic field; it is something (recognisable by certain tests) which you use in your car or in your wireless

347

apparatus, and all that is needed is a recognised name for it. Just so; space-curvature is something found in nature with which we are beginning to be familiar, recognisable by certain tests, for which ordinarily we need not a picture but a name.

It is sometimes said that the difference between the mathematician and the non-mathematician is that the former can picture things in four dimensions. I suppose there is a grain of truth in this, for after working for some time in four or more dimensions one does involuntarily begin to picture them after a fashion. But it has to be added that, although the mathematician visualises four dimensions, his picture is *wrong* in essential particulars — at least mine is. I see our spherical universe like a bubble in four dimensions; length, breadth, and thickness, all lie in the skin of the bubble. Can I picture this bubble rotating? Why, of course I can. I fix on one direction in the four dimensions as axis, and I see the other three dimensions whirling round it. Perhaps I never actually see more than two at a time; but thought flits rapidly from one pair to another, so that all three seem to be hard at it. Can *you* picture it like that? If you fail, it is just as well. For we know by analysis that a bubble in four dimensions does not rotate that way at all. Three dimensions cannot spin round a fourth. They must rotate two round two; that is to say, the bubble does not rotate about a line axis but about a plane. I know that that is true; but I cannot visualise it.

I need scarcely say that our scientific conclusions about the curvature of space are not derived from the false involuntary picture, but by algebraic working out of formulae which, though they may be to some extent illustrated by such pictures, are independent of pictures. In fact, the pictorial conception of space-curvature falls between two stools: it is too abstruse to convey much illumination to the non-mathematician, whilst the mathematician practically ignores it and relies on the more dependable and more powerful algebraic methods of investigating this property of physical space.

Having said so much in disparagement of the picture of our three-dimensional space contorted by curvature in fictitious directions, I must now mention one application in which it is helpful. We are assured by analysis that in one important respect the picture is not misleading. The curvature, or bending round of space, may be sufficient to give a "closed space" — space in which it is impossible to go on indefinitely getting farther and farther from the starting-point. Closed space differs from an open infinite space in the same way that the surface of a sphere differs from a plane infinite surface.

\* \* \*

We may say of the surface of a sphere (1) that it is a *curved* surface, (2) that it is a *closed* surface. Similarly we have to contemplate two possible characteristics of our actual three-dimensional space, *curvature* and *closure*. A closed surface or space must necessarily be curved, but a curved surface or space need not be closed. Thus the idea of closure goes somewhat beyond the idea of curvature; and, for example, it was not contemplated in the first announcement of Einstein's general relativity theory which introduced curved space.

In the ordinary application of Einstein's theory to the solar system and other systems on a similar scale the curvature is small and amounts only to a very slight wrinkling or hummocking. The distortion is local, and does not affect the general character of space as a whole. Our present subject takes us much farther afield, and we have to apply the theory to the great super-system of the galaxies. The small local distortions now have cumulative effect. The new investigations suggest that the curvature actually leads to a complete bending round and closing up of space, so that it becomes a domain of finite extent. It will be seen that this goes beyond the original proposal; and the evidence for it is by no means so secure. But all new exploration passes through a phase of insecurity.

For the purpose of discussion this closed space is generally taken to be spherical. The presence of matter will cause local unevenness; the scale that we are now contemplating is so vast that we scarcely notice the stars, but the galaxies change the curvature locally and so pull the sphere rather out of shape. (Einstein's law of gravitation connects the various components of curvature of space with the density, momentum, and stress of the matter occupying it. I would again remind the reader that space-curvature is the technical name for an observable physical property, so that there is nothing metaphysical in the idea of matter producing curvature any more than in a magnet producing a magnetic field.) The ideal spherical space may be compared to the geoid used to represent the average figure of the earth with the mountains and ocean beds smoothed away. It may be, however, that the irregularity is much greater, and the universe may be pear-shaped or sausage-shaped; the 150 million light-years over which our observational survey extends is only a small fraction of the whole extent of space, so that we are not in a position to dogmatise as to the actual shape. But we can use the spherical world as a typical model, which will illustrate the peculiarities arising from the closure of space.

349

In spherical space, if we go on in the same direction continually, we ultimately reach our starting-point again, having "gone round the world." The same thing happens to a traveller on the earth's surface who keeps straight on bearing neither to the left nor to the right. Thus the closure of space may be thought of as analogous to the closure of a surface, and generally speaking it has the same connection with curvature. The whole area of the earth's surface is finite, and so too the whole volume of spherical space is finite. It is "finite but unbounded"; we never come to a boundary, but owing to the re-entrant property we can never be more than a limited distance away from our starting-point.

In the theory that I am going to describe the galaxies are supposed to be distributed throughout a closed space of this kind. As there is no boundary — no point at which we can enter or leave the closed space — this constitutes a self-contained finite universe.

Perhaps the most elementary characteristics of a spherical universe is that at great distances from us there is not so much room as we should have anticipated. On the earth's surface the area within 2 miles of Charing Cross is very nearly 4 times the area within 1 mile; but at a distance of say 4000 miles this simple progression has broken down badly. Similarly in the universe the volume, or amount of room, within 2 light-years of the sun is very nearly 8 times the volume within 1 light-year; but the volume within 4000 million light-years of the sun is considerably less than 8 times the volume within 2000 million light-years. We have no right to be surprised. How could we have expected to know how much room there would be out there without examining the universe to see? It is a common enough experience that simple rules, which hold well enough for a limited range of trial, break down when pushed too far. There is no juggling with words in these statements; the meaning of distance and volume in surveying the earth or the heavens is not ambiguous; and although there are practical difficulties in measuring these vast distances and volumes there is no uncertainty as to the ideal that is aimed at. I do not suggest that we have checked by direct measurement the falling off of volume at great distances; like many scientific conclusions, it is a very indirect inference. But at least it has been reached by examining the universe; and, however shaky the deduction, it has more weight than a judgment formed without looking at the universe at all.

Much confusion of thought has been caused by the assertion so often made that we can use any kind of space we please (Euclidean or non-Euclidean) for representing physical phenomena, so that it is impossible to disprove Euclidean space observationally. We can graphically represent (or misrepresent) things as we please. It is possible to represent the curved surface of the earth in a flat space as, for example, in maps on Mercator's projection; but this does not render meaningless the labours of geodesists as to the true figure of the earth. Those who *on this ground* defend belief in a flat universe must also defend belief in a flat earth.

There is a widespread impression, which has been encouraged by some scientific writers, that the consideration of spherical space in this subject is an unnecessary mystification, and that we could say all we want to say about the expanding system of the galaxies without using any other conception than that of Euclidean infinite space. It is suggested that talk about expanding space is mere metaphysics, and has no real relevance to the expansion of the material universe itself, which is commonplace and easily comprehensible. This is a mistaken idea. The general phenomenon of expansion, including the explanation provided by relativity theory, can be expounded up to a certain point without any recondite conceptions of space, ... but there are other consequences of the theory which cannot be dealt with so simply. To consider these we have to change the method, and partly transfer our attention from the properties and behaviour of the material system to the properties and behaviour of the space which it occupies. This is necessary because the properties attributed to the material system by the theory are so unusual that they cannot even be described without self-contradiction if we continue to picture the system in flat (i.e. Euclidean) space. This does not constitute an objection to the theory, for there is, of course, no reason for supposing space to be flat unless our observations show it to be flat; and there is no reason why we should be able to picture or describe the system in flat space if it is not in flat space. It is no disparagement to a square peg to say that it will not fit into a round hole.

I will liken the super-system of galaxies (the universe) to a peg which is fitted into a hole — space. ... We were only concerned with a little bit of the peg (the 150 million light-years surveyed) and the question of fit scarcely arose. When we turn to consider the whole peg we find mathematically that, unless something unforeseen occurs beyond the region explored, it is (for the purposes of this analogy) a square peg. Immediately there is an outcry: "That is an impossible sort of peg — not really a peg at all." Our answer is that it is an excellent peg, as good as any on the market, provided that you do not

351

want to fit it into a round hole. "But holes are round. It is the nature of holes to be round. A Greek two thousand years ago said that they are round." And so on. So whether I want it or not, the argument shifts from the peg to the hole — the space into which the material universe is fitted. It is over the hole that the battle has been fought and won; I think now that every authority admits, if only grudgingly, that the square hole — by which I here symbolise closed space — is a physical possibility.

The issue that I am here dealing with is not whether the theory of a closed expanding universe is right or wrong, probable or improbable, but whether, if we hold the theory, spherical space is necessary to the statement of it. I am not here replying to those who disbelieve the theory, but to those who think its strangeness is due to the mystifying language of its exponents. The following will perhaps show that there has been no gratuitous mystification:

I want you to imagine a system of say a billion stars spread approximately uniformly so that each star has neighbours surrounding it on all sides, the distance of each star from its nearest neighbours being approximately the same everywhere. (Lest there be any doubt as to the meaning of *distance*, I define it as the distance found by parallax observation, or by any other astronomical method accepted as equivalent to actually stepping out the distance.) Can you picture this?

— Yes. Except that you forgot to consider that the system will have a boundary; and the stars at the edge will have neighbours on one side only, so that they must be excepted from your condition of having neighbours on all sides.

— No; I meant just what I said. I want *all* the stars to have neighbours surrounding them. If you picture a place where the neighbours are on one side only — what you call a boundary — you are not picturing the system I have in mind.

— But your system is impossible; there must be a boundary.

— Why is it impossible? I could arrange a billion people on the surface of the earth (spread over the whole surface) so that each has neighbours on all sides, and no question of a boundary arises. I only want you to do the same with the stars.

— But that is a distribution over a surface. The stars are to be distributed in three-dimensional space, and space is not like that.

— Then you agree that if space could be "like that" my system would be quite possible and natural?

— I suppose so. But how could space be like that?

— We will discuss space if you wish. But just now when I was trying to explain that according to present theory space does behave "like that," I was told that the discussion of space was an unnecessary mystification, and that if I would stick to a description of my material system it would be seen to be quite commonplace and comprehensible. So I duly described my material system; whereupon *you* immediately raised questions as to the nature of space.

In the spherical universe the character of the material system is as peculiar as the character of the space. The material system, like the space, exhibits *closure*; so that no galaxy is more central than another, and none can be said to be at the outside. Such a distribution is at first sight inconceivable, but that is because we try to conceive it in flat space. The space and the material system have to fit one another. It is no use trying to imagine the system of galaxies contemplated in Einstein's and Lemaître's theories of the universe, if the only kind of space in our minds is one in which such a system cannot exist.

In the foregoing conversation I have credited the reader with a feeling which instinctively rejects the possibility of a spherical space or a closed distribution of galaxies. But spherical space does not contradict our experience of space, any more than the sphericity of the earth contradicts the experience of those who have never travelled far enough to notice the curvature. Apart from our reluctance to tackle a difficult and unfamiliar conception, the only thing that can be urged against spherical space is that more than twenty centuries ago a certain Greek published a set of axioms which (inferentially) stated that spherical space is impossible. He had perhaps more excuse, but no more reason, for his statement than those who repeat it today.

Few scientific men nowadays would reject spherical space as impossible, but there are many who take the attitude that it is an unlikely kind of hypothesis only to be considered as a last resort. Thus, in support of some of the proposed explanations of the motions of the spiral nebulae, it is claimed that they have the "advantage" of not requiring curved space. But what is the supposed disadvantage of curved space? I cannot remember that any disadvantage has ever been pointed out. On the other hand it is well known that the assumption of flat physical space leads to very serious theoretical and logical difficulties. . . .

A closed system of galaxies requires a closed space. If such a system expands, it requires an expanding space. This can be seen at once from the analogy that we have already used, viz. human beings distributed evenly over the surface of the earth; clearly they cannot scatter apart from one another unless the earth's surface expands.

This should make clear how the present theory of the expanding universe stands in relation to (*a*) the expansion of a material system, and (*b*) the expansion of space. The observational phenomenon chiefly concerned (recession of the spiral nebulae) is obviously expansion of a material system; and the onlooker is often puzzled to find theorists proclaiming the doctrine of an expanding space. He suspects that there has been confusion of thought of a rather elementary kind. Why should not the space be there already, and the material system expand into it, as material systems usually do? If the system of galaxies comes to an end not far beyond the greatest distance we have plumbed, then I agree that that is what happens. But the system shows no sign of coming to an end, and, if it goes on much farther, it will alter its character. This change of character is a matter of mathematical computation which cannot be discussed here; I need only say that it is connected with the fact that, if the speed of recession continues to increase outwards, it will ere long approach the speed of light, so that something must break down. The result is that the system becomes a closed system; and we have seen that such a system cannot expand without the space also expanding. That is how *expansion of space* comes in. I daresay that (for historical reasons) expansion of space has often been given too much prominence in expositions of the subject, and readers have been led to think that it is more directly concerned in the explanation of the motions of the nebulae than is actually the case. But if we are to give a full account of the views to which we are led by theory and observation, we must not omit to mention it.

What I have said has been mainly directed towards removing preliminary prejudices against a closed space or a closed system of galaxies. I do not suggest that the reasons for adopting closed space are overwhelmingly strong; but even slight advantages may be of weight when there is nothing to place in the opposite scale. (*Curved* space is fundamental in relativity theory, and the argument for adopting it is generally considered to be overwhelming. It is *closed* space which needs more evidence.) If we adopt open space we encounter certain difficulties (not necessarily insuperable) which closed space entirely avoids; and we do not want to divert the inquiry into a speculation as to the solution of difficulties which need never arise. If we wish to be non-committal, we shall naturally work in terms of a closed universe of finite radius $R$, since we can at any time revert to an infinite universe by making $R$ infinite.

There is one other type of critic to whom a word may be said. He feels that space is not solely a matter that concerns the physicists, and that by their technical definitions and abstractions they are making of it something different from the common man's space. It would be difficult to define precisely what is in his mind. Perhaps he is not thinking especially of space as a measurable constituent of the physical universe, and is imagining a world order transcending the delusions of our sensory organs and the limitations of our micrometers — a space of "things as they really are." It is no part of my present subject to discuss the relation of the world as conceived in physics to a wider interpretation of our experience; I will only say that that part of our conscious experience representable by physical symbols ought not to claim to be the whole. As a conscious being *you* are not one of my symbols; your domain is not circumscribed by my spatial measurements. If, like Hamlet, you count yourself king of an infinite space, I do not challenge your sovereignty. I only invite attention to certain disquieting rumours which have arisen as to the state of Your Majesty's nutshell.

The immediate result of introducing the cosmical term into the law of gravitation was the appearance (in theory) of two universes — the Einstein universe and the de Sitter universe. Both were closed spherical universes; so that a traveller going on and on in the same direction would at last find himself back at the starting-point, having made a circuit of space. Both claimed to be static universes which would remain unchanged for any length of time; thus they provided a permanent framework within which the small scale systems — galaxies and stars — could change and evolve. There were, however, certain points of difference between them. An especially important difference, because it might possibly admit of observational test, was that in de Sitter's universe there would be an apparent recession of remote objects, whereas in Einstein's universe this would not occur. At that time only three radial velocities of spiral nebulae were known, and these somewhat lamely supported de Sitter's universe by a majority of 2 to 1. There the question rested for a time. But in 1922 Prof. V. M. Slipler furnished me with his (then unpublished) measures of 40 spiral nebulae for use in my book *Mathematical Theory of Relativity*. As the majority had become 36 to 4, de Sitter's theory began to appear in a favourable light.

The Einstein and de Sitter universes were two alternatives arising out of the same theoretical basis. To give an analogy — suppose that we are transported to a new star, and that we notice a number of celestial bodies in the neighbourhood. We should know from gravitational theory that their orbits must be either ellipses or hyperbolas; but only observation can decide which. Until the observational test is made there are two alternatives; the objects may have elliptic orbits and constitute a permanent system like the solar system, or they may have hyperbolic orbits and constitute a dispersing system. Actually the question whether the universe would follow Einstein's or de Sitter's model depended on how much matter was present in the universe — a question which could scarcely be settled by theory — and is none too easy to settle by observation.

We have now realised that the changelessness of de Sitter's universe was a mathematical fiction. Taken literally his formulae described a *completely empty* universe; but that was meant to be interpreted generously as signifying that the average density of matter in it, though not zero, was low enough to be neglected in calculating the forces controlling the system. It turned out, however, that the changelessness depended on there being literally no matter present. In fact the "changeless universe" had been invented by the simple expedient of omitting to put into it anything that could exhibit change. We therefore no longer rank de Sitter's as a static universe; and Einstein's is the only form of material universe which is genuinely static or motionless.

The situation has been summed up in the statement that Einstein's universe contains matter but no motion and de Sitter's contains motion but no matter. It is clear that the actual universe containing both matter and motion does not correspond exactly to either of these abstract models. The only question is, Which is the better choice for a first approximation? Shall we put a little motion into Einstein's world of inert matter, or shall we put a little matter into de Sitter's Primum Mobile?

The choice between Einstein's and de Sitter's models is no longer urgent because we are not now restricted to these two extremes; we have available the whole chain of intermediate solutions between motionless matter and matterless motion, from which we can pick out the solution with the right proportion of matter and motion to correspond with what we observe. These solutions were not sought earlier, because their appropriateness was not realized; it was the preconceived idea that a static solution was a necessity in order that everything might be referred to an unchanging background of space. We have seen that this requirement should strictly have barred out de Sitter's solution, but by a fortunate piece of gate-crashing it gained admission; it was the precursor of the other non-static solutions to which attention is now mainly directed.

The deliberate investigation of non-static solutions was carried out by A. Friedmann in 1922. His solutions were rediscovered in 1927 by Abbé G. Lemaître, who brilliantly developed the astronomical theory resulting therefrom. His work was published in a rather inaccessible journal, and seems to have remained unknown until 1930 when attention was called to it by de Sitter and myself. In the meantime the solutions had been discovered for the third time by H. P. Robertson, and through him their interest was beginning to be realised. The astronomical application, stimulated by Hubble and Humason's observational work on the spiral nebulae, was also being rediscovered, but it had not been carried so far as in Lemaître's paper.

The intermediate solutions of Friedmann and Lemaître are "expanding universes." Both the material system and the closed space, in which it exists, are expanding. At one end we have Einstein's universe with no motion and therefore in equilibrium. Then, as we proceed along the series, we have model universes showing more and more rapid expansion until we reach de Sitter's universe at the other end of the series. The rate of expansion increases all the way along the series and the density diminishes; de Sitter's universe is the limit when the average density of celestial matter approaches zero. The series of expanding universes then stops, not because the expansion becomes too rapid, but because there is nothing left to expand.

We can better understand this series of models by starting at the de Sitter end. . . . There are two forces operating, the ordinary Newtonian attraction between the galaxies and the cosmical repulsion. In the de Sitter universe the density of matter is infinitely small so that Newtonian attraction is negligible. The cosmical repulsion acts without check, and we get the greatest possible rate of expansion of the system. When more matter is inserted, the mutual gravitation tends to hold the mass together and opposes the expansion. The more matter put in, the slower the expansion. There will be a particular density at which the Newtonian attraction between the galaxies is just strong enough to counterbalance the cosmical repulsion, so that the expansion is zero. This is Einstein's universe. If we put in still more matter, attraction outweighs repulsion and we obtain a model of a contracting universe.

Primarily this series of models is a series of alternatives, one of which has to be selected to represent our actual universe. But it has a still more interesting application. As time goes on the actual universe travels along the series of models, so that the whole series gives a picture of its life-history. At the present moment the universe corresponds to a particular model; but since it is expanding its density is diminishing. Therefore a million years hence we shall need a model of lower density, i.e. nearer to the de Sitter end of the series.

Tracing this progression as far back as possible we reach the conclusion that the world started as an Einstein universe; it has passed continuously along the series of models having more and more rapid expansion; and it will finish up as a de Sitter universe.

Allusion has been made to the fact that the recession of the galaxies in the present theory of the expanding universe is not precisely the effect foreseen by de Sitter. It may be well to explain the manner of the transition. The phenomenon that is generally called the "de Sitter effect" was a rather mysterious slowing down of time at great distances from the observer; atomic vibrations would be executed more slowly, so that their light would be shifted to the red and imitate the effect of a receding velocity. But besides discovering this, de Sitter examined the equations of motion and noticed that the *real* velocities of distant objects would probably be large; he did not, however, expect these real velocities to favour recession rather than approach. I am not sure when it was first recognised that the complication in the equations of motion was neither more nor less than a repulsive force proportional to the distance; but it must have been before 1922. Summarising the theory at that date, I wrote — "De Sitter's theory gives a double explanation of this motion of recession: first, there is the general tendency to scatter according to the equation

$$d^2r/ds^2 = \tfrac{1}{3} \lambda r;$$

second, there is the general displacement of spectral lines to the red in distant objects due to the slowing down of atomic vibrations which would be erroneously interpreted as a motion of recession." I also pointed out that it was a question of definition whether the latter effect should be regarded as a spurious or a genuine velocity. During the time that its light is travelling to us, the nebula is being accelerated by the cosmical repulsion and acquires an additional outward velocity exceeding the amount in dispute; so that the velocity, which was spurious at the time of emission of the light, has become genuine by the time of its arrival. Inferentially this meant that the slowing down of time had become a very subsidiary effect compared with cosmical repulsion; but this was not so clearly realised as it might have been. The subsequent developments of Friedmann and Lemaître were geometrical and did not allude to anything so crude as "force"; but, examining them to see what has happened, we find that the slowing down of time has been swallowed up in the cosmical repulsion; it was a small portion of the whole effect (a second-order term) which had been artificially detached by the earlier methods of analysis.

An Einstein universe is in equilibrium, but its equilibrium is unstable. The Newtonian attraction and the cosmical repulsion are in exact balance. Suppose that a slight disturbance momentarily upsets the balance; let us say that the Newtonian attraction is slightly weakened. Then repulsion has the upper hand, and a slow expansion begins. The expansion increases the average distance apart of the material bodies so that their attraction on one another is lessened. This widens the difference between attraction and repulsion, and the expansion becomes faster. Thus the balance becomes more and more upset until the universe becomes irrevocably launched on its course of expansion. Similarly if the first slight disturbance were a strengthening of the Newtonian attraction, this would cause a small contraction. The material systems would be brought nearer together and their mutual attraction further increased. The contracting tendency thus becomes more and more reinforced. Einstein's universe is delicately poised so that the slightest disturbance will cause it to topple into a state of ever-increasing expansion or of ever-increasing contraction.

 The original unstable Einstein universe might have turned into an expanding universe or into a contracting universe. Apparently it has chosen expansion. . . .

# GEORGES LEMAÎTRE

■

The Belgian astrophysicist Georges Lemaître (1894–1966) was also a priest, a circumstance that occasioned a certain amount of derisive laughter when he began theorizing about the "big bang" — the fiery event with which the expansion of the universe is thought to have begun. But his perception that the big bang could be examined by means of nuclear physics, explained here in a selection from his 1950 book *The Primeval Atom*, presaged the main theme of cosmology twenty years later.

# *The Primeval Atom*

THE PRIMEVAL ATOM hypothesis is a cosmogonic hypothesis which pictures the present universe as the result of the radioactive disintegration of an atom.

I was led to formulate this hypothesis, some fifteen years ago, from thermodynamic considerations while trying to interpret the law of degradation of energy in the frame of quantum theory. Since then, the discovery of the universality of radioactivity shown by artificially provoked disintegrations, as well as the establishment of the corpuscular nature of cosmic rays, manifested by the force which the Earth's magnetic field exercises on these rays, made more plausible an hypothesis which assigned a radioactive origin to these rays, as well as to all existing matter.

Therefore, I think that the moment has come to present the theory in deductive form. I shall first show how easily it avoids several major objections which would tend to disqualify it from the start. Then I shall strive to deduce its results far enough to account, not only for cosmic rays, but also for the present structure of the universe, formed of stars and gaseous clouds, organized into spiral or elliptical nebulae, sometimes grouped in large clusters of several thousand nebulae which, more often, are composed of isolated nebulae, receding from one another according to the mechanism known by the name of the expanding universe.

For the exposition of my subject, it is indispensable that I recall several elementary geometric conceptions, such as that of the closed space of Riemann, which led to that of space with a variable radius, as well as certain aspects of the theory of relativity, particularly the introduction of the cosmological constant and of the cosmic repulsion which is the result of it.

All partial space is open space. It is comprised in the interior of a surface, its boundary, beyond which there is an exterior region. Our habit of thought about such open regions impels us to think that this is necessarily so, however large the regions being considered may be. It is to Riemann that we are indebted for having demonstrated that total space can be closed. To explain this concept of closed space, the most simple method is to make a small-scale model of it in an open space. Let us imagine, in such a space, a ball in which we are going to represent the whole of closed space. On the spherical surface of the ball, each point of closed space will be supposed to be represented twice, by two points, $A$ and $A'$, which, for example, will be two antipodal points, that is, two extremities of the same diameter. If we join these two points $A$ and $A'$ by a line located in the interior of the ball, this line must be considered as a closed line, since the two extremities $A$ and $A'$ are two distinct representations of the same, single point. The situation is altogether analogous to that which occurs with the Mercator projection, where the points on the 180th meridian are represented twice, at the eastern and western edges of the map. One can thus circulate indefinitely in this space without ever having to leave it.

It is important to notice that the points represented by the outer surface of the ball, in which we have represented all space, are not distinguished by any properties from the other points of space, any more than is the 180th meridian for the geographic map. In order to account for that, let us imagine that we displace the sphere in such a manner that point $A$ is superposed on $B$, and the antipodal point $A'$ on $B'$. We shall then suppose that the entire segment $AB$ and the entire segment $A'B'$ are two representations of a similar segment in closed space. Thus we shall have a portion of space which has already been represented in the interior of the initial sphere which is now represented a second time at the exterior of this sphere. Let us disregard the interior representation as useless; a complete representation of the space in the interior of the new sphere will remain. In this representation, the closed contours will be soldered into a point which is twice represented, namely, by the points $B$ and $B'$, mentioned above, instead of being welded, as they were formerly, to points $A$ and $A'$. Therefore, these latter are not distinguished by an essential property.

Let us notice that when we modify the exterior sphere, it can happen that a closed contour which intersects the first sphere no longer intersects the second, or, more generally, that a contour no longer intersects the finite sphere at the same number of points. Nevertheless, it is evident that the points of intersection can only vary by an even number. Therefore, there are two kinds of closed contours which cannot be continuously distorted within one another. Those of the first kind can be reduced to a point. They do not intersect the outer sphere or they intersect it at an even number of points. The others cannot be reduced to one point; we call them the *odd contours* since they intersect the sphere at an odd number of points.

If, in a closed space, we leave a surface which we can suppose to be horizontal, in going toward the top we can, by going along an odd contour, return to our point of departure from the opposite direction without having deviated to the right or left, backward or forward, without having traversed the horizontal plane passing through the point of departure.

That is the essential of the topology of closed space. It is possible to complete these topological ideas by introducing, as is done in a geographical map, scales which vary from one point to another and from one direction to another. That can be done in such a manner that all the points of space and all the directions in it may be perfectly equivalent. Thus, Riemann's homogeneous space, or elliptical space, is obtained. The straight line is an odd contour of minimum length. Any two points divide it into two segments, the sum of which has a length which is the same for all straight lines and which is called the tour of space.

All elliptical spaces are similar to one another. They can be described by comparison with one among them. The one in which the tour of the straight line is equal to $\pi = 3.1416$ is chosen as the standard elliptical space. In every elliptical space, the distances between two points are equal to the corresponding distances in standard space, multiplied by the number $R$ which is called the radius of elliptical space under consideration. The distances in standard space, called space of unit radius, are termed angular distances. Therefore, the true distances, or linear distances, are the product of the radius of space times the angular distances.

When the radius of space varies with time, space of variable radius is obtained. One can imagine that material points are distributed evenly in it, and that spatio-temporal observations are made on these points. The angular distance of the various observers remains invariant, therefore the linear distances vary proportionally to the radius of space. All the points in space are perfectly equivalent. A displacement can bring any point into the center of the representation. The measurements made by the observers are thus also equivalent, each one of them makes the same map of the universe.

If the radius increases with time, each observer sees all points which surround him receding from him, and that occurs at velocities which become greater as they recede further. It is this which has been observed for the extra-galactic nebulae that surround us. The constant ratio between distance and velocity has been determined by Hubble and Humason. It is equal to $T_H = 2 \times 10^9$ years.

If one makes a graph, plotting as abscissa the values of time and as ordinate the value of radius, one obtains a curve, the sub-tangent of which at the point representing the present instant is precisely equal to $T_H$.

These are the geometric concepts that are indispensable to us. We are now going to imagine that the entire universe existed in the form of an atomic nucleus which filled elliptical space of convenient radius in a uniform manner.

Anticipating that which is to follow, we shall accept that, when the universe had a density of $10^{-27}$ gram per cubic centimeter, the radius of space was about a billion light-years, that is, $10^{27}$ centimeters. Thus the mass of the universe is $10^{54}$ grams. If the universe formerly had a density equal to that of water, its radius was then reduced to $10^{18}$ centimeters, say, one light-year. In it, each proton occupied a sphere of one angstrom, say, $10^{-8}$ centimeter. In an atomic nucleus, the protons are contiguous and their radius is $10^{-13}$, thus about 100,000 times smaller. Therefore, the radius of the corresponding universe is $10^{13}$ centimeters, that is to say, an astronomical unit.

Naturally, too much importance must not be attached to this description of the primeval atom, a description which will have to be modified, perhaps, when our knowledge of atomic nuclei is more perfect.

Cosmogonic theories propose to seek out initial conditions which are ideally simple, from which the present world, in all its complexity, might have resulted, through the natural interplay of known forces. It seems difficult to conceive of conditions which are simpler than those which obtained when all matter was unified in an atomic nucleus. The future of atomic theories will perhaps tell us, some day, how far the atomic nucleus must be considered as a system in which associated particles still retain some individuality of their own. The fact that particles can issue from a nucleus during radioactive transformations certainly does not prove that these particles pre-existed as such. Photons issue from an atom of which they were not constituent parts; electrons appear where they were not previously, and the theoreticians deny them an individual existence in the nucleus. Still more protons or alpha particles exist there, without doubt. When they issue forth, their existence becomes more independent, nevertheless, and their degrees of freedom more numerous. Also, their emergence in the course of radioactive transformations is a typical example of the degradation of energy, with an increase in the number of independent quanta or increase in entropy.

That entropy increases with the number of quanta is evident in the case of electromagnetic radiation in thermodynamic equilibrium. In fact, in black body radiation, the entropy and the total number of photons are both proportional to the third power of the temperature. Therefore, when one mixes radiations of different temperatures and one allows a new statistical equilibrium to be established, the total number of photons has increased. The degradation of energy is manifested as a pulverization of energy. The total quantity of energy is maintained, but it is distributed in an ever larger number of quanta, it becomes broken into fragments which are ever more numerous.

If, therefore, by means of thought, one wishes to attempt to retrace the course of time, one must search in the past for energy concentrated in a lesser number of quanta. The initial condition must be a state of maximum concentration. It was in trying to formulate this condition that the idea of the primeval atom was germinated. Who knows if the evolution of theories of the nucleus will not, some day, permit the consideration of the primeval atom as a single quantum?

# JAMES TREFIL

■

By the 1980s, physicists and astronomers throughout the world were collaborating in order to study the rapid evolution of the universe during the first moments of the big bang, when the largest structures in nature were subatomic particles and the rules of quantum physics applied everywhere. In this 1987 article James Trefil (b. 1938), a professor of physics and a widely read popularizer of science, reports on the ongoing alliance between physics and cosmology.

# *The New Physics and the Universe*

THE MOST EXCITING THING to happen in cosmology in the last decade has been the coming together — some people might call it the shotgun marriage — of elementary particle physics and the study of the evolution of the universe. This may seem at first glance to be a somewhat surprising development — after all, what could be more different than a proton and the whole universe? Yet, as we shall see shortly, there is a kind of logic in this development. In fact, we probably ought to have been able to anticipate it once we knew about the Big Bang and the cosmic microwave background radiation. The upshot of this new alliance between seemingly unconnected branches of science is that we are starting to come very close to answering some very old questions: where the universe came from and how it got to be the way it is.

I suppose we should start by talking about why there is any connection at all between elementary particle physics and cosmology, and maybe back up a little more and talk about what elementary particle physics is. Ever since 1805, when the English chemist John Dalton

365

published his theory that matter was made up of atoms, we have known that the materials we see in the world around us are not themselves elementary but are actually conglomerates of atoms (the word comes from the Greek for "that which cannot be divided"). Dalton based his theory on the regularities that were being uncovered by the new science of chemistry and produced a number of predictions that could be (and were) tested in the laboratory.

But the atom isn't really indivisible, as we all know. In the early years of this century Ernest Rutherford showed that almost all the mass of the atom is concentrated in an unimaginably dense nucleus, around which lightweight electrons move in orbit. Instead of being itself elementary, the atom (and hence all matter) was made of things more elementary still. Through the 1920's and 1930's a simple picture of the world developed, in which all matter was thought to consist of only three particles. These were the proton (with a positive electrical charge), the neutron (with no electrical charge), and the electron (with negative electrical charge). The protons and neutrons, in roughly equal numbers, combined to make the nuclei of atoms, and then the electrons went into their orbits to complete the job. As in Dalton's original scheme, these composite atoms would then come together to make the infinite complexity of molecules that we see in our everyday world.

This simple picture of the fundamental constituents of matter, alas, was not to survive long. By the 1950's, scientists studying the collisions of protons with nuclei began to realize that when a nucleus is torn apart, all sorts of strange and wonderful things can be found in the debris. There were, of course, the protons and neutrons you'd expect to find, but in addition there was a whole collection of new particles. These particles seemed to live inside the nucleus, but when freed from this environment they decayed — came apart — in very short times. The instability of these particles explains why we were not previously aware of them — they can be seen only briefly and under very special conditions such as those that exist in the laboratory.

It quickly became evident, in fact, that there were two classes of particles in nature. There were particles like the proton, the neutron, and the stuff that was seen in the debris of nuclear collisions. These particles seemed to be at home in the nucleus and to contribute in some way to holding it together. The other class of particles were those like the electron — particles that are normally found not inside the nucleus but outside of it. The former class of particles was christened *hadrons* (after the Greek for "strongly interacting ones"). The latter particles were called *leptons,* or weakly interacting ones.

It turned out that going into the nucleus after the hadrons had opened a real Pandora's box. Throughout the 1960's and into the 1970's the number of hadrons being discovered skyrocketed. The last time I looked there were over two hundred, but no one is counting anymore. It's clear that a system of the world with two hundred kinds of "elementary" particles just isn't going to work. Some way of ordering the hadrons had to be found. In the late 1960's such a scheme was proposed, and since then this scheme has come to dominate the study of the basic structure of matter. The scheme is actually very similar to the simple proton-neutron-electron picture of the 1930's. The idea is that just as all nuclei are merely different arrangements of the hadrons, all hadrons are themselves simply different arrangements of things more fundamental still, things for which the name *quarks* was invented. The name actually comes from a line in James Joyce's *Finnegans Wake* that goes "Three quarks for Muster Mark," and it reflects the fact that in its earliest incarnation the quark model had three different kinds of quarks in it.

If you accept for a moment that this model is valid, then it gives us a remarkably simple picture of the structure of matter. There are a small number of quarks, and these quarks are put together in different ways to make the hundreds of hadrons. These hadrons, in turn, combine to form the nuclei of the hundred-plus known chemical elements. The leptons are then added to make up atoms, which then come together to make the molecules.

At the moment, it appears that there are six different kinds of quarks in nature. We have seen hadrons that contain five of them in the laboratory, and it is expected that the next round of new high-energy accelerators (the machines that are popularly called "atom smashers") to come on line will find evidence for the sixth. The quarks are unlike any other particles known. For one thing, they have electrical charges that are either one-third or two-thirds that found on the electron or proton. They are the only particles that are supposed to have this sort of fractional charge. The quarks also have another property, whimsically called *color,* which is analogous to electrical charge. Each type of quark can have one of three possible color charges, just as it can have one of a small number of possible electrical charges.

The six quarks are called, respectively, "up," "down," "strange," "charm," "bottom," and "top." The names are suggested by various properties of the particles and by analogies — if a hadron contains a charm quark, for example, it will survive much longer than you might expect (i.e., lead a "charmed" life). For our purposes, however, we need only know that according to our present notions all hadrons are made from these six quarks.

I should point out one other difference between quarks and hadrons. No one has succeeded in producing unambiguous evidence for the existence of free quarks (that is, quarks not locked into hadrons). Indeed, present theories tell us that quarks cannot be torn loose from hadrons — in the jargon of physics, they are *confined*. The best analogy is to imagine that the quarks are something like the ends of a rubber band. No matter what you do, you can't pull the end of a rubber band off. You can tear the band apart, but this just produces shorter strings with two ends. In just the same way, we believe that you can never pull a quark out of a hadron.

Given this picture of the structure of matter, it becomes easy to see how elementary particle physics and cosmology come together. One thing we know about the Big Bang is that during its early stages the universe was much hotter than it is now. When any material — be it the universe or a pot of boiling water — is heated, its constituent parts move faster. This means that when the inevitable collisions between these constituents occur, they are much more violent in a hot material than in a cold one. As we start to trace the Big Bang back through time, then, we will be encountering situations in which ever more violent (read higher-energy) collisions are taking place at the microscopic level. This is why the things we have come to know about the structure of matter can help us understand the early evolution of the universe.

Let's take atoms as a simple example. The atom is a relatively airy, fragile structure. The electrons, far from the nucleus in their orbits, can be knocked loose in a collision without much difficulty. What this means is that if the temperature of a material is very high, all the electrons will be separated from their nuclei. Matter in this state, in which positively charged nuclei and negatively charged electrons can move around independently of each other, is called a *plasma*. If an electron in a plasma does manage to attach itself to a nucleus to form an atom, it will be torn off in a subsequent collision. When the temperature of a material is too high, then, atoms simply cannot exist.

The temperature needed to create a plasma isn't all that great. It can easily be produced in our laboratories. (Indeed, a large part of the current attempts to reproduce the nuclear fusion that powers the stars depends on being able to produce and confine plasmas.) Since most of the matter in the Sun and other stars is in a plasma state as well, plasma should not be thought of as some sort of exotic and unknown form of matter. In fact, it is by far the most common state of matter in the universe.

From these arguments, it follows that as the universe was expanding and cooling during the early stages of the Big Bang, there must have been a time when the temperature was too high for atoms to exist, followed by a time when the temperature had fallen to the point where they could. In between these two, there must have been a transition during which the atoms formed.

It is useful to consider another, more familiar process in nature which has the same properties. When you lower the temperature of water, there is a sharply defined temperature — 32°F — when it freezes. At this temperature the water changes form, from a liquid to a solid. We say it freezes, and physicists say that it goes through a *phase change*.

We can make the same sort of statement about the formation of atoms in the early universe. It turns out that the "freezing" of the plasma into atoms isn't as sharply defined as the freezing of liquid water into ice, but the two are similar enough to give us a useful analogy. When the universe was about 100,000 years old, the temperature had dropped to the point where electrons that attached themselves to atoms would no longer necessarily be torn loose in collisions. By the time the universe was a million years old, the temperature was so low that the disruption of an atom by collision was a fairly rare event. Matter in the form of atoms was safely established, and we say that the atoms "froze" out of a plasma in the period around 500,000 years.

If we continue our journey backward in time, the next important juncture occurs about three minutes after the Big Bang. This was another freezing, or phase change, but it involved the nuclei themselves rather than the atoms. At around the three-minute mark the temperature had dropped to the point that when two hadrons (a proton and a neutron, for example) came together to form a nucleus, it had a reasonable chance of surviving. Before the universe was three minutes old, any such nucleus that was formed would quickly have been disrupted. Consequently, before this time, matter existed in the form of a sea of hadrons and leptons.

The freezing of hadrons into nuclei at three minutes, like the freezing of the atoms later, was not a simple, instantaneous process. It proceeded in a stepwise fashion, first forming deuterium (one proton and one neutron) and then, by subsequent collisions, tacking on a few more protons and neutrons to form various isotopes of helium and even a little lithium. In this collection of nuclei, it turns out that the simplest — deuterium — is the most easily torn apart in collisions. Therefore, the whole nucleus-building process had to wait until the deuterium was stable before the rest could proceed. Once deuterium was around, the rest of the nuclei formed quickly.

You might be interested to know, as a point of historical interest, that the Big Bang theory was first proposed back in the 1940's as a mechanism by which the nuclei in the periodic table could have been synthesized all at once, at the very beginning. The idea was that the universe created its nuclear "capital" early on, and that it has been living on it ever since. It quickly became obvious, however, that the expansion of the universe would carry protons and neutrons apart from each other in short order, thereby preventing the formation of anything heavier than lithium (three protons, two neutrons). Because of this realization, the Big Bang theory did not really catch on until we understood the formation of elements by stars. Our present understanding is that all nuclei heavier than lithium were created in stars much later in the evolution of the universe.)

Before the three-minute mark, then, the matter in the universe was in the form of a collection of hadrons and leptons, but no nuclei were present. After the first three minutes, hadrons were locked up in the newly formed nuclei.

Resuming our journey backward in time, we pass through a universe where more and more energy is present in collisions, and in which more and more exotic hadrons are created from this energy. The world looks less and less familiar. Finally, when we get back to a time roughly ten microseconds after the beginning, we encounter yet another phase change. At that time, the hadrons were created out of their constituent quarks, or, if you prefer, the quarks froze to create the hadrons. Before this time, matter existed in the most fundamental form we know, as a collection of quarks and leptons. Afterward, the quarks disappeared and were replaced by hadrons.

It is tempting to suppose that this particular phase change took place in just the way the formation of atoms and nuclei did — by a process of adding the pieces together one at a time. In point of fact, however, we now understand that this particular transition (and others we'll discuss shortly) must have been rather different. It was much more like the formation of a magnet out of iron atoms than the freezing of water into ice. In the case of a magnet, at high temperatures all the iron atoms are aligned randomly in space, each one pointing in its own direction. In such a situation the magnetic fields of the individual atoms tend to cancel one another out and the material taken as a whole has no magnetic properties. As the temperature is lowered, however, there comes a point where the individual atoms suddenly line up. The fields now reinforce one another and a magnet is born. The sudden lining up of the atoms is a phase change, even though it doesn't much resemble the freezing of water.

The transition from quarks to hadrons was like the formation of a magnet. At ten microseconds the free, randomly moving quarks suddenly rearranged themselves, just as the individual atoms do in a magnet. After this transition, the quarks were locked in, never to be seen as free particles again.

It would seem, then, that we have reached the point of ultimate simplicity in our backward journey through the history of the universe. Matter has been broken down into its basic constituents — quarks and leptons — and it seems that there's nowhere else to go. This "end of the road" feeling isn't justified, however, because up to now we've been considering only half the story. The basic constituents of all matter may be quarks and leptons, but until we've talked about how these constituents interact with each other, we really haven't given a full description of the state of the universe.

Physicists attack this aspect of the problem by investigating forces. By the end of the nineteenth century we knew of two forces that allowed matter to interact. There was the familiar force of gravity, holding the planets in their orbits and people on the surface of the Earth, and there was the force of electromagnetism, which governed magnets, electrical charges, and other such phenomena. In the twentieth century we added two new forces to this roster. One of these is the so-called *strong force,* the force that holds the hadrons in the nucleus together and locks the quarks into hadrons. The other is the *weak force,* which governs some forms of radioactive decay. Everything that happens in the universe happens because one or more of these forces is acting.

At first glance this categorization of forces appears to be a major achievement. Explaining the infinite number of processes in the universe in terms of just four basic forces seems like an enormous simplification. It is, of course, but that doesn't stop theoretical physicists from wondering if they can't do better. You can understand this point of view by asking the question: "What is the minimum number of forces you need to make a universe?" According to the basic laws of motion that Isaac Newton wrote down in the seventeenth century, any change in an object's state of motion must be caused by the action of a force. A universe with no forces could have no changes — it would be a dull place indeed. A universe in which things happen, therefore, needs at least one force, but it really doesn't need more. The question that naturally arises, then, is whether or not the four forces we see are truly distinct, or whether they are actually no more than different aspects of the same single force.

In the language of theoretical physics, a theory in which two different forces are seen to have an underlying identity is called a *unified field theory*. Let me give you a simple example to show how such a theory might operate: Suppose you were rolling marbles toward two holes in the ground, and that one hole had a slight hill in front of it. If you rolled the marbles slowly, some would stop on that hill and come rolling back toward you. Looking at your results, you would say that the holes were very different — the marbles fell into one but rebounded from the other. In that universe, there would be two kinds of holes.

If, however, you rolled the marbles faster, they would both be able to make it into the region of the holes, and you would see that the two systems were basically identical. The point is that you could see this underlying identity only if the marbles were moving very fast — if they had a lot of energy. At lower speeds the difference would reappear.

The situation with the underlying forces is the same. If particles collide at low energies, the forces look different. If they collide at higher energies, the apparent differences go away and the true underlying identity of the forces becomes manifest. We say that the forces become unified. Since high particle speeds occur only at high temperatures, we expect to see the forces unified early in the life of the universe, when it was very hot. In fact, as we move backward in time from the freezing of the quarks, the major changes we will see will be of this type.

So long as we are thinking about forces, we might take a moment to summarize the current thoughts on what a force is and, in the process, to introduce a new type of particle. When we talk about two particles exerting a force on each other, we do not have in mind something analogous to a collision between billiard balls, where the particles actually come into contact. The picture of the interaction that comes from our modern understanding of physics is more complex. When particles *A* and *B* come near each other, a third particle is kicked out of *A*, travels through the vacuum, and hits *B*. It is the action of this intermediary particle that changes the motion of *A* and *B*, and which we see as producing a force.

Each of the four forces has its own particle to act as intermediary — indeed, we think of the difference in the forces as arising from the differences in the exchanged particles. These force-generating particles are neither quarks nor leptons, but a third class of fundamental entities. This class can be referred to in a number of ways, the most usual being *bosons* or *gauge particles.* The former name comes from the twentieth-century Indian physicist S. N. Bose, who first investigated the general properties of this type of particle; the second refers to the fact that the types of theories that deal with the unification of forces (and hence with the process by which the bosons associated with different forces are seen to be identical) are called gauge theories. This term is something of an historical accident and has nothing to do with the early universe.

With each of the forces, then, we can associate a particle. We can think of this particle as carrying the force. The most familiar of the gauge particles is the *photon,* the particle that makes up ordinary light and other forms of electromagnetic radiation. The exchange of photons is what generates the electromagnetic force. The weak force is generated by a particle called the *intermediate vector boson.* This particle had been predicted by our theories and was seen in the laboratory for the first time in 1983. The strong force is generated by the exchange of things called *gluons* between quarks (they "glue" the quarks together). We believe that the gluons, like the quarks themselves, cannot be seen as independent particles in the laboratory. Finally, the force of gravity is associated with the exchange of a particle called the *graviton,* which has not yet been observed directly.

The idea that the four forces are actually one — or, equivalently, that all the gauge particles are really identical in spite of their apparent differences — is an old one in physics. Albert Einstein spent the latter half of his life trying unsuccessfully to show that gravity and electromagnetism were really identical, a fact that should give you some idea of the difficulty of the problem. In the late 1960's and early 1970's, theorists began to develop the modern version of the unified field theory. It turns out that this unification doesn't happen all at once; we don't suddenly find an energy at which all four forces become identical. Instead, we find that the unification takes place in a stepwise fashion, with each force joining in at a higher energy.

With this background, we can resume our journey back toward the moment of creation. We left off, you will recall, with the freezing of quarks into hadrons at the ten-microseconds mark. At this time, the temperatures were still low enough so that all four forces were seen as distinct and different, and so they remain until our trip takes us back to one ten-billionth of a second after the beginning. At this point, the temperatures were so high that the first unification — which turns out to be the coming together of the electromagnetic and weak forces — could occur. Before the universe was one ten-billionth of a second old, the quarks and leptons interacted with each other through the medium of only three fundamental forces. From this time on, they had the full complement of four.

It may seem strange to you that we can talk with such confidence about what happened so long ago under such unusual conditions. In point of fact, however, today we can re-create the conditions of the universe as it was when it was one ten-billionth of a second old, right here in our particle accelerators. At laboratories such as the European Center for Nuclear Research (CERN) in Geneva, Switzerland, there are machines that take two beams of protons, accelerate them to almost the speed of light, and then let them collide head-on. In some of these collisions the temperature in a volume of space about the size of a proton is raised to the levels it had when the weak and electromagnetic forces unified, and scientists can study what happens under these conditions. Consequently, we have some confidence in our ability to describe the history of the universe at least as far back as this point.

The next important milestone on our way back to the Big Bang occurs at $10^{-35}$ seconds. This is a time so short as to be almost incomprehensible — it is represented by a decimal point, thirty-four zeros, and then a one. At this time the second unification took place. It involved the coming together of the strong force with the newly unified electroweak force. Before $10^{-35}$ seconds there were two forces operating in the universe; afterward there were three.

In many ways, this particular "freezing" was the most important in the entire history of the universe. The theories that describe it are called *grand unified theories,* or GUTs. In addition to describing what happens to the forces themselves, they make some rather striking statements about what else was happening in the universe when the freezing occurred. The most important of these predicted processes is called *inflation*.

You know that when water freezes into ice, it expands. This is why bottles left outside in winter are often broken when their contents freeze. In an analogous process, the GUTs predict that when the strong forces "froze out" at $10^{-35}$ seconds, the universe expanded at a rate much higher than that associated with the normal Big Bang. Although the details of how the rapid expansion occurred are a little complicated, the end result is that in a very, very short time the universe expanded from something with a characteristic size (called its *radius of curvature*) smaller than that of an individual proton to something with a radius of curvature the size of a grapefruit. This rapid ballooning of the size of the universe early in its development is what is called inflation, and a universe that undergoes this process is said to be an *inflationary universe*.

The discovery of inflation solved a very old problem in cosmology. One of the features of the cosmic microwave background is that it is very uniform — no matter which way we look we see essentially the same sort of radiation coming toward us. This means that the regions of the universe that emitted the original radiation must have been at the same temperature when the radiation was emitted. But in the conventional Big Bang scenario, these regions would never have been close enough together to have established a "shared" uniform temperature. This was known, for technical reasons, as the *horizon problem*. With inflation, however, we can get around this difficulty. Before inflation occurred, the universe was much smaller than we might otherwise think, and all its parts would have been in contact with one another. Once a uniform temperature was established, it was maintained through the inflationary expansion and is seen today as the uniformity in the microwaves. This is an elegant solution to a fundamental problem, and it made scientists much more ready to accept the GUTs than they usually are when confronted with new ideas.

The GUTs also predict that quarks and leptons, hitherto thought of as separate and distinct, can actually be transformed into each other if the energies at which they collide are high enough. A quark, in other words, can become a lepton, and vice versa. To a

375

physicist, this is equivalent to saying that the two particles are identical. This means that in the period before the GUT freezing, the universe consisted of just two sorts of particles: the quark/leptons and the gauge particles (or bosons) that are exchanged to generate the forces. In this period, there were also only two forces: the unified strong-electroweak force and gravity.

We believe that the ultimate freezing occurred at $10^{-43}$ seconds, at what is called the *Planck time*. (It's named after the early twentieth-century physicist Max Planck, one of the founders of our modern ideas about the behavior of the subatomic realm, the ideas we call *quantum mechanics*.) At this time, as you probably have already anticipated, the force of gravity becomes unified with the rest of the forces. Before the Planck time, there was only a single unified force operating to produce change in the universe.

The theories that describe this ultimate unification are generally referred to by the term *supersymmetry*, or SUSY. In addition to predicting the unification of gravity with the other forces, these theories indicate that at the temperatures characteristic of the birth of the universe, the two remaining classes of particles — the quark/leptons and the gauge particles — become interchangeable with each other. Consequently, they tell us that in its moment of birth, the universe was as simple as it could possibly be. There was one sort of particle interacting through a single, fully unified force. Each subsequent freezing made the universe more complex and more diverse.

The supersymmetry theories have another property that is very interesting to cosmologists. All versions of the theory predict the existence of a large number of as-yet-undiscovered elementary particles. These new particles would form a sort of mirror image of our ordinary world. For each quark, there would be a supersymmetric quark partner (dubbed, predictably, a *squark*), and so on. In some versions of the theory this mirror-image picture is carried to the extreme of imagining an entire shadow universe made of a new kind of matter, interacting with ours only through the force of gravity. But even more conventional theories suggest that one or another of the new particles predicted by the theories may have survived to the present day.

In fact, if these sorts of particles are present in great numbers, they may even exert enough of a gravitational force to affect the rate of expansion of the universe, and hence its ultimate fate. They might, in fact, be the dark matter whose existence is predicted by our observations of clusters of galaxies (and, we might add, by part of the inflationary-universe theory). In this way, the predictions of supersymmetry and the predictions of the GUTs intertwine with each other.

This, then, is our current picture of the way the universe arrived at its present state. How much confidence can we have in it? Is it dependable, or are scientists talking about GUTs and SUSYs like medieval scholars arguing about how many angels can dance on the head of a pin? As I mentioned, we can have some confidence in the story I've just told, back to the ten-billionth-of-a-second mark, simply because we are capable of reproducing the conditions of the universe at that time in our laboratories. The cosmological implications of the GUTs cannot be tested directly, since we will probably never be able to produce the temperatures that existed at $10^{-35}$ seconds, but we can test other predictions of these theories in the laboratory. The present verdict on the GUT is mixed: Some predictions seem to be borne out, but the most striking prediction — that the proton is not an eternally stable particle but will eventually decay — has yet to be verified. The SUSYs remain firmly in the realm of theoretical physics, with experimentalists just starting to think about finding ways to test them. Consequently, you should probably think of the description of the very early universe I've just given you as something more in the nature of a progress report than a final answer.

# OWEN GINGERICH

■

Our conception of the universe at large traditionally has intertwined with our thoughts about God and creation, though the
latter topics go beyond the reach of science. Here the eminent
historian of science Owen Gingerich (b. 1930) considers some
of the theological and philosophical issues raised by the big-
bang cosmology

# *Let There Be Light: Modern Cosmogony and Biblical Creation*

IN THE EARLY PART of this century, when the Harvard philosophy
department was graced with such luminaries as William James,
Alfred North Whitehead, George Herbert Palmer, and George San-
tayana, the department members consulted with Harvard's President
Eliot about an appropriate motto to be carved on the façade of the
proposed new philosophy building. Prof. Palmer recommended a
quotation from Protagoras, "Man is the measure of all things," but
President Charles Eliot didn't commit himself. When Emerson Hall
was finally completed in 1905 and the motto unveiled, it read "What
is man that thou art mindful of him?" I don't know how those illus-
trious philosophers reacted to this line from Psalm 8, but I do think
it poses a fundamental question for us all. Does man have an essential
place in the universe, or is he a cosmical nonentity?

I shall return eventually to that question, but let me begin with
the first quotation, "Man is the measure of all things." In a curious
dimensional way, which Protagoras could scarcely have intended, we
are a kind of yardstick for the universe, standing between the microscopic and macroscopic worlds. Up and beyond the universe extend

twenty-five orders of magnitude larger than the human body, while down and within, the atomic nucleus lies nearly sixteen orders of magnitude smaller.

In the microscopic world of the atom there are marvels to stagger the imagination. Physicists have become modern alchemists, transforming gold to mercury and uranium to strontium. Submicroscopic particles can annihilate antiparticles, transforming matter into a burst of energy, or, in reverse, pure energy can give birth to matter with exotic, newly described properties of charm, color, and strangeness as well as mass and electrical charge. At the other end of the scale, astronomers plumb the world of the large, delineating our Milky Way as a giant pinwheel galaxy containing over 200 billion stars — about fifty for every man, woman, and child on earth. And beyond our own stellar system, countless other galaxies are scattered out to the fringes of the universe, some 10 to 20 billion light years away.

These are all discoveries of our own century, most of them scarcely fifty years old. Yet none of them is quite as astonishing as the scientific scenario that has now been outlined for the first moments of creation. During the past two decades knowledge of the world of the smallest possible sizes, the domain of particle physics, has been combined with astronomy to describe the universe in its opening stages. The physics ultimately fails as the nucleo-cosmologists push their calculations back to Time Zero, but they get pretty close to the beginning, to $10^{-43}$ second. At that point, at a second split so fine that no clock could measure it, the entire observable universe is compressed within the wavelike blur described by the uncertainty principle, so tiny and compact that it could pass through the eye of a needle. Not just this room, or the earth, or the solar system, but *the entire universe* squeezed into a dense dot of pure energy. And then comes the explosion. "There is no way to express that explosion," writes the poet Robinson Jeffers,

> . . . All that exists
> Roars into flame, the tortured fragments rush away from
>     each other into all the sky, new universes
> Jewel the black breast of night; and far off the outer nebulae
>     like charging spearmen again
> Invade emptiness.

It's an amazing picture, of pure and incredibly energetic light being transformed into matter, and leaving its vestiges behind. It's even more astonishing when we realize that the final fate of the universe, whether it will expand forever or fall back on itself to a future Big Crunch, was determined in that opening moment.

But, you may well ask, what evidence do we have that this wondrous tale is true? Or is it some kind of strange fiction? "Where wast thou when I laid the foundations of the earth?" the Lord asked Job from the whirlwind, and, to be sure, none of us was there. So I must admit that what the scientists in my detective story have devised is an intricate reconstruction, assembled just as the detective in your who-done-it thriller systematically reconstructs the crime. That metaphor is not bad, yet it is not quite adequate to convey the grandeur and extent of modern science. Personally, I believe there is a better metaphor in likening science to a beautiful, panoramic tapestry. It is beautiful in the way the contrasting patterns and themes are organized into a unified, coherent whole. It is panoramic in its scope, the majestic sweep that covers all of nature from the subatomic particles to the vast outer reaches of space and time. Like a tapestry, it is a human artifact, ingeniously and seamlessly woven together. It is not easy to unravel one small part without affecting the whole.

Let me try to weave a few threads of this vast tapestry into place. One part of the design concerns our knowledge of distances, another the time-scale of the universe. Both are woven tightly together.

The ancient Greeks already knew the size of the earth, and by triangulation they got the distance to the moon fairly well. Before the end of the seventeenth century, more precise triangulation, made possible with the telescope and refined measuring scales, allowed astronomers to get distances within the solar system. Then, allowing a moving earth to sweep out a larger baseline, they triangulated to the nearest stars, this delicate feat being accomplished by the middle of the last century. Proxima Centauri, the first star beyond the sun, is 400 trillion kilometers away; its light takes four years to reach us. But triangulation can barely penetrate a hundred times farther, not even one percent of the distance across the Milky Way system. A thousand light years lie beyond the reach of the geometrical methods.

Nevertheless, we can plumb the vaster depth of space with techniques that use the geometrical diminution of light with distance, well expressed by the aphorism "Faintness means farness." To proceed, we must single out classes of objects that always have the same intrinsic brightness, and then by observing the apparent brightness of one of these objects, we can calculate its farness from its observed faintness. In the second decade of this century, Harlow Shapley, a young astronomer at Mount Wilson Observatory, succeeded in calibrating the luminosities of a type of brilliant pulsating star. Because these stars rhythmically varied in light, it was comparatively easy to detect and identify them, and once they were found and measured, their distances could be calculated. In particular, Shapley was studying the immense aggregations of stars called globular clusters, and he found that these stellar congeries were at distances greater than anyone had contemplated before that time. Using the spatial arrangement of globular clusters, Shapley inferred that our Milky Way galaxy was a huge disk of stars and star clouds over a hundred thousand light years across.

It remained for his colleague at Mount Wilson, Edwin Hubble, to apply the same technique to the spiral nebulae and to show that these were also disklike pinwheels of stars, comparable to our own Milky Way, but distributed throughout space beyond our own system. But at a distance of 10 million light years, even the most brilliant pulsating stars are diminished below the limit of visibility for the largest of the Earth-based telescopes.

A class of still brighter objects is needed if we are to explore even greater depths of space, and here the globular clusters themselves, taken as units, offer a candle in the darkness. This step is slightly risky, because not all globular clusters are equally bright, and it takes some averaging and checking with alternative methods to work with any confidence. Still, it does allow distances of perhaps 50 million light years to be fathomed.

To probe further into the inky blackness of space requires shrewd scheming and judicious guesswork. If we assume some uniformity in space and time, then distant galaxies must on the average be like those nearby, and so the galaxies themselves, well chosen, can become the measures for reaching out a few billion light years.

With distances so great, unimaginable vistas of time are now also involved. The problem of ages is closely akin to that of the distances. Around 1917, V. M. Slipher, working at the Lowell Observatory, found that key features in the spectra of the spiral nebulae were shifted toward the red end of the spectral rainbow compared to similar laboratory measurements. When interpreted as a Doppler shift (similar to the change in pitch of a siren as it passes and recedes from us), these red shifts indicated surprisingly high recessional velocities, some exceeding a thousand kilometers per second. Such speeds are high by any terrestrial standard although slow compared to the speed of light. After Edwin Hubble had determined the distances to some of these spirals, it became apparent that the farther the galaxy, the faster it was rushing away from us.

These data arrived on the scene just as the cosmologists had begun to speculate on the large-scale properties of the universe, and out of this confluence of theory and observation arose the concept of the expansion of the universe. Given the rate of expansion and the distances of the galaxies, we can calculate backwards to the time when they were back together, "crushed in one harbor" in Robinson Jeffers's phrase. That comes out about 10 to 20 billion years ago, a time that can be interpreted as the age of the universe itself.

There are two other, quite independent ways to get the ages of some very old things in the universe. I have before me a fragment of the Allende meteorite, a ton of which fell in northern Mexico in 1969. It is probably the oldest object I'll ever touch, dated at 4.6 billion years. How is that age determined? The meteorite contains trace amounts of several radioactive isotopes — such as palladium-107, which very gradually changes to silver. In the early stages of the solar system, certain minerals crystallized out as the meteorite formed and cooled. Once the atoms were locked up in the crystal structure, they couldn't move, but the radioactive palladium would slowly disintegrate at a known rate, increasing the silver-107. But the original amount of the other silver isotopes remains unchanged, so careful analyses of the isotope ratios in several of the minerals can give the time of crystallization. According to this analysis, the minerals in the small white nodules within the Allende meteorite crystallized 4.6 billion years ago.

Not only has it been possible to determine the age of a specific object such as the Allende meteorite but even the elements themselves can be dated. If the atoms were infinitely old, then radioactive uranium and thorium would have turned to lead. Their very existence tells us that they were formed at a finite time past. Long ago a supernova went off in our neighborhood of the Milky Way, spewing forth within its nuclear ashes a wide variety of elements. We are all recycled star stuff. The iron in our blood is a typical sample of reused cosmic wastes. Included among that stellar ejecta is a variety of radioactive atoms, and from their rates of radioactive decay the nuclear physicists can calculate back to the time of the supernova explosion. These radioactive dating methods do not give a highly precise answer, but what seems to me very impressive is that the results fall in the same ball park as the expansion age, that is, roughly 10 billion years.

The third way to get a truly ancient age is to calculate the evolution of stars in a globular star cluster, one of those immense and fundamental units that Shapley used for inferring the vastness of our Milky Way system. The globular clusters are gravitationally very stable, and therefore they can hang together for a long time. Calculations of the ages of old yellow and orange stars in these clusters are complex, and could scarcely be made without high-speed electronic computers, but they do indicate that the globulars are very ancient objects, dating from perhaps 15 billion years ago.

I've described three independent ways for age-dating very old things: globular cluster stars, radioactive isotopes, and the universe itself as established from its expansion. The first two of these indicate that our part of the universe is pretty old, but they don't preclude other parts from being much, much older. Hence we ought to ask if there is any confirming evidence that we live in an evolving universe that has a specific beginning.

The quasars, discovered in the 1960s, furnish evidence that the universe is changing, and was far different 10 billion years ago. As we look out into space, we are looking backwards into time, and from the distribution of quasars by brightness steps, we can conclude that they were actually more abundant in times long past, some billions of years ago. In other words, the universe was different then; the quasars, the most brilliant of the cosmic fireworks, were far more plentiful in the early stages of the universe. Now they are spent, and their dying embers have perhaps vanished into vast black holes. Their distribution thus provides evidence that the universe is changing and evolving on a grand scale. It does not force us to conclude that the universe had a definite beginning, though that seems to be an inviting corollary.

The evidence that the universe had a definite, superdense beginning is somewhat different in nature, and we must approach it more circuitously. Suppose we run time backward in our calculations, to see what the universe is like as its density increases. The total mass and energy remains the same, but the temperature rises as the matter-energy is compressed. Finally the temperature becomes so high, and the mean energy of the components so great, that the presently known laws of physics no longer apply. This happens when we are a split second from squashing the universe into nothingness.

Now let us run the clock forward again. In the first few minutes things happen much faster than we can possibly describe them. In the first microseconds the high-energy photons vastly outnumber particles of matter, but there is a continual interchange between the photons and heavy particles of matter and antimatter. Einstein's famous $E = mc^2$ equation helps describe how the energy of the photons is converted into mass and vice versa. By the end of the first millisecond, the creation of protons and antiprotons is essentially finished, and the vast majority have already been annihilated back into photons. As the universe loses its incredible compression, the average energy per photon drops, and during this first second electrons and antielectrons (called positrons) are repeatedly formed and annihilated, finally leaving about 100 million photons of light for every atom — a ratio that still remains today.

The thermonuclear detonation of the universe is now on its way, and in the next minute fusion reactions take place that build up deuterium and helium nuclei. After the first few minutes the explosive nuclear fireworks are over, but the headlong expansion continues, and the cosmic egg gradually cools. Eventually the photons lose enough of their energy so that they no longer interact so intensively with the atoms; then they fly through space unimpeded, although their color is redshifted by the expansion of the universe. The left-over radiation becomes ours to observe, and those photons have been observed by looking out every direction into space, the fossil evidence of the primeval fireball of the Big Bang. This observed background radiation is one piece of evidence supporting the contemporary scientific picture of creation. The other is the observed abundance of helium and of deuterium, which match well the predicted amounts that would be formed in that cosmic explosion.

The Big Bang makes a thrilling scenario. And its essential framework, of everything springing forth from that blinding flash, bears a striking resonance with those succinct words of Genesis 1:3: "And God said, Let there be light." Who could have guessed even a hundred years ago — not to mention two or three thousand years ago — that a scientific picture would emerge with electromagnetic radiation as the starting point of creation! Astrophysicist Robert Jastrow thinks that the agnostic scientists should sit up and take notice, and even be a little worried. But let us look a little more carefully at the extent of the convergence. Both the contemporary scientific account and the age-old Biblical account assume a beginning. The scientific account concerns only the transformation of everything that now is. Until recently, it has not gone beyond that, to the singularity when there was nothing and then suddenly the inconceivably energetic seed for the universe abruptly came into being. Here science has seemed up against a blank wall. In one memorable passage in his book *God and the Astronomers,* Jastrow writes:

> At this moment it seems as though science will never be able to raise the curtain on the mystery of creation. For the scientist who has lived by his faith in the power of reason, the story ends like a bad dream. He has scaled the mountains of ignorance; he is about to conquer the highest peak; as he pulls himself over the final rock, he is greeted by a band of theologians who have been sitting there for centuries.

The band of theologians has an answer: God did it!

But for either a self-professed agnostic like Jastrow or for a believer, such an answer is unrevealing and even superficial, for it cloaks our ignorance beneath a name and tells us nothing further about God beyond the concept of an omnipotent Creator. Not long ago I happened to be discussing this point with Freeman Dyson, one of the most thoughtful physicists of our day. "It's rather silly," he said, "to think of God's role in creation as just sitting up there on a platform and pushing the switch." Creation is a far broader concept than just the moment of the Big Bang. The very structures of the universe itself, the rules of its operation, its continued maintenance — these are the more important aspects of creation.

If we accept that moment of blazing glory as the whole work of the Creator, we are on rather thin ice. Scientific theories, especially cosmological views, are notoriously subject to change, and cosmologists have taken it as a special challenge to eliminate the singularity of point zero when space and time vanish as the universe becomes infinitely dense. There has recently come on the scene the so-called inflationary universe. I haven't the time to try to explain some of the reasons for supposing that the universe, at a very early stage, rapidly expanded by billions of billions of billions of times in less than the twinkling of an eye. Such an event would solve several nagging problems that have plagued the Big Bang scenario. Now the inflationary universe episode, a tiny moment in the larger picture of the expanding universe, does not so much eliminate the singularity of the first instant of creation as to make it scientifically irrelevant, since the random conditions at the moment of inflation would conceal what had gone before and would control the subsequent Big Bang rather than the conditions in the universe at some earlier split second.

But this was not enough for Stephen Hawking, who, treating time as one of the dimensions of the curved space-time in that opening sequence, made a coordinate transformation that eliminated the origin. In his best-selling book *A Brief History of Time,* Hawking writes:

> The idea that space and time may form a closed surface without boundary also has profound implications for the role of God in the affairs of the universe. With the success of scientific theories in describing events, most people have come to believe that God allows the universe to evolve according to a set of laws and does not intervene to break these laws. However, the laws do not tell us what the universe should have looked like when it started — it would still be up to God to wind up the clockwork and choose how to start it off. So long as the universe had a beginning, we could suppose it had a creator. But if the universe is really completely self-contained, having no boundary or edge, it would have neither beginning nor end: it would simply be. What place, then, for a creator?

From a theistic perspective, the answer to Hawking's question is that God is more than the omnipotence who, in some other space-time dimension, decides when to push the mighty ON switch. God is the Creator in the much larger sense of designer and intender of the universe, the powerful creator with a plan and an intention for the existence of the entire cosmos. I shall return to this idea of design presently. But first I would like to point out that both science and the Bible have something to say about the earlier stages of the universe. In fact, both the scientific and the Biblical accounts portray the latent creativity as the universe begins to unfold, beginning with the grand and simple, and leading to the immediate and complex. The scientific picture sketches the creation of atoms, of galaxies, of stars, of life, even of mankind. Likewise, Genesis speaks of the sun and moon, of plants and animals, and of men and women.

But there is a truly fundamental difference in their viewpoints.

\* \* \*

The great tapestry of science is woven together in a grand and awe-some design with the question How? How can the universe end up with a preponderance of positively charged nuclei? How can fluctuations arise to give birth to galaxies? How can the presence of iron atoms be explained? How can hemoglobin come about? The scientific account starts with our present, everyday universe. Detailed observations of the natural world provide the warp of our tapestry, and the theoretical explanations provide the "how," the weft that holds the picture together.

The Biblical picture also concerns the universe around us, but it addresses an entirely different question, not the interconnections of "how" but the motivations and designs of the "who." The Bible brilliantly affirms that our universe has come to be, not by chance, but by a grand design, and that the designer was Almighty God.

In our country today there is a vocal minority who are confused by the separate roles of the scientific way of building up a world view and the Biblical story of creation. Somehow they feel threatened by the ascendancy of a system of looking at the world that does not explicitly include the designing hand of God in the construction. Science is, by its very nature, godless. It is a mechanistic system, contrived to show how things work, and unable to say anything about the who, the designer.

I can sympathize if a deeply religious person finds this incomplete and unsatisfying, and I can even sympathize mildly with the frustration of the creationists, who wish that some broader philosophical framework could be placed into biology textbooks. But they are mistaken when they take scientific explanations, as such, to be antigod or atheistic, they are wrong when they think that the Genesis account can substitute for the "how" of scientific explanations, and they err when they think that a meaningful tack is to brand evolution as a "mere hypothesis." In a certain sense *all* the theoretical explanations of science, the weft that holds the tapestry together, are hypotheses, and to unthread one section risks destroying the entire fabric.

The conflicts of California, Arkansas, and Louisiana might well have been avoided if teachers of science spent a little less time on the scientific answers and a little more time on the process of inquiry, because the process is the heart of the matter. But perhaps it is naive to suppose that children could grasp some of the distinctions that have escaped their elders.

When I first began to prepare this essay, I received in the mail a colorful flyer offering a $1000 cash reward for "scientific proof-positive that the earth moves." In an accompanying letter, the potential donor, Mr. Elmendorf, wrote that "As an engineer, I am astounded that the question of the earth's motion is apparently not 'all settled' after all these years. I mean, if we don't know *that*, what do we know?"

The question that puzzles Mr. Elmendorf goes back to Copernicus and Galileo and to the birth of modern science. Most of the scholars in the generations immediately following Copernicus assumed that his heliocentric system was merely a hypothetical arrangement for explaining the apparent motions in the heavens, and not that the earth *really* moved. Galileo was intrigued by the Copernican system, but he was only a timid Copernican until he turned the newly devised telescope to the heavens and found there a miniature Copernican system in the satellite system of Jupiter. Galileo found a succession of other instances where the heliocentric system could nicely explain his observations, and so he began to argue that the sun-centered arrangement was not just hypothetical but real. But he was unable to *prove* this in any strict sense: All he could do was make the heliocentric system appear increasingly probable. The Catholic Church objected to Galileo's line of reasoning, pointing out (correctly) that just because certain observations were explained by the Copernican system did not mean that some other hypothesis might not equally well explain them.

Meanwhile, north of the Alps, Johannes Kepler also accepted the physical reality of the Copernican arrangement, and demonstrated that each planet has an elliptical orbit with the sun at one focus. He could not prove the truth of the earth's motion, but without the assumption of a fixed central sun, his orbital system didn't make much sense.

Then along came Isaac Newton, who, with some simple but potent assumptions concerning the nature of matter and motion and with some powerful new mathematics, was able to build up a marvelous system of both explanation and prediction. Still, Newton had no *proof* of the earth's motion. What he had was an elaborate picture of how the physical world worked, so thorough and so probable that most people had no difficulty in accepting it as truth. Everyone knew that if the earth was really moving around the sun, the positions of nearby stars should show an annual displacement, but the failure to find it did not discredit the heliocentric theory. Thus, when the discovery of annual stellar parallax finally took place around 1840, people could hardly get excited about this purported proof of the motion of the earth.

"At the end of the last century," and here I quote from a most perceptive book by Jacob Bronowski, *The Origins of Knowledge and Imagination,* "there were physicists who were perfectly willing to say that there was no need to produce another Newton because there was nothing as fundamental as gravitation for another Newton to discover. . . . Since then," he continues, "the world has fallen about our ears." Why? Because Einstein's special relativity showed us an entirely different way of looking at time and space, and his general relativity taught us another way of looking at gravitation, and brought the scientific community to the realization that there was no way to know when any scientific theory was final and therefore true in some ultimate sense. Science simply does not offer "scientific proofs positive," and that is why it is unlikely that anyone will ever collect Mr. Elmendorf's thousand dollars for a "scientific proof-positive" that the earth moves.

Einstein's work has forcefully awakened us to the provisional nature of our scientific picture. As he himself said of science, "the sense experiences are the given subject-matter. But the theory that shall interpret them is man-made. It is the result of an extremely laborious process of adaptation: hypothetical, never completely final, always subject to question and doubt."

This is why I have continually referred to the scientific world view as a grand tapestry. It is an interlocked and coherent picture, a most workable explanation, but it is not ultimate truth.

I know, just as the Catholic Church in Galileo's day knew, that there can be alternate explanations for certain observations. For example, I accept the Big Bang model of the universe as a working hypothesis, but I know that the evidence I have cited concerning the background radiation and the abundances of the elements might have any number of other explanations. But what I am *not* willing to accept — and this is very important — is some special explanation that does not fit the rest of the picture. All of science is in a sense hypothetical.

But the tapestry is of a piece, and it cannot be shredded easily. That is why I am uninterested in the ad hoc and particular claims made here or there by the advocates of "creation science." Perhaps the ultimate truth is that the world was created only 6,000 years ago, but since the Creator has filled it with wonderful clues pointing back 10 or 20 billion years, I am content to do my science by building a coherent picture of a multi-billion-year-old creation, even though that may be only a grand hypothesis. Why? Because it is the coherency of the picture and the systematic procedures for getting there — *not* the final truth — that science is all about.

I have been describing science as a tapestry with a grand design; let me now turn to the idea of a designer. I am personally persuaded that through the eyes of faith one can see numerous vestiges of the designer's hand in the universe. I have many good friends who cannot see it that way, so I am rather doubtful that you can argue a skeptic into faith with the evidences of science. Nevertheless, I find some of these circumstances of nature impossible to comprehend in the absence of supernatural design.

Consider the complex interaction between the earth's atmosphere and life. From what astronomers have deduced about solar evolution, we believe that the sun was perhaps twenty-five percent less luminous several billion years ago. Today, if the solar luminosity dropped by twenty-five percent, the oceans would freeze solid to the bottom, and it would take a substantial increase beyond the sun's present luminosity to thaw them out again. Life could not have originated on such a frozen globe, so it seems that the earth's surface never suffered such frigid conditions. As it turns out, there is a good reason for this. The earth's early atmosphere, formed from the outgassing of volcanoes, included large amounts of carbon dioxide and water vapor. Such an atmosphere would have produced a strong greenhouse effect, an effect that might be more readily explained with a locked car parked in the sun on a hot summer day than with a greenhouse. When you open the car, it's like an oven inside. The glass lets in the photons of visible light from the sun. Hot as the interior of the car may seem, it's quite cool compared to the sun's surface, so the reradiation from inside the car is in the infrared. The glass is quite opaque for those longer wavelengths, and because the radiation can't get out, the car heats up inside. Similarly, the carbon dioxide and water vapor partially blocked the reradiation from the early earth, raising its surface temperature above the mean freezing point of water.

Over the ages, as the sun's luminosity rose, so did the surface temperature of the earth, and had the atmosphere stayed constant, our planet would now have a runaway greenhouse effect, something like that found on the planet Venus; the earth's oceans would have boiled away, leaving a hot, lifeless globe.

How did our atmosphere change over to oxygen just in the nick of time? Apparently the earliest widely successful life form on earth was the so-called blue-greens, a single-celled prokaryote, which survive to this day as stromatolites. Evidence for them appears in the Precambrian fossil record of a few billion years ago. In the absence of predators, these algae-like organisms covered the oceans, extracting hydrogen from the water and releasing oxygen to the air. Nothing much seems to have happened for over a billion years. However, about a billion years ago the oxygen content of the atmosphere rose rapidly, and then a series of events, quite possibly interrelated, took place:

1. Eukaryotic cells, that is, cells with their genetic information contained within a nucleus, originated, which allowed the invention of sex and the more efficient sharing of genetic material, and hence a more rapid adaptation of life forms to new environments.

2. More complicated organisms breathing oxygen, with its much higher energy yield, developed.

3. The excess carbon dioxide was converted into limestone in the structure of these creatures, thus making the atmosphere more transparent in the infrared and thereby preventing the oceans from boiling away in a runaway greenhouse effect as the sun brightened.

Some critics now argue that the oceans, as a giant reservoir for dissolving carbon dioxide, combined with the crustal movements associated with continental drift, might have brought about the transformation even in the absence of life. But whichever way you cut it, the perfect timing of this complex configuration of circumstances is enough to amaze and bewilder many of my friends who look at all this in purely mechanistic terms — the survival of life on earth seems such a close shave as to border on the miraculous. Can we not see here the designer's hand at work? Could we not perhaps rephrase the description of this concatenation of events, substituting "created" for "originated" and "developed"? Eukaryotic cells were created, which allowed the creation of sex and the more efficient sharing of genetic material and more rapid adaptation of life forms; more complicated organisms breathing oxygen were created; and so on. The first form of these phrases gave the scientific elaboration, the "how." The book of Genesis adds the "who" and "why" to the scientific picture.

Let me touch on one further example of design. Early in this century, after the work of Darwin, which emphasized the fitness of organisms for their various environments, the chemist L. J. Henderson wrote a fascinating book entitled *The Fitness of the Environment,* which pointed out that the organisms themselves would not exist except for certain properties of matter. He argued for the uniqueness of carbon as the chemical basis of life, and everything we have learned since then reinforces his argument. But today it is possible to go still further and to probe the origin of carbon itself, through its synthesis in the nuclear reactions deep inside evolving stars.

Carbon is the fourth most common atom in our galaxy, after hydrogen, helium, and oxygen, but it isn't very abundant. A carbon nucleus can be made by merging three helium nuclei, but a triple collision is tolerably rare. It would be easier if two helium nuclei would stick together to form beryllium, but beryllium is not very stable. Nevertheless, sometimes before the two helium nuclei can come unstuck, a third helium nucleus strikes home, and a carbon nucleus results. And here the internal details of the carbon nucleus become interesting: it turns out that there is precisely the right resonance within the carbon to help this process along. Without it, there would be relatively few carbon atoms. Similarly, the internal details of the oxygen nucleus play a critical role. Oxygen can be formed by combining helium and carbon nuclei, but the corresponding resonance level in the oxygen nucleus is *half a percent too low* for the combination to stay together easily. Had the resonance level in the carbon been 4 percent lower, there would be essentially no carbon. Had that level in the oxygen been only half a percent higher, virtually all of the carbon would have been converted to oxygen. Without that carbon abundance, neither you nor I would be here now.

I am told that Fred Hoyle, who together with William Fowler first noticed the remarkable arrangement of carbon and oxygen nuclear resonances, has said that nothing has shaken his atheism as much as this discovery. In the November 1981 issue of *Engineering and Science,* the Cal Tech alumni magazine, Hoyle writes:

> Would you not say to yourself, "Some supercalculating
> intellect must have designed the properties of the carbon
> atom, otherwise the chance of my finding such an atom

through the blind forces of nature would be utterly minuscule"? Of course you would. . . . A common sense interpretation of the facts suggests that a superintellect has monkeyed with physics, as well as with chemistry and biology, and that there are no blind forces worth speaking about in nature. The numbers one calculates from the facts seem to me so overwhelming as to put this conclusion almost beyond question.

Indeed, some of these circumstances seem so impressive that those scientists who wish to *deny* the role of design have had to take into account its ubiquitous signs. They have even given it a name — the anthropic principle. Briefly stated, they have turned the argument around. Rather than accepting that we are here because of a deliberate supernatural design, they claim that the universe simply must be this way *because* we are here; had the universe been otherwise, we would not be here to observe ourselves, and that is that. As I said, I am doubtful that you can convert a skeptic by the argument of design, and the discussions of the anthropic principle seem to prove the point.

In any event, Sir Fred and I differ about lots of things, but on this we agree: The picture of the universe is far more satisfying if we accept the designing hand of a superintelligence.

But even if natural theology can argue for the existence of God the Creator and Designer, it falls short of revealing the central significance of the Biblical creation story. Without doubt the most crucial sentence of the chapter is verse 27, so quintessential that the idea is immediately repeated lest we miss it: "God created man in his own image, in his own image created he him, male and female created he them." Succinctly put, the stance of the Biblical account is that God is not only Creator and Designer, but there is within us, male and female, a divine creative spark, a touch of the infinite — consciousness, and conscience. It is not so easy to fathom within ourselves those same divine evidences that appear in the external world. As Kepler wrote: "There is nothing I want to find out and long to know with greater urgency than this. Can I find God, whom I can almost grasp with my own hands in looking at the universe, also in myself?"

What Kepler longed to know with such urgency in 1620 is even more urgent for us today. Through science we have grasped the harmony of creation and learned the secrets of the stars, and we have, in a terrifying way, brought those secrets to earth. As Arthur Eddington wrote so presciently in 1920, "If, indeed, the sub-atomic energy in the stars is being freely used to maintain their great furnaces, it seems to bring a little nearer to fulfillment our dream of controlling this latent power for the well-being of the human race — or for its suicide."

I have discussed the integrity of the great tapestry of science, how rational man can fill in the details of the working out of God's incredible design, but for me, the coherency of my own view demands a step further, toward accepting something even more mind-boggling than cosmology or the evolution of intelligent life on earth: God has given us something more, a demonstration of his sacrificial love in the life and death of Jesus. As beings created in the image of God, we are called not to power or personal justice, but to sacrificial love.

I confess that this is not the logical conclusion of my line of argument; indeed, it is the beginning, the point of departure for a way of perceiving science and the universe. But unless we can see the universe in those terms, I believe that we are headed toward nuclear suicide.

When the Psalmist asked, "What is man that thou art mindful of him," I think it not only reminds us of our place in the immensity of the cosmos but it affirms our place in the universe as intelligent beings created in the image of God. I certainly hope that as sentient, moral creatures we have enough intelligence, enough conscience, and even enough sacrificial love to avoid nuclear confrontation and catastrophe.

Concerning these arguments, and in keeping with the image of intelligent, creative, and moral beings, I can do no better than to close with a quotation from the seventeenth-century English virtuoso, Thomas Browne: "The wisdom of God receives small honor from those vulgar heads that rudely stare about, and with a gross rusticity admire his workes; those highly magnifie him whose judicious enquiry into his acts, and deliberate research into his creatures, returne the duty of a devout and learned admiration."

Beginnings and Endings

STEVEN WEINBERG

∎

The American physicist Steven Weinberg (b. 1933) shared the 1979 Nobel Prize in physics for his work on the electroweak theory of nuclear interactions, which described events that may have transpired during the first moments of the expansion of the universe. This passage comes from his 1977 book *The First Three Minutes,* which did much to awaken both the scientific community and the general public to the progress being made in the study of the very early universe.

# The First Three Minutes

WE ARE NOW ABOUT to follow the course of cosmic evolution through its first three minutes. Events move much more swiftly at first than later, so it would not be useful to show pictures spaced at equal time intervals, like an ordinary movie. Instead, I will adjust the speed of our film to the falling temperature of the universe, stopping the camera to take a picture each time that the temperature drops by a factor of about three.

Unfortunately, I cannot start the film at zero time and infinite temperature. Above a threshold temperature of fifteen hundred thousand million degrees Kelvin ($1.5 \times 10^{12}$ °K), the universe would contain large numbers of the particles known as pi mesons, which weigh about one-seventh as much as a nuclear particle. Unlike the electrons, positrons, muons, and neutrinos, the pi mesons interact very strongly with each other and with nuclear particles — in fact, the continual exchange of pi mesons among nuclear particles is responsible for most of the attractive force which holds atomic nuclei together. The presence of large numbers of such strongly interacting particles makes it extraordinarily difficult to calculate the behavior of matter at super-high temperatures, so to avoid such difficult mathematical problems I will start the story in this chapter at about one-hundredth of a second after the beginning, when the temperature had cooled to a mere hundred thousand million degrees Kelvin, safely below the threshold temperatures for pi mesons, muons, and all heavier particles.

With these understandings, let us now start our film.

FIRST FRAME. The temperature of the universe is 100,000 million degrees Kelvin ($10^{11}$ °K). The universe is simpler and easier to describe than it ever will be again. It is filled with an undifferentiated soup of matter and radiation, each particle of which collides very rapidly with the other particles. Thus, despite its rapid expansion, the universe is in a state of nearly perfect thermal equilibrium. The contents of the universe are therefore dictated by the rules of statistical mechanics, and do not depend at all on what went before the first frame. All we need to know is that the temperature is $10^{11}$ °K, and that the conserved quantities — charge, baryon number, lepton number — are all very small or zero.

The abundant particles are those whose threshold temperatures are below $10^{11}$ °K; these are the electron and its antiparticle, the positron, and of course the massless particles, the photon, neutrinos, and antineutrinos. The universe is so dense that even the neutrinos, which can travel for years through lead bricks without being scattered, are kept in thermal equilibrium with the electrons, positrons, and photons by rapid collisions with them and with each other. (Again, I will sometimes simply refer to "neutrinos" when I mean neutrinos and antineutrinos.)

Another great simplification — the temperature of $10^{11}$ °K is far above the threshold temperature for electrons and positrons. It follows that these particles, as well as the photons and neutrinos, are behaving just like so many different kinds of radiation. What is the energy density of these various kinds of radiation? The electrons and positrons together contribute 7/4 as much energy as the photons, and the neutrinos and antineutrinos contribute the same as the electrons and positrons, so the total energy density is greater than the energy density for pure electromagnetic radiation at this temperature, by a factor

$$\frac{7}{4} + \frac{7}{4} + 1 = \frac{9}{2}.$$

The Stefan-Boltzmann law gives the energy density of electromagnetic radiation at a temperature of $10^{11}$ °K as $4.72 \times 10^{44}$ electron volts per liter, so the total energy density of the universe at this temperature was 9/2 as great, or $21 \times 10^{44}$ electron volts per liter. This is equivalent to a mass density of 3.8 thousand million kilograms per liter, or 3.8 thousand million times the density of water under normal terrestrial conditions. (When I speak of a given energy as being equivalent to a given mass, I mean of course that this is the energy that would be released according to the Einstein formula $E = mc^2$, if the mass were converted entirely to energy.) If Mt. Everest were made of matter this dense, its gravitational attraction would destroy the earth.

The universe at the first frame is rapidly expanding and cooling. Its rate of expansion is set by the condition that every bit of the universe is traveling just at escape velocity away from any arbitrary center. At the enormous density of the first frame, the escape velocity is correspondingly high — the characteristic time for expansion of the universe is about 0.02 seconds. (The "characteristic expansion time" can be roughly defined as 100 times the length of time in which the size of the universe would increase 1 percent. To be more precise, the characteristic expansion time at any epoch is the reciprocal of the Hubble "constant" at that epoch. The age of the universe is always less than the characteristic expansion time, because gravitation is continually slowing down the expansion.)

There are a small number of nuclear particles at the time of the first frame, about one proton or neutron for every 1,000 million photons or electrons or neutrinos. In order eventually to predict the abundances of the chemical elements formed in the early universe, we will also need to know the relative proportions of protons and neutrons. The neutron is heavier than the proton, with a mass difference between them equivalent to an energy of 1.293 million electron volts. However, the characteristic energy of the electrons, positrons, and so on, at a temperature of $10^{11}$ °K, is much larger, about 10 million electron volts (Boltzmann's constant times the temperature). Thus, collisions of neutrons or protons with the much more numerous electrons, positrons, and so on, will produce rapid transitions of protons to neutrons and vice versa. The most important reactions are

Antineutrino plus proton yields positron plus neutron
(and vice versa)

Neutrino plus neutron yields electron plus proton
(and vice versa).

Under our assumption that the net lepton number and charge
per photon are very small, there are almost exactly as many neutrinos
as antineutrinos, and as many positrons as electrons, so that the transitions from proton to neutron are just as fast as the transitions from
neutron to proton. (The radioactive decay of the neutron can be
ignored here because it takes about 15 minutes, and we are working
now on a time scale of hundredths of seconds.) Equilibrium thus
requires that the numbers of protons and neutrons be just about
equal at the first frame. These nuclear particles are not yet bound
into nuclei; the energy required to break up a typical nucleus altogether is only six to eight million electron volts per nuclear particle;
this is less than the characteristic thermal energies at $10^{11}$ °K, so complex nuclei are destroyed as fast as they form.

It is natural to ask how large the universe was at very early
times. Unfortunately we do not know, and we are not even sure that
this question has any meaning. The universe may well be infinite
now, in which case it was also infinite at the time of the first frame,
and will always be infinite. On the other hand, it is possible that the
universe now has a finite circumference, sometimes estimated to be
about 125 thousand million light years. (The circumference is the
distance one must travel in a straight line before finding oneself back
where one started. This estimate is based on the present value of the
Hubble constant, under the supposition that the density of the universe is about twice its "critical" value.) Since the temperature of the
universe falls in inverse proportion to its size, the circumference of
the universe at the time of the first frame was less than at present by
the ratio of the temperature then ($10^{11}$ °K) to the present temperature (3 °K); this gives a first-frame circumference of about four light
years. None of the details of the story of cosmic evolution in the first
few minutes will depend on whether the circumference of the universe was infinite or only a few light years.

SECOND FRAME. The temperature of the universe is 30,000 million
degrees Kelvin ($3 \times 10^{10}$ °K). Since the first frame, 0.11 seconds have
elapsed. Nothing has changed qualitatively — the contents of the
universe are still dominated by electrons, positrons, neutrinos, anti-

neutrinos, and photons, all in thermal equilibrium, and all high above their threshold temperatures. Hence the energy density has dropped simply like the fourth power of the temperature, to about 30 million times the energy density contained in the rest mass of ordinary water. The rate of expansion has dropped like the square of the temperature, so that the characteristic expansion time of the universe has now lengthened to about 0.2 seconds. The small number of nuclear particles is still not bound into nuclei, but with falling temperature it is now significantly easier for the heavier neutrons to turn into the lighter protons than vice versa. The nuclear particle balance has consequently shifted to 38 percent neutrons and 62 percent protons.

THIRD FRAME. The temperature of the universe is 10,000 million degrees Kelvin ($10^{10}$ °K). Since the first frame, 1.09 seconds have elapsed. About this time the decreasing density and temperature have increased the mean free time of neutrinos and antineutrinos so much that they are beginning to behave like free particles, no longer in thermal equilibrium with the electrons, positrons, or photons. From now on they will cease to play any active role in our story, except that their energy will continue to provide part of the source of the gravitational field of the universe. Nothing much changes when the neutrinos go out of thermal equilibrium. (Before this "decoupling," the typical neutrino wavelengths were inversely proportional to the temperature, and since the temperature was falling off in inverse proportion to the size of the universe, the neutrino wavelengths were increasing in direct proportion to the size of the universe. After neutrino decoupling the neutrinos will expand freely, but the general red shift will still stretch out their wavelengths in direct proportion to the size of the universe. This shows, incidentally, that it is not too important to determine the precise instant of neutrino decoupling, which is just as well, because it depends on details of the theory of neutrino interactions which are not entirely settled.)

The total energy density is less than it was in the last frame by the fourth power of the ratio of temperatures, so it is now equivalent to a mass density 380,000 times that of water. The characteristic time for expansion of the universe has correspondingly increased to about two seconds. The temperature is now only twice the threshold temperature of electrons and positrons, so they are just beginning to annihilate more rapidly than they can be recreated out of radiation.

It is still too hot for neutrons and protons to be bound into atomic nuclei for any appreciable time. The decreasing temperature has now allowed the proton-neutron balance to shift to 24 percent neutrons and 76 percent protons.

FOURTH FRAME. The temperature of the universe is now 3,000 million degrees Kelvin ($3 \times 10^9$ °K). Since the first frame, 13.82 seconds have elapsed. We are now below the threshold temperature for electrons and positrons, so they are beginning rapidly to disappear as major constituents of the universe. The energy released in their annihilation has slowed down the rate at which the universe cools, so that the neutrinos, which do not get any of this extra heat, are now 8 percent cooler than the electrons, positrons, and photons. From now on, when we refer to the temperature of the universe we will mean the temperature of the *photons*. With electrons and positrons rapidly disappearing, the energy density of the universe is now somewhat less than it would be if it were simply falling off like the fourth power of the temperature.

It is now cool enough for various stable nuclei like helium ($He^4$) to form, but this does not happen immediately. The reason is that the universe is still expanding so rapidly that nuclei can only be formed in a series of fast two-particle reactions. For instance, a proton and a neutron can form a nucleus of heavy hydrogen, or deuterium, with the extra energy and momentum being carried away by a photon. The deuterium nucleus can then collide with a proton or a neutron, forming either a nucleus of the light isotope, helium three ($He^3$), consisting of two protons and a neutron, or the heaviest isotope of hydrogen, called tritium ($H^3$), consisting of a proton and two neutrons. Finally, the helium three can collide with a neutron, and the tritium can collide with a proton, in both cases forming a nucleus of ordinary helium ($He^4$), consisting of two protons and two neutrons. But in order for this chain of reactions to occur, it is necessary to start with the first step, the production of deuterium.

Now, ordinary helium is a tightly bound nucleus, so, as I said, it can indeed hold together at the temperature of the third frame. However, tritium and helium three are much less tightly bound, and deuterium is especially loosely bound. (It takes only a ninth as much energy to pull a deuterium nucleus apart as to pull a single nuclear particle out of a helium nucleus.) At the fourth-frame temperature of $3 \times 10^9$ °K, nuclei of deuterium are blasted apart as soon as they form, so heavier nuclei do not get a chance to be produced. Neutrons are still being converted into protons, although much more slowly than before; the balance now is 17 percent neutrons and 83 percent protons.

FIFTH FRAME. The temperature of the universe is now 1,000 million degrees Kelvin ($10^9$ °K), only about 70 times hotter than the center of the sun. Since the first frame, three minutes and two seconds have elapsed. The electrons and positrons have mostly disappeared, and the chief constituents of the universe are now photons, neutrinos, and antineutrinos. The energy released in electron-positron annihilation has given the photons a temperature 35 percent higher than that of the neutrinos.

The universe is now cool enough for tritium and helium three as well as ordinary helium nuclei to hold together, but the "deuterium bottleneck" is still at work: nuclei of deuterium do not hold together long enough to allow appreciable numbers of heavier nuclei to be built up. The collisions of neutrons and protons with electrons, neutrinos, and their antiparticles have now pretty well ceased, but the decay of the free neutron is beginning to be important; in each 100 seconds, 10 percent of the remaining neutrons will decay into protons. The neutron-proton balance is now 14 percent neutrons, 86 percent protons.

A LITTLE LATER. At some time shortly after the fifth frame, a dramatic event occurs: the temperature drops to the point at which deuterium nuclei can hold together. Once the deuterium bottleneck is passed, heavier nuclei can be built up very rapidly by the chain of two-particle reactions described in the fourth frame. However, nuclei heavier than helium are not formed in appreciable numbers because of other bottlenecks: there are no stable nuclei with five or eight nuclear particles. Hence, as soon as the temperature reaches the point where deuterium can form, almost all of the remaining neutrons are immediately cooked into helium nuclei. The precise temperature at which this happens depends slightly on the number of nuclear particles per photon, because a high particle density would make it a little easier for nuclei to form. (This is why I had to identify this moment imprecisely as "a little later" than the fifth frame.) For 1,000 million photons per nuclear particle, nucleosynthesis will begin at a temperature of 900 million degrees Kelvin ($0.9 \times 10^9$ °K). At this time, three minutes forty-six seconds have passed since the first frame. (The reader will have to forgive my inaccuracy in calling this

*The First Three Minutes.* It sounded better than *The First Three and Three-quarter Minutes.*) Neutron decay would have shifted the neutron-proton balance just before nucleosynthesis began to 13 percent neutrons, 87 percent protons. After nucleosynthesis, the fraction by weight of helium is just equal to the fraction of all nuclear particles that are bound into helium; half of these are neutrons, and essentially all neutrons are bound into helium, so the fraction by weight of helium is simply twice the fraction of neutrons among nuclear particles, or about 26 percent. If the density of nuclear particles is a little higher, nucleosynthesis begins a little earlier, when not so many neutrons would have decayed, so slightly more helium is produced, but probably not more than 28 percent by weight.

We have now reached and exceeded our planned running time, but in order to see better what has been accomplished, let us take a last look at the universe after one more drop in temperature.

SIXTH FRAME. The temperature of the universe is now 300 million degrees Kelvin ($3 \times 10^8$ °K). Since the first frame, 34 minutes and 40 seconds have elapsed. The electrons and positrons are now completely annihilated except for the small (one part in 1,000 million) excess of electrons needed to balance the charge of the protons. The energy released in this annihilation has now given the photons a temperature permanently 40.1 percent higher than the temperature of the neutrinos. The energy density of the universe is now equivalent to a mass density 9.9 percent that of water; of this, 31 percent is in the form of neutrinos and antineutrinos and 69 percent is in the form of photons. This energy density gives the universe a characteristic expansion time of about one and one-quarter hours. Nuclear processes have stopped — the nuclear particles are now for the most part either bound into helium nuclei or are free protons (hydrogen nuclei), with about 22 to 28 percent helium by weight. There is one electron for each free or bound proton, but the universe is still much too hot for stable atoms to hold together.

The universe will go on expanding and cooling, but not much of interest will occur for 700,000 years. At that time the temperature will drop to the point where electrons and nuclei can form stable atoms; the lack of free electrons will make the contents of the universe transparent to radiation; and the decoupling of matter and radiation will allow matter to begin to form into galaxies and stars.

After another 10,000 million years or so, living beings will begin to reconstruct this story.

This account of the early universe has one consequence that can be immediately tested against observation: the material left over from the first three minutes, out of which the stars must originally have formed, consisted of 22–28 percent helium, with almost all the rest hydrogen. As we have seen, this result depends on the assumption that there is a huge ratio of photons to nuclear particles, which in turn is based on the measured 3 °K temperature of the present cosmic microwave radiation background. The first calculation of the cosmological helium production to make use of the measured radiation temperature was carried out by P.J.E. Peebles at Princeton in 1965, shortly after the discovery of the microwave background by Penzias and Wilson. A similar result was obtained independently and at almost the same time in a more elaborate calculation by Robert Wagoner, William Fowler, and Fred Hoyle. This result was a stunning success for the standard model, for there were already at this time independent estimates that the sun and other stars do start their lives as mostly hydrogen, with about 20–30 percent helium!

There is, of course, extremely little helium on earth, but that is just because helium atoms are so light and so chemically inert that most of them escaped the earth ages ago. Estimates of the primordial helium abundance of the universe are based on comparisons of detailed calculations of stellar evolution with statistical analyses of observed stellar properties, plus direct observation of helium lines in the spectra of hot stars and interstellar material. Indeed, as its name indicates, helium was first identified as an element in studies of the spectrum of the sun's atmosphere, carried out in 1868 by J. Norman Lockyer.

During the early 1960s it was noticed by a few astronomers that the abundance of helium in the galaxy is not only large, but does not vary from place to place nearly so much as does the abundance of heavier elements. This of course is just what would be expected if the heavy elements were produced in stars, but helium was produced in the early universe, before any of the stars began to cook. There is still a good deal of uncertainty and variation in estimates of nuclear abundances, but the evidence for a primordial 20–30 percent helium abundance is strong enough to give great encouragement to adherents of the standard model.

In addition to the large amount of helium produced at the end of the first three minutes, there was also a trace of lighter nuclei, chiefly deuterium (hydrogen with one extra neutron) and the light helium isotope He³, that escaped incorporation into ordinary helium nuclei. (Their abundances were first calculated in the 1967 paper of Wagoner, Fowler, and Hoyle.) Unlike the helium abundance, the deuterium abundance is very sensitive to the density of nuclear particles at the time of nucleosynthesis: for higher densities, nuclear reactions proceeded faster, so that almost all deuterium would have been cooked into helium. To be specific, here are the values of the abundance of deuterium (by weight) produced in the early universe, as given by Wagoner, for three possible values of the ratio of photons to nuclear particles:

| *Photons/Nuclear Particle* | *Deuterium Abundance (parts per million)* |
|---|---|
| 100 million | 0.00008 |
| 1,000 million | 16 |
| 10,000 million | 600 |

Clearly, if we could determine the primordial deuterium abundance that existed before stellar cooking began, we could make a precise determination of the photon-to-nuclear particle ratio; knowing the present radiation temperature of 3 °K, we could then determine a precise value for the present nuclear mass density of the universe, and judge whether it is open or closed.

Unfortunately it has been very difficult to determine a truly primordial deuterium abundance. The classic value for the abundance by weight of deuterium in the water on earth is 150 parts per million. (This is the deuterium that will be used to fuel thermonuclear reactors, if thermonuclear reactions can ever be adequately controlled.) However, this is a biased figure; the fact that deuterium atoms are twice as heavy as hydrogen atoms makes it somewhat more likely for them to be bound into molecules of heavy water (HDO), so that a smaller proportion of deuterium than hydrogen would have escaped the earth's gravitational field. On the other hand, spectroscopy indicates a very low abundance of deuterium on the sun's surface — less than four parts per million. This also is a biased figure — deuterium in the outer regions of the sun would have been mostly destroyed by fusing with hydrogen into the light isotope of helium, He³.

Our knowledge of the cosmic deuterium abundance was put on a much firmer basis by ultraviolet observations in 1973 from the artificial Earth satellite *Copernicus*. Deuterium atoms, like hydrogen atoms, can absorb ultraviolet light at certain distinct wavelengths, corresponding to transitions in which the atom is excited from the state of lowest energy to one of the higher states. These wavelengths depend slightly on the mass of the atomic nucleus, so the ultraviolet spectrum of a star whose light passes to us through an interstellar mixture of hydrogen and deuterium will be crossed with a number of dark absorption lines, each split into two components, one from hydrogen and one from deuterium. The relative darkness of any pair of absorption line components then immediately gives the relative abundance of hydrogen and deuterium in the interstellar cloud. Unfortunately, the earth's atmosphere makes it very difficult to do any sort of ultraviolet astronomy from the ground. The satellite *Copernicus* carried an ultraviolet spectrometer which was used to study absorption lines in the spectrum of the hot star 6 Centaurus; from their relative intensities, it was found that the interstellar medium between us and 6 Centaurus contains about 20 parts per million (by weight) of deuterium. More recent observations of ultraviolet absorption lines in the spectra of other hot stars give similar results.

If this 20 parts per million of deuterium was really created in the early universe, then there must have been (and is now) just about 1,100 million photons per nuclear particle (see the table above). At the present cosmic radiation temperature of 3 °K there are 550,000 photons per liter, so there must be now about 500 nuclear particles per million liters. This is considerably less than the minimal density for a closed universe, which is about 3,000 nuclear particles per million liters. The conclusion would then be that the universe is open; that is, the galaxies are moving at above escape velocity, and the universe will expand forever. If some of the interstellar medium has been processed in stars which tend to destroy deuterium (as in the sun), then the cosmologically produced deuterium abundance must have been even greater than the 20 parts per million found by the *Copernicus* satellite, so the density of nuclear particles must be even less than 500 particles per million liters, strengthening the conclusion that we live in an open, eternally expanding universe.

I must say that I personally find this line of argument rather unconvincing. Deuterium is not like helium — even though its abundance seems higher than would be expected for a relatively dense closed universe, deuterium is still extremely rare in absolute terms.

We can imagine that this much deuterium was produced in "recent" astrophysical phenomena — supernovas, cosmic rays, perhaps even quasi-stellar objects. This is not the case for helium; the 20–30 percent helium abundance could not have been created recently without liberating enormous amounts of radiation that we do not observe. It is argued that the 20 parts per million of deuterium found by *Copernicus* could not have been produced by any conventional astrophysical mechanism without also producing unacceptably large amounts of the other rare light elements: lithium, beryllium, and boron. However, I do not see how we are ever going to be sure that this trace of deuterium was not produced by some noncosmological mechanism that no one has thought of yet.

There is one other remnant of the early universe that is present all around us, and yet seems impossible to observe. We saw in the third frame that neutrinos have behaved like free particles since the cosmic temperature dropped below about 10,000 million degrees Kelvin. During this time the neutrino wavelengths have simply expanded in proportion to the size of the universe; the number and energy distribution of the neutrinos have consequently remained the same as they would be in thermal equilibrium, but with a temperature that has dropped in inverse proportion to the size of the universe. This is just about the same as what has happened to photons during this time, even though photons remained in thermal equilibrium far longer than neutrinos. Hence the present neutrino temperature ought to be roughly the same as the present photon temperature. There would therefore be something like 1,000 million neutrinos and antineutrinos for every nuclear particle in the universe.

It is possible to be considerably more precise about this. A little after the universe became transparent to neutrinos, the electrons and positrons began to annihilate, heating the photons but not the neutrinos. In consequence, the present neutrino temperature ought to be a little *less* than the present photon temperature. It is fairly easy to calculate that the neutrino temperature is less than the photon temperature by a factor of the cube root of 4/11, or 71.38 percent; the neutrinos and antineutrinos then contribute 45.42 percent as much energy to the universe as photons. Although I have not said so explicitly, whenever I have quoted cosmic expansion times previously, I have taken this extra neutrino energy density into account.

The most dramatic possible confirmation of the standard model of the early universe would be the detection of this neutrino background. We have a firm prediction of its temperature; it is 71.38 percent of the photon temperature, or just about 2 °K. The only real theoretical uncertainty in the number and energy distribution of neutrinos is in the question of whether the lepton number density is small, as we have been assuming. (Recall that the lepton number is the number of neutrinos and other leptons *minus* the number of anti-neutrinos and other antileptons.) If the lepton number density is as small as the baryon number density, then the number of neutrinos and antineutrinos should be equal to each other, to one part in 1,000 million. On the other hand, if the lepton number density is comparable to the photon number density, then there would be a "degeneracy," an appreciable excess of neutrinos (or antineutrinos) and a deficiency of antineutrinos (or neutrinos). Such a degeneracy would affect the shifting neutron-proton balance in the first three minutes, and hence would change the amounts of helium and deuterium produced cosmologically. Observation of the 2 °K cosmic neutrino and antineutrino background would immediately settle the question of whether the universe has a large lepton number, but much more important, it would prove that the standard model of the early universe is really true.

Alas, neutrinos interact so weakly with ordinary matter that no one has been able to devise any method for observing a 2 °K cosmic neutrino background. It is a truly tantalizing problem: there are some 1,000 million neutrinos and antineutrinos for every nuclear particle, and yet no one knows how to detect them! Perhaps someday someone will.

In following this account of the first three minutes, the reader may feel that he can detect a note of scientific overconfidence. He might be right. However, I do not believe that scientific progress is always best advanced by keeping an altogether open mind. It is often necessary to forget one's doubts and to follow the consequences of one's assumptions wherever they may lead — the great thing is not to be free of theoretical prejudices, but to have the right theoretical prejudices. And always, the test of any theoretical preconception is in where it leads. The standard model of the early universe has scored some successes, and it provides a coherent theoretical framework for future experimental programs. This does not mean that it is true, but it does mean that it deserves to be taken seriously.

Nevertheless, there *is* one great uncertainty that hangs like a dark cloud over the standard model. Underlying all the calculations described in this chapter is the Cosmological Principle, the assumption that the universe is homogeneous and isotropic. (By "homogeneous" we mean that the universe looks the same to any observer who is carried along by the general expansion of the universe, wherever that observer may be located; by "isotropic" we mean that the universe looks the same in all directions to such an observer.) We know from direct observation that the cosmic microwave radiation background is highly isotropic about us, and from this we infer that the universe has been highly isotropic and homogeneous ever since the radiation went out of equilibrium with matter, at a temperature of about 3,000 °K. However, we have no evidence that the Cosmological Principle was valid at earlier times.

It is possible that the universe was initially highly inhomogeneous and anisotropic, but has subsequently been smoothed out by the frictional forces exerted by the parts of the expanding universe on each other. Such a "mixmaster" model has been particularly advocated by Charles Misner of the University of Maryland. It is even possible that the heat generated by the frictional homogenization and isotropization of the universe is responsible for the enormous 1,000 million-to-one present ratio of photons to nuclear particles. However, to the best of my knowledge, no one can say why the universe should have any specific initial degree of inhomogeneity or anisotropy, and no one knows how to calculate the heat produced by its smoothing out.

In my opinion, the appropriate response to such uncertainties is not (as some cosmologists might like) to scrap the standard model, but rather to take it very seriously and to work out its consequences thoroughly, if only in the hope of turning up a contradiction with observation. It is not even clear that a large initial anisotropy and inhomogeneity would have much effect on the story presented in this chapter. It might be that the universe was smoothed out in the first few seconds; in that case the cosmological production of helium and deuterium could be calculated as if the Cosmological Principle were always valid. Even if the anisotropy and inhomogeneity of the universe persisted beyond the era of helium synthesis, the helium and deuterium production in any uniformly expanding clump would depend only on the expansion rate within that clump, and might not be very different from the production calculated in the standard model. It might even be that the whole universe that we can see when we look all the way back to the time of nucleosynthesis is but a homogeneous and isotropic clump within a larger inhomogeneous and anisotropic universe.

The uncertainty surrounding the Cosmological Principle becomes really important when we look back to the very beginning or forward to the final end of the universe. However, it must always be admitted that our simple cosmological models may only describe a small part of the universe, or a limited portion of its history.

# HARALD FRITZSCH

■

The equivalence of matter and energy, revealed in Einstein's special theory of relativity, would have played a major role in the big bang. In this chapter from his 1984 book *The Creation of Matter*, the physicist Harald Fritzsch (b. 1943) describes how particles of matter are thought to have condensed out of raw energy as the early universe expanded and cooled.

# *The Magic Furnace*

YOU MAY HAVE SEEN the hot furnaces used in glass smelting in which glass is heated to extremely high temperatures until it melts. Now, in your mind's eye try to visualize such a furnace whose temperature can be raised to arbitrary high temperatures without itself melting down.

Suppose we had such a furnace with a 100-liter capacity at our disposal. Not fired, its temperature measures approximately 20°C. What does temperature mean? What happens when we heat an object? The atoms or molecules of an object generally are not at rest; they are constantly moving about. Thus every atom in the furnace wall oscillates around a specific point. The prevailing temperature determines the extent of these oscillations. As the temperature rises the oscillations increase in size, and vice versa. Temperature is simply a measure of the kinetic energy of atoms or molecules.

It is possible to cool objects so that their atoms cease to move altogether. When that occurs we speak of the absolute zero of temperature. Normally one uses the Celsius or Fahrenheit scale to indicate temperature. On the Celsius scale the zero point coincides with the temperature at which water freezes under normal conditions. This, however, is an arbitrary point, not related to absolute zero, which is equivalent to −273°C. The atoms of an element are at rest at a temperature of $-273°C$. Lower temperatures cannot be achieved.

Since the absolute zero of temperature obviously plays a special role it was decided to begin measuring temperature at that point. The Kelvin scale of temperature, named after the English physicist Lord Kelvin, does this. Zero degrees Kelvin hence means $-273°C$ (strictly speaking, absolute zero lies at $-273.15°C$, but for our purposes here we will round it off). Consequently, $293°K$ is equivalent to $+20°C$, the average room temperature. Under normal conditions water boils at $100°C$, or $373°K$. Henceforth, when we speak of temperature, we mean absolute temperature.

Let us assume that we succeeded in extracting all air from our furnace. It now contains nothing. Inside the furnace there is a vacuum. When I say "nothing," this is not strictly speaking true. The constantly moving atoms of the furnace wall continue to emit electromagnetic radiation, just as a hot radiator does. The radiation fills up the furnace's interior with photons.

One finds (in accordance with the physical laws of thermodynamics) that the properties of this radiation are determined by the temperature. The higher the furnace's temperature, the greater the average energy of the photons. Temperature and average energy are in proportion. The number of photons also depends on the temperature. It increases by the third power of the temperature, measured from absolute zero. It thus follows that the total energy of the photon gas in the furnace, that is to say the sum of all photon energies, increases by the fourth power of the absolute temperature. When the temperature is raised by a factor of two, energy contained in the photon gas rises by a factor of $2^4 = 16$.

This law, incidentally, also applies to the radiation of an oven or radiator. A radiator in a room emits a stream of electromagnetic radiation, hence photons. The heat radiation of the radiator rises by the fourth power of the absolute temperature. It is easy to establish that a radiator at a temperature of $100°C$ emits 1.57 times as much energy in the form of heat radiation as one at a temperature of $60°C$: $1.57 = (100 + 273)^4/(60 + 273)^4$.

What interests us primarily is the energy density of photon gas. It, too, follows the laws of quantum physics and thermodynamics. The figure we arrive at is easy to remember. At a temperature of $1°K$, the energy density, or the energy per unit of volume, of photon gas in our furnace amounts to 4.72 eV/liter, or 0.00472 eV/cubic centimeter. To find the energy density at a higher temperature all we have to do is multiply the above figure with the fourth power of the temperature. For example: if the furnace is fired to a temperature of 1000°K, the energy per liter of the photon gas in the kiln will be $4.72 \times (1000)^4$ eV $= 4.72 \times 10^{12}$ eV $= 4720$ GeV. This energy is equivalent to the rest energy, or the mass, of almost 5000 protons.

Another magnitude of interest to us is the mean energy of a photon in photon gas. It is proportional to the temperature. For our purposes we can apply this simple rule: if we raise the temperature of the furnace by 1°C, the mean energy of a photon increases by 0.00008617 eV. If the temperature in the furnace measures only 1°K (−272°C), the mean energy of the photons equals 0.00008617 eV. In this case we are dealing with electromagnetic radio waves. If the furnace has a temperature of 1000°K the mean energy of the photons will be 0.086 eV.

Now that we have laid the groundwork we can proceed with our thought experiment. In that context we will assume that the walls of our furnace are absolutely impenetrable and temperature-resistant, conditions which, of course, could not be found in reality.

We will now fire our furnace. For a long time nothing happens. The energy of the photon gas inside it continues to increase by the fourth power of the temperature. When the temperature climbs to about 6 billion degrees ($6 \times 10^9$ degrees Kelvin) something strange occurs. As you will recall, the annihilation of a positron and an electron produces two photons. The same process can also take place in reverse; when two photons meet, an electron and a positron can come into being out of nothing, provided that the photons possess sufficient energy. The energy of each of the photons must be at least equal to the mass of the electron expressed in units of energy, that is, 0.511 MeV. As soon as the photons in our furnace reach a mean energy of a little more than 0.5 MeV, electron-positron pairs come into being. The temperature needed for this to occur can easily be figured out. It is about 6 billion degrees. At that temperature the photons in the furnace cease to predominate. They will exist side by side with electrons and positrons.

Incidentally, temperatures of that magnitude cannot normally be found in the universe. In the interior of stars temperatures reach "only" a few million degrees.

Suppose we raise the temperature in our imaginary furnace still more. As soon as it is sufficiently high compared to 6 billion degrees, the electrons, positrons, and photons begin to balance each other.

There will be an equal number of electrons, positrons, and photons per unit of space. (This is only an approximation because of subtle differences between leptons — electrons and positrons — and photons.) The leptons and photons carry the same mean energy. Electrons and positrons are continuously being produced by the collision of two photons, and they continuously annihilate each other into two photons. On the average, however, there are as many photons as there are electrons and positrons.

Generally, but not always, the annihilation of an electron and positron gives rise to two photons. There also exists a subtle process called the weak interaction. (This interaction, which exists alongside electromagnetic and strong interactions in nature, is responsible for neutron decay.) It involves the annihilation of an electron and positron into a neutrino-antineutrino pair:

$$e^- + e^+ \rightarrow \nu_e + \bar{\nu}_e.$$

By this process neutrinos and antineutrinos are constantly forming in the furnace's bubbling "electron-positron-photon-stew." Let us assume that these neutrinos and antineutrinos are being deflected by the furnace walls, that they cannot escape. (Considering the highly unrealistic assumptions we have already made with regard to the nature of these walls, one more such assumption should not matter.) As a result we not only obtain a balance between the electrons, positrons, and photons, but the neutrinos turn into fully fledged equal partners. The furnace is now filled with electrons, positrons, neutrinos, antineutrinos, and photons.

Our heating process continues. The energy of the particles flying about in the furnace increases further, until it reaches a mean of 1 GeV, or a temperature of about $10^{13}$ degrees. At that temperature it is possible to create proton-antiproton or neutron-antineutron pairs. The collision of two photons, or of an electron and a positron, produces a proton-antiproton or a neutron-antineutron pair. Such processes are not mere theoretical speculations; they can be observed in accelerator laboratories like PETRA in Hamburg or PEP in Stanford.

When the temperature in our furnace begins to climb above $10^{13}$ degrees, the strongly interacting particles — the π mesons as well as the protons and neutrons — join in our game. From here on things become more complicated, for there exists a whole series of such particles. If I fail to go into this in greater detail it is because,

413

as we shall find out, it would be pointless. At still higher temperatures the story again becomes quite simple.

Let us continue to raise the temperature in the furnace, this time to $10^{14}$ degrees. The particles in it now possess a mean energy of about 10 GeV. At this temperature something astonishing takes place. Protons and neutrons have disappeared from the furnace; instead, in addition to the photons, electrons, positrons, and neutrinos, we now find quarks and gluons. The mean energy of the particles has increased so greatly that protons and neutrons can no longer exist as independent entities. They disintegrate into their components — into quarks and gluons. These quarks and gluons are so dense that the mean distance between them is far less than $10^{-13}$ centimeters. In effect, the quarks and antiquarks and gluons inside the furnace behave like free particles, like electrons or neutrinos. All the particles are in a state of equilibrium. Electrons and positrons continue their reciprocal annihilation into photons or quark-antiquark pairs, while the latter annihilate themselves into gluons, into electron-positron or neutrino-antineutrino pairs. As a result of these continuous processes, the number of electrons, positrons, neutrinos, antineutrinos, photons, $u$ quarks, $d$ quarks, and gluons remains equal. (Subtle differences, such as those caused by the color characteristics of the quarks, do exist, but these amount at most to a factor of three in the various particle densities. Because for the time being I am concerned only with qualitative aspects, I am ignoring these factors here.)

And still we continue to raise the temperature in our furnace. Up to this point our imaginary experiment can be followed experimentally in our accelerator laboratories. The reactions relevant to my speculations, such as the annihilation of an electron and a positron into a quark and antiquark, are being closely studied in these labs. If I now raise the temperature in our furnace to more than $10^{14}$ degrees, I can no longer base myself on experiments, but only on theory. However, since in an imaginary experiment this is entirely permissible let us proceed.

For a long time, so the theory says, nothing much happens. At a temperature of, say, $10^{20}$ degrees, the particles in the furnace have a mean temperature of about ten million GeV. But the number of particles of all types remains equal. Finally we get up to a temperature of $10^{28}$ degrees, equivalent to a mean energy of $10^{15}$ GeV. As you undoubtedly recall, we came across this energy in the unified theories of leptons and quarks. It is determined by the mass of the hypothetical $X$ particles, and it plays a role in joining leptons and quarks into a unit. As soon as we exceed the temperature of $10^{28}$ degrees the $X$ particles also join in our game. At this point it becomes impossible to distinguish leptons from quarks.

Finally we find that our furnace contains a mixture of leptons, quarks, photons, gluons, and $X$ particles. As a matter of fact, it is no longer possible to distinguish between leptons and quarks, nor between photons, gluons, and $X$ particles. The former as well as the latter form a unit. At temperatures of more than $10^{28}$ degrees particles lose their individuality.

It might be useful to figure out the energy density in our furnace at a temperature of $10^{28}$ degrees. We find that it amounts to about $10^{103}$ GeV per liter, an energy density beyond the grasp even of physicists. The number of quarks and electrons in the observable universe is estimated to be about $10^{80}$. If all these particles were compressed into a volume the size of a liter, the corresponding energy density would still be minuscule compared to the density in our initially empty furnace. Our experiment obviously is an imaginary one and must remain so forever. Still, that does not mean that we should stop. Therefore let us raise the temperature still higher, to about $10^{32}$ degrees. At that point the average energy of the particles will be somewhat greater than $10^{19}$ GeV. We now leave the realm of theoretical physics and enter unexplored, virgin terrain. At an energy level of $10^{19}$ GeV normal concepts of space and time collapse — at least according to quantum theory. We do not know what takes their place, nor do we know what happens at temperatures of $10^{32}$ degrees or more. At these high temperatures or energy densities we probably can no longer speak of particles. The concept of particles, like the concepts of space and time, of energy and temperature, simply evanesces.

In our thought experiment we raised the temperature of the furnace from $0°K$ to $10^{32}$ degrees. When we began, the furnace was empty; by the time we finished, it was filled with highly intensive radiation composed of leptons, quarks, photons, gluons, and $X$ particles (and still other particles which, for the sake of simplicity, we ignored).

We can also let this process take place in reverse. Suppose, beginning with a temperature of $10^{32}$ degrees, we slowly let the furnace cool off until it reaches a temperature of $0°K$, absolute zero. At temperatures below $10^{28}$ degrees, the $X$ particles disappear from the furnace. What is left is a mixture of leptons, quarks, photons, gluons, and various other particles. When the temperature falls below $10^{13}$ degrees, quarks and gluons disappear, and all that is left are leptons and photons. Finally we reach the absolute zero of temperature. We look into the furnace and find that it is empty. During the process of cooling off, everything disappeared. Leptons and quarks annihilated each other; the surviving photons subsequently disappeared as we approached absolute zero.

Let us once more return to the hot furnace (which shows a temperature of more than $10^{28}$ degrees). At this temperature it contains leptons, quarks, photons, gluons, and, above all, $X$ particles. We know that these $X$ particles decay immediately upon birth, possibly into positrons and antiquarks:

$$X \rightarrow e^+ \, \overline{d}.$$

The anti-$X$ particles correspondingly decay into electrons and $d$ quarks:

$$\overline{X} \rightarrow e^- \, d.$$

A continuous transformation process among leptons, $q$ quarks, and $X$ particles is taking place in the furnace. Let us take a closer look at these processes.

Strange asymmetries between particles and antiparticles can be found in nature. In some processes particles and antiparticles behave a little bit differently. One such effect was discovered by the physicists James W. Cronin and Val Logsdon Fitch, among others, in 1964, and earned Cronin and Fitch the 1980 Nobel Prize in Physics. Yet at the time, no one suspected that this discovery might hold a clue to our understanding of the universe.

Once again let us take a look at $X$ particle decay. Some $X$ particles can decay both into $e^+\overline{d}$ as well as $uu$; the corresponding antiparticles decay into $e^-d$ and $\overline{uu}$. One expects these decays to differ slightly from each other. For example, the $\overline{X} \rightarrow e^-d$ process should be more rapid than the $X \rightarrow e^+\overline{d}$ process. In the extreme, only the $\overline{X} \rightarrow e^-d$ and $X \rightarrow uu$ process might perhaps be possible. In that case an $X\overline{X}$ pair would decay into $e^-uud$, that is, into one electron and three quarks which at the appropriate low temperature could easily form a proton. With that, the decay of the $X\overline{X}$ pair leads directly to the production of hydrogen.

The radiation in our furnace is composed of a variety of particles and antiparticles. Thus there exists an equal number of $X$ and $\overline{X}$ particles which can be grouped into $X\overline{X}$ pairs. Suppose we look at the furnace at a specific moment. We find that soon after, some of the $X$ particles will have decayed. If every $X\overline{X}$ pair decays into $e^-uud$ we can expect suddenly to find more quarks than antiquarks in the furnace. We have created quarks out of nothing, out of energy. Does that mean that we have created matter?

The answer to this unfortunately is no, we have not. Upon closer inspection it turns out that we have forgotten one important aspect. True, the $X$ particles in the furnace continue to decay, but at the same time new ones are forming. We are faced with a situation which in physics is called "thermal equilibrium." Whenever we find such a balance we also find complete symmetry between particles and antiparticles. More specifically, the number of quarks and antiquarks in the furnace remains constant. The only way we could create more quarks than antiquarks in the furnace is by destroying its thermal equilibrium.

This brings us to the last and most important stage of our thought experiment. We raise the temperature in our furnace to more than $10^{28}$ degrees and then try to cool it down as rapidly as possible. Since the furnace exists only in our imagination we are able to move it anywhere we want, to the remotest possible place, even interstellar space. And a good thing, too, because our next experiment is not altogether without danger. For what we are about to do is remove the walls of the furnace and leave the hot plasma with its temperature of about $10^{30}$ degrees standing free.

A furnace with an internal temperature of more than $10^{28}$ degrees must be able to weather a great many things. Its walls must not only be heat-resistant, but also be able to withstand enormous pressure. If we suddenly remove its walls the particles in the furnace no longer are confined. The hot plasma flies apart with the speed of light; an explosion takes place. An explosion is a physical phenomenon in which thermal equilibrium no longer exists. The particles squeezed together in the furnace now fly off in all directions. Each particle is on its own. Above all, the $X$ and $\overline{X}$ particles decay after a short time, with practically no new ones being created, the thermal equilibrium having been destroyed. After these decays have taken place we count the quarks and antiquarks. The result does not come as a surprise: quarks outnumber antiquarks.

After the explosion of our furnace all unstable particles decay. Eventually the quarks and antiquarks unite and form either mesons, which also decay, or protons and neutrons and their respective anti-particles (the neutrons, however, in turn decay into protons and leptons). But at the end there will be more quarks than antiquarks, which means more protons than antiprotons. If we bother to gather up the residue left in the universe after the explosion, the antiprotons and their respective protons will annihilate themselves. Finally we will be left with some surviving protons that will unite with wandering electrons to form hydrogen atoms.

We have thus succeeded in producing matter out of "nothing," out of pure energy. It is the reverse process of the proton decay touched on earlier. In proton decay matter is transformed into radiation. Here we have done the reverse — we have made matter out of radiation.

Our mission is completed. We started with an empty furnace and heated it to temperatures of more than $10^{28}$ degrees. In the consequent explosion matter formed and we got hydrogen atoms.

The matter of the universe consists largely of hydrogen. What would seem more logical than to look at this hydrogen as the residue of an explosion that took place at temperatures of more than $10^{28}$ degrees? Are galaxies, stars, planets, and human beings composed of the ash of such an explosion?

ALAN LIGHTMAN

■

Stephen Hawking has endeavored to investigate the origin of the universe by combining concepts from relativity and quantum mechanics. Alan Lightman (b. 1948), a physicist at Smithsonian Astrophysical Observatory and professor of astronomy at Harvard, published these musings on Hawking's work in 1986.

# *The Origin of the Universe*

FOR THE LAST COUPLE of decades, physicists have been pushing their theories of matter and energy backward in time, closer and closer to the primal explosion that started the universe, some 10 billion years ago. Indeed, it is common these days for scientists to hold forth on "The Origin of the Universe." When you go to such lectures, however, you soon learn that you're not getting The Origin, but a billionth or a trillionth of a second later.

And so it was that I casually took my seat at yet another lecture titled "The Origin of the Universe," given at Harvard in the spring of 1984. The lecturer was the British scientist Stephen Hawking. The hall was packed. Hawking, then forty-two, has become one of the seminal theoretical physicists of our time. He has also suffered for years from a worsening motor neuron disease, which has ravaged his body but spared his mind. On this afternoon, as Hawking sat in his wheelchair, laboring to utter a series of sounds that were translated into words by a student, I gradually realized what I was hearing: Hawking had traveled back the whole distance. For the first time, a preeminent scientist was tackling the *initial* condition of the universe — not a split second after the Big Bang, as I'd heard about before, but the very beginning, the instant of creation, the pristine pattern of matter and energy that would later form atoms and galaxies and planets.

In any other circumstance, physicists would hardly think twice about discussing initial conditions. The "initial conditions" along with the "laws" are the two essential parts of every model of nature. The initial conditions tell how the particles and forces of nature are arranged at the beginning of an experiment. The laws tell what happens next. Any predictions rest on both parts. Set a pendulum swinging, for example, and its motion will be determined by the initial height where your hand let go, as well as by the laws of gravity and mechanics. But Hawking's pendulum is the entire universe. And he is attempting to reason out what theologians and scientists alike have previously assumed as a given. He is attempting to *calculate* where the hand let go. Hawking's equations for the initial state of the universe, together with the laws of nature, could predict the complete outcome of the universe. They could tell us whether our universe will expand forever or reach a maximum size and then collapse. They might explain the existence of planets, or of time.

How could anyone know whether Hawking's equations are right? Could the human mind even grasp Creation? And, just as puzzling, how did science arrive at such outrageous self-confidence? I asked these questions to myself and a dazed colleague as we wandered out of the lecture hall and away from the campus, walking past cars and children in mittens.

Physicists today are not modest — and with some reason. Just in this century, they have discovered and successfully tested a new law for gravity, a theory for the strong nuclear force, and a unified theory of the electromagnetic and weak nuclear forces. They have proposed further laws that might unify all of the forces of nature. Physicists have demonstrated that time doesn't flow at a uniform rate and that subatomic particles seem to occupy several places at once. These victories, often in territory far removed from human sensory perception, have created a strong sense of confidence.

In many of the more heady advances, theory has outdistanced observation, let alone application. For example, the unified theory of the electromagnetic and weak nuclear forces, developed in the 1960s, predicted the existence of new particles that weren't discovered in the laboratory until the 1980s. Super-dense stars, fifteen miles in diameter, were predicted in the 1930s — more than thirty years before they were first observed in space, thousands of light years from earth. Einstein's general theory of relativity predicted that a ray of starlight passing near the sun should be deflected five ten-thousandths of a degree by the sun's gravity. When a novel experiment confirmed this minuscule effect several years later and Einstein seemed blasé, a student asked him what he would have done if his prediction had been refuted. He answered that then he would have been sorry for the dear Lord, because "the theory *is* correct."

With such confidence, physicists have grown accustomed to extrapolating their theories to situations that could not possibly be witnessed by human beings. Hawking's work on the beginning of the universe is an extreme example of this kind. In the evolution of the universe as a whole, gravity is the dominant force to be reckoned with. Hawking has extrapolated Einstein's theory of gravity back to an epoch that was not simply prior to life, but prior to atoms. Stranger still, the early universe was of such high density that its entire contents, including the geometry of space itself, behaved in the hazy, hard-to-pin-down manner of subatomic particles. The methodology needed to describe such behavior is called quantum mechanics, and the application of this methodology to gravity is called quantum gravity. According to the theory of quantum gravity, it was possible for the entire universe to appear out of nothing.

Hawking has mathematically investigated the kind of universe that could have appeared out of nothing. Would the infant universe be finite or infinite in extent? Would it curve in on itself? Would it look the same in all directions? Would it be expanding rapidly or slowly? The answers to these questions are buried in a difficult equation. That equation will likely take a long time to solve, an even longer time to test against its predictions, and it may be plain wrong. Nevertheless, it reflects an extreme confidence in the power of human reason to reveal the natural world. Hawking, like Darwin, has ventured into regions previously forbidden to human beings, and to scientists in particular. Hawking's work, right or wrong, is a celebration of human power and entitlement to knowledge.

Men and women have always longed to understand and control their world, but they have constantly met with obstacles. In different ages and cultures, they have tried different means to clear their path — magic in primitive cultures, religion and science in more evolved societies. Primitive man believes in his power to control nature and other men through magic. He believes he can make rain by climbing up a fir tree and drumming a bowl to imitate thunder. He believes he can bring a cool breeze by wrapping a horsehair around a stick and waving it in the air. But with growing experience, man realizes that these methods have their limitations. Rain and cool breezes don't always come when requested. At this stage of development, as the anthropologist Sir James Frazer says in *The Golden*

*Bough,* man stops relying on himself and throws himself at the mercy of higher beings. Thus begins religion. And a surrender of personal power. My rabbi once told me that man has always made of God what he wished to be himself.

But, as man increases his knowledge, this new reckoning with the universe also needs revision. For the gods, reflecting the ignorance and superstition of man, have often been given human personality along with their powers. The gods get drunk, as the Babylonian deities did on the night before Marduk went out to do battle against Chaos. They are jealous and spiteful, like Hera, who destroyed the Trojan race because she had placed runner-up in a beauty contest judged by a Trojan. If natural phenomena are controlled by such gods, then those phenomena should be subject to whim and to passion. Yet the more man studies nature, the more he finds evidence for regular laws. Seasons repeat, stars move on course, and stones fall predictably. The study of these regularities marks the method of science. Through science, man regains much of his primitive confidence in self-power, with control now replaced by knowledge. Knowledge is power. Man might not be able to control the weather, but he can try to predict it.

With the beginning of modern science in Europe, people measured eclipses, they dissected cadavers, they observed mountains on the moon with the new telescopes, they peered at lake water through microscopes, they studied magnets and electricity. Copernicus declared that the earth went around the sun. Paracelsus announced that disease was caused by agents outside of the body, not by internal humors. Galileo pointed out that moving bodies maintained their motion unless acted on by external forces.

Yet flowing deep through human culture remained the idea that some areas of understanding were off limits or beyond mortal grasp. Adam and Eve were punished because they ate from the forbidden tree of knowledge, which opened their eyes and made them "as gods." In *Paradise Lost,* Adam asks the angel Raphael to explain Creation. Raphael reveals a little, and then says:

> . . . the rest
> From Man or Angel the great Architect
> Did wisely to conceal, and not divulge
> His secrets to be scann'd by them who ought
> Rather admire. . . .

Doctor Faustus approached other authorities for knowledge and had to pay with his soul. People also questioned to what extent the universe was subject to human rationality. Descartes had likened the world to a giant machine, but many viewed such reductions as threats to the power of God. In his Condemnation of 1277, the Bishop of Paris made it clear that no amount of human logic could hinder God's freedom to do what He wanted. Even Isaac Newton, master logician and reductionist, surveyor of all natural phenomena, comes to the end of *The Principia*, the General Scholium, where he lets down his hair and confesses that the synchronized performance of moons and planets could never be explained by "mere mechanical causes," but requires "the counsel and dominion of an intelligent and powerful Being." Furthermore, it would be impossible for mortal man to fathom the art of that divine balancing act: "As a blind man has no idea of colors, so we have no idea of the manner by which the all-wise God perceives and understands all things." Newton, both scientist and believer, was caught between his own power of calculation and the unknowable power of God.

But the unknowable continued to beckon, and man, although fearful of lifting the veils, was still driven to try. After Newton, a great debate arose over whether the solar system could be explained on a rational basis. Similar debates echoed through later centuries. In the eighteenth and nineteenth centuries, geologists argued over whether changes in the earth came about through gradual transformations, obeying natural law, or through sudden catastrophes, ordered by a tampering God. The mode of thinking in the late 1800s, just before Madame Curie found that the sacred atom could be splintered, was described by Henry Adams in this way: ". . . since Bacon and Newton, English thought had gone on impatiently protesting that no one must try to know the unknowable at the same time that every one went on thinking about it."

For Henry Adams, the unknowable that became known, was the atom. For modern biologists, it is the structure of DNA and possibly the creation of life. For modern astronomers, it is the distance to the galaxies and the shape of the cosmos; for modern physicists, perhaps the grand unified force and the birth of the universe. Layer by layer, the unknowable has been peeled away, and examined, and made rational. Scientists today, humbled by their dwindling size in the cosmos but emboldened by their success at adjusting, have staked out all of the physical universe as their rightful territory. And they intend to let their theories and equations take them to places they cannot go with their bodies. In the introduction to one of his recent papers,

Hawking says: "Many people would claim that the [initial] conditions [of the universe] are not part of physics but belong to metaphysics or religion. They would claim that nature had complete freedom to start the universe off any way it wanted. . . . Yet all the evidence is that [the universe] evolves in a regular way according to certain laws. It would therefore seem reasonable to suppose that there are also laws governing the [initial] conditions."

To me, Hawking's work, although strikingly bold, is a natural extension of what science has been doing for the last five hundred years. But the question remains: After physics has reduced the birth of the universe to an equation, is there room left for God? I asked this to a colleague who has done significant calculations on the origin of the universe and is also a devout believer in God. He answered that while physics can describe what is created, Creation itself lies outside physics. But with your equations, I said, you're not giving God any freedom. And he answered, "But that's His choice."

# JOHN D. BARROW
# AND JOSEPH SILK

■

So far as we can tell, the laws of nature have remained the same since the beginning of time. Their stability emboldens scientists to use them to predict the distant future — as described in this passage from *The Left Hand of Creation,* by the English astronomer John D. Barrow (b. 1952) of the University of Sussex and the American astronomer Joseph Silk (b. 1942) of the University of California at Berkeley.

# *How Will the World End?*

"SOME SAY THE WORLD will end in fire. Some say in ice." Robert Frost's well-known words still sum up the possibilities fairly accurately. If the universe contains sufficient material then, eventually, the expansion will be reversed into contraction. Redshifts will turn into blueshifts, and the cosmos will plunge into a singularity that will differ in several respects from the one from whence it originated. Whereas the beginning appears to have been regular and quiescent to a high degree, the final state will be chaotic and violent. The present moderate irregularities in the universe will amplify dramatically during the catastrophic approach to the singularity. Furthermore there was a period of "inflation" during the initial expansion, but there will be no "deflationary" phase during the recollapse. The following describes what will happen.

First as the singularity approaches and the temperature rises, all galaxies, stars, and atoms will dissolve into nuclei and radiation. Then the nuclei will be dismembered into protons and neutrons. They, in turn, will be squeezed until the quarks confined within them are liberated into a huge cosmic soup of freely interacting quarks and leptons. At first, there will be more quarks than antiquarks because of the present matter-antimatter asymmetry of the universe,

425

but as the singularity approaches . . . complete symmetry will be restored. The final plunge into the unknown quantum gravitational era is about $10^{-43}$ second before the singularity occurs when the density exceeds $10^{96}$ that of water.

If you do not fancy such a gruesome prospect, even though it lies at least ten billion years in the future, there is an alternative long-range forecast. The following scenario is possible if astronomers fail to find enough cosmic matter to close the universe and make its temporal future finite. Today, just ten billion years after the "big bang," we sit on a hospitable planet in stable orbit around a middle-aged and reliable star. But after another ten billion years elapse, the sun's fuel will be close to exhaustion, and it will expand to encompass the orbit of the earth. Even if we were ingeniously to evade that catastrophe, we would find ourselves evicted from the solar system after about $10^{15}$ years by the close passage of a neighboring star. Likewise, the sun and its associates will probably be dispatched from our Milky Way galaxy after $10^{19}$ years. Any stars remaining in galaxies will have completed their steady slide into an all-consuming black hole at the galactic nucleus after about $10^{24}$ years. Any beings with the resilience and ingenuity to survive all this will still have to cross their greatest hurdle — the decay of all matter. After about $10^{32}$ years, we expect all protons and neutrons and nuclei to have decayed away. All that will survive are leptons and light, and slowly evaporating black holes. Only after a fantastic $10^{100}$ years will the black holes that were once galaxies evaporate away, leaving behind unpredictable naked singularities and a sea of inert particles and light. Throughout these long aeons of lingering decay, the shape of the cosmos may change as radically as its contents. The last vestiges of geometrical symmetry will be lost.

If life, in any shape or form, is to survive this ultimate environmental crisis, then the universe must satisfy certain basic requirements. The basic prerequisite for intelligence to survive is a source of energy. Such a source could be present even in the indefinite future, if there were a deviation from complete uniformity in temperature and some degree of disorder. The potential for this does seem to exist. The anisotropies in the cosmic expansion, the evaporating black holes, the remnant naked singularities are all life preservers of a sort. Even when the black holes have all dissolved and the naked singularities are few and far between, irregularities may still grow on a cosmic scale and provide a source of heat as they eventually are smoothed out. An infinite amount of information is potentially available in an open universe, and its assimilation would be the principal goal of any surviving noncorporeal intelligence. As the temperature approaches absolute zero, never quite arriving there, the remaining aeons seem doomed to eternal tedium. But where there is quantum theory there is hope. We can never be completely sure this cosmic heat death will occur because we can never predict the future of a quantum universe with complete certainty; for in an infinite quantum future anything that can happen, will eventually. . . .

# PART THREE

## *The Cosmos of Numbers*

Irony is delicious, and ubiquitous too. Higher mathematics, once considered to be the most impractical of all human activities, has become an indispensable tool in comprehending the workings of the real world. The fancies of non-euclidean geometry are employed to chart the shoals of curved space where black holes reside. The austere equations of fractal geometry yield replicas of natural structures from clusters of galaxies to terrestrial coastlines. Computers built to do drudge work wind up prompting us to wonder whether our minds are but calculating machines. Nor has this invasion of worldly wisdom left the halls of mathematics untouched. Old certitudes have fallen, the once-comforting sense of academic detachment has been abridged, and mathematicians have been brought face to face with the terrible possibility that as wild as their fantasies have been — their uncountable infinities, their irrational numbers and imaginary numbers — the real world may be wilder still.

# About Mathematics

## GODFREY HAROLD HARDY

■

At his peak, Godfrey Harold Hardy (1877–1947) was said to be the best mathematician in England. He published his *Apology*, a short book from which this brief excerpt is taken, in 1940, and it soon came to be regarded as a classic. C. P. Snow called it "the most beautiful statement of the creative mind ever written or ever likely to be written."

# A Mathematician's Apology

. . . I CANNOT REMEMBER ever having wanted to be anything but a mathematician. I suppose that it was always clear that my specific abilities lay that way, and it never occurred to me to question the verdict of my elders. I do not remember having felt, as a boy, any *passion* for mathematics, and such notions as I may have had of the career of a mathematician were far from noble. I thought of mathematics in terms of examinations and scholarships: I wanted to beat other boys, and this seemed to be the way in which I could do so most decisively.

I was about fifteen when (in a rather odd way) my ambitions took a sharper turn. There is a book by "Alan St. Aubyn"[1] called *A Fellow of Trinity*, one of a series dealing with what is supposed to be Cambridge college life. I suppose that it is a worse book than most of Marie Corelli's; but a book can hardly be entirely bad if it fires a clever boy's imagination. There are two heroes, a primary hero called Flowers, who is almost wholly good, and a secondary hero, a much

1. "Alan St. Aubyn" was Mrs. Frances Marshall, wife of Matthew Marshall.

weaker vessel, called Brown. Flowers and Brown find many dangers in university life, but the worst is a gambling saloon in Chesterton[2] run by the Misses Bellenden, two fascinating but extremely wicked young ladies. Flowers survives all these troubles, is Second Wrangler and Senior Classic, and succeeds automatically to a Fellowship (as I suppose he would have done then). Brown succumbs, ruins his parents, takes to drink, is saved from delirium tremens during a thunderstorm only by the prayers of the Junior Dean, has much difficulty in obtaining even an Ordinary Degree, and ultimately becomes a missionary. The friendship is not shattered by these unhappy events, and Flowers's thoughts stray to Brown, with affectionate pity, as he drinks port and eats walnuts for the first time in Senior Combination Room.

Now Flowers was a decent enough fellow (so far as "Alan St. Aubyn" could draw one), but even my unsophisticated mind refused to accept him as clever. If he could do these things, why not I? In particular, the final scene in Combination Room fascinated me completely, and from that time, until I obtained one, mathematics meant to me primarily a Fellowship of Trinity.

I found at once, when I came to Cambridge, that a Fellowship implied "original work," but it was a long time before I formed any definite idea of research. I had of course found at school, as every future mathematician does, that I could often do things much better than my teachers; and even at Cambridge I found, though naturally much less frequently, that I could sometimes do things better than the College lecturers. But I was really quite ignorant, even when I took the Tripos, of the subjects on which I have spent the rest of my life; and I still thought of mathematics as essentially a "competitive" subject. My eyes were first opened by Professor Love, who taught me for a few terms and gave me my first serious conception of analysis. But the great debt which I owe to him — he was, after all, primarily an applied mathematician — was his advice to read Jordan's famous *Cours d'analyse;* and I shall never forget the astonishment with which I read that remarkable work, the first inspiration for so many mathematicians of my generation, and learnt for the first time as I read it what mathematics really meant. From that time onwards I was in my way a real mathematician, with sound mathematical ambitions and a genuine passion for mathematics.

I wrote a great deal during the next ten years, but very little of

2. Actually, Chesterton lacks picturesque features.

any importance; there are not more than four or five papers which I can still remember with some satisfaction. The real crises of my career came ten or twelve years later, in 1911, when I began my long collaboration with Littlewood, and in 1913, when I discovered Ramanujan. All my best work since then has been bound up with theirs, and it is obvious that my association with them was the decisive event of my life. I still say to myself when I am depressed, and find myself forced to listen to pompous and tiresome people, "Well, I have done one thing *you* could never have done, and that is to have collaborated with both Littlewood and Ramanujan on something like equal terms." It is to them that I owe an unusually late maturity: I was at my best at a little past forty, when I was a professor at Oxford. Since then I have suffered from that steady deterioration which is the common fate of elderly men and particularly of elderly mathematicians. A mathematician may still be competent enough at sixty, but it is useless to expect him to have original ideas.

It is plain now that my life, for what it is worth, is finished, and that nothing I can do can perceptibly increase or diminish its value. It is very difficult to be dispassionate, but I count it a "success"; I have had more reward and not less than was due to a man of my particular grade of ability. I have held a series of comfortable and "dignified" positions. I have had very little trouble with the duller routine of universities. I hate "teaching," and have had to do very little, such teaching as I have done having been almost entirely supervision of research; I love lecturing, and have lectured a great deal to extremely able classes; and I have always had plenty of leisure for the researches which have been the one great permanent happiness of my life. I have found it easy to work with others, and have collaborated on a large scale with two exceptional mathematicians; and this has enabled me to add to mathematics a good deal more than I could reasonably have expected. I have had my disappointments, like any other mathematician, but none of them has been too serious or has made me particularly unhappy. If I had been offered a life neither better nor worse when I was twenty, I would have accepted without hesitation.

It seems absurd to suppose that I could have "done better." I have no linguistic or artistic ability, and very little interest in experimental science. I might have been a tolerable philosopher, but not one of a very original kind. I think that I might have made a good lawyer; but journalism is the only profession, outside academic life, in which I should have felt really confident of my chances. There is no doubt that I was right to be a mathematician, if the criterion is to be what is commonly called success.

My choice was right, then, if what I wanted was a reasonably comfortable and happy life. But solicitors and stockbrokers and bookmakers often lead comfortable and happy lives, and it is very difficult to see how the world is the richer for their existence. Is there any sense in which I can claim that my life has been less futile than theirs? It seems to me again that there is only one possible answer: yes, perhaps, but, if so, for one reason only.

I have never done anything "useful." No discovery of mine has made, or is likely to make, directly or indirectly, for good or ill, the least difference to the amenity of the world. I have helped to train other mathematicians, but mathematicians of the same kind as myself, and their work has been, so far at any rate as I have helped them to it, as useless as my own. Judged by all practical standards, the value of my mathematical life is nil; and outside mathematics it is trivial anyhow. I have just one chance of escaping a verdict of complete triviality, that I may be judged to have created something worth creating. And that I have created something is undeniable: the question is about its value.

The case for my life, then, or for that of any one else who has been a mathematician in the same sense in which I have been one, is this: that I have added something to knowledge, and helped others to add more; and that these somethings have a value which differs in degree only, and not in kind, from that of the creations of the great mathematicians, or of any of the other artists, great or small, who have left some kind of memorial behind them.

# ALFRED ADLER

∎

What makes for creative aptitude in mathematics? Why do a few
people have the gift, most not? Why, for that matter, are "nearly
all mathematicians . . . eldest sons"? The German mathematician Alfred Adler (b. 1930) contemplates these and other mysteries of his discipline in this essay, written for *The New Yorker*
in 1972.

# *Mathematics and Creativity*

MATHEMATICS, LIKE CHESS, requires too direct and personal a confrontation to allow graceful defeat. There is no element of luck;
there are no partners to share the blame for mistakes; the nature of
the discipline places it precisely at the center of the intellectual being,
where true cerebral power waits to be tested. A loser must admit that
in some very important way he is the intellectual inferior of a winner.
Both mathematics and chess spread before the participant a vast
domain of confrontation of intellect with strong opposition, together
with extreme purity, elegance of form, and an infinitude of possibilities. Mathematics is pure language — the language of science. It
is unique among languages in its ability to provide precise expression
for every thought or concept that can be formulated in its terms. (In
a spoken language, there exist words, like "happiness," that defy definition.) It is also an art — the most intellectual and classical of the
arts. And almost no one is capable of doing significant mathematics.
There are no acceptably good mathematicians. Each generation has
its few great mathematicians, and mathematics would not even notice
the absence of the others. They are useful as teachers, and their
research harms no one, but it is of no importance at all. A mathematician is great or he is nothing. Perhaps that is where the purity
of the discipline begins. In mathematics, no one can be fooled; there
is never the remotest chance of great work's suddenly being done

by a previously uninspired man; everyone recognizes mathematical genius immediately, and without bitterness or reservation. So it is tempting to approach the nature of mathematical creativity by describing the nature of creative mathematicians. But that is the wrong way to go about it. For example, the approach cannot explain the fact, amusing at first and then puzzling, that nearly all mathematicians are eldest sons. The few exceptions tend to be mathematicians whose older brothers are mathematicians, too. Mathematical gifts cannot be unique to oldest sons. It must be that the discipline itself is crucial — that it provides a domain in which mental power and creativity can be wielded in a manner congenial to the special qualities of men who for a time, at least, were the only sons in their families.

The logical way to begin is to describe what mathematics is and what mathematicians do. But this is probably impossible. Almost certainly, any two mathematicians would disagree in significant ways if they were asked to give such a description, and a day later each would disagree with what he had said the day before. It can be said that mathematics is whatever mathematicians are doing. This sounds circular, and in some respects it is. But there is never any doubt about who is and who is not a creative mathematician, so all that is required is to keep track of the activities of these few men. A stronger case can be made for the assertion that mathematics is in large part the discovery and study of analogies. Mathematicians will consider two mathematical objects and ask whether they are similar in some significant way, and whether there are other objects that are like those two in the same significant way, and then, perhaps, whether there are among these objects further similarities, which have not yet been discovered. It is almost impossible to give examples of this. Mathematics is at a stage where the apparently simple and easily posed problems are too difficult to approach, so it is necessary to construct elaborate mathematical objects with a great variety of special properties before there is any possibility of discovering fundamental relationships between the objects. Then, there are the branches of mathematics concerned with foundations. At one level, they deal with the nature of logic itself. At another level, they deal with numbers — the basic units of mathematics, and perhaps of all human thought. Often, the problems can be simply stated and yet are profoundly difficult — in fact, usually impossible even to begin to solve. Here is an example of the kind of question that is asked. A number is called a prime if it is divisible only by the number itself and the number one. So the numbers two, three, five, seven, eleven are primes, and the number six is not a prime. It seems apparent that all composite numbers are products of primes, as twelve is the product of two and two and three. It is another matter, but not a difficult one, to prove that this is so. Primes, then, are the building blocks of the whole number system, and mathematicians pursue their special properties. How many primes are there? Are they finite in number, or can primes of arbitrarily large size be found? A shrewd guess will not do. What is required is an assertion, together with its proof. Perhaps the reader can supply it. If so, he will have proved a theorem of Euclid, and will have discovered a fundamental property of the number system.

There are two simple and ruthless ethical standards by which the purity of any discipline can be determined. Mathematics has at times seemed almost alone in the attainment of these standards. What is required is, first, an institutionalized indifference to men whose work has been completed — a disregard or contempt for those who have accomplished much but who have lost the will to create and whose major accomplishments are of the past. This applies perfectly to mathematics. There is in mathematics no repressive establishment of older men, no traditional way to do mathematics and so to become an important member of the mathematical community, no shortage of room for the rapid progress of gifted men, none of the slavish behavior toward academic superiors which is common in many other disciplines. What is also required is the institutionalized conviction that accomplishment is important only if it advances the discipline in some significant way. Competition must exist for creative achievement only — with and for the discipline itself, rather than with competitors. This, too, applies perfectly. Plagiarism among mathematicians is essentially nonexistent. There is little haste to be credited with results before someone else is, because it simply does not matter who achieves results as long as someone does achieve them. This is not at all in the spirit of "The Double Helix," a book that, perhaps unfairly, indicts biochemistry as a field in which a major concomitant of success is a cynical attempt to keep others unaware of what is being done, and, when this attempt fails, to mislead and misdirect them. Science simply cannot function in this manner, and mathematics is a science.

The mathematical life of a mathematician is short. Work rarely improves after the age of twenty-five or thirty. If little has been accomplished by then, little will ever be accomplished. If greatness has been attained, good work may continue to appear, but the level of accomplishment will fall with each decade. Perhaps this is due to an early failure of the nerve for excellence. It is easy to believe that life is long and one's gifts are vast — easy at the beginning, that is. But the limits of life grow more evident; it becomes clear that great work can be done rarely, if at all. Moreover, there are family responsibilities and there are professional sinecures. Hard work can certainly continue. But creativity requires more than steady, hard, regular, capable work. It requires total commitment over years, with the likelihood of failure at the end, and so the likelihood of a total waste of those years. It requires work of truly immense concentration. Such consuming commitment can rarely be continued into middle and old age, and mathematicians after a time do minor work. In addition, mathematics is continually generating new concepts, which seem profound to the older men and must be painstakingly studied and learned. The young mathematicians absorb these concepts in their university studies and find them simple. What is agonizingly difficult for their teachers appears only natural to them. The students begin where the teachers have stopped; the teachers become scholarly observers.

One quality, that of absolute skepticism, is organically a part of mathematics more than of most other disciplines. Not that the world is deficient in this quality; every thinking person possesses it. But where most of life's encounters invite skepticism, mathematics, by its nature, forces skepticism on its students as a first requirement. This can be put more strongly: Mathematics is a field in which much that appears obviously true is in fact false. (A natural corollary would be that mathematics is a field in which very little is known, and this is indeed so, but the statement is equally true of every other domain of human thought.) The ambient uncertainty is apparent even at an elementary level. For example, nothing could seem more reasonable than the assertion that every surface — every page of a book, say — has two sides. It is reasonable, but it is false. The Möbius band, a figure produced by twisting a strip of paper and then gluing the ends together, has only one side. This is no trick, no joke, no semantic sport. It is an aspect of reality that cannot be dismissed, and therefore no proof of anything can ever make use of the falsehood that all surfaces have two sides. Another example: It is easy to sketch a curve with an angle in it (for instance, the letter *v*). It is just as easy, if more time-consuming, to sketch one with three angles (the letter *w*), or twenty, or ten million. The process becomes absurd, however, when one is asked to construct a curve consisting entirely of angles, or corners. Absurd, and yet it can be done. About a century ago, the mathematician Weierstrass gave an example of a curve consisting of angles, or corners, and nothing else. It cannot be sketched, but its equation can be written down, and can be comprehended by most college juniors and seniors. The examples proliferate. They cannot be avoided, for they lurk in every branch of mathematics, and the mathematician learns early to accept no fact, to believe no statement, however apparently reasonable or obvious or trivial, until it has been proved, rigorously and totally, by a series of steps proceeding from universally accepted first principles.

And even the first principles have been exposed as uncertain. Again, it is not a matter of semantics. Nor is it a matter of the pronouncement that it is fallible human beings who are doing the thinking and hence the product of their thought is fallible. This is true, but too obvious to be very fruitful — not a statement likely to force us to take a large piece of our picture of ourselves and of our world and alter it radically. Goedel's Theorem does precisely that. It proves that there exist meaningful mathematical statements that are neither provable nor disprovable, now or ever — neither provable nor disprovable, that is, not simply because human thought or knowledge is insufficiently advanced but because the very nature of logic renders them incapable of resolution, no matter how long the human race survives or how wise it becomes. There is no way to escape this conundrum. It is not a question of sophistry of any kind. The theorem itself was proved some decades ago, and recently the first example of an undecidable mathematical statement was found. Called the Continuum Hypothesis, it is the assertion that in a set-theoretic sense there exist no sets with more elements than the integers but fewer than the set of all real numbers. So there it stands — an assertion that to the end of time cannot be proved and yet cannot be disproved. The philosophical implications are devastating.

The mathematician's disciplined skepticism and creative imagination might seem enough to guarantee success in the non-mathematical world. Mathematicians are often expected to manage brilliantly in the fields of business and finance. Of course, they do nothing of the kind. Their non-mathematical efforts are, on the whole, pitifully inept. The qualities embedded in the mind of the mathematician by the discipline of mathematics fail to extend beyond the boundaries of mathematics. It appears to be mathematics itself, rather than any inner constraint, that anchors the mathematician to caution and rational thought in his professional work — a measure of the astonishing power of the discipline. For example, departmental and mathematics-society meetings are occupied mainly with talk — aimless and pedantic talk, billowing with Latinisms. Little of substance is ever accomplished, or even intended. The financial adventures of mathematicians consist of wildly speculative stock-market excursions — always exciting but usually unsuccessful — rationalized by elaborate but irrelevant formulas and systems. Almost no mathematicians are to be found in banking or finance, the two fields to which one might suppose them to be ideally suited. Nor do mathematicians distinguish themselves by their political activities, at any level. They have been known for their radical positions, usually on the left but sometimes on the right — positions defended emotionally and often irrationally. There is no reason to expect mathematicians to do well in politics, true. However, their aversion to any but extreme and speculative positions has caused them to forfeit even the modest power and influence due them in bureaucratic affairs, where, consequently, mathematics at all levels is much less influential than any of the other sciences and much less influential than its scientific importance and its procedural virtues warrant. All this contrasts vividly with the achievements of mathematicians when they do mathematics: meaningful results of brevity and simplicity, accomplished by an insistence on total rationality both in hypotheses and in proofs. The professional restraints are so severe that the reaction is too powerful. As soon as the bonds are loosened, mathematicians adopt careless procedures that, together with a vast self-esteem and a conviction of intellectual superiority, cause them to overlook crucial aspects of whatever they are doing — to lose the mental self-control essential to almost every successful human effort.

Why do otherwise rational men fail to see what is happening to them? It may be because the nonacademic world makes it easy for them to succeed in very minor ways — to fulfill small ambitions and a greed for acclaim. Perhaps mathematicians, lacking the imagination to appreciate the scope and sophistication of the outside world, confuse minor success with real achievement and are satisfied with it. Then, too, they seldom recognize failure when they are confronted by it; rather, they tend to think of it as simply one more betrayal by a society that usually patronizes them while elevating armies of patently inferior claimants. In the academic world, on the other hand, mathematicians often enjoy rewards that they do not merit. They are engulfed by admirers from the departments of philosophy and the social sciences, disciplines that suffer from a dangerous confusion of thought; namely, that the presence and the casual contributions of scientists certify them as sciences. Mathematicians are too vain to assess such admiration at its true worth.

Mathematicians as teachers are another matter entirely. There is no reason to expect good mathematicians to be good teachers, any more than to expect them to be good financiers, or even good philosophers. These subjects all rely to a large extent on mathematical reasoning and techniques but involve other talents as well. Nevertheless, almost every good mathematician is also a good teacher, while almost no mediocre mathematician can teach the subject adequately even at an elementary level. This phenomenon is easier to recognize than to explain. Students, even though in most cases they do not know what constitutes good mathematics or which are the best mathematicians they have encountered, will unfailingly pick out the best mathematicians when asked to identify their best mathematics teachers. Love of mathematics and active involvement in its development forge ties between the teacher and his students; the latter are rarely fooled by style or dramatic effect. The usual confusions are absent: confusions between content and presentation, between the subject and the man, between profound inspiration and trivial manipulation — in short, those confusions common in the classrooms of so many other subjects, and common, in fact, in so great a part of life. There is no such thing as a man who does not create mathematics and yet is a fine mathematics teacher. Textbooks, course material — these do not approach in importance the communication of what mathematics is really about, of where it is going, and of where it currently stands with respect to the specific branch of it being taught. What really matters is the communication of the spirit of mathematics. It is a spirit that is active rather than contemplative — a spirit of disciplined search for adventures of the intellect. Only an adventurer can really tell of adventures.

There is no Nobel Prize in mathematics. In view of the record of the Nobel committee, this doesn't matter very much. But it is an indication of what all mathematicians eventually learn to expect, and even take for granted — that successful research and teaching are the only rewards they will ever receive. Money? Salaries are adequate but never have been and never will be generous enough to permit more than a modest middle-class way of life. Power? It is almost ridic-

441

ulous to speak of any relationship between mathematics and secular power. Of course, money and power are only superficial rewards, and it is possible to live without them; less tangible rewards, those of recognition and understanding, are what make effort and accomplishment rich and fulfilling when things are going well, and effort and failure bearable when things are going badly. But mathematicians cannot expect these, either. For example, it would be astonishing if the reader could identify more than two of the following names: Gauss, Cauchy, Euler, Hilbert, Riemann. It would be equally astonishing if he should be unfamiliar with the names of Mann, Stravinsky, de Kooning, Pasteur, John Dewey. The point is not that the first five are the mathematical equivalents of the second five. They are not. They are the mathematical equivalents of Tolstoy, Beethoven, Rembrandt, Darwin, Freud. The geometry of relativity — the work of Riemann — has had consequences as profound as psychoanalysis. The mathematical equivalents of Mann, Stravinsky, and the others would be Bochner, Thom, Serre, Cartan, and Weil, and it is all but certain that the reader is unfamiliar with every one of these names. There is really no reason he should be. But that is precisely the cruelty of the situation. All professions reward accomplishment in part with admiration by peers, but mathematics can reward it with admiration of no other kind. It is, in fact, impossible for a mathematician even to talk intelligently to non-mathematicians about his mathematical work. In the company of friends, writers can discuss their books, economists the state of the economy, lawyers their latest cases, and businessmen their recent acquisitions, but mathematicians cannot discuss their mathematics at all. And the more profound their work, the less understandable it is; a spirited high-school teacher can regale his audience with puzzles and magic squares, but there is no way for the serious mathematician to talk to the non-mathematician about his latest results on the homotopy groups of spheres. Few laymen are really interested enough to distinguish between real mathematicians and fools who can multiply six-digit numbers in their heads. Even a well-educated layman is generally willing to grant, at most, an hour's time to the consideration of the implications of the last half century of mathematical discovery, and then only if he is in a benign humor and the explicant is eloquent and talks philosophy rather than mathematics. The listener will then almost certainly leave without any understanding of what is going on in mathematics, because he has not cared to expend any effort on understanding mathematics by first learning a bit about it; that is, by learning it as a language. The most rudimentary requirement for comprehending a language is a knowledge of its vocabulary, and acquiring a vocabulary demands some hard work. The mathematician can take it for granted that acquiring the vocabulary of mathematics is simply out of the question for his friends, acquaintances — nearly anyone.

And yet the language of mathematics is so natural and so simple in comparison to the spoken languages that the resistance it encounters is difficult to understand. All children learn to count, and most seem to enjoy it. Perhaps incompetent teaching spoils mathematics for them early. Elementary-school teachers know all about arithmetical and algebraic techniques, but not many are able to alert their students to the power and beauty of mathematics, at which these techniques can only hint. Algebra, geometry, calculus — some of the most profound residues of human thought — are taught simply as trades. The operational methods of mathematics become a calculating aid. Yet it is operational methods that distinguish one language from another, that give each language its particular "feel" and, in many ways, endow nationalities with their divergent patterns of thought. The operational methods of mathematics are designed to refine and simplify the search for analogies. Their purposes are to abstract, to generalize, to compress, to isolate, and to expose.

The first abstraction, that of number, is perhaps the most profound. There exist cultures that have not discovered numbers — cultures that can speak of "one" and "many" but of nothing between. Most attempts at a definition of the concept of number seem reasonable at first but prove unsatisfactory on close inspection. Not until this century, in fact, was the first logically acceptable definition found. A further abstraction is that of algebra. This requires speaking of numbers as a whole, rather than of specific numbers, so letters — called variables — are introduced to take the place of particular numbers. The letter $t$ might represent time — not one second or thirty hours but any amount of time whatever. The letter $h$ might represent height — any height — and the letter $w$ any weight. In this way, the concept of numbers is abstracted to that of variables. And it is possible to go much further. A first generalization is accomplished by the recognition that the variables $t$, $h$, and $w$ are simply special cases of a more general phenomenon, that of mathematical functions. A second generalization occurs with the observation that functions need not be restricted to real values, such as lengths or widths, but can be given meaning for other mathematical objects as well. One can consider collections of functions, and then speak of

443

functions of functions, and so on. In short, by abstracting and generalizing first the concept of number, then the concept of the variable, and then the concept of function, it is possible to arrive at an extremely abstract and general level of thought, and so to isolate the kernel of subtlety and difficulty that the less abstract levels obscure.

Then, there is the method of accommodation. The mathematical language is continually being altered to fit new results, to simplify new techniques. In this respect, it differs from the spoken languages, which usually resist accommodation. It is a commonplace that human thought is often bent by the accumulated meanings of words, instead of bending the words to denote refinements of their old meanings. Mathematics does not suffer from this weakness. In the calculus, for instance, the derivative was at one time denoted by a Newtonian symbol — adequate but uninspired. Another notation, introduced by Leibniz, denoted the derivative by the symbol $dy/dx$, which looks like a fraction but is nothing of the kind. It is only a symbol, and the two $d$s cannot be cancelled, as they could if the $dy/dx$ were a fraction. The Leibniz notation nevertheless suggests that in some sense derivatives can be expected to behave like fractions, even though they are not fractions. But can they really? Do they satisfy some of the usual rules for fractions? It became the task of mathematicians to find out. They can and do. The appropriate notation thus led directly to a collection of results that had been hidden by inappropriate notation. One good definition is worth three theorems.

Finally, there are the methods of compression and referral, typified by the evolution of the theory of tensors. The original theory, still in use among some scientists, emphasizes the manipulation of tensors, which are enormously complicated variables, involving elaborate arrays of subscripts and superscripts. The modern theory derives from the observation that these complications are not essential — that the theory would remain intact, and, in fact, would be greatly advanced, if the complicated objects (tensors) representing something simple (numbers) were replaced by simple objects (called forms) representing something complicated (called matrices). It does not really matter whether the reader is familiar with the technical concepts; the significant thing is the method. What has been accomplished is a compression of the notation into one that is easy to manipulate mathematically, together with a referral of the difficulties to the one place where they really belong (the theory of multilinear algebra) and away from the subscripts and superscripts, which serve only to obscure the basic problems.

It is all very simple: Each of these methods results in the isolation and, finally, the exposure of some essential — which is to say difficult — mathematical problem. It is axiomatic among mathematicians that only difficult tasks are worth doing. A difficult problem will lead to important results; a truly intractable problem is one that is worth any amount of time and effort. The most exasperating questions promise the most significant mathematical answers. For example, one of the first results of the theory of integrals is a formula for powers of $x$. The formula, which is simple, is valid for all powers except the power $-1$. Therefore, precisely this power must be the important one. Why does the formula break down? Where is one to look for a resolution of the special case? Is there some mathematical object whose importance has never been fully appreciated and which is suddenly signalling its hidden meanings? The answer is known: the formula for the case $-1$ requires the introduction of the logarithm — the logarithm of high-school algebra, which in high school is usually relegated to a computational role. Why should it appear among the formulas of the theory of integrals, when it really belongs to algebra? What in the world is the logarithm doing in so diverse a pair of domains? These are among the next questions that must be asked about this most important mathematical object. It is difficult to believe that anyone would not wish to know the answers.

Nothing ventured, nothing gained. It is time now to propose the existence of a pattern in all that has been said: The essential feature of mathematical creativity is the exploration, under the pressure of powerful implosive forces, of difficult problems for whose validity and importance the explorer is eventually held accountable by reality. The reality is the physical world. Mathematics allows great speculative freedom, but in the end each mathematical theory must be relevant to physical reality, either directly or by relevance to the main body of mathematics, which, in turn, has direct physical origins. There is perhaps no other way to force the distinction between style and substance, and so to purify a discipline periodically of stylistic eccentricities and of trivial but fashionable diversions. The forces that exert the pressure have been described. There is the constant awareness of time, of the certainty that mathematical creativity ends early in life, so that important work must begin early and proceed quickly if it is to be completed. There is the focus on problems of great difficulty, because the discipline is unforgiving in its contempt for the solution of easy problems and in its indifference to the solution of almost any problems but the most profound and difficult ones. There is the pressure of the nature of mathematics itself — of the elusiveness of truth, of the ever-present necessity for skepticism. And, finally, there is the non-mathematical world, in which the mathematician appears unable to find success, and which at almost all points accords the mathematician a monolithic indifference. So there is no way out for mathematicians; there is no place for them to turn but to other mathematicians and inward on themselves. The insanity and suicide levels among mathematicians are probably the highest in any of the professions. But the rewards are proportionately great. A new mathematical result, entirely new, never before conjectured or understood by anyone, nursed from the first tentative hypothesis through labyryinths of false attempted proofs, wrong approaches, unpromising directions, and months or years of difficult and delicate work — there is nothing, or almost nothing, in the world that can bring a joy and a sense of power and tranquility to equal those of its creator. And a great new mathematical edifice is a triumph that whispers of immortality. What is more, mathematics generates a momentum, so that any significant result points automatically to another new result, or perhaps to two or three other new results. And so it goes — goes, until the momentum all at once dissipates. Then the mathematical career is, essentially, over; the frustrations remain, but the satisfactions have vanished. It has been said that no man should become a philosopher before the age of forty. Perhaps no man should remain a mathematician after the age of forty. The world is, after all, full of worlds to conquer.

# BENOIT B. MANDELBROT

■

We think of geometry as ancient — as the science of Euclid and of the Egyptian surveyors — but it is also a fertile, living discipline. The potential of its undiscovered riches was demonstrated anew in the 1970s with the discovery, by Benoit B. Mandelbrot (b. 1924), of *fractals,* a new field in geometry capable of generating and interpreting structures the complexity and beauty of which rival that of nature's own. Disinclined to underestimate his own abilities (he had taught economics at Harvard, engineering at Yale, and physiology at the Einstein School of Medicine), Mandelbrot proclaimed the news of fractals in a heroic tone that prompted one reviewer of his book *The Fractal Geometry of Nature* to remark approvingly, "I like people who write for glory and not just for money."

# *How Long Is the Coast of Britain?*

## *Fractal Geometry*

Why is geometry often described as "cold" and "dry"? One reason lies in its inability to describe the shape of a cloud, a mountain, a coastline, or a tree. Clouds are not spheres, mountains are not cones, coastlines are not circles, and bark is not smooth, nor does lightning travel in a straight line.

More generally, I claim that many patterns of Nature are so irregular and fragmented, that, compared with *Euclid* — a term used in this work to denote all of standard geometry — Nature exhibits not simply a higher degree but an altogether different level of complexity. The number of distinct scales of length of natural patterns is for all practical purposes infinite.

The existence of these patterns challenges us to study those forms that Euclid leaves aside as being "formless," to investigate the morphology of the "amorphous." Mathematicians have disdained this challenge, however, and have increasingly chosen to flee from nature by devising theories unrelated to anything we can see or feel.

Responding to this challenge, I conceived and developed a new geometry of nature and implemented its use in a number of diverse fields. It describes many of the irregular and fragmented patterns around us, and leads to full-fledged theories, by identifying a family of shapes I call *fractals*. The most useful fractals involve *chance,* and both their regularities and their irregularities are statistical. Also, the shapes described here tend to be *scaling,* implying that the degree of their irregularity and/or fragmentation is identical at all scales. The concept of *fractal* (Hausdorff) *dimension* plays a central role in this work.

Some fractal sets are curves or surfaces, others are disconnected "dusts," and yet others are so oddly shaped that there are no good terms for them in either the sciences or the arts.

\* \* \* \* \*

F. J. Dyson has given an eloquent summary of this theme of mine:
"*Fractal* is a word invented by Mandelbrot to bring together under one heading a large class of objects that have [played] . . . an historical role . . . in the development of pure mathematics. A great revolution of ideas separates the classical mathematics of the 19th century from the modern mathematics of the 20th. Classical mathematics had its roots in the regular geometric structures of Euclid and the continuously evolving dynamics of Newton. Modern mathematics began with Cantor's set theory and Peano's space-filling curve. Historically, the revolution was forced by the discovery of mathematical structures that did not fit the patterns of Euclid and Newton. These new structures were regarded . . . as 'pathological,' . . . as a 'gallery of monsters,' kin to the cubist painting and atonal music that were upsetting established standards of taste in the arts at about the same time. The mathematicians who created the monsters regarded them as important in showing that the world of pure mathematics contains a richness of possibilities going far beyond the simple structures that they saw in Nature. Twentieth-century mathematics flowered in the belief that it had transcended completely the limitations imposed by its natural origins.

"Now, as Mandelbrot points out, . . . Nature has played a joke on the mathematicians. The 19th-century mathematicians may have been lacking in imagination, but Nature was not. The same pathological structures that the mathematicians invented to break loose from 19th-century naturalism turn out to be inherent in familiar objects all around us."[1]

In brief, I have confirmed Blaise Pascal's observation that imagination tires before Nature. . . .

Fractal geometry reveals that some of the most austerely formal chapters of mathematics had a hidden face: a world of pure plastic beauty unsuspected till now. . . .

## How Long Is the Coast of Britain?

To introduce a first category of fractals, namely curves whose fractal dimension is greater than 1, consider a stretch of coastline. It is evident that its length is at least equal to the distance measured along a straight line between its beginning and its end. However, the typical coastline is irregular and winding, and there is no question it is much longer than the straight line between its end points.

There are various ways of evaluating its length more accurately, and this chapter analyzes several of them. The result is most peculiar: coastline length turns out to be an elusive notion that slips between the fingers of one who wants to grasp it. All measurement methods ultimately lead to the conclusion that the typical coastline's length is very large and so ill determined that it is best considered infinite. Hence, if one wishes to compare different coastlines from the viewpoint of their "extent," length is an inadequate concept.

This chapter seeks an improved substitute, and in doing so finds it impossible to avoid introducing various forms of the fractal concepts of dimension, measure, and curve.

### MULTIPLICITY OF ALTERNATIVE METHODS OF MEASUREMENT

METHOD A: Set dividers to a prescribed opening $\varepsilon$, to be called the yardstick length, and walk these dividers along the coastline, each new step starting where the previous step leaves off. The number of steps multiplied by $\varepsilon$ is an approximate length $L(\varepsilon)$. As the dividers'

1. From "Characterizing Irregularity" by Freeman Dyson, *Science*, May 12, 1978, vol. 200, no. 4342, pp. 677–678. Copyright © 1978 by the American Association for the Advancement of Science.

opening becomes smaller and smaller, and as we repeat the operation, we have been taught to expect $L(\epsilon)$ to settle rapidly to a well-defined value called the *true length*. But in fact what we expect does not happen. In the typical case, the observed $L(\epsilon)$ tends to increase without limit.

The reason for this behavior is obvious: When a bay or peninsula noticed on a map scaled to 1/100,000 is reexamined on a map at 1/10,000, subbays and subpeninsulas become visible. On a 1/1,000 scale map, sub-subbays and sub-subpeninsulas appear, and so forth. Each adds to the measured length.

Our procedure acknowledges that a coastline is too irregular to be measured directly by reading it off in a catalog of lengths of simple geometric curves. Therefore, METHOD A replaces the coastline by a sequence of broken lines made of straight intervals, which are curves we know how to handle.

METHOD B: Such "smoothing out" can also be accomplished in other ways. Imagine a man walking along the coastline, taking the shortest path that stays no farther from the water than the prescribed distance $\epsilon$. Then he resumes his walk after reducing his yardstick, then again, after another reduction; and so on, until $\epsilon$ reaches, say 50 cm. Man is too big and clumsy to follow any finer detail. One may further argue that this unreachable fine detail (a) is of no direct interest to Man and (b) varies with the seasons and the tides so much that it is altogether meaningless. We take up argument (a) later on in this chapter. In the meantime, we can neutralize argument (b) by restricting our attention to a rocky coastline observed when the tide is low and the waves are negligible. In principle, Man could follow such a curve down to finer details by harnessing a mouse, then an ant, and so forth. Again, as our walker stays increasingly closer to the coastline, the distance to be covered continues to increase with no limit.

METHOD C: An asymmetry between land and water is implied in METHOD B. To avoid it, Cantor suggests, in effect, that one should view the coastline with an out-of-focus camera that transforms every point into a circular blotch of radius $\epsilon$. In other words, Cantor considers all the points of both land and water for which the distance to the coastline is no more than $\epsilon$. These points form a kind of sausage or tape of width $2\epsilon$. . . . Measure the area of the tape and divide it by $2\epsilon$. If the coastline were straight, the tape would be a rectangle, and the above quotient would be the actual length. With actual coastlines, we have an estimated length $L(\varepsilon)$. As $\varepsilon$ decreases, this estimate increases without limit.

METHOD D: Imagine a map drawn in the manner of pointillist painters using circular blotches of radius $\varepsilon$. Instead of using circles centered on the coastline, as in METHOD C, let us require that the blotches that cover the entire coastline be as few in number as possible. As a result, they may well lie mostly inland near the capes and mostly in the sea near the bays. Such a map's area, divided by $2\varepsilon$, is an estimate of the length. This estimate also "misbehaves."

### ARBITRARINESS OF THE RESULTS OF MEASUREMENT

To summarize the preceding section, the main finding is always the same. As $\varepsilon$ is made smaller and smaller, every approximate length tends to increase steadily without bound.

In order to ascertain the meaning of this result, let us perform analogous measurements on a standard curve from Euclid. For an interval of straight line, the approximate measurements are essentially identical and define the length. For a circle, the approximate measurements increase but converge rapidly to a limit. The curves for which a length is thus defined are called *rectifiable*.

An even more interesting contrast is provided by the results of measurement on a coastline that Man has tamed, say the coast at Chelsea as it is today. Since very large features are unaffected by Man, a very large yardstick again yields results that increase as $\varepsilon$ decreases.

However, there is an intermediate zone of $\varepsilon$'s in which $L(\varepsilon)$ varies little. This zone may go from 20 meters down to 20 centimeters (but do not take these values too strictly). But $L(\varepsilon)$ increases again after $\varepsilon$ becomes less than 20 centimeters and measurements become affected by the irregularity of the stones. Thus, if we trace the curves representing $L(\varepsilon)$ as a function of $\varepsilon$, there is little doubt that the length exhibits, in the zone of $\varepsilon$'s between $\varepsilon = 20$ meters and $\varepsilon = 20$ centimeters, a flat portion that was not observable before the coast was tamed.

Measurements made in this zone are obviously of great practical use. Since boundaries between different scientific disciplines are largely a matter of conventional division of labor between scientists, one might restrict geography to phenomena above Man's reach, for example, on scales above 20 meters. This restriction would yield a well-defined value of geographical length. The Coast Guard may well choose to use the same ε for untamed coasts, and encyclopedias and almanacs could adopt the corresponding $L(\varepsilon)$.

However, the adoption of the same ε by all the agencies of a government is hard to imagine, and its adoption by all countries is all but inconceivable. For example (Richardson 1961), the lengths of the common frontiers between Spain and Portugal, or Belgium and Netherlands, as reported in these neighbors' encyclopedias, differ by 20%. The discrepancy must in part result from different choices of ε. An empirical finding to be discussed soon shows that it suffices that the ε differ by a factor of 2, and one should not be surprised that a small country (Portugal) measures its borders more accurately than its big neighbor.

The second and more significant reason against deciding on an arbitrary ε is philosophical and scientific. Nature does exist apart from Man, and anyone who gives too much weight to any specific ε and $L(\varepsilon)$ lets the study of Nature be dominated by Man, either through his typical yardstick size or his highly variable technical reach. If coastlines are ever to become an object of scientific inquiry, the uncertainty concerning their lengths cannot be legislated away. In one manner or another, the concept of geographic length is not as inoffensive as it seems. It is not entirely "objective." The observer inevitably intervenes in its definition.

\* \* \* \* \*

### THE RICHARDSON EFFECT

The variation of the approximate length $L(\varepsilon)$ obtained by METHOD A has been studied empirically in Richardson 1961, a reference that chance (or fate) put in my way. I paid attention because I knew of Lewis Fry Richardson as a great scientist whose originality mixed with eccentricity. . . . We are indebted to him for some of the most profound and most durable ideas regarding the nature of turbulence, notably the notion that turbulence involves a self-similar cascade. He also concerned himself with other difficult problems, such as the nature of armed conflict between states. His experiments were of classic simplicity, but he never hesitated to use refined concepts when he deemed them necessary.

. . . [Richardson concluded] that there are two constants, which we shall call λ and $D$, such that — to approximate a coastline by a broken line — one needs roughly $F\varepsilon^{-D}$ intervals of length $\varepsilon$, adding up to the length

$$L(\varepsilon) \sim F\varepsilon^{1-D}.$$

The value of the exponent $D$ seems to depend upon the coastline that is chosen, and different pieces of the same coastline, if considered separately, may produce different values of $D$. To Richardson, the $D$ in question was a simple exponent of no particular significance. However, its value seems to be independent of the method chosen to estimate the length of a coastline. Thus $D$ seems to warrant attention.

### A COASTLINE'S FRACTAL DIMENSION

Having unearthed Richardson's work, I proposed that, despite the fact that the exponent $D$ is not an integer, it can and should be interpreted as a dimension, namely, as a fractal dimension. Indeed, I recognized that all the above listed methods of measuring $L(\varepsilon)$ correspond to nonstandard generalized definitions of dimension already used in pure mathematics. The definition of length based on the coastline being covered by the smallest number of blotches of radius $\varepsilon$ is used in Pontrjagin & Schnirelman 1932 to define the covering dimension. The definition of length based on the coastline being covered by a tape of width $2\varepsilon$ implements an idea of Cantor and Minkowski, and the corresponding dimension is due to Bouligand. Yet these two examples only hint at the many dimensions (most of them known only to a few specialists) that star in diverse specialized chapters of mathematics. . . .

Why did mathematicians introduce this plethora of distinct definitions? Because in some cases they yield distinct values. Luckily, however, such cases are never encountered in this Essay, and the list of possible alternative dimensions can be reduced to two that I have not yet mentioned. The older and best investigated one dates back to Hausdorff and serves to define fractal dimension; we come to it momentarily. The simpler one is similarity dimension: it is less general, but in many cases is more than adequate. . . .

Clearly, I do not propose to present a mathematical proof that Richardson's $D$ is a dimension. No such proof is conceivable in any natural science. The goal is merely to convince the reader that the notion of length poses a conceptual problem, and that $D$ provides a manageable and convenient answer. Now that fractal dimension is injected into the study of coastlines, even if specific reasons come to be challenged, I think we shall never return to the stage when $D = 1$ was accepted thoughtlessly and naively. He who continues to think that $D = 1$ has to argue his case. . . .

### HAUSDORFF FRACTAL DIMENSION

If we accept that various natural coasts are really of infinite length and that the length based on an anthropocentric value of ε gives only a partial idea of reality, how can different coastlines be compared to each other? Since infinity equals four times infinity, every coastline is four times longer than each of its quarters, but this is not a useful conclusion. We need a better way to express the sound idea that the entire curve must have a "measure" that is four times greater than each of its fourths.

A most ingenious method of reaching this goal has been provided by Felix Hausdorff. It is intuitively motivated by the fact that the linear measure of a polygon is calculated by adding its sides' lengths without transforming them in any way. One may say (the reason for doing so will soon become apparent) that these lengths are raised to the power $D = 1$, the Euclidean dimension of a straight line. The surface measure of a closed polygon's interior is similarly calculated by paving it with squares, and adding the squares' sides raised to the power $D = 2$, the Euclidean dimension of a plane. When, on the other hand, the "wrong" power is used, the result gives no specific information: the area of every closed polygon is zero, and the length of its interior is infinite.

Let us proceed likewise for a polygonal approximation of a coastline made up of small intervals of length ε. If their lengths are raised to the power $D$, we obtain a quantity we may call tentatively an "approximate measure in the dimension $D$." Since according to Richardson the number of sides is $N = F\varepsilon^{-D}$, said approximate measure takes the value $F\varepsilon^{D}\varepsilon^{-D} = F$.

Thus, *the approximate measure in the dimension* D *is independent of* ε. With actual data, we simply find that this approximate measure varies little with ε.

In addition, the fact that the length of a square is infinite has a simple counterpart and generalization: a coastline's approximate measure evaluated in any dimension $d$ smaller than $D$ tends to $\infty$ as $\varepsilon \to 0$. Similarly, the area and the volume of a straight line are zero. And when $d$ takes any value larger than $D$, the corresponding approximate measure of a coastline tends to $O$ as $\varepsilon \to O$. The approximate measure behaves reasonably if and only if $d = D$.

### A CURVE'S FRACTAL DIMENSION MAY EXCEED 1; FRACTAL CURVES

By design, the Hausdorff dimension preserves the ordinary dimension's role as exponent in defining a *measure*.

But from another viewpoint, $D$ is very odd indeed: it is a fraction! In particular, it exceeds 1, which is the intuitive dimension of curves and which may be shown rigorously to be their topological dimension $D_T$.

I propose that curves for which the fractal dimension exceeds the topological dimension 1 be called *fractal curves*. And the present chapter can be summarized by asserting that, within the scales of interest to the geographer, coastlines can be modeled by fractal curves. Coastlines are *fractal patterns*.

# JAMES GLEICK

■

Whereas fractal geometry made complex natural structures more accessible to mathematics, the emerging science of chaos brought other complexities within the potential grasp of science — from changes in the weather to predictions of the rise and fall of the stock market. James Gleick (b. 1954), a reporter for the *New York Times*, reported on this new development in his best-selling book *Chaos*, published in 1987.

# *Chaos*

THE POLICE IN THE SMALL TOWN of Los Alamos, New Mexico worried briefly in 1974 about a man seen prowling in the dark, night after night, the red glow of his cigarette floating along the back streets. He would pace for hours, heading nowhere in the starlight that hammers down through the thin air of the mesas. The police were not the only ones to wonder. At the national laboratory some physicists had learned that their newest colleague was experimenting with twenty-six-hour days, which meant that his waking schedule would slowly roll in and out of phase with theirs. This bordered on strange, even for the Theoretical Division.

In the three decades since J. Robert Oppenheimer chose this unworldly New Mexico landscape for the atomic bomb project, Los Alamos National Laboratory had spread across an expanse of desolate plateau, bringing particle accelerators and gas lasers and chemical plants, thousands of scientists and administrators and technicians, as well as one of the world's greatest concentrations of supercomputers. Some of the older scientists remembered the wooden buildings rising hastily out of the rimrock in the 1940s, but to most of the Los Alamos staff, young men and women in college-style corduroys and work shirts, the first bombmakers were just ghosts. The laboratory's locus of purest thought was the Theoretical Division, known as T division, just as computing was C division and weapons was X division. More than a hundred physicists and mathematicians worked in T division, well paid and free of academic pressures to teach and publish. These scientists had experience with brilliance and with eccentricity. They were hard to surprise.

But Mitchell Feigenbaum was an unusual case. He had exactly one published article to his name, and he was working on nothing that seemed to have any particular promise. His hair was a ragged mane, sweeping back from his wide brow in the style of busts of German composers. His eyes were sudden and passionate. When he spoke, always rapidly, he tended to drop articles and pronouns in a vaguely middle European way, even though he was a native of Brooklyn. When he worked, he worked obsessively. When he could not work, he walked and thought, day or night, and night was best of all. The twenty-four-hour day seemed too constraining. Nevertheless, his experiment in personal quasiperiodicity came to an end when he decided he could no longer bear waking to the setting sun, as had to happen every few days.

At the age of twenty-nine he had already become a savant among the savants, an ad hoc consultant whom scientists would go to see about any especially intractable problem, when they could find him. One evening he arrived at work just as the director of the laboratory, Harold Agnew, was leaving. Agnew was a powerful figure, one of the original Oppenheimer apprentices. He had flown over Hiroshima on an instrument plane that accompanied the Enola Gay, photographing the delivery of the laboratory's first product.

"I understand you're real smart," Agnew said to Feigenbaum. "If you're so smart, why don't you just solve laser fusion?"

Even Feigenbaum's friends were wondering whether he was ever going to produce any work of his own. As willing as he was to do impromptu magic with their questions, he did not seem interested in devoting his own research to any problem that might pay off. He thought about turbulence in liquids and gases. He thought about time — did it glide smoothly forward or hop discretely like a sequence of cosmic motion-picture frames? He thought about the eye's ability to see consistent colors and forms in a universe that physicists knew to be a shifting quantum kaleidoscope. He thought about clouds, watching them from airplane windows (until, in 1975, his scientific travel privileges were officially suspended on grounds of overuse) or from the hiking trails above the laboratory.

In the mountain towns of the West, clouds barely resemble the

457

sooty indeterminate low-flying hazes that fill the Eastern air. At Los Alamos, in the lee of a great volcanic caldera, the clouds spill across the sky, in random formation, yes, but also not-random, standing in uniform spikes or rolling in regularly furrowed patterns like brain matter. On a stormy afternoon, when the sky shimmers and trembles with the electricity to come, the clouds stand out from thirty miles away, filtering the light and reflecting it, until the whole sky starts to seem like a spectacle staged as a subtle reproach to physicists. Clouds represented a side of nature that the mainstream of physics had passed by, a side that was at once fuzzy and detailed, structured and unpredictable. Feigenbaum thought about such things, quietly and unproductively.

To a physicist, creating laser fusion was a legitimate problem; puzzling out the spin and color and flavor of small particles was a legitimate problem; dating the origin of the universe was a legitimate problem. Understanding clouds was a problem for a meteorologist. Like other physicists, Feigenbaum used an understated, tough-guy vocabulary to rate such problems. *Such a thing is obvious,* he might say, meaning that a result could be understood by any skilled physicist after appropriate contemplation and calculation. *Not obvious* described work that commanded respect and Nobel Prizes. For the hardest problems, the problems that would not give way without long looks into the universe's bowels, physicists reserved words like *deep.* In 1974, though few of his colleagues knew it, Feigenbaum was working on a problem that was deep: chaos.

Where chaos begins, classical science stops. For as long as the world has had physicists inquiring into the laws of nature, it has suffered a special ignorance about disorder in the atmosphere, the turbulent sea, in the fluctuations of wildlife populations, in the oscillations of the heart and the brain. The irregular side of nature, the discontinuous and erratic side — these have been puzzles to science, or worse, monstrosities.

But in the 1970s a few scientists in the United States and Europe began to find a way through disorder. They were mathematicians, physicists, biologists, chemists, all seeking connections between different kinds of irregularity. Physiologists found a surprising order in the chaos that develops in the human heart, the prime cause of sudden, unexplained death. Ecologists explored the rise and fall of gypsy moth populations. Economists dug out old stock price data and tried a new kind of analysis. The insights that emerged led directly into the natural world — the shapes of clouds, the paths of lightning, the microscopic intertwining of blood vessels, the galactic clustering of stars.

When Mitchell Feigenbaum began thinking about chaos at Los Alamos, he was one of a handful of scattered scientists, mostly unknown to one another. A mathematician in Berkeley, California, had formed a small group dedicated to creating a new study of "dynamical systems." A population biologist at Princeton University was about to publish an impassioned plea that all scientists should look at the surprisingly complex behavior lurking in some simple models. A geometer working for IBM was looking for a new word to describe a family of shapes — jagged, tangled, splintered, twisted, fractured — that he considered an organizing principle in nature. A French mathematical physicist had just made the disputatious claim that turbulence in fluids might have something to do with a bizarre, infinitely tangled abstraction that he called a strange attractor.

A decade later, chaos has become a shorthand name for a fast-growing movement that is reshaping the fabric of the scientific establishment. Chaos conferences and chaos journals abound. Government program managers in charge of research money for the military, the Central Intelligence Agency, and the Department of Energy have put ever greater sums into chaos research and set up special bureaucracies to handle the financing. At every major university and every major corporate research center, some theorists ally themselves first with chaos and only second with their nominal specialties. At Los Alamos, a Center for Nonlinear Studies was established to coordinate work on chaos and related problems; similar institutions have appeared on university campuses across the country.

Chaos has created special techniques of using computers and special kinds of graphic images, pictures that capture a fantastic and delicate structure underlying complexity. The new science has spawned its own language, an elegant shop talk of *fractals* and *bifurcations, intermittencies* and *periodicities, folded-towel diffeomorphisms* and *smooth noodle maps*. These are the new elements of motion, just as, in traditional physics, quarks and gluons are the new elements of matter. To some physicists chaos is a science of process rather than state, of becoming rather than being.

Now that science is looking, chaos seems to be everywhere. A rising column of cigarette smoke breaks into wild swirls. A flag snaps back and forth in the wind. A dripping faucet goes from a steady

459

pattern to a random one. Chaos appears in the behavior of the weather, the behavior of an airplane in flight, the behavior of cars clustering on an expressway, the behavior of oil flowing in underground pipes. No matter what the medium, the behavior obeys the same newly discovered laws. That realization has begun to change the way business executives make decisions about insurance, the way astronomers look at the solar system, the way political theorists talk about the stresses leading to armed conflict.

Chaos breaks across the lines that separate scientific disciplines. Because it is a science of the global nature of systems, it has brought together thinkers from fields that had been widely separated. "Fifteen years ago, science was heading for a crisis of increasing specialization," a Navy official in charge of scientific financing remarked to an audience of mathematicians, biologists, physicists, and medical doctors. "Dramatically, that specialization has reversed because of chaos." Chaos poses problems that defy accepted ways of working in science. It makes strong claims about the universal behavior of complexity. The first chaos theorists, the scientists who set the discipline in motion, shared certain sensibilities. They had an eye for pattern, especially pattern that appeared on different scales at the same time. They had a taste for randomness and complexity, for jagged edges and sudden leaps. Believers in chaos — and they sometimes call themselves believers, or converts, or evangelists — speculate about determinism and free will, about evolution, about the nature of conscious intelligence. They feel that they are turning back a trend in science toward reductionism, the analysis of systems in terms of their constituent parts: quarks, chromosomes, or neurons. They believe that they are looking for the whole.

The most passionate advocates of the new science go so far as to say that twentieth-century science will be remembered for just three things: relativity, quantum mechanics, and chaos. Chaos, they contend, has become the century's third great revolution in the physical sciences. Like the first two revolutions, chaos cuts away at the tenets of Newton's physics. As one physicist put it: "Relativity eliminated the Newtonian illusion of absolute space and time; quantum theory eliminated the Newtonian dream of a controllable measurement process; and chaos eliminates the Laplacian fantasy of deterministic predictability." Of the three, the revolution in chaos applies to the universe we see and touch, to objects at human scale. Everyday experience and real pictures of the world have become legitimate targets for inquiry. There has long been a feeling, not always expressed openly, that theoretical physics has strayed far from human intuition about the world. Whether this will prove to be fruitful heresy or just plain heresy, no one knows. But some of those who thought physics might be working its way into a corner now look to chaos as a way out.

Within physics itself, the study of chaos emerged from a backwater. The mainstream for most of the twentieth century has been particle physics, exploring the building blocks of matter at higher and higher energies, smaller and smaller scales, shorter and shorter times. Out of particle physics have come theories about the fundamental forces of nature and about the origin of the universe. Yet some young physicists have grown dissatisfied with the direction of the most prestigious of sciences. Progress has begun to seem slow, the naming of new particles futile, the body of theory cluttered. With the coming of chaos, younger scientists believed they were seeing the beginnings of a course change for all of physics. The field had been dominated long enough, they felt, by the glittering abstractions of high-energy particles and quantum mechanics.

The cosmologist Stephen Hawking, occupant of Newton's chair at Cambridge University, spoke for most of physics when he took stock of his science in a 1980 lecture titled "Is the End in Sight for Theoretical Physics?"

"We already know the physical laws that govern everything we experience in everyday life. . . . It is a tribute to how far we have come in theoretical physics that it now takes enormous machines and a great deal of money to perform an experiment whose results we cannot predict."

Yet Hawking recognized that understanding nature's laws on the terms of particle physics left unanswered the question of how to apply those laws to any but the simplest of systems. Predictability is one thing in a cloud chamber where two particles collide at the end of a race around an accelerator. It is something else altogether in the simplest tub of roiling fluid, or in the earth's weather, or in the human brain.

Hawking's physics, efficiently gathering up Nobel Prizes and big money for experiments, has often been called a revolution. At times it seemed within reach of that grail of science, the Grand Unified Theory or "theory of everything." Physics had traced the development of energy and matter in all but the first eyeblink of the universe's history. But was postwar particle physics a revolution? Or was it just the fleshing out of the framework laid down by Einstein, Bohr, and the other fathers of relativity and quantum mechanics? Certainly, the achievements of physics, from the atomic bomb to the transistor, changed the twentieth-century landscape. Yet if anything, the scope of particle physics seemed to have narrowed. Two generations had passed since the field produced a new theoretical idea that changed the way nonspecialists understand the world.

The physics described by Hawking could complete its mission without answering some of the most fundamental questions about nature. How does life begin? What is turbulence? Above all, in a universe ruled by entropy, drawing inexorably toward greater and greater disorder, how does order arise? At the same time, objects of everyday experience like fluids and mechanical systems came to seem so basic and so ordinary that physicists had a natural tendency to assume they were well understood. It was not so.

As the revolution in chaos runs its course, the best physicists find themselves returning without embarrassment to phenomena on a human scale. They study not just galaxies but clouds. They carry out profitable computer research not just on Crays but on Macintoshes. The premier journals print articles on the strange dynamics of a ball bouncing on a table side by side with articles on quantum physics. The simplest systems are now seen to create extraordinarily difficult problems of predictability. Yet order arises spontaneously in those systems — chaos and order together. Only a new kind of science could begin to cross the great gulf between knowledge of what one thing does — one water molecule, one cell of heart tissue, one neuron — and what millions of them do.

Watch two bits of foam flowing side by side at the bottom of a waterfall. What can you guess about how close they were at the top? Nothing. As far as standard physics was concerned, God might just as well have taken all those water molecules under the table and shuffled them personally. Traditionally, when physicists saw complex results, they looked for complex causes. When they saw a random relationship between what goes into a system and what comes out, they assumed that they would have to build randomness into any realistic theory, by artificially adding noise or error. The modern study of chaos began with the creeping realization in the 1960s that quite simple mathematical equations could model systems every bit as violent as a waterfall. Tiny differences in input could quickly become overwhelming differences in output — a phenomenon given the name "sensitive dependence on initial conditions." In weather, for example, this translates into what is only half-jokingly known as the Butterfly Effect — the notion that a butterfly stirring the air today in Peking can transform storm systems next month in New York.

\* \* \* \* \*

Two decades ago . . . most practicing scientists shared a set of beliefs about complexity. They held these beliefs so closely that they did not need to put them into words. Only later did it become possible to say what these beliefs were and to bring them out for examination.

*Simple systems behave in simple ways.* A mechanical contraption like a pendulum, a small electrical circuit, an idealized population of fish in a pond — as long as these systems could be reduced to a few perfectly understood, perfectly deterministic laws, their long-term behavior would be stable and predictable.

*Complex behavior implies complex causes.* A mechanical device, an electrical circuit, a wildlife population, a fluid flow, a biological organ, a particle beam, an atmospheric storm, a national economy — a system that was visibly unstable, unpredictable, or out of control must either be governed by a multitude of independent components or subject to random external influences.

*Different systems behave differently.* A neurobiologist who spent a career studying the chemistry of the human neuron without learning anything about memory or perception, an aircraft designer who used wind tunnels to solve aerodynamic problems without understanding the mathematics of turbulence, an economist who analyzed the psychology of purchasing decisions without gaining an ability to forecast large-scale trends — scientists like these, knowing that the components of their disciplines were different, took it for granted that the complex systems made up of billions of these components must also be different.

Now all that has changed. In the intervening twenty years, physicists, mathematicians, biologists, and astronomers have created an alternative set of ideas. Simple systems give rise to complex behavior. Complex systems give rise to simple behavior. And most important, the laws of complexity hold universally, caring not at all for the details of a system's constituent atoms.

For the mass of practicing scientists — particle physicists or neurologists or even mathematicians — the change did not matter immediately. They continued to work on research problems within their disciplines. But they were aware of something called chaos.

They knew that some complex phenomena had been explained, and they knew that other phenomena suddenly seemed to need new explanations. A scientist studying chemical reactions in a laboratory or tracking insect populations in a three-year field experiment or modeling ocean temperature variations could not respond in the traditional way to the presence of unexpected fluctuations or oscillations — that is, by ignoring them. For some, that meant trouble. On the other hand, pragmatically, they knew that money was available from the federal government and from corporate research facilities for this faintly mathematical kind of science. More and more of them realized that chaos offered a fresh way to proceed with old data, forgotten in desk drawers because they had proved too erratic. More and more felt the compartmentalization of science as an impediment to their work. More and more felt the futility of studying parts in isolation from the whole. For them, chaos was the end of the reductionist program in science.

Uncomprehension; resistance; anger; acceptance. Those who had promoted chaos longest saw all of these. Joseph Ford of the Georgia Institute of Technology remembered lecturing to a thermodynamics group in the 1970s and mentioning that there was a chaotic behavior in the Duffing equation, a well-known textbook model for a simple oscillator subject to friction. To Ford, the presence of chaos in the Duffing equation was a curious fact — just one of those things he knew to be true, although several years passed before it was published in *Physical Review Letters*. But he might as well have told a gathering of paleontologists that dinosaurs had feathers. They knew better.

"When I said that? Jee-sus Christ, the audience began to bounce up and down. It was, 'My daddy played with the Duffing equation, and my granddaddy played with the Duffing equation, and nobody seen anything like what you're talking about.' You would really run across resistance to the notion that nature is complicated. What I didn't understand was the hostility."

Comfortable in his Atlanta office, the winter sun setting outside, Ford sipped soda from an oversized mug with the word CHAOS painted in bright colors. His younger colleague Ronald Fox talked about his own conversion, soon after buying an Apple II computer for his son, at a time when no self-respecting physicist would buy such a thing for his work. Fox heard that Mitchell Feigenbaum had discovered universal laws guiding the behavior of feedback functions, and he decided to write a short program that would let him see the behavior on the Apple display. He saw it all painted across the screen — pitchfork bifurcations, stable lines breaking in two, then four, then eight; the appearance of chaos itself; and within the chaos, the astonishing geometric regularity. "In a couple of days you could redo all of Feigenbaum," Fox said. Self-teaching by computing persuaded him and others who might have doubted a written argument.

Some scientists played with such programs for a while and then stopped. Others could not help but be changed. Fox was one of those who had remained conscious of the limits of standard linear science. He knew he had habitually set the hard nonlinear problems aside. In practice a physicist would always end up saying, *This is a problem that's going to take me to the handbook of special functions, which is the last place I want to go, and I'm sure as hell not going to get on a machine and do it, I'm too sophisticated for that.*

"The general picture of nonlinearity got a lot of people's attention — slowly at first, but increasingly," Fox said. "Everybody that looked at it, it bore fruit for. You now look at any problem you looked at before, no matter what science you're in. There was a place where you quit looking at it because it became nonlinear. Now you know how to look at it and you go back."

Ford said, "If an area begins to grow, it has to be because some clump of people feel that there's something it offers *them* — that if they modify their research, the rewards could be very big. To me chaos is like a dream. It offers the possibility that, if you come over and play this game, you can strike the mother lode."

Still, no one could quite agree on the word itself.

Philip Holmes, a white-bearded mathematician and poet from Cornell by way of Oxford: *The complicated, aperiodic, attracting orbits of certain (usually low-dimensional) dynamical systems.*

Hao Bai-Lin, a physicist in China who assembled many of the historical papers of chaos into a single reference volume: *A kind of order without periodicity.* And: *A rapidly expanding field of research to which mathematicians, physicists, hydrodynamicists, ecologists and many others have all made important contributions.* And: *A newly recognized and ubiquitous class of natural phenomena.*

H. Bruce Stewart, an applied mathematician at Brookhaven National Laboratory on Long Island: *Apparently random recurrent behavior in a simple deterministic (clockwork-like) system.*

Roderick V. Jensen of Yale University, a theoretical physicist exploring the possibility of quantum chaos: *The irregular, unpredictable behavior of deterministic, nonlinear dynamical systems.*

James Crutchfield of the Santa Cruz collective: *Dynamics with positive, but finite, metric entropy. The translation from mathese is: behavior that produces information (amplifies small uncertainties), but is not utterly unpredictable.*

And Ford, self-proclaimed evangelist of chaos: *Dynamics freed at last from the shackles of order and predictability. . . . Systems liberated to randomly explore their every dynamical possibility. . . . Exciting variety, richness of choice, a cornucopia of opportunity.*

John Hubbard, exploring iterated functions and the infinite fractal wildness of the Mandelbrot set, considered chaos a poor name for his work, because it implied randomness. To him, the overriding message was that simple processes in nature could produce magnificent edifices of complexity *without* randomness. In nonlinearity and feedback lay all the necessary tools for encoding and then unfolding structures as rich as the human brain.

To other scientists, like Arthur Winfree, exploring the global topology of biological systems, chaos was too narrow a name. It implied simple systems, the one-dimensional maps of Feigenbaum and the two- or three- (and a fraction) dimensional strange attractors of Ruelle. Low-dimensional chaos was a special case, Winfree felt. He was interested in the laws of many-dimensional complexity — and he was convinced that such laws existed. Too much of the universe seemed beyond the reach of low-dimensional chaos.

The journal *Nature* carried a running debate about whether the earth's climate followed a strange attractor. Economists looked for recognizable strange attractors in stock market trends but so far had not found them. Dynamicists hoped to use the tools of chaos to explain fully developed turbulence. Albert Libchaber, now at the University of Chicago, was turning his elegant experimental style to the service of turbulence, creating a liquid-helium box thousands of times larger than his tiny cell of 1977. Whether such experiments, liberating fluid disorder in both space and time, would find simple attractors, no one knew. As the physicist Bernardo Huberman said, "If you had a turbulent river and put a probe in it and said, 'Look, here's a low-dimensional strange attractor,' we would all take off our hats and look."

Chaos was the set of ideas persuading all these scientists that they were members of a shared enterprise. Physicist or biologist or mathematician, they believed that simple, deterministic systems could breed complexity; that systems too complex for traditional mathematics could yet obey simple laws; and that, whatever their particular field, their task was to understand complexity itself.

"Let us again look at the laws of thermodynamics," wrote James E. Lovelock, author of the Gaia hypothesis. "It is true that at first sight they read like the notice at the gate of Dante's Hell . . ." But.

The Second Law is one piece of technical bad news from science that has established itself firmly in the nonscientific culture. Everything tends toward disorder. Any process that converts energy from one form to another must lose some as heat. Perfect efficiency is impossible. The universe is a one-way street. *Entropy must always increase in the universe and in any hypothetical isolated system within it.* However expressed, the Second Law is a rule from which there seems no appeal. In thermodynamics that is true. But the Second Law has had a life of its own in intellectual realms far removed from science, taking the blame for disintegration of societies, economic decay, the breakdown of manners, and many other variations on the decadent theme. These secondary, metaphorical incarnations of the Second Law now seem especially misguided. In our world, complexity flourishes, and those looking to science for a general understanding of nature's habits will be better served by the laws of chaos.

Somehow, after all, as the universe ebbs toward its final equilibrium in the featureless heat bath of maximum entropy, it manages to create interesting structures. Thoughtful physicists concerned with the workings of thermodynamics realize how disturbing is the question of, as one put it, "how a purposeless flow of energy can wash life and consciousness into the world." Compounding the trouble is the slippery notion of entropy, reasonably well-defined for thermodynamic purposes in terms of heat and temperature, but devilishly hard to pin down as a measure of *disorder*. Physicists have trouble enough measuring the degree of order in water, forming crystalline structures in the transition to ice, energy bleeding away all the while. But thermodynamic entropy fails miserably as a measure of the changing degree of form and formlessness in the creation of amino acids, of microorganisms, of self-reproducing plants and animals, of complex information systems like the brain. Certainly these evolving islands of order must obey the Second Law. The important laws, the creative laws, lie elsewhere.

Nature forms patterns. Some are orderly in space but disorderly in time, others orderly in time but disorderly in space. Some patterns are fractal, exhibiting structures self-similar in scale. Others give rise to steady states or oscillating ones. Pattern formation has become a branch of physics and of materials science, allowing scientists to model the aggregation of particles into clusters, the fractured spread of electrical discharges, and the growth of crystals in ice and metal alloys. The dynamics seem so basic — shapes changing in space and time — yet only now are the tools available to understand them. It is a fair question now to ask a physicist, "Why are all snowflakes different?"

Ice crystals form in the turbulent air with a famous blending of symmetry and chance, the special beauty of six-fold indeterminacy. As water freezes, crystals send out tips; the tips grow, their boundaries becoming unstable, and new tips shoot out from the sides. Snowflakes obey mathematical laws of surprising subtlety, and it was impossible to predict precisely how fast a tip would grow, how narrow it would be, or how often it would branch. Generations of scientists sketched and cataloged the variegated patterns: plates and columns, crystals and polycrystals, needles and dendrites. The treatises treated crystal formation as a classification matter, for lack of a better approach.

Growth of such tips, dendrites, is now known as a highly nonlinear unstable free boundary problem, meaning that models need to track a complex, wiggly boundary as it changes dynamically. When solidification proceeds from outside to inside, as in an ice tray, the boundary generally remains stable and smooth, its speed controlled by the ability of the walls to draw away the heat. But when a crystal solidifies outward from an initial seed — as a snowflake does, grabbing water molecules while it falls through the moisture-laden air — the process becomes unstable. Any bit of boundary that gets out ahead of its neighbors gains an advantage in picking up new water molecules and therefore grows that much faster — the "lightning-rod effect." New branches form, and then subbranches.

One difficulty was in deciding which of the many physical forces involved are important and which can safely be ignored. Most important, as scientists have long known, is the diffusion of the heat released when water freezes. But the physics of heat diffusion cannot completely explain the patterns researchers observe when they look at snowflakes under microscopes or grow them in the laboratory. Recently scientists worked out a way to incorporate another process: surface tension. The heart of the new snowflake model is the essence of chaos: a delicate balance between forces of stability and forces of instability; a powerful interplay of forces on atomic scales and forces on everyday scales.

Where heat diffusion tends to create instability, surface tension creates stability. The pull of surface tension makes a substance prefer smooth boundaries like the wall of a soap bubble. It costs energy to make surfaces that are rough. The balancing of these tendencies depends on the size of the crystal. While diffusion is mainly a large-scale, macroscopic process, surface tension is strongest at the microscopic scales.

Traditionally, because the surface tension effects are so small, researchers assumed that for practical purposes they could disregard them. Not so. The tiniest scales proved crucial; there the surface effects proved infinitely sensitive to the molecular structure of a solidifying substance. In the case of ice, a natural molecular symmetry gives a built-in preference for six directions of growth. To their surprise, scientists found that the mixture of stability and instability manages to amplify this microscopic preference, creating the near-fractal lacework that makes snowflakes. The mathematics came not from atmospheric scientists but from theoretical physicists, along with metallurgists, who had their own interest. In metals the molecular symmetry is different, and so are the characteristic crystals, which help determine an alloy's strength. But the mathematics are the same: the laws of pattern formation are universal.

Sensitive dependence on initial conditions serves not to destroy but to create. As a growing snowflake falls to earth, typically floating in the wind for an hour or more, the choices made by the branching tips at any instant depend sensitively on such things as the temperature, the humidity, and the presence of impurities in the atmosphere. The six tips of a single snowflake, spreading within a millimeter space, feel the same temperatures, and because the laws of growth are purely deterministic, they maintain a near-perfect symmetry. But the nature of turbulent air is such that any pair of snowflakes will experience very different paths. The final flake records the history of all the changing weather conditions it has experienced, and the combinations may as well be infinite.

Snowflakes are nonequilibrium phenomena, physicists like to say. They are products of imbalance in the flow of energy from one piece of nature to another. The flow turns a boundary into a tip, the tip into an array of branches, the array into a complex structure never before seen. As scientists have discovered such instability obeying the universal laws of chaos, they have succeeded in applying the same methods to a host of physical and chemical problems, and, inevitably, they suspect that biology is next. In the back of their minds, as they look at computer simulations of dendrite growth, they see algae, cell walls, organisms budding and dividing.

From microscopic particles to everyday complexity, many paths now seem open. In mathematical physics the bifurcation theory of Feigenbaum and his colleagues advances in the United States and Europe. In the abstract reaches of theoretical physics scientists probe other new issues, such as the unsettled question of quantum chaos: Does quantum mechanics admit the chaotic phenomena of classical mechanics? In the study of moving fluids Libchaber builds his giant liquid-helium box, while Pierre Hohenberg and Günter Ahlers study the odd-shaped traveling waves of convection. In astronomy chaos experts use unexpected gravitational instabilities to explain the origin of meteorites — the seemingly inexplicable catapulting of asteroids from far beyond Mars. Scientists use the physics of dynamical systems to study the human immune system, with its billions of components and its capacity for learning, memory, and pattern recognition, and they simultaneously study evolution, hoping to find universal mechanisms of adaptation. Those who make such models quickly see structures that replicate themselves, compete, and evolve by natural selection.

"Evolution is chaos with feedback," Joseph Ford said. The universe is randomness and dissipation, yes. But randomness with direction can produce surprising complexity. And as Lorenz discovered so long ago, dissipation is an agent of order.

"God plays dice with the universe," is Ford's answer to Einstein's famous question. "But they're loaded dice. And the main objective of physics now is to find out by what rules were they loaded and how can we use them for our own ends."

Such ideas help drive the collective enterprise of science forward. Still, no philosophy, no proof, no experiment ever seems quite enough to sway the individual researchers for whom science must first and always provide a way of working. In some laboratories, the traditional ways falter. Normal science goes astray, as Kuhn put it; a piece of equipment fails to meet expectations; "the profession can no longer evade anomalies." For any one scientist the ideas of chaos could not prevail until the method of chaos became a necessity.

Every field had its own examples. In ecology, there was William M. Schaffer, who trained as the last student of Robert MacArthur, the dean of the field in the fifties and sixties. MacArthur built a conception of nature that gave a firm footing to the idea of *natural balance*. His models supposed that equilibriums would exist and that populations of plants and animals would remain close to them. To MacArthur, balance in nature had what could almost be called a moral quality — states of equilibrium in his models entailed the most efficient use of food resources, the least waste. Nature, if left alone, would be good.

Two decades later MacArthur's last student found himself realizing that ecology based on a sense of equilibrium seems doomed to fail. The traditional models are betrayed by their linear bias. Nature is more complicated. Instead he sees chaos, "both exhilarating and a bit threatening." Chaos may undermine ecology's most enduring assumptions, he tells his colleagues. "What passes for fundamental concepts in ecology is as mist before the fury of the storm — in this case, a full, nonlinear storm."

Schaffer is using strange attractors to explore the epidemiology of childhood diseases such as measles and chicken pox. He has collected data, first from New York City and Baltimore, then from Aberdeen, Scotland, and all England and Wales. He has made a dynamical model, resembling a damped, driven pendulum. The diseases are driven each year by the infectious spread among children returning to school, and damped by natural resistance. Schaffer's model predicts strikingly different behavior for these diseases. Chicken pox should vary periodically. Measles should vary chaotically. As it happens, the data show exactly what Schaffer predicts. To a traditional epidemiologist the yearly variations in measles seemed inexplicable — random and noisy. Schaffer, using the techniques of phase-space reconstruction, shows that measles follow a strange attractor, with a fractal dimension of about 2.5.

Schaffer computed Lyapunov exponents and made Poincaré maps. "More to the point," Schaffer said, "if you look at the pictures it jumps out at you, and you say, 'My God, this is the same thing.'" Although the attractor is chaotic, some predictability becomes possible in light of the deterministic nature of the model. A year of high measles infection will be followed by a crash. After a year of medium infection, the level will change only slightly. A year of low infection produces the greatest unpredictability. Schaffer's model also pre-

dicted the consequences of damping the dynamics by mass inocula-
tion programs — consequences that could not be predicted by standard epidemiology.

On the collective scale and on the personal scale, the ideas of
chaos advance in different ways and for different reasons. For
Schaffer, as for many others, the transition from traditional science
to chaos came unexpectedly. He was a perfect target for Robert
May's evangelical plea in 1975; yet he read May's paper and discarded it. He thought the mathematical ideas were unrealistic for the
kinds of systems a practicing ecologist would study. Oddly, he knew
too much about ecology to appreciate May's point. These were one-
dimensional maps, he thought — what bearing could they have on
continuously changing systems? So a colleague said, "Read Lorenz."
He wrote the reference on a slip of paper and never bothered to
pursue it.

Years later Schaffer lived in the desert outside of Tucson, Arizona, and summers found him in the Santa Catalina mountains just
to the north, islands of chaparral, merely hot when the desert floor
is roasting. Amid the thickets in June and July, after the spring
blooming season and before the summer rain, Schaffer and his graduate students tracked bees and flowers of different species. This ecological system was easy to measure despite all its year-to-year
variation. Schaffer counted the bees on every stalk, measured the
pollen by draining flowers with pipettes, and analyzed the data mathematically. Bumblebees competed with honeybees, and honeybees
competed with carpenter bees, and Schaffer made a convincing
model to explain the fluctuations in population.

By 1980 he knew that something was wrong. His model broke
down. As it happened, the key player was a species he had overlooked: ants. Some colleagues suspected unusual winter weather;
others unusual summer weather. Schaffer considered complicating
his model by adding more variables. But he was deeply frustrated.
Word was out among the graduate students that summer at 5,000
feet with Schaffer was hard work. And then everything changed.

He happened upon a preprint about chemical chaos in a complicated laboratory experiment, and he felt that the authors had
experienced exactly his problem: the impossibility of monitoring
dozens of fluctuating reaction products in a vessel matched the
impossibility of monitoring dozens of species in the Arizona mountains. Yet they had succeeded where he had failed. He read about
reconstructing phase space. He finally read Lorenz, and Yorke, and

others. The University of Arizona sponsored a lecture series on "Order in Chaos." Harry Swinney came, and Swinney knew how to talk about experiments. When he explained chemical chaos, displaying a transparency of a strange attractor, and said, "That's real data," a chill ran up Schaffer's spine.

"All of a sudden I knew that that was my destiny," Schaffer said. He had a sabbatical year coming. He withdrew his application for National Science Foundation money and applied for a Guggenheim Fellowship. Up in the mountains, he knew, the ants changed with the season. Bees hovered and darted in a dynamical buzz. Clouds skidded across the sky. He could not work the old way any more.

# Artificial Intelligence and All That

## LEWIS THOMAS

■

Lewis Thomas (b. 1913), director of the Sloan-Kettering Cancer Center in New York and a former dean of the Yale and New York University medical schools, was sixty years old when he published his first collection of essays, *The Lives of a Cell,* in 1974. (His inspiration was Montaigne, who published *his* first book at the age of forty-seven.) The book established Thomas as one of the most perspicacious essayists of our time; the brevity and originality of the following selection suggests why.

## Computers

YOU CAN MAKE COMPUTERS that are almost human. In some respects they are superhuman; they can beat most of us at chess, memorize whole telephone books at a glance, compose music of a certain kind and write obscure poetry, diagnose heart ailments, send personal invitations to vast parties, even go transiently crazy. No one has yet programmed a computer to be of two minds about a hard problem, or to burst out laughing, but that may come. Sooner or later, there will be real human hardware, great whirring, clicking cabinets intelligent enough to read magazines and vote, able to think rings around the rest of us.

Well, maybe, but not for a while anyway. Before we begin organizing sanctuaries and reservations for our software selves, lest we vanish like the whales, here is a thought to relax with.

Even when technology succeeds in manufacturing a machine as

475

big as Texas to do everything we recognize as human, it will still be, at best, a single individual. This amounts to nothing, practically speaking. To match what we can do, there would have to be 3 billion of them with more coming down the assembly line, and I doubt that anyone will put up the money, much less make room. And even so, they would all have to be wired together, intricately and delicately, as we are, communicating with each other, talking incessantly, listening. If they weren't *at* each other this way, all their waking hours, they wouldn't be anything like human, after all. I think we're safe, for a long time ahead.

It is in our collective behavior that we are most mysterious. We won't be able to construct machines like ourselves until we've understood this, and we're not even close. All we know is the phenomenon: we spend our time sending messages to each other, talking and trying to listen at the same time, exchanging information. This seems to be our most urgent biological function; it is what we do with our lives. By the time we reach the end, each of us has taken in a staggering store, enough to exhaust any computer, much of it incomprehensible, and we generally manage to put out even more than we take in. Information is our source of energy; we are driven by it. It has become a tremendous enterprise, a kind of energy system on its own. All 3 billion of us are being connected by telephones, radios, television sets, airplanes, satellites, harangues on public-address systems, newspapers, magazines, leaflets dropped from great heights, words got in edgewise. We are becoming a grid, a circuitry around the earth. If we keep at it, we will become a computer to end all computers, capable of fusing all the thoughts of the world into a syncytium.

Already, there are no closed, two-way conversations. Any word you speak this afternoon will radiate out in all directions, around town before tomorrow, out and around the world before Tuesday, accelerating to the speed of light, modulating as it goes, shaping new and unexpected messages, emerging at the end as an enormously funny Hungarian joke, a fluctuation in the money market, a poem, or simply a long pause in someone's conversation in Brazil.

We do a lot of collective thinking, probably more than any other social species, although it goes on in something like secrecy. We don't acknowledge the gift publicly, and we are not as celebrated as the insects, but we do it. Effortlessly, without giving it a moment's thought, we are capable of changing our language, music, manners, morals, entertainment, even the way we dress, all around the earth in a year's turning. We seem to do this by general agreement, without voting or even polling. We simply think our way along, pass information around, exchange codes disguised as art, change our minds, transform ourselves.

Computers cannot deal with such levels of improbability, and it is just as well. Otherwise, we might be tempted to take over the control of ourselves in order to make long-range plans, and that would surely be the end of us. It would mean that some group or other, marvelously intelligent and superbly informed, undoubtedly guided by a computer, would begin deciding what human society ought to be like, say, over the next five hundred years or so, and the rest of us would be persuaded, one way or another, to go along. The process of social evolution would then grind to a standstill, and we'd be stuck in today's rut for a millennium.

Much better we work our way out of it on our own without governance. The future is too interesting and dangerous to be entrusted to any predictable, reliable agency. We need all the fallibility we can get. Most of all, we need to preserve the absolute unpredictability and total improbability of our connected minds. That way we can keep open all the options, as we have in the past.

It would be nice to have better ways of monitoring what we're up to so that we could recognize change while it is occurring, instead of waking up as we do now to the astonished realization that the whole century just past wasn't what we thought it was, at all. Maybe computers can be used to help in this, although I rather doubt it. You can make simulation models of cities, but what you learn is that they seem to be beyond the reach of intelligent analysis; if you try to use common sense to make predictions, things get more botched up than ever. This is interesting, since a city is the most concentrated aggregation of humans, all exerting whatever influence they can bring to bear. The city seems to have a life of its own. If we cannot understand how this works, we are not likely to get very far with human society at large.

Still, you'd think there would be some way in. Joined together, the great mass of human minds around the earth seems to behave like a coherent, living system. The trouble is that the flow of information is mostly one-way. We are all obsessed by the need to feed information in, as fast as we can, but we lack sensing mechanisms for getting anything much back. I will confess that I have no more sense of what goes on in the mind of mankind than I have for the mind of an ant. Come to think of it, this might be a good place to start.

477

# JOHN VON NEUMANN

■

The Hungarian mathematician John von Neumann (1903–1957) is, along with Alan Turing, regarded as one of the architects of modern computer science. His book *The Computer and the Brain*, published in 1958, marked a milestone in the effort to understand human thought in terms of its relationship to artificial intelligence.

# *The Computer and the Brain*

SINCE I AM NEITHER a neurologist nor a psychiatrist, but a mathematician, the work that follows requires some explanation and justification. It is an approach toward the understanding of the nervous system from the mathematician's point of view. However, this statement must immediately be qualified in both of its essential parts.

First, it is an overstatement to describe what I am attempting here as an "approach toward the understanding"; it is merely a somewhat systematized set of speculations as to how such an approach ought to be made. That is, I am trying to guess which of the — mathematically guided — lines of attack seem, from the hazy distance in which we see most of them, a priori promising, and which ones have the opposite appearance. I will also offer some rationalizations of these guesses.

Second, the "mathematician's point of view," as I would like to have it understood in this context, carries a distribution of emphases that differs from the usual one: apart from the stress on the general mathematical techniques, the logical and the statistical aspects will be in the foreground. Furthermore, logics and statistics should be primarily, although not exclusively, viewed as the basic tools of "information theory." Also, that body of experience which has grown up around the planning, evaluating, and coding of complicated logical

478

and mathematical automata will be the focus of much of this information theory. The most typical, but not the only, such automata are, of course, the large electronic computing machines.

Let me note, in passing, that it would be very satisfactory if one could talk about a "theory" of such automata. Regrettably, what at this moment exists — and to what I must appeal — can as yet be described only as an imperfectly articulated and hardly formalized "body of experience."

Lastly, my main aim is actually to bring out a rather different aspect of the matter. I suspect that a deeper mathematical study of the nervous system — "mathematical" in the sense outlined above — will affect our understanding of the aspects of mathematics itself that are involved. In fact, it may alter the way in which we look on mathematics and logics proper. . . .

I begin by discussing some of the principles underlying the systematics and the practice of computing machines.

Existing computing machines fall into two broad classes: "analog" and "digital." This subdivision arises according to the way in which the numbers, on which the machine operates, are represented in it.

In an analog machine each number is represented by a suitable physical quantity, whose value, measured in some pre-assigned unit, is equal to the number in question. This quantity may be the angle by which a certain disk has rotated, or the strength of a certain current, or the amount of a certain (relative) voltage, etc. To enable the machine to compute, i.e. to operate on these numbers according to a predetermined plan, it is necessary to provide organs (or components) that can perform on these representative quantities the basic operations of mathematics.

These basic operations are usually understood to be the "four species of arithmetic": addition (the operation $x + y$), subtraction ($x - y$), multiplication ($xy$), division ($x/y$).

Thus it is obviously not difficult to add or to subtract two currents (by merging them in parallel or in antiparallel directions). Multiplication (of two currents) is more difficult, but there exist various kinds of electrical componentry which will perform this operation. The same is true for division (of one current by another). (For multiplication as well as for division — but not for addition and subtrac-

479

tion — of course the unit in which the current is measured is relevant.)

A rather remarkable attribute of some analog machines, on which I will have to comment a good deal further, is this. Occasionally the machine is built around other "basic" operations than the four species of arithmetic mentioned above. Thus the classical "differential analyzer," which expresses numbers by the angles by which certain disks have rotated, proceeds as follows. Instead of addition, $x + y$, and subtraction, $x - y$, the operations $(x \pm y)/2$ are offered, because a readily available, simple component, the "differential gear" (the same one that is used on the back axle of an automobile), produces these. Instead of multiplication, $xy$, an entirely different procedure is used: In the differential analyzer all quantities appear as functions of time, and the differential analyzer makes use of an organ called the "integrator," which will, for two such quantities $x(t)$, $y(t)$ form the ("Stieltjes") integral $z(t) \equiv \int^t x(t)\, dy(t)$.

The point in this scheme is threefold:

First: the three above operations will, in suitable combinations, reproduce three of the four usual basic operations, namely addition, subtraction, and multiplication.

Second: in combination with certain "feedback" tricks, they will also generate the fourth operation, division. I will not discuss the feedback principle here, except by saying that while it has the appearance of a device for solving implicit relations, it is in reality a particularly elegant short-circuited iteration and successive approximation scheme.

Third, and this is the true justification of the differential analyzer: its basic operations $(x \pm y)/2$ and integration are, for wide classes of problems, more economical than the arithmetical ones $(x + y, x - y, xy, x/y)$. More specifically: any computing machine that is to solve a complex mathematical problem must be "programmed" for this task. This means that the complex operation of solving that problem must be replaced by a combination of the basic operations of the machine. Frequently it means something even more subtle: approximation of that operation — to any desired (prescribed) degree — by such combinations. Now for a given class of problems one set of basic operations may be more efficient, i.e. allow the use of simpler, less extensive, combinations, than another such set. Thus, in particular, for systems of total differential equations — for which the differential analyzer was primarily designed — the above-mentioned basic operations of that machine are more efficient

480

than the previously mentioned arithmetical basic operations $(x + y, x - y, xy, x/y)$.

Next, I pass to the digital class of machines.

In a decimal digital machine each number is represented in the same way as in conventional writing or printing, i.e. as a sequence of decimal digits. Each decimal digit, in turn, is represented by a system of "markers."

A marker which can appear in ten different forms suffices by itself to represent a decimal digit. A marker which can appear in two different forms only will have to be used so that each decimal digit corresponds to a whole group. (A group of three two-valued markers allows 8 combinations; this is inadequate. A group of four such markers allows 16 combinations; this is more than adequate. Hence, groups of at least four markers must be used per decimal digit. There may be reasons to use larger groups; see below.) An example of a ten-valued marker is an electrical pulse that appears on one of ten pre-assigned lines. A two-valued marker is an electrical pulse on a pre-assigned line, so that its presence or absence conveys the information (the marker's "value"). Another possible two-valued marker is an electrical pulse that can have positive or negative polarity. There are many other equally valid marker schemes.

I will make one more observation on markers: The above-mentioned ten-valued marker is clearly a group of ten two-valued markers, in other words, highly redundant in the sense noted above. The minimum group, consisting of four two-valued markers, can also be introduced within the same framework. Consider a system of four pre-assigned lines, such that (simultaneous) electrical pulses can appear on any combination of these. This allows for 16 combinations, any 10 of which can be stipulated to correspond to the decimal digits.

Note that these markers, which are usually electrical pulses (or possibly electrical voltages or currents, lasting as long as their indication is to be valid), must be controlled by electrical gating devices.

In the course of the development up to now, electromechanical relays, vacuum tubes, crystal diodes, ferromagnetic cores, and transistors have been successively used — some of them in combination with others, some of them preferably in the memory organs of the machine (cf. below), and others preferably outside the memory (in the "active" organs) — giving rise to as many different species of digital machines.

Now a number in the machine is represented by a sequence of

ten-valued markers (or marker groups), which may be arranged to appear simultaneously, in different organs of the machine — in *parallel* — or in temporal succession, in a single organ of the machine — in *series*. If the machine is built to handle, say, twelve-place decimal numbers, e.g. with six places "to the left" of the decimal point, and six "to the right," then twelve such markers (or marker groups) will have to be provided in each information channel of the machine that is meant for passing numbers. (This scheme can be — and is in various machines — made more flexible in various ways and degrees. Thus, in almost all machines, the position of the decimal point is adjustable. However, I will not go into these matters here any further.)

The operations of a digital machine have so far always been based on the four species of arithmetic. Regarding the well-known procedures that are being used, the following should be said:

First, on addition: in contrast to the physical processes that mediate this process in analog machines (cf. above), in this case rules of strict and logical character control this operation — how to form digital sums, when to produce a carry, and how to repeat and combine these operations. The logical nature of the digital sum becomes even clearer when the binary (rather than decimal) system is used. Indeed, the binary addition table $(0 + 0 = 00, \ 0 + 1 = 1 + 0 = 01, 1 + 1 = 10)$ can be stated thus: The sum digit is 1 if the two addend digits differ, otherwise it is 0; the carry digit is 1 if both addend digits are 1, otherwise it is 0. Because of the possible presence of a carry digit, one actually needs a binary addition table for three terms $(0 + 0 + 0 = 00, \ 0 + 0 + 1 = 0 + 1 + 0 = 1 + 0 + 0 = 01, 0 + 1 + 1 = 1 + 0 + 1 = 1 + 1 + 0 = 10, 1 + 1 + 1 = 11)$, and this states: The sum digit is 1, if the number of 1's among the addend (including the carry) digits is odd (1 or 3), otherwise it is 0; the carry digit is 1 if the 1's among the addend (including the carry) digits form a majority (2 or 3), otherwise it is 0.

Second, on subtraction: the logical structure of this is very similar to that one of addition. It can even be — and usually is — reduced to the latter by the simple device of "complementing" the subtrahend.

Third, on multiplication: the primarily logical character is even more obvious — and the structure more involved — than for addition. The products (of the multiplicand) with each digit of the multiplier are formed (usually preformed for all possible decimal digits, by various addition schemes), and then added together (with suitable shifts). Again, in the binary system the logical character is even more transparent and obvious. Since the only possible digits are 0 and 1, a (multiplier) digital product (of the multiplicand) is omitted for 0 and it is the multiplicand itself for 1.

All of this applies to products of positive factors. When both factors may have both signs, additional logical rules control the four situations that can arise.

Fourth, on division: the logical structure is comparable to that of the multiplication, except that now various iterated, trial-and-error subtraction procedures intervene, with specific logical rules (for the forming of the quotient digits) in the various alternative situations that can arise, and that must be dealt with according to a serial, repetitive scheme.

To sum up: all these operations now differ radically from the physical processes used in analog machines. They all are patterns of alternative actions, organized in highly repetitive sequences, and governed by strict and logical rules. Especially in the cases of multiplication and division these rules have a quite complex logical character. (This may be obscured by our long and almost instinctive familiarity with them, but if one forces oneself to state them fully, the degree of their complexity becomes apparent.)

Beyond the capability to execute the basic operations singly, a computing machine must be able to perform them according to the. sequence — or rather, the logical pattern — in which they generate the solution of the mathematical problem that is the actual purpose of the calculation in hand. In the traditional analog machines — typified by the "differential analyzer" — this "sequencing" of the operation is achieved in this way. There must be a priori enough organs present in the machine to perform as many basic operations as the desired calculation calls for — i.e. enough "differential gears" and "integrators" (for the two basic operations $(x \pm y)/2$ and $\int^t x(t) \, dy(t)$, respectively, cf. above). These — i.e. their "input" and "output" disks (or, rather, the axes of these) — must then be so connected to each other (by cogwheel connections in the early models, and by electrical follower-arrangements ["selsyns"] in the later ones) as to constitute a replica of the desired calculation. It should be noted that this connection-pattern can be set up at will — indeed, this is the means by which the problem to be solved, i.e. the intention of the user, is impressed on the machine. This "setting up" occurred in the early (cogwheel-connected, cf. above) machines by mechanical means, while in the later (electrically connected, cf. above) machines it was done by plugging. Nevertheless, it was in all these types always a fixed setting for the entire duration of a problem. . . .

The discussion up to this point has provided the basis for the comparison that is the objective of this work. I have described, in some detail, the nature of modern computing machines and the broad alternative principles around which they can be organized. It is now possible to pass on to the other term of the comparison, the human nervous system. I will discuss the points of similarity and dissimilarity between these two kinds of "automata." Bringing out the elements of similarity leads over well-known territory. There are elements of dissimilarity, too, not only in rather obvious respects of size and speed but also in certain much deeper-lying areas: These involve the principles of functioning and control, of over-all organization, etc. My primary aim is to develop some of these. However, in order to appreciate them properly, a juxtaposition and combination with the points of similarity, as well as with those of more superficial dissimilarity (size, speed; cf. above) are also required. Hence the discussion must place considerable emphasis on these, too.

The most immediate observation regarding the nervous system is that its functioning is *prima facie* digital. It is necessary to discuss this fact, and the structures and functions on which its assertion is based, somewhat more fully.

The basic component of this system is the *nerve cell,* the *neuron,* and the normal function of a neuron is to generate and to propagate a *nerve impulse.* This impulse is a rather complex process, which has a variety of aspects — electrical, chemical, and mechanical. It seems, nevertheless, to be a reasonably uniquely defined process, i.e. nearly the same under all conditions; it represents an essentially reproducible, unitary response to a rather wide variety of stimuli.

Let me discuss this — i.e. those aspects of the nerve impulse that seem to be the relevant ones in the present context — in somewhat more detail.

The nerve cell consists of a *body* from which originate, directly or indirectly, one or more branches. Such a branch is called an *axon* of the cell. The nerve impulse is a continuous change, propagated — usually at a fixed speed, which may, however, be a function of the nerve cell involved — along the (or rather, along each) axon. As mentioned above, this condition can be viewed under multiple aspects. One of its characteristics is certainly that it is an electrical disturbance; in fact, it is most frequently described as being just that.

This disturbance is usually an electrical potential of something like 50 millivolts and of about a millisecond's duration. Concurrently with this electrical disturbance there also occur chemical changes along the axon. Thus, in the area of the axon over which the pulse-potential is passing, the ionic constitution of the intracellular fluid changes, and so do the electrical-chemical properties (conductivity, permeability) of the wall of the axon, the *membrane*. At the endings of the axon the chemical character of the change is even more obvious; there, specific and characteristic substances make their appearance when the pulse arrives. Finally, there are probably mechanical changes as well. Indeed, it is very likely that the changes of the various ionic permeabilities of the cell membrane (cf. above) can come about only by reorientation of its molecules, i.e. by mechanical changes involving the relative positions of these constituents.

It should be added that all these changes are reversible. In other words, when the impulse has passed, all conditions along the axon, and all its constituent parts, resume their original states.

Since all these effects occur on a molecular scale — the thickness of the cell membrane is of the order of a few tenth-microns (i.e. $10^{-5}$ cm.), which is a molecular dimension for the large organic molecules that are involved here — the above distinctions between electrical, chemical, and mechanical effects are not so definite as it might first appear. Indeed, on the molecular scale there are no sharp distinctions between all these kinds of changes: every chemical change is induced by a change in intramolecular forces which determine changed relative positions of the molecules, i.e. it is mechanically induced. Furthermore, every such intramolecular mechanical change alters the electrical properties of the molecule involved, and therefore induces changed electrical properties and changed relative electrical potential levels. To sum up: on the usual (macroscopic) scale, electrical, chemical, and mechanical processes represent alternatives between which sharp distinctions can be maintained. However, on the near-molecule level of the nerve membrane, all these aspects tend to merge. It is, therefore, not surprising that the nerve impulse turns out to be a phenomenon which can be viewed under any one of them.

As I mentioned before, the fully developed nerve impulses are comparable, no matter how induced. Because their character is not an unambiguously defined one (it may be viewed electrically as well as chemically, cf. above), its induction, too, can be alternatively attributed to electrical or to chemical causes. Within the nervous system, however, it is mostly due to one or more other nerve impulses. Under such conditions, the process of its induction — the *stimulation* of a nerve impulse — may or may not succeed. If it fails, a passing disturbance arises at first, but after a few milliseconds, this dies out. Then no disturbances propagate along the axon. If it succeeds, the disturbance very soon assumes a (nearly) standard form, and in this form it spreads along the axon. That is to say, as mentioned above, a standard nerve impulse will then move along the axon, and its appearance will be reasonably independent of the details of the process that induced it.

The stimulation of the nerve impulse occurs normally in or near the body of the nerve cell. Its propagation, as discussed above, occurs along the axon.

I can now return to the digital character of this mechanism. The nervous pulses can clearly be viewed as (two-valued) markers, in the sense discussed previously: the absence of a pulse then represents one value (say, the binary digit 0), and the presence of one represents the other (say, the binary digit 1). This must, of course, be interpreted as an occurrence on a specific axon (or, rather, on all the axons of a specific neuron), and possibly in a specific time relation to other events. It is, then, to be interpreted as a marker (a binary digit 0 or 1) in a specific, logical role.

As mentioned above, pulses (which appear on the axons of a given neuron) are usually stimulated by other pulses that are impinging on the body of the neuron. This stimulation is, as a rule, conditional, i.e. only certain combinations and synchronisms of such primary pulses stimulate the secondary pulse in question — all others will fail to so stimulate. That is, the neuron is an organ which accepts and emits definite physical entities, the pulses. Upon receipt of pulses in certain combinations and synchronisms it will be stimulated to emit a pulse of its own, otherwise it will not emit. The rules which describe to which groups of pulses it will so respond are the rules that govern it as an active organ.

This is clearly the description of the functioning of an organ in a digital machine, and of the way in which the role and function of a digital organ has to be characterized. It therefore justifies the original assertion, that the nervous system has a *prima facie* digital character.

Let me add a few words regarding the qualifying "prima facie." The above description contains some idealizations and simplifications, which will be discussed subsequently. Once these are taken into account, the digital character no longer stands out quite so clearly and unequivocally. Nevertheless, the traits emphasized in the above are the primarily conspicuous ones. It seems proper, therefore, to begin the discussion as I did here, by stressing the digital character of the nervous system.

Before going into this, however, some orienting remarks on the size, energy requirements, and speed of the nerve cell are in order. These will be particularly illuminating when stated in terms of comparisons with the main "artificial" competitors: the typical active organs of modern logical and computing machines. These are, of course, the vacuum tube and (more recently) the transistor.

I stated above that the stimulation of the nerve cell occurs normally on or near its body. Actually, a perfectly normal stimulation is possible along an axon, too. That is, an adequate electrical potential or a suitable chemical stimulant in adequate concentration, when applied at a point of the axon, will start there a disturbance which soon develops into a standard pulse, traveling both up and down the axon, from the point stimulated. Indeed, the "usual" stimulation described above mostly takes place on a set of branches extending from the body of the cell for a short distance, which, apart from their smaller dimensions, are essentially axons themselves, and it propagates from these to the body of the nerve cell (and then to the regular axons). By the way, these stimulation-receptors are called *dendrites*. The normal stimulation, when it comes from another pulse (or pulses) emanates from a special ending of the axon (or axons) that propagated the pulse in question. This ending is called a *synapse*. (Whether a pulse can stimulate only through a synapse, or whether, in traveling along an axon, it can stimulate directly another, exceptionally close-lying axon, is a question that need not be discussed here. The appearances are in favor of assuming that such a short-circuited process is possible.) The time of trans-synaptic stimulation amounts to a few times $10^{-4}$ seconds, this time being defined as the duration between the arrival of a pulse at a synapse and the appearance of the stimulated pulse on the nearest point of an axon of the stimulated neuron. However, this is not the most significant way to define the reaction time of a neuron, when viewed as an active organ in a logical machine. The reason for this is that immediately after the stimulated pulse has become evident, the stimulated neuron has not yet reverted to its original, prestimulation condition. It is *fatigued*, i.e. it could not immediately accept stimulation by another pulse and respond in the standard way. From the point of view of machine economy, it is a more important measure of speed to state after how much time a stimulation that induced a standard response can be followed by another stimulation that will also induce a standard response. This duration is about 1.5 times $10^{-2}$ seconds. It is clear from these figures that only one or two per cent of this time is needed for the actual trans-synaptic stimulation, the remainder representing recovery time, during which the neuron returns from its fatigued, immediate post-stimulation condition to its normal, prestimulation one. It should be noted that this recovery from fatigue is a gradual one — already at a certain earlier time (after about .5 times $10^{-2}$ seconds) the neuron can respond in a nonstandard way, namely it will produce a standard pulse, but only in response to a stimulus which is significantly stronger than the one needed under standard conditions. This circumstance has somewhat broad significance, and I will come back to it later on.

Thus the reaction time of a neuron is, depending on how one defines it, somewhere between $10^{-4}$ and $10^{-2}$ seconds, but the more significant definition is the latter one. Compared to this, modern vacuum tubes and transistors can be used in large logical machines at reaction times between $10^{-6}$ and $10^{-7}$ seconds. (Of course, I am allowing here, too, for the complete recovery time; the organ in question is, after this duration, back to its prestimulation condition.) That is, our artifacts are, in this regard, well ahead of the corresponding natural components, by factors like $10^4$ to $10^5$.

With respect to size, matters have a rather different aspect. There are various ways to evaluate size, and it is best to take these up one by one.

The linear size of a neuron varies widely from one nerve cell to the other, since some of these cells are contained in closely integrated large aggregates and have, therefore, very short axons, while others conduct pulses between rather remote parts of the body and may, therefore, have linear extensions comparable to those of the entire human body. One way to obtain an unambiguous and significant comparison is to compare the logically active part of the nerve cell with that of a vacuum tube, or transistor. For the former this is the cell membrane, whose thickness as mentioned before is of the order of a few times $10^{-5}$ cm. For the latter it is as follows: in the case of the vacuum tube, it is the grid-to-cathode distance, which varies from $10^{-1}$ to a few times $10^{-2}$ cm.; in the case of the transistor, it is the distance between the so-called "whisker electrodes" (the nonohmic electrodes — the "emitter" and the "control-electrode"), about 3 folded in order to account for the immediate, active environment of these subcomponents, and this amounts to somewhat less than $10^{-2}$ cm. Thus, with regard to linear size, the natural components seem to lead our artifacts by a factor like $10^3$.

Next, a comparison with respect to volume is possible. The central nervous system occupies a space of the order magnitude of a liter (in the brain), i.e. of $10^3$ cm.$^3$ The number of neurons contained in this system is usually estimated to be of the order of $10^{10}$, or somewhat higher. This would allow about $10^{-7}$ cm.$^3$ per neuron.

The density with which vacuum tubes or transistors can be packed can also be estimated — although not with absolute unambiguity. It seems clear that this packing density is (on either side of the comparison) a better measure of size efficiency than the actual volume of a single component. With present-day techniques, aggregates of a few thousand vacuum tubes will certainly occupy several times 10 ft.$^3$; for transistors the same may be achieved, in something like one, or a few, ft.$^3$ Using the figure of the latter order as a measure of the best that can be done today, one obtains something like $10^5$ cm.$^3$ for a few times $10^3$ active organs, i.e. about 10 to $10^2$ cm.$^3$ per active organ. Thus the natural components lead the artificial ones with respect to volume requirements by factors like $10^8$ to $10^9$. In comparing this with the estimates for the linear size, it is probably best to consider the linear size-factor as being on one footing with the cube root of the volume-factor. The cube root of the above $10^8$ to $10^9$ is .5 to 1 times $10^3$. This is in good accord with the $10^3$ arrived at above by a direct method.

Finally, a comparison can be made with respect to energy consumption. An active logical organ does not, by its nature, do any work: the stimulated pulse that it produces need not have more energy than the prorated fraction of the pulses which stimulate it — and in any case there is no intrinsic and necessary relationship between these energies. Consequently, the energy involved is almost entirely dissipated, i.e. coverted into heat without doing relevant mechanical work. Thus the energy consumed is actually energy dissipated, and one might as well talk about the energy dissipation of such organs.

The energy dissipation in the human central nervous system (in the brain) is of the order of 10 watts. Since, as pointed out above, the order of $10^{10}$ neurons are involved here, this means a dissipation of $10^{-9}$ watts per neuron. The typical dissipation of a vacuum tube is of the order of 5 to 10 watts. The typical dissipation of a transistor may be as little as $10^{-1}$ watts. Thus the natural components lead the artificial ones with respect to dissipation by factors like $10^8$ to $10^9$ — the same factors that appeared above with respect to volume requirements.

Summing up all of this, it appears that the relevant comparison-factor with regard to size is about $10^8$ to $10^9$ in favor of the natural componentry versus the artificial one. This factor is obtained from the cube of a linear comparison, as well as by a volume-comparison and an energy-dissipation comparison. Against this there is a factor of about $10^4$ to $10^5$ on speed in favor of the artificial componentry versus the natural one.

On these quantitative evaluations certain conclusions can be based. It must be remembered, of course, that the discussion is still moving very near to the surface, so that conclusions arrived at at this stage are very much subject to revision in the light of the further progress of the discussion. It seems nevertheless worth while to formulate certain conclusions at this point. They are the following ones.

First: in terms of the number of actions that can be performed by active organs of the same total size (defined by volume or by energy dissipation) in the same interval, the natural componentry is a factor $10^4$ ahead of the artificial one. This is the quotient of the two factors obtained above, i.e. of $10^8$ to $10^9$ by $10^4$ to $10^5$.

Second: the same factors show that the natural componentry favors automata with more, but slower, organs, while the artificial one favors the reverse arrangement of fewer, but faster, organs. Hence it is to be expected that an efficiently organized large natural automaton (like the human nervous system) will tend to pick up as many logical (or informational) items as possible simultaneously, and process them simultaneously, while an efficiently organized large artificial automaton (like a large modern computing machine) will be more likely to do things successively — one thing at a time, or at any rate not so many things at a time. That is, large and efficient natural automata are likely to be highly *parallel*, while large and efficient artificial automata will tend to be less so, and rather to be *serial*. (Cf. some earlier remarks on parallel versus serial arrangements.)

Third: it should be noted, however, that parallel and serial operation are not unrestrictedly substitutable for each other — as would be required to make the first remark above completely valid, with its simple scheme of dividing the size-advantage factor by the speed-disadvantage factor in order to get a single (efficiency) "figure of merit." More specifically, not everything serial can be immediately paralleled — certain operations can only be performed after certain others, and not simultaneously with them (i.e. they must use the results of the latter). In such a case, the transition from a serial scheme to a parallel one may be impossible, or it may be possible but only concurrently with a change in the logical approach and organization of the procedure. Conversely, the desire to serialize a parallel procedure may impose new requirements on the automaton. Specifically, it will almost always create new memory requirements, since the results of the operations that are performed first must be stored while the operations that come after these are performed. Hence the logical approach and structure in natural automata may be expected to differ widely from those in artificial automata.

Also, it is likely that the memory requirements of the latter will turn out to be systematically more severe than those of the former.

# ALAN TURING

■

In this closely reasoned essay the English mathematician and logician Alan Turing (1912–1954) describes the "Turing test," a method for deciding whether a computer can be said to have achieved human intelligence. The Turing test is still widely discussed, to the illumination of those attempting to better understand what is meant by such terms as *reason* and *intelligence*.

# Can a Machine Think?

I PROPOSE TO CONSIDER the question, "Can machines think?" This should begin with definitions of the meaning of the terms "machine" and "think." The definitions might be framed so as to reflect so far as possible the normal use of the words, but this attitude is dangerous. If the meaning of the words "machine" and "think" are to be found by examining how they are commonly used it is difficult to escape the conclusion that the meaning and the answer to the question, "Can machines think?" is to be sought in a statistical survey such as a Gallup poll. But this is absurd. Instead of attempting such a definition I shall replace the question by another, which is closely related to it and is expressed in relatively unambiguous words.

The new form of the problem can be described in terms of a game which we call the "imitation game." It is played with three people, a man (A), a woman (B), and an interrogator (C) who may be of either sex. The interrogator stays in a room apart from the other two. The object of the game for the interrogator is to determine which of the other two is the man and which is the woman. He knows them by labels X and Y, and at the end of the game he says either "X is A and Y is B" or "X is B and Y is A." The interrogator is allowed to put questions to A and B thus:

C: Will X please tell me the length of his or her hair?

492

Now suppose *X* is actually *A*, then *A* must answer. It is *A*'s object in the game to try and cause *C* to make the wrong identification. His answer might therefore be, "My hair is shingled, and the longest strands are about nine inches long."

In order that tones of voice may not help the interrogator the answers should be written, or better still, typewritten. The ideal arrangement is to have a teleprinter communicating between the two rooms. Alternatively the question and answers can be repeated by an intermediary. The object of the game for the third player (*B*) is to help the interrogator. The best strategy for her is probably to give truthful answers. She can add such things as "I am the woman, don't listen to him!" to her answers, but it will avail nothing as the man can make similar remarks.

We now ask the question, "What will happen when a machine takes the part of *A* in this game?" Will the interrogator decide wrongly as often when the game is played like this as he does when the game is played between a man and a woman? These questions replace our original, "Can machines think?"

As well as asking, "What is the answer to this new form of the question," one may ask, "Is this new question a worthy one to investigate?" This latter question we investigate without further ado, thereby cutting short an infinite regress.

The new problem has the advantage of drawing a fairly sharp line between the physical and the intellectual capacities of a man. No engineer or chemist claims to be able to produce a material which is indistinguishable from the human skin. It is possible that at some time this might be done, but even supposing this invention available we should feel there was little point in trying to make a "thinking machine" more human by dressing it up in such artificial flesh. The form in which we have set the problem reflects this fact in the condition which prevents the interrogator from seeing or touching the other competitors, or hearing their voices. Some other advantages of the proposed criterion may be shown up by specimen questions and answers. Thus:

Q: Please write me a sonnet on the subject of the Forth Bridge.
A: Count me out on this one. I never could write poetry.
Q: Add 34957 to 70764.

A: (Pause about 30 seconds and then give as answer)
   105621.
Q: Do you play chess?
A: Yes.
Q: I have K at my K1, and no other pieces. You have
   only K at K6 and R at R1. It is your move. What do
   you play?
A: (After a pause of 15 seconds) R-R8 mate.

The question and answer method seems to be suitable for introducing almost any one of the fields of human endeavour that we wish to include. We do not wish to penalise the machine for its inability to shine in beauty competitions, nor to penalise a man for losing in a race against an aeroplane. The conditions of our game make these disabilities irrelevant. The "witnesses" can brag, if they consider it advisable, as much as they please about their charms, strength or heroism, but the interrogator cannot demand practical demonstrations.

The game may perhaps be criticised on the ground that the odds are weighted too heavily against the machine. If the man were to try and pretend to be the machine he would clearly make a very poor showing. He would be given away at once by slowness and inaccuracy in arithmetic. May not machines carry out something which ought to be described as thinking but which is very different from what a man does? This objection is a very strong one, but at least we can say that if, nevertheless, a machine can be constructed to play the imitation game satisfactorily, we need not be troubled by this objection.

It might be urged that when playing the "imitation game" the best strategy for the machine may possibly be something other than imitation of the behaviour of a man. This may be, but I think it is unlikely that there is any great effect of this kind. In any case there is no intention to investigate here the theory of the game, and it will be assumed that the best strategy is to try to provide answers that would naturally be given by a man.

The question which we put [at the outset] will not be quite definite until we have specified what we mean by the word "machine." It is natural that we should wish to permit every kind of engineering technique to be used in our machines. We also wish to allow the possibility that an engineer or team of engineers may construct a machine which works, but whose manner of operation cannot be satisfactorily described by its constructors because they have applied a method which is largely experimental. Finally, we wish to exclude from the machines men born in the usual manner. It is difficult to frame the definitions so as to satisfy these three conditions. One might for instance insist that the team of engineers should be all of one sex, but this would not really be satisfactory, for it is probably possible to rear a complete individual from a single cell of the skin (say) of a man. To do so would be a feat of biological technique deserving of the very highest praise, but we would not be inclined to regard it as a case of "constructing a thinking machine." This prompts us to abandon the requirement that every kind of technique should be permitted. We are the more ready to do so in view of the fact that the present interest in "thinking machines" has been aroused by a particular kind of machine, usually called an "electronic computer" or "digital computer." Following this suggestion we only permit digital computers to take part in our game.

This restriction appears at first sight to be a very drastic one. I shall attempt to show that it is not so in reality. To do this necessitates a short account of the nature and properties of these computers.

It may also be said that this identification of machines with digital computers, like our criterion for "thinking," will only be unsatisfactory if (contrary to my belief) it turns out that digital computers are unable to give a good showing in the game.

There are already a number of digital computers in working order, and it may be asked, "Why not try the experiment straight away? It would be easy to satisfy the conditions of the game. A number of interrogators could be used, and statistics compiled to show how often the right identification was given." The short answer is that we are not asking whether all digital computers would do well in the game nor whether the computers at present available would do well, but whether there are imaginable computers which would do well. But this is only the short answer. We shall see this question in a different light later.

The idea behind digital computers may be explained by saying that these machines are intended to carry out any operations which could be done by a human computer. The human computer is supposed to be following fixed rules; he has no authority to deviate from them in any detail. We may suppose that these rules are supplied in a book, which is altered whenever he is put on to a new job. He has also an unlimited supply of paper on which he does his calculations. He may also do his multiplications and additions on a "desk machine," but this is not important.

If we use the above explanation as a definition we shall be in danger of circularity of argument. We avoid this by giving an outline of the means by which the desired effect is achieved. A digital computer can usually be regarded as consisting of three parts:

(*i*) Store.
(*ii*) Executive unit.
(*iii*) Control.

The store is a store of information, and corresponds to the human computer's paper, whether this is the paper on which he does his calculations or that on which his book of rules is printed. In so far as the human computer does calculations in his head a part of the store will correspond to his memory.

The executive unit is the part which carries out the various individual operations involved in a calculation. What these individual operations are will vary from machine to machine. Usually fairly lengthy operations can be done such as "Multiply 3540675445 by 7076345687," but in some machines only very simple ones such as "Write down 0" are possible.

We have mentioned that the "book of rules" supplied to the computer is replaced in the machine by a part of the store. It is then called the "table of instructions." It is the duty of the control to see that these instructions are obeyed correctly and in the right order. The control is so constructed that this necessarily happens.

The information in the store is usually broken up into packets of moderately small size. In one machine, for instance, a packet might consist of ten decimal digits. Numbers are assigned to the parts of the store in which the various packets of information are stored, in some systematic manner. A typical instruction might say, "Add the number stored in position 6809 to that in 4302 and put the result back into the latter storage position."

Needless to say it would not occur in the machine expressed in English. It would more likely be coded in a form such as 6809430217. Here 17 says which of various possible operations is to be performed on the two numbers. In this case the operation is that described above, *viz.* "Add the number. . . ." It will be noticed that the instruction takes up 10 digits and so forms one packet of information, very conveniently. The control will normally take the instructions to be obeyed in the order of the positions in which they are stored, but occasionally an instruction such as "Now obey the instruction stored in position 5606, and continue from there" may be encountered, or again "If position 4505 contains 0 obey next the instruction stored in 6707, otherwise continue straight on."

Instructions of these latter types are very important because they make it possible for a sequence of operations to be repeated over and over again until some condition is fulfilled, but in doing so to obey, not fresh instructions on each repetition, but the same ones over and over again. To take a domestic analogy. Suppose Mother wants Tommy to call at the cobbler's every morning on his way to school to see if her shoes are done, she can ask him afresh every morning. Alternatively she can stick up a notice once and for all in the hall which he will see when he leaves for school and which tells him to call for the shoes, and also to destroy the notice when he comes back if he has the shoes with him.

The reader must accept it as a fact that digital computers can be constructed, and indeed have been constructed, according to the principles we have described, and that they can in fact mimic the actions of a human computer very closely.

The book of rules which we have described our human computer as using is of course a convenient fiction. Actual human computers really remember what they have got to do. If one wants to make a machine mimic the behaviour of the human computer in some complex operation one has to ask him how it is done, and then translate the answer into the form of an instruction table. Constructing instruction tables is usually described as "programming." To "programme a machine to carry out the operation $A$" means to put the appropriate instruction table into the machine so that it will do $A$.

An interesting variant on the idea of a digital computer is a "digital computer with a random element." These have instructions involving the throwing of a die or some equivalent electronic process; one such instruction might for instance be, "Throw the die and put the resulting number into store 1000." Sometimes such a machine is described as having free will (though I would not use this phrase myself). It is not normally possible to determine from observing a machine whether it has a random element, for a similar effect can be produced by such devices as making the choices depend on the digits of the decimal for π.

Most actual digital computers have only a finite store. There is no theoretical difficulty in the idea of a computer with an unlimited store. Of course only a finite part can have been used at any one time. Likewise only a finite amount can have been constructed, but we can imagine more and more being added as required. Such computers have special theoretical interest and will be called infinitive capacity computers.

The idea of a digital computer is an old one. Charles Babbage, Lucasian Professor of Mathematics at Cambridge from 1828 to 1839, planned such a machine, called the Analytical Engine, but it was never completed. Although Babbage had all the essential ideas, his machine was not at that time such a very attractive prospect. The speed which would have been available would be definitely faster than a human computer but something like 100 times slower than the Manchester machine, itself one of the slower of the modern machines. The storage was to be purely mechanical, using wheels and cards.

The fact that Babbage's Analytical Engine was to be entirely mechanical will help us to rid ourselves of a superstition. Importance is often attached to the fact that modern digital computers are electrical, and that the nervous system also is electrical. Since Babbage's machine was not electrical, and since all digital computers are in a sense equivalent, we see that this use of electricity cannot be of theoretical importance. Of course electricity usually comes in where fast signalling is concerned, so that it is not surprising that we find it in both these connections. In the nervous system chemical phenomena are at least as important as electrical. In certain computers the storage system is mainly acoustic. The feature of using electricity is thus seen to be only a very superficial similarity. If we wish to find such similarities we should look rather for mathematical analogies of function.

The digital computers considered in the last section may be classified amongst the "discrete state machines." These are the machines which move by sudden jumps or clicks from one quite definite state to another. These states are sufficiently different for the possibility of confusion between them to be ignored. Strictly speaking there are no such machines. Everything really moves continuously. But there are many kinds of machine which can profitably be *thought of* as being discrete state machines. For instance in considering the switches for a lighting system it is a convenient fiction that each switch must be definitely on or definitely off. There must be intermediate positions, but for most purposes we can forget about them. As an example of a discrete state machine we might consider a wheel which clicks round through 120° once a second, but may be stopped by a lever which can be operated from outside; in addition a lamp is to light in one of the positions of the wheel. This machine could be described abstractly as follows. The internal state of the machine (which is described by the position of the wheel) may be $q_1$, $q_2$ or $q_3$. There is an input signal $i_0$ or $i_1$ (position of lever). The internal state at any moment is determined by the last state and input signal according to the table

|  |  | Last State |  |  |
|--|--|--|--|--|
|  |  | $q_1$ | $q_2$ | $q_3$ |
|  | $i_0$ | $q_2$ | $q_3$ | $q_1$ |
| Input |  |  |  |  |
|  | $i_1$ | $q_1$ | $q_2$ | $q_3$ |

The output signals, the only externally visible indication of the internal state (the light) are described by the table

| State | $q_1$ | $q_2$ | $q_3$ |
|--|--|--|--|
| Output | $o_0$ | $o_0$ | $o_1$ |

This example is typical of discrete state machines. They can be described by such tables provided they have only a finite number of possible states.

It will seem that given the initial state of the machine and the input signals it is always possible to predict all future states. This is reminiscent of Laplace's view that from the complete state of the universe at one moment of time, as described by the positions and velocities of all particles, it should be possible to predict all future states. The prediction which we are considering is, however, rather nearer to practicability than that considered by Laplace. The system of the "universe as a whole" is such that quite small errors in the initial conditions can have an overwhelming effect at a later time. The displacement of a single electron by a billionth of a centimetre at one moment might make the difference between a man being killed by an avalanche a year later, or escaping. It is an essential property of the mechanical systems which we have called "discrete state machines" that this phenomenon does not occur. Even when we consider the actual physical machines instead of the idealised machines, reasonably accurate knowledge of the state at one moment yields reasonably accurate knowledge any number of steps later.

As we have mentioned, digital computers fall within the class of discrete state machines. But the number of states of which such a machine is capable is usually enormously large. For instance, the number for the machine now working at Manchester is about $2^{165,000}$, *i.e.*, about $10^{50,000}$. Compare this with our example of the clicking wheel described above, which had three states. It is not difficult to see why the number of states should be so immense. The computer includes a store corresponding to the paper used by a human computer. It must be possible to write into the store any one of the combinations of symbols which might have been written on the paper. For simplicity suppose that only digits from 0 to 9 are used as symbols. Variations in handwriting are ignored. Suppose the computer is allowed 100 sheets of paper each containing 50 lines each with room for 30 digits. Then the number of states is $10^{100 \times 50 \times 30}$, *i.e.*, $10^{150,000}$. This is about the number of states of three Manchester machines put together. The logarithm to the base two of the number of states is usually called the "storage capacity" of the machine. Thus the Manchester machine has a storage capacity of about 165,000 and the wheel machine of our example about 1.6. If two machines are put together their capacities must be added to obtain the capacity of the resultant machine. This leads to the possibility of statements such as "The Manchester machine contains 64 magnetic tracks each with a capacity of 2560, eight electronic tubes with a capacity of 1280. Miscellaneous storage amounts to about 300 making a total of 174,380."

Given the table corresponding to a discrete state machine it is possible to predict what it will do. There is no reason why this calculation should not be carried out by means of a digital computer. Provided it could be carried out sufficiently quickly the digital computer could mimic the behaviour of any discrete state machine. The imitation game could then be played with the machine in question (as *B*) and the mimicking digital computer (as *A*), and the interrogator would be unable to distinguish them. Of course the digital computer must have an adequate storage capacity as well as working sufficiently fast. Moreover, it must be programmed afresh for each new machine which it is desired to mimic.

This special property of digital computers, that they can mimic any discrete state machine, is described by saying that they are *universal* machines. The existence of machines with this property has the important consequence that, considerations of speed apart, it is unnecessary to design various new machines to do various computing processes. They can all be done with one digital computer, suitably programmed for each case. It will be seen that as a consequence of this all digital computers are in a sense equivalent.

We may now consider again the point raised [earlier]. It was suggested tentatively that the question, "Can machines think?" should be replaced by "Are there imaginable digital computers which would do well in the imitation game?" If we wish we can make this superficially more general and ask "Are there discrete state machines which would do well?" But in view of the universality property we see that either of these questions is equivalent to this, "Let us fix our attention on one particular digital computer $C$. Is it true that by modifying this computer to have an adequate storage, suitably increasing its speed of action, and providing it with an appropriate programme, $C$ can be made to play satisfactorily the part of $A$ in the imitation game, the part of $B$ being taken by a man?"

We may now consider the ground to have been cleared and we are ready to proceed to the debate on our question, "Can machines think?" and the variant of it quoted at the end of the last section. We cannot altogether abandon the original form of the problem, for opinions will differ as to the appropriateness of the substitution and we must at least listen to what has to be said in this connexion.

It will simplify matters for the reader if I explain first my own beliefs in the matter. Consider first the more accurate form of the question. I believe that in about fifty years' time it will be possible to programme computers, with a storage capacity of about $10^9$, to make them play the imitation game so well that an average interrogator will not have more than 70 per cent chance of making the right identification after five minutes of questioning. The original question, "Can machines think?" I believe to be too meaningless to deserve discussion. Nevertheless I believe that at the end of the century the use of words and general educated opinion will have altered so much that one will be able to speak of machines thinking without expecting to be contradicted. I believe further that no useful purpose is served by concealing these beliefs. The popular view that scientists proceed inexorably from well-established fact to well-established fact, never being influenced by any unproved conjecture, is quite mistaken. Provided it is made clear which are proved facts and which are conjectures, no harm can result. Conjectures are of great importance since they suggest useful lines of research.

I now proceed to consider opinions opposed to my own.

(1) *The Theological Objection.* Thinking is a function of man's immortal soul. God has given an immortal soul to every man and woman, but not to any other animal or to machines. Hence no animal or machine can think.[1]

I am unable to accept any part of this, but will attempt to reply in theological terms. I should find the argument more convincing if animals were classed with men, for there is a greater difference, to my mind, between the typical animate and the inanimate than there is between man and the other animals. The arbitrary character of the orthodox view becomes clearer if we consider how it might appear to a member of some other religious community. How do Christians regard the Moslem view that women have no souls? But let us leave this point aside and return to the main argument. It appears to me that the argument quoted above implies a serious restriction of the omnipotence of the Almighty. It is admitted that there are certain things that He cannot do such as making one equal to two, but should we not believe that He has freedom to confer a soul on an elephant if He sees fit? We might expect that He would only exercise this power in conjunction with a mutation which provided the elephant with an appropriately improved brain to minister to the needs of this soul. An argument of exactly similar form may be made for the case of machines. It may seem different because it is more difficult to "swallow." But this really only means that we think it would be less likely that He would consider the circumstances suitable for conferring a soul. The circumstances in question are discussed in the rest of this paper. In attempting to construct such machines we should not be irreverently usurping His power of creating souls, any more than we are in the procreation of children: rather we are, in either case, instruments of His will providing mansions for the souls that He creates.

However, this is mere speculation. I am not very impressed with theological arguments whatever they may be used to support. Such arguments have often been found unsatisfactory in the past. In the time of Galileo it was argued that the texts "And the sun stood still ... and hasted not to go down about a whole day" (Joshua x. 13) and "He laid the foundations of the earth, that it should not move

---

1. Possibly this view is heretical. St. Thomas Aquinas (*Summa Theologica*, quoted by Bertrand Russell, *A History of Western Philosophy*, Simon and Schuster, New York, 1945, p. 458) states that God cannot make a man to have no soul. But this may not be a real restriction on His powers, but only a result of the fact that men's souls are immortal, and therefore indestructible.

at any time" (Psalm cv. 5) were an adequate refutation of the Co-pernican theory. With our present knowledge such an argument appears futile. When that knowledge was not available it made a quite different impression.

(2) *The "Heads in the Sand" Objection.* "The consequences of machines thinking would be too dreadful. Let us hope and believe that they cannot do so."

This argument is seldom expressed quite so openly as in the form above. But it affects most of us who think about it at all. We like to believe that Man is in some subtle way superior to the rest of creation. It is best if he can be shown to be *necessarily* superior, for then there is no danger of him losing his commanding position. The popularity of the theological argument is clearly connected with this feeling. It is likely to be quite strong in intellectual people, since they value the power of thinking more highly than others, and are more inclined to base their belief in the superiority of Man on this power.

I do not think that this argument is sufficiently substantial to require refutation. Consolation would be more appropriate: perhaps this should be sought in the transmigration of souls.

(3) *The Mathematical Objection.* There are a number of results of mathematical logic which can be used to show that there are limitations to the powers of discrete-state machines. The best known of these results is known as Gödel's theorem, and shows that in any sufficiently powerful logical system statements can be formulated which can neither be proved nor disproved within the system, unless possibly the system itself is inconsistent. There are other, in some respects similar, results due to Church, Kleene, Rosser, and Turing. The latter result is the most convenient to consider, since it refers directly to machines, whereas the others can only be used in a comparatively indirect argument: for instance if Gödel's theorem is to be used we need in addition to have some means of describing logical systems in terms of machines, and machines in terms of logical systems. The result in question refers to a type of machine which is essentially a digital computer with an infinite capacity. It states that there are certain things that such a machine cannot do. If it is rigged up to give answers to questions as in the imitation game, there will be some questions to which it will either give a wrong answer, or fail to give an answer at all however much time is allowed for a reply. There may, of course, be many such questions, and questions which cannot be answered by one machine may be satisfactorily answered by another. We are of course supposing for the present that the ques-

tions are of the kind to which an answer "Yes" or "No" is appropriate, rather than questions such as "What do you think of Picasso?" The questions that we know the machines must fail on are of this type: "Consider the machine specified as follows. . . . Will this machine ever answer 'Yes' to any question?" The dots are to be replaced by a description of some machine in a standard form. . . . When the machine described bears a certain comparatively simple relation to the machine which is under interrogation, it can be shown that the answer is either wrong or not forthcoming. This is the mathematical result: it is argued that it proves a disability of machines to which the human intellect is not subject.

The short answer to this argument is that although it is established that there are limitations to the powers of any particular machine, it has only been stated, without any sort of proof, that no such limitations apply to the human intellect. But I do not think this view can be dismissed quite so lightly. Whenever one of these machines is asked the appropriate critical question, and gives a definite answer, we know that this answer must be wrong, and this gives us a certain feeling of superiority. Is this feeling illusory? It is no doubt quite genuine, but I do not think too much importance should be attached to it. We too often give wrong answers to questions ourselves to be justified in being very pleased at such evidence of fallibility on the part of the machines. Further, our superiority can only be felt on such an occasion in relation to the one machine over which we have scored our petty triumph. There would be no question of triumphing simultaneously over *all* machines. In short, then, there might be men cleverer than any given machine, but then again there might be other machines cleverer again, and so on.

Those who hold to the mathematical argument would, I think, mostly be willing to accept the imitation game as a basis for discussion. Those who believe in the two previous objections would probably not be interested in any criteria.

(4) *The Argument from Consciousness.* This argument is very well expressed in Professor Jefferson's Lister Oration for 1949, from which I quote: "Not until a machine can write a sonnet or compose a concerto because of thoughts and emotions felt, and not by the chance fall of symbols, could we agree that machine equals brain — that is, not only write it but know that it had written it. No mechanism could feel (and not merely artificially signal, an easy contrivance) pleasure at its successes, grief when its valves fuse, be warmed by

flattery, be made miserable by its mistakes, be charmed by sex, be angry or depressed when it cannot get what it wants."

This argument appears to be a denial of the validity of our test. According to the most extreme form of this view the only way by which one could be sure that a machine thinks is to *be* the machine and to feel oneself thinking. One could then describe these feelings to the world, but of course no one would be justified in taking any notice. Likewise according to this view the only way to know that a *man* thinks is to be that particular man. It is in fact the solipsist point of view. It may be the most logical view to hold but it makes communication of ideas difficult. *A* is liable to believe "*A* thinks but *B* does not" whilst *B* believes "*B* thinks but *A* does not." Instead of arguing continually over this point it is usual to have the polite convention that everyone thinks.

I am sure that Professor Jefferson does not wish to adopt the extreme and solipsist point of view. Probably he would be quite willing to accept the imitation game as a test. The game (with the player *B* omitted) is frequently used in practice under the name of *viva voce* to discover whether some one really understands something or has "learnt it parrot fashion." Let us listen in to a part of such a *viva voce*:

Interrogator: In the first line of your sonnet which reads "Shall I compare thee to a summer's day," would not "a spring day" do as well or better?

Witness: It wouldn't scan.

Interrogator: How about "a winter's day." That would scan all right.

Witness: Yes, but nobody wants to be compared to a winter's day.

Interrogator: Would you say Mr. Pickwick reminded you of Christmas?

Witness: In a way.

Interrogator: Yet Christmas is a winter's day, and I do not think Mr. Pickwick would mind the comparison.

Witness: I don't think you're serious. By a winter's day one means a typical winter's day, rather than a special one like Christmas.

And so on. What would Professor Jefferson say if the sonnet-writing machine was able to answer like this in the *viva voce*? I do not know whether he would regard the machine as "merely artificially signalling" these answers, but if the answers were as satisfactory and sustained as in the above passage I do not think he would describe it as "an easy contrivance." This phrase is, I think, intended to cover such devices as the inclusion in the machine of a record of someone reading a sonnet, with appropriate switching to turn it on from time to time.

In short then, I think that most of those who support the argument from consciousness could be persuaded to abandon it rather than be forced into the solipsist position. They will then probably be willing to accept our test.

I do not wish to give the impression that I think there is no mystery about consciousness. There is, for instance, something of a paradox connected with any attempt to localise it. But I do not think these mysteries necessarily need to be solved before we can answer the question with which we are concerned in this paper.

(5) *Arguments from Various Disabilities*. These arguments take the form, "I grant you that you can make machines do all the things you have mentioned but you will never be able to make one to do *X*." Numerous features *X* are suggested in this connexion. I offer a selection:

> Be kind, resourceful, beautiful, friendly, have initiative, have a sense of humour, tell right from wrong, make mistakes, fall in love, enjoy strawberries and cream, make some one fall in love with it, learn from experience, use words properly, be the subject of its own thought, have as much diversity of behaviour as a man, do something really new.

No support is usually offered for these statements. I believe they are mostly founded on the principle of scientific induction. A man has seen thousands of machines in his lifetime. From what he sees of them he draws a number of general conclusions. They are ugly, each is designed for a very limited purpose, when required for a minutely different purpose they are useless, the variety of behaviour of any one of them is very small, etc., etc. Naturally he concludes that these are necessary properties of machines in general. Many of these limitations are associated with the very small storage capacity of most machines. (I am assuming that the idea of storage capacity is extended in some way to cover machines other than discrete-state machines. The exact definition does not matter as no mathematical accuracy is claimed in the present discussion.) A few years ago, when very little had been heard of digital computers, it was possible to elicit much incredulity concerning them, if one mentioned their properties without describing their construction. That was presumably due to a similar application of the principle of scientific induction. These applications of the principle are of course largely unconscious. When a burnt child fears the fire and shows that he fears it by avoiding it, I should say that he was applying scientific induction. (I could of course also describe his behaviour in many other ways.) The works and customs of mankind do not seem to be very suitable material to which to apply scientific induction. A very large part of space-time must be investigated, if reliable results are to be obtained. Otherwise we may (as most English children do) decide that everybody speaks English, and that it is silly to learn French.

There are, however, special remarks to be made about many of the disabilities that have been mentioned. The inability to enjoy strawberries and cream may have struck the reader as frivolous. Possibly a machine might be made to enjoy this delicious dish, but any attempt to make one do so would be idiotic. What is important about this disability is that it contributes to some of the other disabilities, *e.g.*, to the difficulty of the same kind of friendliness occurring between man and machine as between white man and white man, or between black man and black man.

The claim that "machines cannot make mistakes" seems a curious one. One is tempted to retort, "Are they any the worse for that?" But let us adopt a more sympathetic attitude, and try to see what is really meant. I think this criticism can be explained in terms of the imitation game. It is claimed that the interrogator could distinguish the machine from the man simply by setting them a number of problems in arithmetic. The machine would be unmasked because of its deadly accuracy. The reply to this is simple. The machine (programmed for playing the game) would not attempt to give the *right* answers to the arithmetic problems. It would deliberately introduce mistakes in a manner calculated to confuse the interrogator. A mechanical fault would probably show itself through an unsuitable decision as to what sort of a mistake to make in the arithmetic. Even this interpretation of the criticism is not sufficiently sympathetic. But we cannot afford the space to go into it much further. It seems to me that this criticism depends on a confusion between two kinds of mistake. We may call them "errors of functioning" and "errors of conclusion." Errors of functioning are due to some mechanical or electrical fault which causes the machine to behave otherwise than it was designed to do. In philosophical discussions one likes to ignore the possibility of such errors; one is therefore discussing "abstract machines." These abstract machines are mathematical fictions rather than physical objects. By definition they are incapable of errors of functioning. In this sense we can truly say that "machines can never make mistakes." Errors of conclusion can only arise when some meaning is attached to the output signals from the machine. The machine might, for instance, type out mathematical equations, or sentences in English. When a false proposition is typed we say that the machine has committed an error of conclusion. There is clearly no reason at all for saying that a machine cannot make this kind of mistake. It might do nothing but type out repeatedly "$0 = 1$." To take a less perverse example, it might have some method for drawing conclusions by scientific induction. We must expect such a method to lead occasionally to erroneous results.

The claim that a machine cannot be the subject of its own thought can of course only be answered if it can be shown that the machine has *some* thought with *some* subject matter. Nevertheless, "the subject matter of a machine's operations" does seem to mean something, at least to the people who deal with it. If, for instance, the machine was trying to find a solution of the equation $x^2 - 40x - 11 = 0$ one would be tempted to describe this equation as part of the machine's subject matter at that moment. In this sort of sense a machine undoubtedly can be its own subject matter. It may be used to help in making up its own programmes, or to predict the effect of alterations in its own structure. By observing the results of its own behaviour it can modify its own programmes so as to achieve some purpose more effectively. These are possibilities of the near future, rather than Utopian dreams.

The criticism that a machine cannot have much diversity of behaviour is just a way of saying that it cannot have much storage capacity. Until fairly recently a storage capacity of even a thousand digits was very rare.

The criticisms that we are considering here are often disguised forms of the argument from consciousness. Usually if one maintains that a machine *can* do one of these things, and describes the kind of method that the machine could use, one will not make much of an impression. It is thought that the method (whatever it may be, for it must be mechanical) is really rather base. . . .

(6) *Lady Lovelace's Objection.* Our most detailed information of Babbage's Analytical Engine comes from a memoir by Lady Lovelace. In it she states, "The Analytical Engine has no pretensions to *originate* anything. It can do *whatever we know how to order it* to perform" (her italics). This statement is quoted by Hartree who adds: "This does not imply that it may not be possible to construct electronic equipment which will 'think for itself,' or in which, in biological terms, one could set up a conditioned reflex, which would serve as a basis for 'learning.' Whether this is possible in principle or not is a stimulating and exciting question, suggested by some of these recent developments. But it did not seem that the machines constructed or projected at the time had this property."

I am in thorough agreement with Hartree over this. It will be noticed that he does not assert that the machines in question had not got the property, but rather that the evidence available to Lady Lovelace did not encourage her to believe that they had it. It is quite possible that the machines in question had in a sense got this property. For suppose that some discrete-state machine has the property. The Analytical Engine was a universal digital computer, so that, if its storage capacity and speed were adequate, it could by suitable programming be made to mimic the machine in question. Probably this argument did not occur to the Countess or to Babbage. In any case there was no obligation on them to claim all that could be claimed.

This whole question will be considered again [in the next section].

A variant of Lady Lovelace's objection states that a machine can "never do anything really new." This may be parried for a moment with the saw, "There is nothing new under the sun." Who can be certain that "original work" that he has done was not simply the growth of the seed planted in him by teaching, or the effect of following well-known general principles. A better variant of the objection says that a machine can never "take us by surprise." This statement is a more direct challenge and can be met directly. Machines take me by surprise with great frequency. This is largely because I do not do sufficient calculation to decide what to expect them to do, or rather because, although I do a calculation, I do it in a hurried, slipshod fashion, taking risks. Perhaps I say to myself, "I suppose the voltage here ought to be the same as there: anyway let's assume it is." Naturally I am often wrong, and the result is a surprise for me for by the time the experiment is done these assumptions have been forgotten. These admissions lay me open to lectures on the subject of my vicious ways, but do not throw any doubt on my credibility when I testify to the surprises I experience.

I do not expect this reply to silence my critic. He will probably say that such surprises are due to some creative mental act on my part, and reflect no credit on the machine. This leads us back to the argument from consciousness, and far from the idea of surprise. It is a line of argument we must consider closed, but it is perhaps worth remarking that the appreciation of something as surprising requires as much of a "creative mental act" whether the surprising event originates from a man, a book, a machine or anything else.

The view that machines cannot give rise to surprises is due, I believe, to a fallacy to which philosophers and mathematicians are particularly subject. This is the assumption that as soon as a fact is presented to a mind all consequences of that fact spring into the mind simultaneously with it. It is a very useful assumption under many circumstances, but one too easily forgets that it is false. A natural consequence of doing so is that one then assumes that there is no virtue in the mere working out of consequences from data and general principles.

(7) *Argument from Continuity in the Nervous System.* The nervous system is certainly not a discrete-state machine. A small error in the information about the size of a nervous impulse impinging on a neuron may make a large difference to the size of the outgoing impulse. It may be argued that, this being so, one cannot expect to be able to mimic the behaviour of the nervous system with a discrete-state system.

It is true that a discrete-state machine must be different from a continuous machine. But if we adhere to the conditions of the imitation game, the interrogator will not be able to take any advantage of this difference. The situation can be made clearer if we consider some other simpler continuous machine. A differential analyser will do very well. (A differential analyser is a certain kind of machine not of the discrete-state type used for some kinds of calculation.) Some of these provide their answers in a typed form, and so are suitable for taking part in the game. It would not be possible for a digital computer to predict exactly what answers the differential analyser would give to a problem, but it would be quite capable of giving the right sort of answer. For instance, if asked to give the value of π (actually about 3.1416) it would be reasonable to choose at random between the values 3.12, 3.13, 3.14, 3.15, 3.16 with the probabilities of 0.05, 0.15, 0.55, 0.19, 0.06 (say). Under these circumstances it would be very difficult for the interrogator to distinguish the differential analyser from the digital computer.

(8) *The Argument from Informality of Behaviour.* It is not possible to produce a set of rules purporting to describe what a man should do in every conceivable set of circumstances. One might for instance have a rule that one is to stop when one sees a red traffic light, and to go if one sees a green one, but what if by some fault both appear together? One may perhaps decide that it is safest to stop. But some further difficulty may well arise from this decision later. To attempt to provide rules of conduct to cover every eventuality, even those arising from traffic lights, appears to be impossible. With all this I agree.

From this it is argued that we cannot be machines. I shall try to reproduce the argument, but I fear I shall hardly do it justice. It seems to run something like this. "If each man had a definite set of rules of conduct by which he regulated his life he would be no better than a machine. But there are no such rules, so men cannot be machines." The undistributed middle is glaring. I do not think the argument is ever put quite like this, but I believe this is the argument used nevertheless. There may however be a certain confusion between "rules of conduct" and "laws of behaviour" to cloud the issue. By "rules of conduct" I mean precepts such as "Stop if you see red lights," on which one can act, and of which one can be conscious. By "laws of behaviour" I mean laws of nature as applied to a man's body such as "if you pinch him he will squeak." If we substitute "laws of behaviour which regulate his life" for "laws of conduct by which he regulates his life" in the argument quoted the undistributed middle is no longer insuperable. For we believe that it is not only true that being regulated by laws of behaviour implies being some sort of machine (though not necessarily a discrete-state machine), but that conversely being such a machine implies being regulated by such laws. However, we cannot so easily convince ourselves of the absence of complete laws of behaviour as of complete rules of conduct. The only way we know of for finding such laws is scientific observation, and we certainly know of no circumstances under which we could say, "We have searched enough. There are no such laws."

We can demonstrate more forcibly that any such statement would be unjustified. For suppose we could be sure of finding such laws if they existed. Then given a discrete-state machine it should certainly be possible to discover by observation sufficient about it to predict its future behaviour, and this within a reasonable time, say a thousand years. But this does not seem to be the case. I have set up on the Manchester computer a small programme using only 1000 units of storage, whereby the machine supplied with one sixteen figure number replies with another within two seconds. I would defy anyone to learn from these replies sufficient about the programme to be able to predict any replies to untried values.

(9) *The Argument from Extra-Sensory Perception.* I assume that the reader is familiar with the idea of extra-sensory perception, and the meaning of the four items of it, *viz.* telepathy, clairvoyance, precognition and psycho-kinesis. These disturbing phenomena seem to deny all our usual scientific ideas. How we should like to discredit them! Unfortunately the statistical evidence, at least for telepathy, is overwhelming. It is very difficult to rearrange one's ideas so as to fit these new facts in. Once one has accepted them it does not seem a very big step to believe in ghosts and bogies. The idea that our bodies move simply according to the known law of physics, together with some others not yet discovered but somewhat similar, would be one of the first to go.

This argument is to my mind quite a strong one. One can say in reply that many scientific theories seem to remain workable in practice, in spite of clashing with E.S.P.; that in fact one can get along very nicely if one forgets about it. This is rather cold comfort, and one fears that thinking is just the kind of phenomenon where E.S.P. may be especially relevant.

A more specific argument based on E.S.P. might run as follows: "Let us play the imitation game, using as witnesses a man who is good as a telepathic receiver, and a digital computer. The interrogator can ask such questions as 'What suit does the card in my right hand belong to?' The man by telepathy or clairvoyance gives the right answer 130 times out of 400 cards. The machine can only guess at random, and perhaps gets 104 right, so the interrogator makes the right identification." There is an interesting possibility which opens here. Suppose the digital computer contains a random number generator. Then it will be natural to use this to decide what answer to give. But then the random number generator will be subject to the psycho-kinetic powers of the interrogator. Perhaps this psycho-kinesis might cause the machine to guess right more often than would be expected on a probability calculation, so that the interrogator might still be unable to make the right identification. On the other hand, he might be able to guess right without any questioning, by clairvoyance. With E.S.P. anything may happen.

If telepathy is admitted it will be necessary to tighen our test up. The situation could be regarded as analogous to that which would occur if the interrogator were talking to himself and one of the competitors was listening with his ear to the wall. To put the competitors into a "telepathy-proof room" would satisfy all requirements.

The reader will have anticipated that I have no very convincing arguments of a positive nature to support my views. If I had I should not have taken such pains to point out the fallacies in contrary views. Such evidence as I have I shall now give.

Let us return for a moment to Lady Lovelace's objection, which stated that the machine can only do what we tell it to do. One could say that a man can "inject" an idea into the machine, and that it will respond to a certain extent and then drop into quiescence, like a piano string struck by a hammer. Another simile would be an atomic pile of less than critical size: an injected idea is to correspond to a neutron entering the pile from without. Each such neutron will cause a certain disturbance which eventually dies away. If, however, the size of the pile is sufficiently increased, the disturbance caused by such an incoming neutron will very likely go on and on increasing until the whole pile is destroyed. Is there a corresponding phenomenon for minds, and is there one for machines? There does seem to be one for the human mind. The majority of them seem to be "sub-critical," *i.e.*, to correspond in this analogy to piles of subcritical size. An idea presented to such a mind will on average give rise to less than one idea in reply. A smallish proportion are super-critical. An idea presented to such a mind may give rise to a whole "theory" consisting of secondary, tertiary and more remote ideas. Animals' minds seem to be very definitely sub-critical. Adhering to this analogy we ask, "Can a machine be made to be super-critical?"

The "skin of an onion" analogy is also helpful. In considering the functions of the mind or the brain we find certain operations which we can explain in purely mechanical terms. This we say does not correspond to the real mind: it is a sort of skin which we must strip off if we are to find the real mind. But then in what remains we find a further skin to be stripped off, and so on. Proceeding in this way do we ever come to the "real" mind, or do we eventually come to the skin which has nothing in it? In the latter case the whole mind is mechanical. (It would not be a discrete-state machine however. We have discussed this.)

These last two paragraphs do not claim to be convincing arguments. They should rather be described as "recitations tending to produce belief."

The only really satisfactory support that can be given for the view expressed [elsewhere] will be that provided by waiting for the end of the century and then doing the experiment described. But what can we say in the meantime? What steps should be taken now if the experiment is to be successful?

As I have explained, the problem is mainly one of programming. Advances in engineering will have to be made too, but it seems unlikely that these will not be adequate for the requirements. Estimates of the storage capacity of the brain vary from $10^{10}$ to $10^{15}$ binary digits. I incline to the lower values and believe that only a very small fraction is used for the higher types of thinking. Most of it is probably used for the retention of visual impressions. I should be surprised if more than $10^9$ was required for satisfactory playing of the imitation game, at any rate against a blind man. (Note — The capacity of the *Encyclopædia Britannica*, 11th edition, is $2 \times 10^9$.) A storage capacity of $10^7$ would be a very practicable possibility even by present techniques. It is probably not necessary to increase the speed of operations of the machines at all. Parts of modern machines which can be regarded as analogues of nerve cells work about a thousand times faster than the latter. This should provide a "margin of safety" which could cover losses of speed arising in many ways. Our problem then is to find out how to programme these machines to play the game. At my present rate of working I produce about a thousand digits of programme a day, so that about sixty workers, working steadily through the fifty years, might accomplish the job, if nothing went into the waste-paper basket. Some more expeditious method seems desirable.

In the process of trying to imitate an adult human mind we are bound to think a good deal about the process which has brought it to the state that it is in. We may notice three components,

(*i*) The initial state of the mind, say at birth,
(*ii*) The education to which it has been subjected,
(*iii*) Other experience, not to be described as education, to which it has been subjected.

Instead of trying to produce a programme to simulate the adult mind, why not rather try to produce one which simulates the child's? If this were then subjected to an appropriate course of education one would obtain the adult brain. Presumably the child-brain is something like a note-book as one buys it from the stationers. Rather little mechanism, and lots of blank sheets. (Mechanism and writing are from our point of view almost synonymous.) Our hope is that there is so little mechanism in the child-brain that something like it can be easily programmed. The amount of work in the education we can assume, as a first approximation, to be much the same as for the human child.

We have thus divided our problem into two parts. The child-programme and the education process. These two remain very closely connected. We cannot expect to find a good child-machine at the first attempt. One must experiment with teaching one such machine and see how well it learns. One can then try another and see if it is better or worse. There is an obvious connection between this process and evolution, by the identifications

Structure of the child-machine = Hereditary material
Changes     "      "      "      = Mutations
Natural selection             = Judgment of the experimenter

One may hope, however, that this process will be more expeditious than evolution. The survival of the fittest is a slow method for measuring advantages. The experimenter, by the exercise of intelligence, should be able to speed it up. Equally important is the fact that he is not restricted to random mutations. If he can trace a cause for some weakness he can probably think of the kind of mutation which will improve it.

It will not be possible to apply exactly the same teaching process to the machine as to a normal child. It will not, for instance, be provided with legs, so that it could not be asked to go out and fill the coal scuttle. Possibly it might not have eyes. But however well these deficiencies might be overcome by clever engineering, one could not send the creature to school without the other children making excessive fun of it. It must be given some tuition. We need not be too concerned about the legs, eyes, etc. The example of Miss Helen Keller shows that education can take place provided that communication in both directions between teacher and pupil can take place by some means or other.

We normally associate punishments and rewards with the teaching process. Some simple child-machines can be constructed or programmed on this sort of principle. The machine has to be so constructed that events which shortly preceded the occurrence of a punishment-signal are unlikely to be repeated, whereas a reward-signal increased the probability of repetition of the events which led up to it. These definitions do not presuppose any feelings on the part of the machine. I have done some experiments with one such child-machine, and succeeded in teaching it a few things, but the teaching method was too unorthodox for the experiment to be considered really successful.

The use of punishments and rewards can at best be a part of the teaching process. Roughly speaking, if the teacher has no other means of communicating to the pupil, the amount of information which can reach him does not exceed the total number of rewards and punishments applied. By the time a child has learnt to repeat "Casabianca" he would probably feel very sore indeed, if the text could only be discovered by a "Twenty Questions" technique, every "NO" taking the form of a blow. It is necessary therefore to have some other "unemotional" channels of communication. If these are available it is possible to teach a machine by punishments and rewards to obey orders given in some language, *e.g.*, a symbolic language. These orders are to be transmitted through the "unemotional" channels. The use of this language will diminish greatly the number of punishments and rewards required.

Opinions may vary as to the complexity which is suitable in the child-machine. One might try to make it as simple as possible consistently with the general principles. Alternatively one might have a complete system of logical inference "built in."[2] In the latter case the store would be largely occupied with definitions and propositions. The propositions would have various kinds of status, *e.g.*, well-established facts, conjectures, mathematically proved theorems, statements given by an authority, expressions having the logical form of proposition but not belief-value. Certain propositions may be described as "imperatives." The machine should be so constructed that as soon as an imperative is classed as "well-established" the appropriate action automatically takes place. To illustrate this, suppose the teacher says to the machine, "Do your homework now." This may

2. Or rather "programmed in," for our child-machine will be programmed in a digital computer. But the logical system will not have to be learnt.

cause "Teacher says 'Do your homework now'" to be included amongst the well-established facts. Another such fact might be, "Everything that teacher says is true." Combining these may eventually lead to the imperative, "Do your homework now," being included amongst the well-established facts, and this, by the construction of the machine, will mean that the homework actually gets started, but the effect is very satisfactory. The processes of inference used by the machine need not be such as would satisfy the most exacting logicians. There might for instance be no hierarchy of types. But this need not mean that type fallacies will occur, any more than we are bound to fall over unfenced cliffs. Suitable imperatives (expressed *within* the systems, not forming part of the rules *of* the system) such as "Do not use a class unless it is a subclass of one which has been mentioned by teacher" can have a similar effect to "Do not go too near the edge."

The imperatives that can be obeyed by a machine that has no limbs are bound to be of a rather intellectual character, as in the example (doing homework) given above. Important amongst such imperatives will be ones which regulate the order in which the rules of the logical system concerned are to be applied. For at each stage when one is using a logical system, there is a very large number of alternative steps, any of which one is permitted to apply, so far as obedience to the rules of the logical system is concerned. These choices make the difference between a brilliant and a footling reasoner, not the difference between a sound and a fallacious one. Propositions leading to imperatives of this kind might be "When Socrates is mentioned, use the syllogism in Barbara" or "If one method has been proved to be quicker than another, do not use the slower method." Some of these may be "given by authority," but others may be produced by the machine itself, *e.g.*, by scientific induction.

The idea of a learning machine may appear paradoxical to some readers. How can the rules of operation of the machine change? They should describe completely how the machine will react whatever its history might be, whatever changes it might undergo. The rules are thus quite time-invariant. This is quite true. The explanation of the paradox is that the rules which get changed in the learning process are of a rather less pretentious kind, claiming only an ephemeral validity. The reader may draw a parallel with the Constitution of the United States.

An important feature of a learning machine is that its teacher will often be very largely ignorant of quite what is going on inside, although he may still be able to some extent to predict his pupil's behaviour. This should apply most strongly to the later education of a machine arising from a child-machine of well-tried design (or programme). This is in clear contrast with normal procedure when using a machine to do computation: one's object is then to have a clear mental picture of the state of the machine at each moment in the computation. This object can only be achieved with a struggle. The view that "the machine can only do what we know how to order it to do" appears strange in face of this. Most of the programmes which we can put into the machine will result in its doing something that we cannot make sense of at all, or which we regard as completely random behaviour. Intelligent behaviour presumably consists in a departure from the completely disciplined behaviour involved in computation, but a rather slight one, which does not give rise to random behaviour, or to pointless repetitive loops. Another important result of preparing our machine for its part in the imitation game by a process of teaching and learning is that "human fallibility" is likely to be omitted in a rather natural way, *i.e.*, without special "coaching." Processes that are learnt do not produce a hundred per cent certainty of result; if they did they could not be unlearnt.

It is probably wise to include a random element in a learning machine. A random element is rather useful when we are searching for a solution of some problem. Suppose for instance we wanted to find a number between 50 and 200 which was equal to the square of the sum of its digits, we might start at 51 then try 52 and go on until we got a number that worked. Alternatively we might choose numbers at random until we got a good one. This method has the advantage that it is unnecessary to keep track of the values that have been tried, but the disadvantage that one may try the same one twice, but this is not very important if there are several solutions. The systematic method has the disadvantage that there may be an enormous block without any solutions in the region, which has to be investigated first. Now the learning process may be regarded as a search for a form of behaviour which will satisfy the teacher (or some other criterion). Since there is probably a very large number of satisfactory solutions the random method seems to be better than the systematic. It should be noticed that it is used in the analogous process of evolution. But there the systematic method is not possible. How could one keep track of the different genetical combinations that had been tried, so as to avoid trying them again?

We may hope that machines will eventually compete with men in all purely intellectual fields. But which are the best ones to start with? Even this is a difficult decision. Many people think that a very abstract activity, like the playing of chess, would be best. It can also be maintained that it is best to provide the machine with the best sense organs that money can buy, and then teach it to understand and speak English. This process could follow the normal teaching of a child. Things would be pointed out and named, etc. Again I do not know what the right answer is, but I think both approaches should be tried.

We can only see a short distance ahead, but we can see plenty there that needs to be done.

# Math Angst

## MORRIS KLINE

∎

Mathematics was once thought to be perfectly precise, and the fact that mathematical equations could predict the outcome of natural phenomena was taken to mean that nature, too, was precisely predictable. But nature turned out to be contaminated by chance (see, for example, Werner Heisenberg's "The Copenhagen Interpretation of Quantum Theory"), and mathematics, too, proved to be less certain than had been supposed. Morris Kline (b. 1908) sounded the alarm in his widely read book *Mathematics: The Loss of Certainty,* published in 1980.

# *The Loss of Certainty*

. . . FROM THE VERY BIRTH of mathematics as an independent body of knowledge, fathered by the classical Greeks, and for a period of over two thousand years, mathematicians pursued truth. Their accomplishments were magnificent. The vast body of theorems about number and geometric figures offered in itself what appeared to be an almost endless vista of certainty.

Beyond the realm of mathematics proper, mathematical concepts and derivations supplied the essence of remarkable scientific theories. Though the knowledge obtained through the collaboration of mathematics and science employed physical principles, these seemed to be as secure as the principles of mathematics proper because the predictions in the mathematical theories of astronomy, mechanics, optics, and hydrodynamics were in remarkably accurate accord with observation and experiment. Mathematics, then, provided a firm grip on the workings of nature, an understanding which dissolved mystery and replaced it by law and order. Man could pridefully survey the world about him and boast that he had grasped many of the secrets of the universe, which in essence were a series of mathematical laws. The conviction that mathematicians were securing truths is epitomized in Laplace's remark that Newton was a most fortunate man because there is just one universe and Newton had discovered its laws.

To achieve its marvelous and powerful results, mathematics relied upon a special method, namely, deductive proof from self-evident principles called axioms, the methodology we still learn, usually in high school geometry. Deductive reasoning, by its very nature, guarantees the truth of what is deduced if the axioms are truths. By utilizing this seemingly clear, infallible, and impeccable logic, mathematicians produced apparently indubitable and irrefutable conclusions. This feature of mathematics is still cited today. Whenever someone wants an example of certitude and exactness of reasoning, he appeals to mathematics.

The successes mathematics achieved with its methodology attracted the greatest intellectuals. Mathematics had demonstrated the capacities, resources, and strengths of human reason. Why should not this methodology be employed, they asked, to secure truths in fields dominated by authority, custom, and habit, fields such as philosophy, theology, ethics, aesthetics, and the social sciences? Man's reason, so evidently effective in mathematics and mathematical physics, could surely be the arbiter of thought and action in these other fields and obtain for them the beauty of truths and the truths of beauty. And so, during the period called the Enlightenment or the Age of Reason, mathematical methodology and even some mathematical concepts and theorems were applied to human affairs.

The most fertile source of insight is hindsight. Creations of the early 19th century, strange geometries and strange algebras, forced mathematicians, reluctantly and grudgingly, to realize that mathematics proper and the mathematical laws of science were not truths. They found, for example, that several differing geometries fit spatial experience equally well. All could not be truths. Apparently mathematical design was not inherent in nature, or if it was, man's mathematics was not necessarily the account of that design. The key to reality had been lost. This realization was the first of the calamities to befall mathematics.

The creation of these new geometries and algebras caused mathematicians to experience a shock of another nature. The conviction that they were obtaining truths had entranced them so much that they had rushed impetuously to secure these seeming truths at the cost of sound reasoning. The realization that mathematics was not a body of truths shook their confidence in what they had created, and they undertook to reexamine their creations. They were dismayed to find that the logic of mathematics was in sad shape.

In fact mathematics had developed illogically. Its illogical development contained not only false proofs, slips in reasoning, and inadvertent mistakes which with more care could have been avoided. Such blunders there were aplenty. The illogical development also involved inadequate understanding of concepts, a failure to recognize all the principles of logic required, and an inadequate rigor of proof; that is, intuition, physical arguments, and appeal to geometrical diagrams had taken the place of logical arguments.

However, mathematics was still an effective description of nature. And mathematics itself was certainly an attractive body of knowledge and in the minds of many, the Platonists especially, a part of reality to be prized in and for itself. Hence mathematicians decided to supply the missing logical structure and to rebuild the defective portions. During the latter half of the 19th century the movement often described as the rigorization of mathematics became the outstanding activity.

By 1900 the mathematicians believed they had achieved their goal. Though they had to be content with mathematics as an approximate description of nature and many even abandoned the belief in the mathematical design of nature, they did gloat over their reconstruction of the logical structure of mathematics. But before they had finished toasting their presumed success, contradictions were discovered in the reconstructed mathematics. Commonly these contradictions were referred to as paradoxes, a euphemism that avoids facing the fact that contradictions vitiate the logic of mathematics.

The resolution of the contradictions was undertaken almost immediately by the leading mathematicians and philosophers of the times. In effect four different approaches to mathematics were conceived, formulated, and advanced, each of which gathered many adherents. These foundational schools all attempted not only to resolve the known contradictions but to ensure that no new ones could ever arise, that is, to establish the consistency of mathematics. Other issues arose in the foundational efforts. The acceptability of some axioms and some principles of deductive logic also became bones of contention on which the several schools took differing positions.

As late as 1930 a mathematician might perhaps have been content with accepting one or another of the several foundations of mathematics and declared that his mathematical proofs were at least in accord with the tenets of that school. But disaster struck again in the form of a famous paper by Kurt Gödel in which he proved, among other significant and disturbing results, that the logical principles accepted by the several schools could not prove the consistency of mathematics. This, Gödel showed, cannot be done without involving logical principles so dubious as to question what is accomplished. Gödel's theorems produced a debacle. Subsequent developments brought further complications. For example, even the axiomatic-deductive method so highly regarded in the past as *the* approach to exact knowledge was seen to be flawed. The net effect of these newer developments was to add to the variety of possible approaches to mathematics and to divide mathematicians into an even greater number of differing factions.

The current predicament of mathematics is that there is not one but many mathematics and that for numerous reasons each fails to satisfy the members of the opposing schools. It is now apparent that the concept of a universally accepted, infallible body of reasoning — the majestic mathematics of 1800 and the pride of man — is a grand illusion. Uncertainty and doubt concerning the future of mathematics have replaced the certainties and complacency of the past. The disagreements about the foundations of the "most certain" science are both surprising and, to put it mildly, disconcerting. The present state of mathematics is a mockery of the hitherto deep-rooted and widely reputed truth and logical perfection of mathematics.

There are mathematicians who believe that the differing views on what can be accepted as sound mathematics will some day be reconciled. Prominent among these is a group of leading French mathematicians who write under the pseudonym of Nicholas Bourbaki:

Since the earliest times, all critical revisions of the principles of mathematics as a whole, or of any branch of it, have almost invariably followed periods of uncertainty, where contradictions did appear and had to be resolved. . . . There are now twenty-five centuries during which the mathematicians have had the practice of correcting their errors and thereby seeing their science enriched, not impoverished; this gives them the right to view the future with serenity.

However, many more mathematicians are pessimistic. Hermann Weyl, one of the greatest mathematicians of this century, said in 1944:

> The question of the foundations and the ultimate meaning of mathematics remains open; we do not know in what direction it will find its final solution or even whether a final objective answer can be expected at all. "Mathematizing" may well be a creative activity of man, like language or music, of primary originality, whose historical decisions defy complete objective rationalization.

In the words of Goethe, "The history of a science is the science itself."

The disagreements concerning what correct mathematics is and the variety of differing foundations affect seriously not only mathematics proper but most vitally physical science. As we shall see, the most well-developed physical theories are entirely mathematical. (To be sure, the conclusions of such theories are interpreted in sensuous or truly physical objects, and we hear voices over our radios even though we have not the slightest physical understanding of what a radio wave is.) Hence scientists, who do not personally work on foundational problems, must nevertheless be concerned about what mathematics can be confidently employed if they are not to waste years on unsound mathematics.

The loss of truth, the constantly increasing complexity of mathematics and science, and the uncertainty about which approach to mathematics is secure have caused most mathematicians to abandon science. With a "plague on all your houses" they have retreated to specialties in areas of mathematics where the methods of proof seem to be safe. They also find problems concocted by humans more appealing and manageable than those posed by nature.

The crises and conflicts over what sound mathematics is have also discouraged the application of mathematical methodology to many areas of our culture such as philosophy, political science, ethics, and aesthetics. The hope of finding objective, infallible laws and standards has faded. The Age of Reason is gone.

Despite the unsatisfactory state of mathematics, the variety of approaches, the disagreements on acceptable axioms, and the danger that new contradictions, if discovered, would invalidate a great deal of mathematics, some mathematicians are still applying mathematics to physical phenomena and indeed extending the applied fields to economics, biology, and sociology. The continuing effectiveness of mathematics suggests two themes. The first is that effectiveness can be used as the criterion of correctness. Of course such a criterion is provisional. What is considered correct today may prove wrong in the next application.

The second theme deals with a mystery. In view of the disagreements about what sound mathematics is, why is it effective at all? Are we performing miracles with imperfect tools? If man has been deceived, can nature also be deceived into yielding to man's mathematical dictates? Clearly not. Yet, do not our successful voyages to the moon and our explorations of Mars and Jupiter, made possible by technology which itself depends heavily on mathematics, confirm mathematical theories of the cosmos? How can we, then, speak of the artificiality and varieties of mathematics? Can the body live on when the mind and spirit are bewildered? Certainly this is true of human beings and it is true of mathematics. It behooves us therefore to learn why, despite its uncertain foundations and despite the conflicting theories of mathematicians, mathematics has proved to be so incredibly effective.

# EUGENE P. WIGNER

■

The question with which Morris Kline concludes the preceding
essay — Why has mathematics proved to be so effective? — is
investigated in this 1960 essay by the Hungarian mathematician
Eugene Paul Wigner (b. 1902). Wigner's mathematical insights
have aided the progress of physics in general and relativity and
quantum mechanics in particular. He was awarded the 1963
Nobel Prize in physics for his contributions to the study of atomic
nuclei.

# *The Unreasonable Effectiveness of Mathematics in the Natural Sciences*

THERE IS A STORY about two friends, who were classmates in high
school, talking about their jobs. One of them became a statistician
and was working on population trends. He showed a reprint to his
former classmate. The reprint started, as usual, with the Gaussian
distribution and the statistician explained to his former classmate the
meaning of the symbols for the actual population, for the average
population, and so on. His classmate was a bit incredulous and was
not quite sure whether the statistician was pulling his leg. "How can
you know that?" was his query. "And what is this symbol here?" "Oh,"
said the statistician, "this is π." "What is that?" "The ratio of the circumference of the circle to its diameter." "Well, now you are pushing
your joke too far," said the classmate, "surely the population has
nothing to do with the circumference of the circle."

Naturally, we are inclined to smile about the simplicity of the
classmate's approach. Nevertheless, when I heard this story, I had to
admit to an eerie feeling because, surely, the reaction of the classmate

betrayed only plain common sense. I was even more confused when, not many days later, someone came to me and expressed his bewilderment[1] with the fact that we make a rather narrow selection when choosing the data on which we test our theories. "How do we know that, if we made a theory which focuses its attention on phenomena we disregard and disregards some of the phenomena now commanding our attention, that we could not build another theory which has little in common with the present one but which, nevertheless, explains just as many phenomena as the present theory?" It has to be admitted that we have not definite evidence that there is no such theory.

The preceding two stories illustrate the two main points which are the subjects of the present discourse. The first point is that mathematical concepts turn up in entirely unexpected connections. Moreover, they often permit an unexpectedly close and accurate description of the phenomena in these connections. Secondly, just because of this circumstance, and because we do not understand the reasons for their usefulness, we cannot know whether a theory formulated in terms of mathematical concepts is uniquely appropriate. We are in a position similar to that of a man who was provided with a bunch of keys and who, having to open several doors in succession, always hit on the right key on the first or second trial. He became skeptical concerning the uniqueness of the coordination between keys and doors.

Most of what will be said on these questions will not be new; it has probably occurred to most scientists in one form or another. My principal aim is to illuminate it from several sides. The first point is that the enormous usefulness of mathematics in the natural sciences is something bordering on the mysterious and that there is no rational explanation for it. Second, it is just this uncanny usefulness of mathematical concepts that raises the question of the uniqueness of our physical theories. In order to establish the first point, that mathematics plays an unreasonably important role in physics, it will be useful to say a few words on the question "What is mathematics?" then, "What is physics?" then, how mathematics enters physical theories, and last, why the success of mathematics in its role in physics appears so baffling. Much less will be said on the second point: the uniqueness of the theories of physics. A proper answer to this ques-

---

1. The remark to be quoted was made by F. Werner when he was a student in Princeton.

tion would require elaborate experimental and theoretical work which has not been undertaken to date.

Somebody once said that philosophy is the misuse of a terminology which was invented just for this purpose.[2] In the same vein, I would say that mathematics is the science of skillful operations with concepts and rules invented just for this purpose. The principal emphasis is on the invention of concepts. Mathematics would soon run out of interesting theorems if these had to be formulated in terms of the concepts which already appear in the axioms. Furthermore, whereas it is unquestionably true that the concepts of elementary mathematics and particularly elementary geometry were formulated to describe entities which are directly suggested by the actual world, the same does not seem to be true of the more advanced concepts, in particular the concepts which play such an important role in physics. Thus, the rules for operations with pairs of numbers are obviously designed to give the same results as the operations with fractions which we first learned without reference to "pairs of numbers." The rules for the operations with sequences, that is with irrational numbers, still belong to the category of rules which were determined so as to reproduce rules for the operations with quantities which were already known to us. Most more advanced mathematical concepts, such as complex numbers, algebras, linear operators, Borel sets — and this list could be continued almost indefinitely — were so devised that they are apt subjects on which the mathematician can demonstrate his ingenuity and sense of formal beauty. In fact, the definition of these concepts, with a realization that interesting and ingenious considerations could be applied to them, is the first demonstration of the ingeniousness of the mathematician who defines them. The depth of thought which goes into the formation of the mathematical concepts is later justified by the skill with which these concepts are used. The great mathematician fully, almost ruthlessly, exploits the domain of permissible reasoning and skirts the impermissible. That his recklessness does not lead him into a morass of contradictions is a miracle in itself: certainly it is hard to believe that our reasoning power was brought, by Darwin's process of natural selection, to the perfection which it seems to possess. However, this is not our present subject. The principal point

2. This statement is quoted here from W. Dubislav's *Die Philosophie der Mathematik in der Gegenwart.* Junker und Dunnhaupt Verlag, Berlin, 1932, p. 1.

which will have to be recalled later is that the mathematician could formulate only a handful of interesting theorems without defining concepts beyond those contained in the axioms and that the concepts outside those contained in the axioms are defined with a view of permitting ingenious logical operations which appeal to our aesthetic sense both as operations and also in their results of great generality and simplicity.[3]

The complex numbers provide a particularly striking example for the foregoing. Certainly, nothing in our experience suggests the introduction of these quantities. Indeed, if a mathematician is asked to justify his interest in complex numbers, he will point, with some indignation, to the many beautiful theorems in the theory of equations, of power series and of analytic functions in general, which owe their origin to the introduction of complex numbers. The mathematician is not willing to give up his interest in these most beautiful accomplishments of his genius.[4]

The physicist is interested in discovering the laws of inanimate nature. In order to understand this statement, it is necessary to analyze the concept "law of nature."

The world around us is of baffling complexity and the most obvious fact about it is that we cannot predict the future. Although the joke attributes only to the optimist the view that the future is uncertain, the optimist is right in this case: the future is unpredictable. It is, as Schrödinger (1933) has remarked, a miracle that in spite of the baffling complexity of the world, certain regularities in the events could be discovered. One such regularity, discovered by Galileo, is that two rocks, dropped at the same time from the same height, reach the ground at the same time. The laws of nature are concerned with such regularities. Galileo's regularity is a prototype of a large class of regularities. It is a surprising regularity for three reasons.

The first reason that it is surprising is that it is true not only in

3. M. Polanyi, in his *Personal Knowledge,* University of Chicago Press, 1958, says: "All these difficulties are but consequences of our refusal to see that mathematics cannot be defined without acknowledging its most obvious feature: namely, that it is interesting" (page 188).

4. The reader may be interested, in this connection, in Hilbert's rather testy remarks about intuitionism which "seeks to break up and to disfigure mathematics," Abh. Math. Sem. Univ. Hamburg, Vol. 157, 1922, or *Gesammelte Werke,* Springer, Berlin, 1935, page 188.

Pisa, and in Galileo's time, it is true everywhere on the Earth, was always true, and will always be true. This property of the regularity is a recognized invariance property and, as I had occasion to point out some time ago (Wigner, 1949), without invariance principles similar to those implied in the preceding generalization of Galileo's observation, physics would not be possible. The second surprising feature is that the regularity which we are discussing is independent of so many conditions which could have an effect on it. It is valid no matter whether it rains or not, whether the experiment is carried out in a room or from the Leaning Tower, no matter whether the person who drops the rocks is a man or a woman. It is valid even if the two rocks are dropped, simultaneously and from the same height, by two different people. There are, obviously, innumerable other conditions which are all immaterial from the point of view of the validity of Galileo's regularity. The irrelevancy of so many circumstances which *could* play a role in the phenomenon observed has also been called an invariance (Wigner, 1949). However, this invariance is of a different character than the preceding one since it cannot be formulated as a general principle. The exploration of the conditions which do, and which do not, influence a phenomenon is part of the early experimental exploration of a field. It is the skill and ingenuity of the experimenter which shows him phenomena which depend on a relatively narrow set of relatively easily realizable and reproducible conditions.[5] In the present case, Galileo's restriction of his observations to relatively heavy bodies was the most important step in this regard. Again, it is true that if there were no phenomena which are independent of all but a manageably small set of conditions, physics would be impossible.

The preceding two points, though highly significant from the point of view of the philosopher, are not the ones which surprised Galileo most, nor do they contain a specific law of nature. The law of nature is contained in the statement that the length of time which it takes for a heavy object to fall from a given height is independent of the size, material and shape of the body which drops. In the framework of Newton's second "law," this amounts to the statement that the gravitational force which acts on the falling body is propor-

5. See, in this connection, the graphic essay of M. Deutsch, *Daedalus*, Vol. 87, 1958, page 86. A. Shimony has called my attention to a similar passage in C. S. Peirce's *Essays in the Philosophy of Science,* The Liberal Arts Press, New York, 1957 (page 237).

tional to its mass but independent of the size, material and shape of the body which falls.

The preceding discussion is intended to remind, first, that it is not at all natural that "laws of nature" exist, much less that man is able to discover them.[6] The present writer had occasion, some time ago, to call attention to the succession of layers of "laws of nature," each layer containing more general and more encompassing laws than the previous one and its discovery constituting a deeper penetration into the structure of the universe than the layers recognized before (Wigner, 1950). However, the point which is most significant in the present context is that all these laws of nature contain, in even their remotest consequences, only a small part of our knowledge of the inanimate world. All the laws of nature are conditional statements which permit a prediction of some future events on the basis of the knowledge of the present, except that some aspects of the present state of the world, in practice the overwhelming majority of the determinants of the present state of the world, are irrelevant from the point of view of the prediction. The irrelevancy is meant in the sense of the second point in the discussion of Galileo's theorem.[7]

As regards the present state of the world, such as the existence of the earth on which we live and on which Galileo's experiments were performed, the existence of the sun and of all our surroundings, the laws of nature are entirely silent. It is in consonance with this, first, that the laws of nature can be used to predict future events only under exceptional circumstances — when all the relevant determinants of the present state of the world are known. It is also in consonance with this that the construction of machines, the functioning of which he can foresee, constitutes the most spectacular accomplishment of the physicist. In these machines, the physicist creates a situation in which all the relevant coordinates are known so that the behavior of the machine can be predicted. Radars and nuclear reactors are examples of such machines.

The principal purpose of the preceding discussion is to point out that the laws of nature are all conditional statements and they relate only to a very small part of our knowledge of the world. Thus,

6. E. Schrödinger, in his *What Is Life,* Cambridge University Press, 1945, says that this second miracle may well be beyond human understanding (page 31).

7. The writer feels sure that it is unnecessary to mention that Galileo's theorem, as given in the text, does not exhaust the content of Galileo's observations in connection with the laws of freely falling bodies.

classical mechanics, which is the best known prototype of a physical theory, gives the second derivatives of the positional coordinates of all bodies, on the basis of the knowledge of the positions, etc., of these bodies. It gives no information on the existence, the present positions, or velocities of these bodies. It should be mentioned, for the sake of accuracy, that we have learned about thirty years ago that even the conditional statements cannot be entirely precise: that the conditional statements are probability laws which enable us only to place intelligent bets on future properties of the inanimate world, based on the knowledge of the present state. They do not allow us to make categorical statements, not even categorical statements conditional on the present state of the world. The probabilistic nature of the "laws of nature" manifests itself in the case of machines also, and can be verified, at least in the case of nuclear reactors, if one runs them at very low power. However, the additional limitation of the scope of the laws of nature which follows from their probabilistic nature will play no role in the rest of the discussion.

Having refreshed our minds as to the essence of mathematics and physics, we should be in a better position to review the role of mathematics in physical theories.

Naturally, we do use mathematics in everyday physics to evaluate the results of the laws of nature, to apply the conditional statements to the particular conditions which happen to prevail or happen to interest us. In order that this be possible, the laws of nature must already be formulated in mathematical language. However, the role of evaluating the consequences of already established theories is not the most important role of mathematics in physics. Mathematics, or, rather, applied mathematics, is not so much the master of the situation in this function: it is merely serving as a tool.

Mathematics does play, however, also a more sovereign role in physics. This was already implied in the statement, made when discussing the role of applied mathematics, that the laws of nature must be already formulated in the language of mathematics to be an object for the use of applied mathematics. The statement that the laws of nature are written in the language of mathematics was properly made three hundred years ago;[8] it is now more true than ever before. In order to show the importance which mathematical concepts possess in the formulation of the laws of physics, let us recall, as an

8. It is attributed to Galileo.

example, the axioms of quantum mechanics as formulated, explicitly, by the great mathematician, von Neumann (1955), or, implicitly, by the great physicist, Dirac (1947). There are two basic concepts in quantum mechanics: states and observables. The states are vectors in Hilbert space, the observables self-adjoint operators on these vectors. The possible values of the observations are the characteristic values of the operators — but we had better stop here lest we engage in a listing of the mathematical concepts developed in the theory of linear operators.

It is true, of course, that physics chooses certain mathematical concepts for the formulation of the laws of nature, and surely only a fraction of all mathematical concepts are used in physics. It is true also that the concepts which were chosen were not selected arbitrarily from a listing of mathematical terms but were developed, in many if not most cases, independently by the physicist and recognized then as having been conceived before by the mathematician. It is not true, however, as is so often stated, that this had to happen because mathematics uses the simplest possible concepts and these were bound to occur in any formalism. As we saw before, the concepts of mathematics are not chosen for their conceptual simplicity — even sequences of pairs of numbers are far from being the simplest concepts — but for their amenability to clever manipulations and to striking, brilliant arguments. Let us not forget that the Hilbert space of quantum mechanics is the complex Hilbert space, with a Hermitean scalar product. Surely to the unpreoccupied mind, complex numbers are far from natural or simple and they cannot be suggested by physical observations. Furthermore, the use of complex numbers is in this case not a calculational trick of applied mathematics but comes close to being a necessity in the formulation of the laws of quantum mechanics. Finally, it now begins to appear that not only numbers but so-called analytic functions are destined to play a decisive role in the formulation of quantum theory. I am referring to the rapidly developing theory of dispersion relations.

It is difficult to avoid the impression that a miracle confronts us here, quite comparable in its striking nature to the miracle that the human mind can string a thousand arguments together without getting itself into contradictions or to the two miracles of the existence of laws of nature and of the human mind's capacity to divine them. The observation which comes closest to an explanation for the mathematical concepts' cropping up in physics which I know is Einstein's statement that the only physical theories which we are willing to accept are the beautiful ones. It stands to argue that the concepts of mathematics, which invite the exercise of so much wit, have the quality of beauty. However, Einstein's observation can at best explain properties of theories which we are willing to believe and has no reference to the intrinsic accuracy of the theory. We shall, therefore, turn to this latter question.

A possible explanation of the physicist's use of mathematics to formulate his laws of nature is that he is a somewhat irresponsible person. As a result, when he finds a connection between two quantities which resembles a connection well-known from mathematics, he will jump at the conclusion that the connection *is* that discussed in mathematics simply because he does not know of any other similar connection. It is not the intention of the present discussion to refute the charge that the physicist is a somewhat irresponsible person. Perhaps he is. However, it is important to point out that the mathematical formulation of the physicist's often crude experience leads in an uncanny number of cases to an amazingly accurate description of a large class of phenomena. This shows that the mathematical language has more to commend it than being the only language which we can speak; it shows that it is, in a very real sense, the correct language. Let us consider a few examples.

The first example is the oft quoted one of planetary motion. The laws of falling bodies became rather well established as a result of experiments carried out principally in Italy. These experiments could not be very accurate in the sense in which we understand accuracy today partly because of the effect of air resistance and partly because of the impossibility, at that time, to measure short time intervals. Nevertheless, it is not surprising that as a result of their studies, the Italian natural scientists acquired a familiarity with the ways in which objects travel through the atmosphere. It was Newton who then brought the law of freely falling objects into relation with the motion of the moon, noted that the parabola of the thrown rock's path on the earth, and the circle of the moon's path in the sky, are particular cases of the same mathematical object of an ellipse and postulated the universal law of gravitation, on the basis of a single, and at that time very approximate, numerical coincidence. Philosophically, the law of gravitation as formulated by Newton was repugnant to his time and to himself. Empirically, it was based on very scanty observations. The mathematical language in which it was formulated contained the concept of a second derivative and those of us who have tried to draw an osculating circle to a curve know that the second derivative is not a very immediate concept. The law of gravity which Newton reluctantly established and which he could verify with an accuracy of about 4 percent has proved to be accurate to less than a ten thousandth of a percent and became so closely associated with the idea of absolute accuracy that only recently did physicists become again bold enough to inquire into the limitations of its accuracy. Certainly, the example of Newton's law, quoted over and over again, must be mentioned first as a monumental example of a law, formulated in terms which appear simple to the mathematician, which has proved accurate beyond all reasonable expectation. Let us just recapitulate our thesis on this example: first, the law, particularly since a second derivative appears in it, is simple only to the mathematician, not to common sense or to non-mathematically-minded freshmen; second, it is a conditional law of very limited scope. It explains nothing about the earth which attracts Galileo's rocks, or about the circular form of the moon's orbit, or about the planets of the sun. The explanation of these initial conditions is left to the geologist and the astronomer, and they have a hard time with them.

The second example is that of ordinary, elementary quantum mechanics. This originated when Max Born noticed that some rules of computation, given by Heisenberg, were formally identical with the rules of computation with matrices, established a long time before by mathematicians. Born, Jordan and Heisenberg then proposed to replace by matrices the position and momentum variables of the equations of classical mechanics (1925, 1926). They applied the rules of matrix mechanics to a few highly idealized problems and the results were quite satisfactory. However, there was, at that time, no rational evidence that their matrix mechanics would prove correct under more realistic conditions. Indeed, they say "if the mechanics as here proposed should already be correct in its essential traits." As a matter of fact, the first application of their mechanics to a realistic problem, that of the hydrogen atom, was given several months later, by Pauli. This application gave results in agreement with experience. This was satisfactory but still understandable because Heisenberg's rules of calculation were abstracted from problems which included the old theory of the hydrogen atom. The miracle occurred only when matrix mechanics, or a mathematically equivalent theory, was applied to problems for which Heisenberg's calculating rules were meaningless. Heisenberg's rules presupposed that the classical equations of motion had solutions with certain periodicity properties; and the equations of motion of the two electrons of the helium atom, or of the even greater number of electrons of heavier atoms, simply do not have these properties, so that Heisenberg's rules cannot be applied to these cases. Nevertheless, the calculation of the lowest energy level of helium, as carried out a few months ago by Kinoshita at Cornell and by Bazley at the Bureau of Standards, agree with the experimental data within the accuracy of the observations, which is one part in ten millions. Surely in this case we "got something out" of the equations that we did not put in.

The same is true of the qualitative characteristics of the "complex spectra," that is the spectra of heavier atoms. I wish to recall a conversation with Jordan who told me, when the qualitative features of the spectra were derived, that a disagreement of the rules derived from quantum mechanical theory, and the rules established by empirical research, would have provided the last opportunity to make a change in the framework of matrix mechanics. In other words, Jordan felt that we would have been, at least temporarily, helpless had an unexpected disagreement occurred in the theory of the helium atom. This was, at that time, developed by Kellner and by Hilleraas. The mathematical formalism was too clear and unchangeable so that, had the miracle of helium which was mentioned before not occurred, a true crisis would have arisen. Surely, physics would have overcome that crisis in one way or another. It is true, on the other hand, that physics as we know it today would not be possible without a constant recurrence of miracles similar to the one of the helium atom which is perhaps the most striking miracle that has occurred in the course of the development of elementary quantum mechanics, but by far not the only one. In fact, the number of analogous miracles is limited, in our view, only by our willingness to go after more similar ones. Quantum mechanics had, nevertheless, many almost equally striking successes which gave us the firm conviction that it is, what we call, correct.

The last example is that of quantum electrodynamics, or the theory of the Lamb shift. Whereas Newton's theory of gravitation still had obvious connections with experience, experience entered the formulation of matrix mechanics only in the refined or sublimated form of Heisenberg's prescriptions. The quantum theory of the Lamb shift, as conceived by Bethe and established by Schwinger, is a purely mathematical theory, and the only direct contribution of experiment was to show the existence of a measurable effect. The agreement with calculation is better than one part in a thousand.

The preceding three examples, which could be multiplied almost indefinitely, should illustrate the appropriateness and accuracy of the mathematical formulation of the laws of nature in terms of concepts chosen for their manipulability, the "laws of nature" being of almost fantastic accuracy but of strictly limited scope. I propose to refer to the observation which these examples illustrate as the empirical law of epistemology. Together with the laws of invariance of physical theories, it is an indispensable foundation of these theories. Without the laws of invariance the physical theories could have been given no foundation of fact; if the empirical law of epistemology were not correct, we would lack the encouragement and reassurance which are emotional necessities without which the "laws of nature" could not have been successfully explored. Dr. R. G. Sachs, with whom I discussed the empirical law of epistemology, called it an article of faith of the theoretical physicist, and it is surely that. However, what he called our article of faith can be well supported by actual examples — many examples in addition to the three which have been mentioned.

The empirical nature of the preceding observation seems to me to be self-evident. It surely is not a "necessity of thought" and it should not be necessary, in order to prove this, to point to the fact that it applies only to a very small part of our knowledge of the inanimate world. It is absurd to believe that the existence of mathematically simple expressions for the second derivative of the position is self-evident, when no similar expressions for the position itself or for the velocity exist. It is therefore surprising how readily the wonderful gift contained in the empirical law of epistemology was taken for granted. The ability of the human mind to form a string of 1000 conclusions and still remain "right," which was mentioned before, is a similar gift.

Every empirical law has the disquieting quality that one does not know its limitations. We have seen that there are regularities in the events in the world around us which can be formulated in terms of mathematical concepts with an uncanny accuracy. There are, on the other hand, aspects of the world concerning which we do not believe in the existence of any accurate regularities. We call these initial conditions. The question which presents itself is whether the different regularities, that is the various laws of nature which will be discovered, will fuse into a single consistent unit, or at least asymptotically approach such a fusion. Alternately, it is possible that there always will be some laws of nature which have nothing in common with each other. At present, this is true, for instance, of the laws of heredity and of physics. It is even possible that some of the laws of nature will be in conflict with each other in their implications, but each convincing enough in its own domain so that we may not be willing to abandon any of them. We may resign ourselves to such a state of affairs or our interest in clearing up the conflict between the various theories may fade out. We may lose interest in the "ultimate truth," that is in a picture which is a consistent fusion into a single unit of the little pictures, formed on the various aspects of nature.

It may be useful to illustrate the alternatives by an example. We now have, in physics, two theories of great power and interest: the theory of quantum phenomena and the theory of relativity. These two theories have their roots in mutually exclusive groups of phenomena. Relativity theory applies to macroscopic bodies, such as stars. The event of coincidence, that is in ultimate analysis of collision, is the primitive event in the theory of relativity and defines a point in space-time, or at least would define a point if the colliding particles were infinitely small. Quantum theory has its roots in the microscopic world and, from its point of view, the event of coincidence, or of collision, even if it takes place between particles of no spatial extent, is not primitive and not at all sharply isolated in space-time. The two theories operate with different mathematical concepts — the four dimensional Riemann space and the infinite dimensional Hilbert space, respectively. So far, the two theories could not be united, that is, no mathematical formulation exists to which both of these theories are approximations. All physicists believe that a union of the two theories is inherently possible and that we shall find it. Nevertheless, it is possible also to imagine that no union of the two theories can be found. This example illustrates the two possibilities, of union and of conflict, mentioned before, both of which are conceivable.

In order to obtain an indication as to which alternative to expect ultimately, we can pretend to be a little more ignorant than we are and place ourselves at a lower level of knowledge than we actually possess. If we can find a fusion of our theories on this lower level of intelligence, we can confidently expect that we will find a fusion of our theories also at our real level of intelligence. On the other hand, if we would arrive at mutually contradictory theories at a somewhat lower level of knowledge, the possibility of the permanence of conflicting theories cannot be excluded for ourselves either. The level of knowledge and ingenuity is a continuous variable and it is unlikely that a relatively small variation of this continuous variable changes the attainable picture of the world from inconsistent to consistent.[9]

Considered from this point of view, the fact that some of the theories which we know to be false give such amazingly accurate results is an adverse factor. Had we somewhat less knowledge, the group of phenomena which these "false" theories explain would appear to us to be large enough to "prove" these theories. However, these theories are considered to be "false" by us just for the reason that they are, in ultimate analysis, incompatible with more encompassing pictures and, if sufficiently many such false theories are discovered, they are bound to prove also to be in conflict with each other. Similarly, it is possible that the theories, which we consider to be "proved" by a number of numerical agreements which appears to be large enough for us, are false because they are in conflict with a possible more encompassing theory which is beyond our means of discovery. If this were true, we would have to expect conflicts between our theories as soon as their number grows beyond a certain point and as soon as they cover a sufficiently large number of groups of phenomena. In contrast to the article of faith of the theoretical physicist mentioned before, this is the nightmare of the theorist.

Let us consider a few examples of "false" theories which give, in view of their falseness, alarmingly accurate descriptions of groups of phenomena. With some goodwill, one can dismiss some of the evidence which these examples provide. The success of Bohr's early and pioneering ideas on the atom was always a rather narrow one and the same applies to Ptolemy's epicycles. Our present vantage point gives an accurate description of all phenomena which these more primitive theories can describe. The same is not true any more of the so-called free-electron theory which gives a marvelously accurate picture of many, if not most, properties of metals, semiconductors and insulators. In particular, it explains the fact, never properly understood on the basis of the "real theory," that insulators show a specific resistance to electricity which may be $10^{26}$ times greater than

9. This passage was written after a great deal of hesitation. The writer is convinced that it is useful, in epistemological discussions, to abandon the idealization that the level of human intelligence has a singular position on an absolute scale. In some cases it may even be useful to consider the attainment which is possible at the level of the intelligence of some other species. However, the writer also realizes that his thinking along the lines indicated in the text was too brief and not subject to sufficient critical appraisal to be reliable.

that of metals. In fact, there is no experimental evidence to show that the resistance is not infinite under the conditions under which the free-electron theory would lead us to expect an infinite resistance. Nevertheless, we are convinced that the free-electron theory is a crude approximation which should be replaced, in the description of all phenomena concerning solids, by a more accurate picture.

If viewed from our real vantage point, the situation presented by the free-electron theory is irritating but is not likely to forebode any inconsistencies which are unsurmountable for us. The free-electron theory raises doubts as to how much we should trust numerical agreement between theory and experiment as evidence for the correctness of the theory. We are used to such doubts.

A much more difficult and confusing situation would arise if we could, some day, establish a theory of the phenomena of consciousness, or of biology, which would be as coherent and convincing as our present theories of the inanimate world. Mendel's laws of inheritance and the subsequent work on genes may well form the beginning of such a theory as far as biology is concerned. Furthermore, it is quite possible that an abstract argument can be found which shows that there is a conflict between such a theory and the accepted principles of physics. The argument could be of such abstract nature that it might not be possible to resolve the conflict, in favor of one or of the other theory, by an experiment. Such a situation would put a heavy strain on our faith in our theories and on our belief in the reality of the concepts which we form. It would give us a deep sense of frustration in our search for what I called the "ultimate truth." The reason that such a situation is conceivable is that, fundamentally, we do not know why our theories work so well. Hence their accuracy may not prove their truth and consistency. Indeed, it is this writer's belief that something rather akin to the situation which was described above exists if the present laws of heredity and of physics are confronted.

Let me end on a more cheerful note. The miracle of the appropriateness of the language of mathematics for the formulation of the laws of physics is a wonderful gift which we neither understand nor deserve. We should be grateful for it and hope that it will remain valid in future research and that it will extend, for better or for worse, to our pleasure even though perhaps also to our bafflement, to wide branches of learning. . . .

JOHN D. BARROW

■

Philosophers have long pondered the question of whether the beautiful relationships discovered in mathematics are but fabrications of the human mind, or, instead, really exist in the outer world — whether, in other words, they are invented by mathematicians or discovered by them. This survey of the main currents of thought on the matter was written by John D. Barrow (b. 1952). It comes from his book *The World Within the World,* published in 1988.

# *What Is Mathematics?*

IT IS NOT EASY TO SAY what mathematics is, but "I know it when I see it" is the most likely response of anyone to whom this question is put. The most striking thing about mathematics is that it is very different to science, and this compounds the problem of why it should be found so useful in describing and predicting how the Universe works. Whereas science is like a long text that is constantly being redrafted, updated, and edited, mathematics is entirely cumulative. Contemporary science is going to be proven wrong, but mathematics is not. The scientists of the past were well justified in holding naïve and erroneous views about physical phenomena in the context of the civilizations in which they lived, but there can never be any justification for establishing erroneous mathematical results. The mechanics of Aristotle is wrong, but the geometry of Euclid is, was, and always will be correct. Right and wrong mean different things in science and mathematics. In the former, "right" means correspondence with reality; in mathematics it means logical consistency.

Before we can draw any conclusions from our good fortune in finding our experience of reality to be so well described by mathematics, we need to have some understanding of what mathematicians think mathematics is, or at least what they think it could be. We have learnt some mathematics; we can write down some mathematical formulae which are "true": what is it that makes them so?

There are essentially four interpretations of mathematics, and the view to which one subscribes will determine to a large extent the assessment one makes of the remarkable effectiveness of mathematics in describing Nature. Conversely, the natural applicability of mathematics to the world of experience might enter as a key witness in deciding the merits of the competing interpretations of mathematics. Each interpretation is a possible view of what is meant by the statement that some mathematical statement is "true." We shall call the four options *Platonism, Conceptualism, Formalism,* and *Intuitionism.* Their *curricula vitae* read like this:

*Platonism:* This is the view that mathematicians discover mathematics rather than invent it. All the concepts they arrive at and find useful, like groups and sets, triangles and points, infinities, and even numbers, really exist "out there" independently of you and me. There would exist mathematical quantities even if there were no mathematicians; they are not creations of the human mind but manifestations of the intrinsic character of reality. "Pi" really is in the sky. These mathematical objects do not exist in the space and time we experience. They are abstract entities, and mathematical truth means correspondence between the properties of these abstract objects and our systems of symbols.

The reason for the Platonic association with this notion should be clear. . . . Mathematical ideas like the number "seven" are regarded as immaterial and immutable ideas that really exist in some abstract realm, whereas our observations are of specific secondary realizations like seven dwarfs, seven brides, or seven brothers.

On this view we could use mathematical entities as a language with which to communicate with alien beings from other worlds, and be confident that they would have discovered many of the same mathematical structures that we have. They would not, almost certainly, be using the same symbols or language as ourselves to represent those mathematical entities, but none the less they would be expected to have representations of the same idea, just as surely as the words "seven," "sieben," and "sept" code the same information to English, German, and French speakers. Of no other part of our human experience can this be said. Our art and ethics, our forms of government and styles of literature would probably be unintelligible to alien beings because they do not describe something independent of our minds. Platonists would be confident of using mathematics as a universal language because they hold it to be a description of something ethereal and absolute. The reason why mathematics is so accurate in capturing the workings of Nature is, therefore, attributed to the simple but inexplicable fact that Nature really *is* mathematical, and is in fact the basic source of our mathematics. Despite being idealists, mathematical Platonists often regard themselves as realists simply because they hold the most straightforward interpretation of mathematics. They regard mathematics as an absolute truth — "God is a mathematician" — whatever your definition of God.

This has given rise to a mystical offshoot dubbed "neo-Platonism," one of whose Soviet proponents has likened mathematics to the composition of a cosmic symphony by independent contributors, each moving it towards some grand final synthesis. This goal cannot, he claims, be something so mundane as a description of the world or the application of mathematics to the solution of practical problems. It has an other-worldly quality because completely independent mathematical discoveries by different mathematicians working in different cultures so often turn out to be identical. But few professional mathematicians would seek to draw such cosmic conclusions. Most work as though Platonism were true, and seek to "discover" new mathematical structures that are interesting because they are rich in internal properties and possess unexpected connections with other, superficially unrelated, branches of mathematics. Talk of "existence" proofs of particular mathematical objects is a reflection of this subconscious attitude. However, if pressed to defend this common-sense approach, few pure mathematicians would seek to do so. On the other hand, the applied mathematician and other "consumers" of mathematics (physicists and economists, for instance) provided "ready-packaged" for them by others generally regard mathematics as a useful "black-box": a tool for obtaining answers to specific problems whose effectiveness is promised by the innumerable successes in the past.

*Conceptualism:* This is the complete antithesis of Platonism, and is popular with sociologists rather than mathematicians and scientists, most of whom would instinctively reject it. It maintains that we create an array of mathematical structures, symmetries, and patterns, and then force the world into this mould because we find it so compelling. Ultimately, the choice of what mathematics we construct is culturally derived. We invent mathematics, we do not discover it. Mathematics is what mathematicians do. The suspicion that there is some truth in this view has led to a gradual and often unnoticed shift in the way

543

applied mathematicians describe what they are doing. Whereas the classical mathematicians of yesteryear would write treatises or give lecture courses on "the mathematical theory of . . . ," today there is a growing emphasis upon the less grandiose term "mathematical modelling." This illustrates the fact that the conceptualist does not regard the Universe as being intrinsically mathematical: God is not a mathematician, but what He does can be fairly accurately described by mathematical "models." Mathematics is entirely a product of the human mind. Conceptualists would not expect to be able to communicate with the Andromedans by using our mathematical concepts.

This viewpoint might have significant consequences for the physicist. It could, for instance, be argued that our so-called constants of Nature (quantities like Newton's gravitational constant $G$), which arise as theoretically undetermined constants of proportionality in our mathematical equations, are solely artefacts of the particular mathematical representation we have chosen to use for the gravitational force. In that sense "$G$" is seen as a cultural creation. It reflects our inclination to express natural phenomena in a particular way. It also reflects the views of Kant regarding the innate categories of thought whereby all our experience is ordered by our minds. Whether or not the "thing in itself" is intrinsically mathematical (as the Platonist believes), we can experience it in no other way than the mathematical. Thus, our minds imprint mathematical ideas upon experience. It also displays something of the opposing belief that cultural elements completely condition our mathematical experience of Nature, because Nature has impressed mathematics onto our minds during the course of our evolutionary adaption. We honed our ability to formulate and manipulate abstract symbols most effectively where they were based upon things that actually exist in the real world. This latter viewpoint has far-reaching consequences, for it maintains that mathematics has been shaped by our experience. If this is so, then we should not expect it to hold good when confronted with newly discovered phenomena outside of everyday human experience. It could be said that many aspects of the world of physics, be they to do with astronomy or particle physics, turn out to be unfamiliar in scale, but not in the type of mathematical reasoning involved, so we should not necessarily be impressed by the fact that our human, evolved mathematical sense provides a good description. However, there are places where no past evolutionary history can have produced the requisite concepts; the peculiar aspects of

544

quantum reality ... may be signalling to us that it is not just our physical theory that needs improving if we are to deal with the micro-world. Indeed, mathematics may not even be the language that most naturally describes what is happening. There have been investigations into the possibility that classical logic (where statements are either true or false) does not apply at the quantum level, but is replaced by a three-valued "quantum logic" which allows the additional status of "undecided" to be associated with a statement; thus a statement that is not true need not be false. In this way a different answer can be given to the question "what slit does the particle go through in the two-slit experiment?" However, the solution to the problems of quantum reality may be far more radical. It may require some other type of description: a new language that does for logic what logic does for mathematics.

Conceptualism is a form of anti-realism directed against theories about, rather than the existence of, a bedrock of underlying reality.

One can certainly detect specific cultural biases in the way mathematics has been and is being done. The British style eschews general formalism for the sake of elegance alone, is biased towards applications, and motivated by the desire to solve practical problems. The French, by contrast, are attracted by formalism and abstraction, as epitomized by the encyclopaedic project of the Bourbaki group. Are these national styles a harmless irrelevance, or do they indicate a deep-seated subjectivity that colours the development of all human mathematics, and thereby dictates the possible laws and explanations that are available to scientists in their representations of Nature?

*Formalism:* The next of the "isms" is one that grew up primarily near the turn of the century. Logicians had uncovered a number of embarrassing logical paradoxes, and there had begun to appear mathematical proofs which established the existence of particular objects but offered no way of constructing them explicitly in a finite number of steps (in this respect they might be compared with certain legitimate configurations which can be laid down in John Conway's board game "Life," but which cannot be reached from some starting state in a finite number of moves). These logical entities arose as a consequence of properties of infinite collections of objects, when it was assumed that infinite collections of things were governed by the same common-sense logic that applies to finite collections (like the assumption that the collection either possesses a certain property or it does not). Some mathematicians were nervous about such a step of faith, because infinite sets were not physically realizable. In response to these doubts David Hilbert proposed a programme to eradicate such ambiguities. The underlying philosophy of Hilbert's formalist programme was to define mathematics as nothing more than the manipulation of symbols according to specified rules. The resultant paper edifice has no special meaning at all. It should, if the manipulations were performed correctly, result in a vast collection of tautological statements: an embroidery of logical connections. Sometimes the word "model" is used in this approach, but with a different meaning to that intended by the conceptualists: here a "model" for a set of axioms is some collection of conceptual entities that satisfy them, and hence to which the resulting tautologies (which we call theorems) derivable from the axioms apply. The focus of attention is upon the relations between entities and the rules governing them, rather than the question of whether the objects being manipulated have any intrinsic meaning. The connection between the world of Nature and the structure of mathematics is totally irrelevant to the formalists. One of the aims of this approach was to avoid having to worry about the *meaning* of non-intuitive objects like infinite sets. Attention was focused upon relationships between concepts rather than on the concepts themselves. The only goal of mathematical investigation was to show particular sets of axioms to be self-consistent, and hence acceptable as starting-points for the logical network of symbols.

From what we have said one might consider Euclid to be an archetypical formalist. In retrospect this appears so, but we have to remember that he abstracted his axioms from observation of the real world. All his theorems were visualizable by drawing points and lines in the sand and measuring angles. Modern mathematics does not require its sets of axioms to possess visualizable or self-evident properties. It is sufficient simply that they be self-consistent. This view of mathematics — that it is a logical game, like chess, with lists of pieces and rules — is opposed to the Platonic picture, for it regards the mathematical rules and axioms as entirely our own creations. They have no independent meaning except through their interrelationships. We determine these relationships by setting the rules of the game. Formulae exist: mathematical objects do not. For the formalist, the utility of mathematics in describing Nature is a curiosity which had nothing to do with mathematics. For the formalist, a mathematical theory is only intelligible if it is meaningless.

*Intuitionism:* This interpretation was also a reaction to the use of non-intuitive concepts in mathematical proofs. To avoid founding whole areas of mathematics upon the assumption that infinite sets share the "obvious" properties possessed by finite ones, it was proposed that only quantities that can be constructed from the natural numbers 1, 2, 3, ... in a finite number of logical steps should be regarded as proven true (prior to Cantor's work on infinite sets mathematicians had not made use of an actual infinity, but only exploited the existence of quantities that could be made arbitrarily large or small — this idea forms the essence of the rigorous definition of a "limit" introduced in the nineteenth century by Cauchy and Weierstrauss). Each step must unambiguously specify the next logical step to be taken. For this reason it is also referred to as *constructivism*. The name "intuitionism" reflects the idea that only the simplest intuitive ideas could be used. Anything outside our experience must be constructed from the simplest ingredients by a sequence of intuitively familiar steps. This approach is analogous to the operationalist stance . . . ; whereas the operationalist restricts attention to measurable quantities in order to avoid introducing "obvious" concepts like simultaneity which may turn out to be experimentally meaningless, the intuitionist retains the "obvious" to avoid arriving at the meaningless.

There is a parallel between the goal of the intuitionists and that of some interpreters of the quantum theory. . . . Both Bohr and the intuitionists were trying to introduce new views of physical and mathematical quantities that divorced them from objective reality. A quantum measurement reflects one's state of knowledge about physical reality. A mathematical formula, according to the intuitionists, describes only the set of computations that has been carried out to arrive at it. It is not a representation of any reality existing independently of the act of computation.

Intuitionists exclude all arguments that prove the existence of something without providing the recipe for constructing it. Of course, if we can prove that there exists *some* special member of a *finite* set of objects then, although this result is not acceptable to the intuitionist, it can be made so by working through the finite collection of objects (perhaps by searching through the mathematical possibilities using a fast computer, as in the recent proof of the famous "four-colour theorem") so as to isolate the special member explicitly. This is a legitimate constructive procedure. However, if we had proved only the existence of a special member of an *infinite* set, then this result would be judged unacceptable because it could not be checked constructively in a finite number of steps using the computer. It is very interesting that the famous singularity theorems of Hawking and Penrose, which establish the existence of a beginning to space and time in the Universe's past if a number of observationally checkable assumptions are met, do *not* meet the intuitionists' requirements. They predict the existence of some path (or paths) through space and time which inevitably comes to an end a finite time ago, but do not construct it explicitly. The best we could do is to find explicit solutions of Einstein's equations of general relativity that possess initial singularities. However, the only such solutions that we have been clever enough to find are rather special cases.

The early enthusiasts for the constructivist approach, like Kronecker and Brouwer, proposed rebuilding the whole of mathematics constructively, avoiding the use of non-intuitive entities like infinite sets. Not surprisingly, this overly positivist proposal did not meet with great enthusiasm. It would have decimated mathematics (remember that depressing feeling when it was suggested that your high-school essay or scientific paper be completely rewritten?!). Hilbert believed that Brouwer's programme would be a disaster, even if it succeeded. He claimed that after the constructivists had finished with mathematics, "compared with the immense expanse of modern mathematics, what would the wretched remnant mean, a few isolated results, incomplete and unrelated, that the intuitionists have obtained."

Brouwer's dogmatic approach actually created quite a rumpus within the world of mathematics. He was one of the editors of the German journal *Mathematische Annalen*, the leading mathematics journal of the day, and declared war on mathematicians who did not accept his constructivist philosophy by rejecting all papers submitted to the journal that employed non-constructive notions like infinite sets or his favourite *bête noire*, the Aristotelian law of the excluded middle, which states that something is either true or false. This created a crisis which the other members of the editorial board resolved by resigning and then re-electing a new editorial board — only Brouwer found that he was not on it! The Dutch government viewed this as an insult to their distinguished countryman, and responded by creating a rival journal with Brouwer as editor.

In practice, the intuitionists did not regard all mathematical statements as either true or false. They stipulated a third category: *undecidable*. This three-fold logic is reminiscent of Scottish courts of law, where the verdict of "not proven" may be returned, whereas English courts require a verdict of either "guilty" or "not guilty." The "undecided" state of limbo was the fate of statements that could not be demonstrated true or false in a finite number of constructive logical steps.

The intuitionist does not admit as valid arguments any that begin with statements like "in the infinite decimal expansion of pi there either exists a run of one hundred consecutive odd numbers or there does not." Such properties of pi are in the undecidable limbo. This restricted logic also outlaws a classic method of proof called *reductio ad absurdum*, because it no longer follows that the negation of the negation of some statement *S* implies that *S* is true. The intuitionists are therefore working with a system of logic that lacks one of the most useful procedures for generating new true statements. They are fighting, as it were, with one hand tied behind their backs. Of course, anything true in this reduced system of logic will also be true in ordinary logic, but not vice versa. The intuitionists have a much more demanding criterion of mathematical truth.

Few mathematicians are intuitionists, but recently there has been a revival of interest in this approach. Like formalism, intuitionism is decidedly non-Platonic. It regards mathematics as invented, not discovered. Moreover, it is constructed by *human* manipulations from a set of basic, intuitively obvious notions — but intuitively obvious to *us*. Whereas the formalists were unperturbed by the presence of non-intuitive concepts like actual infinities in their logical procedures, the intuitionists excluded them by fiat, but both philosophies reject as meaningless the idea of mathematics without mathematicians. One obvious drawback of the intuitionist programme is that the scope of the subject is not well defined. There is no precise definition of what the constructive methods are. We do not know if someone will come along tomorrow and construct results that you believed to be unconstructable by finite intuitive steps. In fact, not long ago some results involving properties of infinite sets proved first by Cantor, and which provoked the original intuitionist revolt, were added to the body of results provable by finite constructive steps.

There exists an indirect connection between the intuitionists' logic and the search for laws of Nature. . . . We discussed the dilemma of quantum reality and the issue of whether a neutron "really" went through one slit or the other, or whether curiosity killed

Schrödinger's Cat.* The resolution of this logical impasse was the adoption of a radically new picture of reality. An alternative approach to the resolution of the quantum measurement puzzle was the adoption of the three-valued logic of the intuitionists, but not the constructivist methodology that motivated it. In this context it is referred to as "quantum logic." Thus, it is no longer necessary to conclude that the neutron went through one slit *or* the other. Schrödinger's Cat is not necessarily dead if it is not alive, and Bell's theorem can no longer be proved. There now exists a further intermediate logical state. The adoption of this quantum logic can provide an "explanation" of sorts for the world of quantum strangeness, but only at the expense of giving up the logic that applies to everything else. Most physicists regard this as an unacceptable schizophrenia. After all, one has to use ordinary logic to argue for the application of quantum logic.

There is no way of deciding which of these approaches towards mathematics is right or wrong. There is suggestive evidence for some and against others. The best that one can do is to recognize how the adoption of one view or another influences conclusions one might draw about the structure of the Universe. We should also appreciate that the intuitionist position is of a slightly different nature to the others. It is not solely a view of what mathematics is, but an attempt to confine it in a rigid operational fashion. The required definition appears to be quite restrictive, and results in a subject considerably smaller than conventional mathematics. To my knowledge no one has attempted to indicate what would be left of the mathematical sciences if only the intuitionists' corpus of mathematics were regarded as true. The three non-Platonic stances must each come to terms with the effectiveness of mathematics in describing Nature. Why is it that abstract mathematical concepts, devised and explored in the distant past with no apparent application, so often turn out to be key elements in the description of some new area of discovery in physics? The following imaginary dialogue between a Platonist and a mathematician who maintains that mathematics is a human invention covers some of the pros and cons of their respective positions.

Whenever you tell me that mathematics is just a human invention like the game of chess I would like to believe you. It would make things so much less mysterious. But I keep returning to the same

---

*See Heinz R. Pagels, "Schrödinger's Cat," above. [editor's note]

problem. Why does the mathematics we have discovered in the past so often turn out to describe the workings of the Universe? Surely this cannot be an accident?

*It certainly isn't an accident. When we set out to describe Nature we have to use the only tools that are available to us. For all we know there may be a better language for that purpose than mathematics. We derived most of our mathematics from Nature in the first place, so it would be rather surprising if we couldn't then describe Nature with it.*

What about Riemannian geometry? That was developed as a branch of pure mathematics by Riemann and others long before Einstein found that it describes the structure of space-time.

*That's an unfortunate example. You see, Riemann's study of the geometrical properties of curved surfaces arose from his interest in a very practical problem: the distortion of sheets of metal when they are heated. The effect is not dissimilar to the distortion of space-time geometry by mass and energy according to Einstein's theory. In fact, some physicists have even used the heated metal sheet as a heuristic to explain Einstein's theory to the general public.*

What about group structures and symmetries? The interactions of elementary particles, indeed the *existence* of particular elementary particles, appear to be dictated by mathematical symmetries, and these are described by groups. All these groups were discovered and classified by pure mathematicians more than a hundred years ago, oblivious of modern physics. These properties are exact as well; they predict that there exists a certain number of particles of a certain type and no more. It's not a case of the mathematics being a pretty good approximation. It's either right or wrong, and experience shows us that it's usually right. One might also question your assumption that if some aspect of mathematics is derived from the natural world then this makes it culturally or humanly derived. Quite the opposite, I would have thought. It just seems to reinforce the view that the world is intrinsically mathematical.

*I think the world just is. I can't see why I should go any further and say it is mathematical. And I don't think there is anything especially mathematical about symmetry. In fact, mathematicians and physicists were among the last to latch on to the importance of symmetry. Architects and artists had identified and appreciated it long before them. Mathematicians derived it from Nature. Mathematicians' interest in group theory is probably just a sophisticated version of the human attraction to symmetry and pattern, an attraction that was acquired from Nature as a result of natural selection in the first place. I'll give you a good example. The draughtsman Maurits Escher pro-*

*duced designs with no knowledge of mathematics at all, but subsequently math-*
*ematicians showed that his pictures contain deep mathematical symmetries and*
*constructions of a complicated sort. The whole Universe could have been fash-*
*ioned in this way, with no mathematics built into it, only aesthetics. Then*
*along we come with our mathematics and proclaim the Universe to be intrins-*
*ically mathematical. You think God is a mathematician, but on the basis of*
*the same sort of evidence you would have claimed Escher was a mathemati-*
*cian: but you would have been wrong. I might also mention that your impli-*
*cation that the usefulness and applicability of mathematics in the real world*
*supports the Platonic view is a bit of a cheat, because your view can't offer*
*any explanation as to why that numinous collection of abstract mathematical*
*objects should have anything at all to do with our everyday physical world.*
*There are all sorts of abstract Platonic entities, like unicorns and centaurs,*
*that are not useful, and which don't seem to exist in the real world: why are*
*the abstract mathematical ones not also of this useless non-existing variety?*
*Lacking an answer to this I think you are begging the question. And anyway,*
*even if I could offer no explanation at all for the usefulness of mathematics*
*it wouldn't strengthen your position one iota.*

But how can you defend the idea that mathematics is culturally
derived? This seems absurd. Different mathematicians in different
cultures living in different circumstances have invariably come up
with the same mathematics. Pythagoras' theorem says the same thing
whether found by the Greeks, or the Indians, or whoever. It contains
a core of truth that transcends individual biases. The whole reason
for adopting mathematics as the language of science was, after all,
precisely because it was culturally independent and lacking any subjective element. I agree there are fashions in mathematics, national
styles even, but they simply dictate the direction in which investigations move or the style in which results are presented; they don't
determine what mathematical results will be found true or false. Five
is a prime number whether you like it or not!

*There must be some contribution by our minds. It's unavoidable I would*
*have thought.*

But it's a harmless simplification to ignore this. If you don't then
you end up studying an image of a reality which is unknowable by
our minds, so why not treat the image as the only reality in the first
place? It's the only reality that is relevant.

*Then everything is subjective you are saying?*

No, I just want to assume our cognition does not distort what
is really out there. You don't. That's the crux of the matter.

*Math Angst*

*I think we shall find there is a little more of a distinction between us than that.*

If, as you say, we invent mathematics rather than discover it, why do we find it so hard to understand? Why do mathematical objects like sets and groups seem so meaningful to us, if they are just convenient descriptions of some patterns that turn up in our minds — so meaningful, in fact, that we are led to call some of their properties "true"? Your view must face the same awkward problem that confronts the solipsist. If everything is our subjective creation, why is some of it so hard to understand, and why does it all have so little connection with us in other respects?

*Chess is a human invention, but that doesn't stop us failing to solve difficult chess puzzles. These things often defeat us because we are not very smart, that's all.*

But I still have this strong conviction that mathematics is altogether different to games like chess. They share common elements of course, but mathematics describes the way the world works: chess doesn't.

*Suppose we apply our viewpoints to some other symbolic language — like music, for example. Is a Beethoven symphony invented or discovered? No musician would dream of taking the Platonic view to claim that it was discovered. It is an invention of the composer. It embodies aspects of his personality. Why is mathematics any different?*

Although music is a symbolic language it differs from mathematics in a significant way: it does not contain an unbending built-in logic. There are rules of composition, but they are rules to be broken. The reason you regard Beethoven's Fifth Symphony as a creation rather than a discovery is because you cannot believe that somebody else would have written it if Beethoven hadn't. But if Pythagoras had not discovered his theorem somebody else would have. In fact, somebody else did! There are lots of examples of this multiple discovery: there are no examples of multiple creation in the arts. Newton and Leibniz both discovered the calculus. Gauss, Lobachevski, and Bolyai all appear to have had the same ideas about non-Euclidean geometry. There are not two Hamlets! This feature demonstrates the difference between the sciences and the arts. The products of the latter, being almost totally subjective in form and content, are necessarily unique; but because mathematics is the discovery of something that already exists, it can be — and often is — independently duplicated.

*Of course, this fact that very different individuals discover the same mathematical results could be put as evidence for the view that mathematics is determined by a universal human trait rather than particular cultural ones, but I must confess that sounds a bit too much like special pleading.*

I agree. It has often been claimed that primitive mathematical notions are innate to the human mind. Poincaré regarded geometries and continuous groups of symmetries as concepts that pre-exist in the human mind. The concept of "number" is probably the most basic intuition we could attribute in this way. It is often claimed by psychologists that children have the abstract concept of, say, the number "three" before they ever understand the concrete examples of it — like three little pigs or three kings. However, recently archaeologists have made some remarkable discoveries which shed light upon how abstract mathematical notions evolved amongst ancient Sumerian societies. It appears that around 8000 B.C. the Sumerians had trading tokens which did not recognize the unity of the concept of a number. They distinguished two sheep from two goats, but they resisted the addition of different types of thing. Lists of their numbers would always include the description of what the numbered things were. However, by 3100 B.C. they had evolved the concept of number independent of the objects being enumerated. There now existed separate tokens to distinguish numbers of things from the identity of the things being counted.

*How can mathematical concepts like "points," infinitesimally small quantities, or irrational numbers be anything but products of our minds? After all, they do not really exist do they? If these mathematical entities exist "out there," where are they? It can't be in the space and time of our Universe I'm afraid. You tell me how many dimensions of space there are, and I'll tell you about spatial geometry in twice as many dimensions, even though it doesn't exist.*

I'm not claiming that mathematical objects exist in the space and time of our Universe.

*You mean these concepts have an existence independent of particular examples of them?*

Yes, but I don't know "where" they exist. All I want to argue is that the Universe is intrinsically mathematical.

*Just now you complained at my creating a shadow world of sense impressions of reality one step removed from it. Now you're doing the same. You have created another world stocked with all the mathematical furnishings you require in this one. You can hardly claim that your view that mathematical*

554

*objects are real is supported by simply inventing another world of real mathematical abstractions.*

My other world is unique, and explains why we all see the same mathematical structures, but your other worlds are as many as there are mathematicians, and make it seem distinctly odd that we all detect the same mathematics if it doesn't originate from a common source.

*But you don't seem to have an explanation of the effectiveness of mathematics in describing the workings of this world anymore. Your mathematical entities live in some other world, and you have to explain why it just so happens that they describe what goes on in our space and time. And I don't see how you can claim to know anything about your abstract world even if it does exist. You admit that these abstract mathematical entities do not sit in our space and time, but that surely means we cannot know about them. They don't interact with human beings, because our cognition is limited to things in the Universe of space and time.*

Well maybe this is just a problem created by our limited picture of what it means to "know" about something. I'm certainly not claiming that my view is complete yet. It needs a lot of development. And I should add that I am content with the view that the Universe just *is* intrinsically mathematical. I can then avoid all mention of the "other world" of abstract entities. It might be that mathematics is identical to what you would call logic. Although we think it mysterious that the world is well described by mathematics, I don't hear many people expressing surprise that the world is described and governed by logic.

*Maybe intelligent beings cannot exist unless the world is governed by logic and mathematics?*

Ah! You've been reading that damned fat square book by Bipler and Tarrow, or whatever their names are. Do you really believe that there are other worlds which are not describable by mathematics? Even the Many Worlds of Everett obey the mathematics of quantum mechanics.

*It's not impossible. But I don't think it is going to settle our disagreement either way. It would be enough to know why mathematics describes the structure of any world.*

There is one difficulty for us both that I would like to mention. We have both been blithely discussing "mathematics" as if one word will do to describe the entire thing — whatever it is. Although I believe that Nature is invariably mathematical, I don't think I could claim that *all* mathematics is used in Nature. Perhaps, being British, we must seek a compromise, and maintain that some mathematics is discovered whilst the rest is invented and derived from the former in some way?

*That's an interesting idea. It doesn't affect my argument of course, but I notice that it partially undermines yours — and it gives you the awkward problem of deciding where to draw the line between the two different sorts of mathematics.*

Well you would say that wouldn't you? Actually, I think it strengthens my position.

*You mean because it allows you to shift all the problematic examples which undermine your case into the non-Platonic category? I see a more subtle problem here. Take classical mechanics for example. There are two ways of using mathematics to determine the trajectories of particles moving under the influence of a force, gravity say: either we use differential equations, and determine the future state in terms of the present one in a causal way, or we can use a variational principle which determines the actual trajectory taken by the particle to be that path which minimizes a certain quantity connecting the initial and final state. In this second approach the future partially determines the past. Philosophically, there is a world of difference between these two ideas, but mathematically they turn out to be completely equivalent. So the Platonist has the dilemma of deciding which of the two descriptions is the "real" one.*

Or which is the right one. They may not turn out to be fully equivalent. Incidentally, why do mathematicians regard mathematics as both "beautiful" and useful?

*We are attracted to certain mathematical structures because they are elegant or "beautiful." By that mathematicians and scientists mean that they exhibit a deep underlying unity in the face of superficial diversity. They grow into the largest and most intricate logical structures from the simplest beginnings, and exhibit unexpected relationships with other branches of mathematics despite their apparent dissimilarity with them. We might imagine that it is structures like these that are most likely to find application in the real world simply because their applicability requires a minimum of special circumstances to hold.*

This little exchange shows the problem we have in determining both the meaning of mathematics and the reason for its special effectiveness in science. There is something emotionally attractive about the Platonic stance. It offers such an encitingly simple explanation for everything. But the arguments cited against it are very persuasive. The final point of the dialogue about different types of mathematics is probably a vital one. We could think of mathematics in the way we think of a language like English. It originates as an expedient way to communicate. It is useful. Sometimes it is even necessary to invent new words. Applied mathematics begins like this. But then along come the grammarians who want to put the whole morass of practice on a firm and logical footing, building their house from the roof downwards. They partially succeed. Next there arise writers and poets who love the language itself. They do not only want to use it for mundane practical purposes. They are attracted by its inherent rhythm and rhyme, and the scope of its grammatical structure for succinct expression. They are conscious of style and form, of the difference between poetry and prose. They are like the pure mathematicians, who pursue mathematics for its own intrinsic structure. They mould the rugged corner-stone into a beautiful sculpture. If this mathematics turns out to be useful so much the better.

Yet this sociological analysis still provides no explanation for the effectiveness of the corner-stones in supporting the scientific description of Nature. The issue is clouded yet further by the fact that there can exist several different mathematical solutions to a physical problem, some of which cannot apply to reality. Thus, some additional correspondence principle must be used to judge whether a piece of mathematics applies to the real world. A good illustration of this ambiguity is provided by the "Coconut Puzzle" which emerged as a challenge problem in pre-war Cambridge. It can be stated as follows:

Five men find themselves shipwrecked on an island, with nothing edible in sight but coconuts, plenty of these, and a monkey. They agree to split the coconuts into five equal integer lots, any remainder going to the monkey.

Man 1 suddenly feels hungry in the middle of the night, and decides to take his share of coconuts at that very moment. He finds the remainder to be one after division by five, so he gives this remaining coconut to the monkey and takes his fifth of the rest, lumping the coconuts that remain back together. A while later, Man 2 wakes up hungry too, and does exactly the same — takes a fifth of the coconuts, gives the monkey the remainder, which is again one, and leaves the rest behind. So do men 3, 4, and 5. In the morning they all get up, and no one mentions anything about his coconut-affair the previous night. So they share the remaining lot in five equal parts finding,

once again, a remainder of one left for the monkey. Find
the initial number of coconuts.

There are in fact an infinite number of solutions to this
problem, but we would like the smallest whole number of coconuts.
It is 15621. However, soon after this problem was posed Paul Dirac
gave another solution: −4 coconuts! This is clearly a solution. Each
time a man arrives at the coconut store he finds −4 coconuts, gives
+1 to the monkey leaving −5. His one-fifth share of −5 is −1,
which he takes, leaving −4 behind either for the next man or the
final share-out!

Here we see an example of a perfectly good *mathematical* solution which (since we cannot realize a "negative coconut") does not
have a realist out-working (however, there is a tradition that Dirac's
negative solution played some role in his thinking which led to the
introduction of the concept of antimatter). In this situation it is easy
to exercise a criterion which eliminates Dirac's solution as an unrealistic one, but in more esoteric areas of mathematical physics it may
not be so easy to decide upon criteria by which to reject some mathematical predictions as unrealistic.

# PHILIP J. DAVIS AND REUBEN HERSH

■

Mathematics is so useful in describing nature that we may sometimes imagine that equations can express *all* reality. But the
mathematicians Philip J. Davis (b. 1923) and Reuben Hersh
(b. 1927) remind us that mathematics, however beguiling, is but
a small part of the whole.

# *The Limits of Mathematics*

BUT WAIT! Can everything be mathematized? Is there anything in
the world which can never become the subject of a mathematical
theory? Certainly in the physical world we do not believe there is
anything un-mathematizable. There may be phenomena, such as turbulence, whose mathematical descriptions are so complex we are
unable to analyze them or compute them in any reasonably effective
sense. We are confident, however, that physics can encompass any
physical phenomenon, and do so by means of a mathematical formalism, whether it be the old, familiar one of differential equations
with initial and boundary conditions,, or the up-to-date one of map-
pings between high-dimensional or infinite-dimensional non-linear
differentiable manifolds.

To find things that cannot be mathematized, then, we must look
away from the physical world. What other world is there? If you are
a sufficiently fanatical mechanical materialist you may say none.
Period. Discussion concluded.

If you are more of a human being, you will be aware that there
are such things as emotions, beliefs, attitudes, dreams, intentions,
jealousy, envy, yearning, regret, longing, anger, compassion, and
many others. These things — the inner world of human life — can
never be mathematized.

True, some psychologists and sociologists have come around
with their questionnaires and chi-square statistics, purporting to

study the human mind quantitatively, but most such investigations are so remote from the target that the critic need hardly say, "Pooh!" They fall over of their own absurdity and pomposity.

I don't mean to say that it is only the inner life of the individual that is beyond mathematics. Even more so is the "inner life" of society, of civilization itself, for example, literature, music, politics, the tides and currents of history, the stuff and nonsense that fill the daily newspaper. All this falls outside the computer, outside any equations or inequalities. And a good thing, too.

# PART FOUR

## *The Ways of Science*

Scientific research is a thoroughly human activity, one that involves not only scientists' knowledge and intellect but also their personalities, their psychologies, their cultural outlooks, and their hopes and dreams. And the results have a potential impact not only on science but on the wider society. In Part Four we present profiles of leading scientists, reflections by some on how they go about their work, and examples of the influence of scientific ideas on society, poetry, and philosophy.

# Scientists' Lives and Works

## JOHN ARCHIBALD WHEELER

■

Our cultural memory of Albert Einstein, the most celebrated scientist of the twentieth century, is to some degree distorted by his formidable myth. Few writers have both the technical competence to understand Einstein's research and the sensitivity to appreciate the subtleties of his personality. One of the exceptions is the physicist, essayist, and philosopher of science John Archibald Wheeler (b. 1911), as this graceful sketch demonstrates.

# Albert Einstein

ALBERT EINSTEIN WAS BORN in Ulm, Germany on March 14, 1879. After education in Germany, Italy, and Switzerland, and professorships in Bern, Zurich, and Prague, he was appointed Director of Kaiser Wilhelm Institute for Physics in Berlin in 1914. He became a professor in the School of Mathematics at the Institute for Advanced Study in Princeton beginning the fall of 1933, became an American citizen in the summer of 1936, and died in Princeton, New Jersey on April 18, 1955. In the Berlin where in 1900 Max Planck discovered the quantum, Einstein fifteen years later explained to us that gravitation is not something foreign and mysterious acting through space, but a manifestation of space geometry itself. He came to understand that the universe does not go on from everlasting to everlasting, but begins with a big bang. Of all the questions with which the great thinkers have occupied themselves in all lands and all centuries, none has ever claimed greater primacy than the origin of the universe, and no contributions to this issue ever made by any man anytime have proved themselves richer in illuminating power than those that Einstein made.

Einstein's 1915 geometrical and still standard theory of gravity provides a prototype unsurpassed even today for what a physical theory should be and do, but for him it was only an outlying ridge in the arduous climb to a greater goal that he never achieved. Scale the greatest Everest that there is or ever can be, uncover the secret of existence — that was what Einstein struggled for with all the force of his life.

How the mountain peak magnetized his attention he told us over and over. "Out yonder," he wrote, "lies this huge world, which exists independently of us human beings and which stands before us like a great, eternal riddle. . . ."[1] And again, "The most incomprehensible thing about the world is that it is comprehensible."[2] And yet again, "All of these endeavors are based on the belief that existence should have a completely harmonious structure. Today we have less ground than ever before for allowing ourselves to be forced away from this wonderful belief."[3]

When the climber laboring toward the Everest peak comes to the summit of an intermediate ridge, he stops at the new panorama of beauty for a new fix on the goal of his life and a new charting of the road ahead; but he knows that he is at the beginning, not at the end of his travail. What Einstein did in spacetime physics, in statistical mechanics, and in quantum physics, he viewed as such intermediate ridges, such way stations, such panoramic points for planning further advance, not as achievements in themselves. Those way stations were not his goals. They were not even preplanned means to his goal. They were catch-as-catch-can means to his goal.

Those who know physicists and mountaineers know the traits they have in common: a "dream-and-drive" spirit, a bulldog tenacity of purpose, and an openness to try any route to the summit. Who does not know Einstein's definition of a scientist as "an unscrupulous opportunist;"[4] or his words on another occasion, "But the years of

---

1. A. Einstein, "Autobiographical Notes," in *Einstein: Philosopher-Scientist,* ed. P. A. Schilpp (Evanston, Ill.: Library of Living Philosophers, 1949), p. 4.

2. B. Hoffmann, *Albert Einstein: Creator and Rebel* (New York: Viking, 1972), p. 18.

3. A. Einstein, *Essays in Science* (New York: Philosophical Library, 1934), p. 114.

4. A. Einstein, "Reply to Criticisms," in *Einstein: Philosopher-Scientist,* ed. P. A. Schilpp (Evanston, Ill.: Library of Living Philosophers, 1949), p. 648.

anxious searching in the dark, with their intense longing, their alternations of confidence and exhaustion, and the final emergence into the light — only those who have experienced it can understand that."[5] For such a man there are not goals. There is only *the* goal, that distant peak.

Who was this climber? How did he come to be bewitched by the mountain? Where did he learn to climb so well? Who were his companions? What were some of his adventures? And how far did he get?

I first saw and heard Einstein in the fall of 1933, shortly after he had come to Princeton to take up his long-term residence there. It was a small, quiet, unpublicized seminar. Unified field theory was to be the topic, it became clear, when Einstein entered the room and began to speak. His English, though a little accented, was beautifully clear and slow. His delivery was spontaneous and serious, with every now and then a touch of humor. I was not familiar with his subject at that time, but I could sense that he had his doubts about the particular version of unified field theory he was then discussing. It was clear on this first encounter that Einstein was following very much his own line, independent of the interest in nuclear physics then at high tide in the United States.

There was one extraordinary feature of Einstein the man I glimpsed that day, and came to see ever more clearly each time I visited his house, climbed to his upstairs study, and we explained to each other what we did not understand. Over and above his warmth and considerateness, over and above his deep thoughtfulness, I came to see, he had a unique sense of the world of man and nature as one harmonious and someday understandable whole, with all of us feeling our way forward through the darkness together.

Our last time together came twenty-one years later, on April 14, 1954, when Einstein kindly accepted an invitation to speak at my relativity seminar. It was the last talk he ever gave, almost exactly a year before his death. He not only reviewed how he looked at general relativity and how he had come to general relativity, he also spoke as strongly as ever of his discomfort with the probabilistic features that the quantum had brought into the description of nature. "When a person such as a mouse observes the universe," he asked feelingly,

5. M. J. Klein, *Einstein, The Life and Times*, R. W. Clark, book review, *Science*, 174: 1315.

"does that change the state of the universe?"[6] He also commented in the course of the seminar that the laws of physics should be simple. One of us asked, "But what if they are not simple?" "Then I would not be interested in them,"[7] he replied.

How Einstein the boy became Einstein the man is a story told in more than one biography, but nowhere better than in Einstein's own sketch of his life, so well known as to preclude repetition here. Who does not remember him in difficulty in secondary school, antagonized by his teacher's determination to stuff knowledge down his throat, and in turn antagonizing the teacher? Who that takes the fast train from Bern to Zürich does not feel a lift of the heart as he flashes through the little town of Aarau? There, we recall, Einstein was sent to a special school because he could not get along in the ordinary school. There, guided by a wise and kind teacher, he could work with mechanical devices and magnets as well as books and paper. Einstein was fascinated. He grew. He succeeded in entering the Züricher Polytechnikum. One who was a rector there not long ago told me that during his period of rectorship he had taken the record book from Einstein's year off the shelf. He discovered that Einstein had not been the bottom student, but next to the bottom student. And how had he done in the laboratory? Always behind. He still did not hit it off with his teachers, excellent teachers as he himself said. His professor, Minkowski, later to be one of the warmest defenders of Einstein's ideas, was nevertheless turned off by Einstein the student. Einstein frankly said he disliked lectures and examinations. He liked to read. If one thinks of him as lonesome, one makes a great mistake. He had close colleagues. He talked and walked and walked and talked.

To Einstein's development, his few close student colleagues meant much; but even more important were the older colleagues he met in books. Among them were Leibniz and Newton, Hume and Kant, Faraday and Helmholtz, Hertz and Maxwell, Kirchhoff and Mach, Boltzmann and Planck. Through their influence, he turned from mathematics to physics, from a subject where there are dismayingly multitudinous directions for dizzy man to choose between, to a subject where this one and only physical world directs our endeavors.

6. J. A. Wheeler, "Mercer Street and Other Memories," in *Albert Einstein, His Influence on Physics, Philosophy, and Politics*, ed. P. C. Aichelburg and R. U. Sexl (Braunscheig: Vieweg, 1979), p. 202.

7. *Ibid.*, p. 204.

Of all heroes, Spinoza was Einstein's greatest. No one expressed more strongly than he a belief in the harmony, the beauty, and — most of all — the ultimate comprehensibility of nature. In a letter to his old and close friend, Maurice Solovine, Einstein wrote, "I can understand your aversion to the use of the term 'religion' to describe an emotional and psychological attitude which shows itself most clearly in Spinoza. [But] I have not found a better expression than 'religious' for the trust in the rational nature of reality that is, at least to a certain extent, accessible to human reason."[8] In later years, Einstein was asked to do a life on Spinoza. He excused himself from writing the biography itself on the ground that it required "exceptional purity, imagination and modesty,"[9] but he did write the introduction. If it is true, as Thomas Mann tells us, that each one of us models his or her life consciously or unconsciously on someone who has gone before, then who was closer to being role-creator for Einstein than Spinoza?

Search out the simple central principles of this physical world — that was becoming Einstein's goal. But how? Many a man in the street thinks of Einstein as a man who could only make headway in his work by dint of pages of complicated mathematics; the truth is the direct opposite. As Hilbert put it, "Every boy in the streets of our mathematical Göttingen understands more about four-dimensional geometry than Einstein. Yet, despite that, Einstein did the work and not the mathematicians."[10] Time and again, in the photoelectric effect, in relativity, in gravitation, the amateur grasped the simple point that had eluded the expert. Where did Einstein acquire this ability to sift the essential from the non-essential?

The management consultant firm of Booz, Allen & Hamilton, which does so much today to select leaders of great enterprises, has a word of advice: What a young man does and who he works with in his first job has more effect on his future than anything else one can easily analyze. What was Einstein's first job? In the view of many, the position of clerk in the Swiss patent office was no proper job at all, but it was the best job available to anyone with his unpromising university record. He served in the Bern office for seven years, from

8. A. Einstein, *Lettres à Maurice Solovine* (Paris: Gauthier-Villars, 1956), p. 102 (January 1, 1956).

9. B. Hoffmann, *Albert Einstein: Creator and Rebel* (New York: Viking, 1972), p. 95.

10. P. Frank, *Einstein, Sein Leben und seine Zeit* (München: Paul List Verlag, 1949), p. 335.

June 23, 1902 to July 6, 1909. Every morning he faced his quota of patent applications. Those were the days when a patent application had to be accompanied by a working model. Over and above the applications and the models was the boss, a kind man, a strict man, and a wise man. He gave strict instructions: explain very briefly, if possible in a single sentence, why the device will work or why it won't; why the application should be granted or why it should be denied. Day after day Einstein had to distill the central lesson out of objects of the greatest variety that man has power to invent. Who knows a more marvelous way to acquire a sense of what physics is and how it works? It is no wonder that Einstein always delighted in the machinery of the physical world — from the action of a compass needle to the meandering of a river, and from the perversities of a gyroscope to the drive of Flettner's rotor ship.

Whoever asks how Einstein won his unsurpassed power of expression, let him turn back to the days in the patent office and the boss who, "More severe than my father . . . taught me to express myself correctly."[11] The writings of Galileo are studied in secondary schools in Italy today, not for their physics, but for their clarity and power of expression. Let the secondary school student of our day take up the writings of Einstein if he would see how to make in the pithiest way a telling point.

From Bern, fate took Einstein to Zürich, to Prague, and then to the Berlin where his genius flowered. Colleagueship never meant more in his life than it did during his 19 years there, and never did he have greater colleagues: Max Planck, James Franck, Walter Nernst, Max von Laue, and others. Colleagueship did not mean chat; it meant serious consultation on troubling issues. No tool of colleagueship was more useful than the seminar. James Franck explained to me the democracy of this trial by jury. The professor, he emphasized, stood on no pinnacle, beyond question by any student. On the contrary, the student had both the right and the obligation to question and to speak up.

If the writing of letters is a test of colleagueship, let no one question Einstein's power to give and to receive. Consider his enormous correspondence. Look at the postcards he sent over the years to the closest in spirit of all his colleagues, Paul Ehrenfest in Leyden. They

11. "Errinerungen an Albert Einstein, 1902–1909," Bureau Fédéral de la Propriété Intellectuelle (Berne, Switzerland), as quoted in: R. W. Clark, *Einstein, The Life and Times* (New York: The World Publishing Co., 1971), p. 75.

deal with the issues nearest to his heart at the moment, whether the direction of time in statistical mechanics, or quantum fluctuations in radiation, or a problem of general relativity. Or examine his correspondence with Max Born, or Maurice Solovine, or with everyday people. To a schoolgirl who mentioned among many other things her problems with mathematics, he replied, "Do not worry about your difficulties in mathematics; I can assure you that mine are greater."[12] Why did Einstein correspond so much with people that you and I would call outsiders? Did he not feel that the amateur brings a freshness of outlook unmatched by the specialist with his narrow view?

The benefits of colleagueship with Einstein I experienced more than once, but never with greater immediate benefit than in statistical mechanics. In a discussion of radiation damping, he referred me to a published dialogue of 1909 between himself and Walter Ritz. The two men agreed to disagree and stated their opposing positions in this single clear sentence: "Ritz treats the limitations to retarded potentials as one of the foundations of the second law of thermodynamics, while Einstein believes that the irreversibility of radiation depends exclusively on considerations of probability."[13] In accord with the position of Einstein, Richard Feynman and I found that the one-sidedness in time of radiation reaction can be understood as originating in the one-sidedness in time of the conditions imposed on the far-away absorber particles, and not at all in the elementary law of interaction between particle and particle. I joined the ranks of what I can only call "the worriers" — those like Boltzmann, Ehrenfest, and Einstein himself, and many, many others — who ask, why initial conditions? Why not final conditions? Or why not some mixture of the two? And most of all, why thus and such initial conditions and no other? No one who knows of Einstein's lifelong concern with such issues can fail to have a new sense of appreciation on reading his great early papers on statistical mechanics, and not least among them the famous 1905 paper on the theory of the Brownian motion. Surely the perspective he won from these worries will someday help show us the way to Everest.

Best known of Einstein's great trio of 1905 papers, however, is that on special relativity. "Henceforth," as Minkowski put the lesson

12. H. Dukas and B. Hoffmann, eds., *Albert Einstein; The Human Side: New Glimpses from His Archives* (Princeton, N.J.: Princeton Univ. Press, 1979), p. 8.
13. A. Einstein and W. Ritz, *Physikalisches Zeitschrift*, 10 (1909): 323–34.

of Einstein, "space by itself and time by itself, are doomed to fade away into mere shadows, and only a kind of union of the two will preserve an independent reality."[14] Historians of science can tell us that if Einstein had not come to this version of spacetime it would have been achieved by Lorentz, or Poincaré, or another, who would also have come eventually to that famous equation $E = mc^2$, with all its consequences. But it still comes to us as a miracle that the patent office clerk was the one to deduce this greatest of lessons about space-time from clues on the surface so innocent as those afforded by electricity and magnetism. Miracle? Would it not have been a greater miracle if anyone but a patent office clerk had discovered relativity? Who else could have distilled this simple central point from all the clutter of electromagnetism than someone whose job it was over and over each day to extract simplicity out of complexity?

If others could have given us special relativity, who else but Einstein, sixty-four years ago, could have given us general relativity? Who else knew out of the welter of facts to fasten on that which is absolutely central? Did the central point come to him, as legend has it, from talking to a housepainter who had fallen off a roof and reported feeling weightless during the fall? We all know that he called that 1908 insight the "happiest thought of my life"[15] — the idea that there is no such thing as gravitation, only free-fall. By thus giving up gravitation, Einstein won back gravitation as a manifestation of a warp in the geometry of space. His 1915 and still standard geometric theory of gravitation can be summarized, we know today, in a single, simple sentence: "Space tells matter how to move and matter tells space how to curve."[16] Through his insight that there is no such thing as gravity, he had had the creative imagination to bring together two great currents of thought out of the past. Riemann had stressed that geometry is not a God-given perfection, but a part of physics; and Mach had argued that acceleration makes no sense except with respect to frame determined by the other masses in the universe.

It is unnecessary to recall the three famous early tests of Einstein's geometric theory of gravitation: the bending of light by the sun, the red-shift of light from the sun, and the precession of the

14. C. Reid, *Hilbert* (Berlin: Springer, 1970), p. 12.
15. A. Einstein, "The Fundamental Idea of General Relativity in Its Original Form," unpublished essay, 1919 (excerpts, *New York Times*, 28 March 1972), p. L.32.
16. J. A. Wheeler, University of Texas, lecture of 2 March 1979.

orbit of the planet Mercury going around the sun. Neither is it necessary to expound the important insights that have come and continue to come out of general relativity. Einstein showed that the law for the motion of a mass in space and time does not have to be made a separate item in the conceptual structure of physics. Instead, it comes straight out of geometric law as applied to the space immediately surrounding the mass in question. Moreover, the geometry that he had freed from slavery to Euclid, and that he had assigned to carry gravitation force, could throw off its chains, become a free agent, and, under the name of "gravitational radiation," carry energy from place to place over and above any energy carried by electromagnetic waves — an effect for which Joseph H. Taylor, L. A. Fowler, P. M. McCulloch, and their Arecibo Observatory colleagues in December 1978 announced impressive evidence.[17]

One does not need to go into the theory of gravitationally collapsed objects or the evidence we have today, some impressive, some less convincing, for black holes: one of some ten solar masses in the constellation Cygnus; others in the range of a hundred or a thousand solar masses at the centers of five of the star clusters in our galaxy; one about four million times as massive as the sun at the center of the Milky Way; and one with a mass of about five billion suns in the center of the galaxy M87.

The collapse at the center of a black hole marks a third "gate of time,"[18] additional to the big bang and the big crunch. Einstein tried to escape all three. Two years after general relativity, Einstein was already applying it to cosmology. He gave reasons to regard the universe as closed and qualitatively similar to a three sphere, the three-dimensional generalization of the surface of a rubber balloon. To his surprise, he found that the universe is dynamic and not static.

Einstein could not accept this result. First, he found fault with Alexander Friedmann's mathematics. Then he retracted this criticism, and looked for the fault in his own theory of gravitation. It turned out there was no natural way to change that theory. The arguments of simplicity and correspondence in the appropriate limit with the Newtonian theory of gravitation left no alternative. There being no natural way to change the theory, he looked for the least

17. L. A. Fowler, P. M. McCulloch, and J. H. Taylor, "Measurement of General Relativistic Effects in the Binary Pulsar PSR 1913 + 16," *Nature*, 277 (8 February 1979) 437–40.

18. J. A. Wheeler, "Genesis and Observership," in *Foundational Problems in the Special Sciences*, ed. R. E. Butts and K. J. Hintikka (Reidel: Dordrecht, 1977), p. 11.

unnatural way he could find to alter it. He introduced a so-called "cosmological term" with the sole point and purpose to hold the universe static. A decade later, Edwin Hubble, working at Mount Wilson Observatory, gave convincing evidence that the universe is actually expanding. Thereafter, Einstein remarked that the cosmological term "was the biggest blunder of my life."[19] Today, looking back, we can forgive him his blunder and give him the credit for the theory of gravitation that predicted the expansion. Of all the great predictions that science has ever made over the centuries, each of us has his own list of spectaculars, but among them all was there ever one greater than this, to predict, and predict correctly, and predict against all expectation, a phenomenon so fantastic as the expansion of the universe? When did nature ever grant man greater encouragement to believe he will someday understand the mystery of existence?

Why did Einstein in the beginning reject his own greatest discovery? Why did he feel that the universe should go on from everlasting to everlasting, when to all brought up in the Judeo-Christian tradition an original creation is the natural concept? I am indebted to Professor Hans Küng for suggesting an important influence on Einstein from his hero Spinoza. Why was twenty-four-year-old Spinoza excommunicated in 1656 from the synagogue in Amsterdam? Because he denied the doctrine of an original creation. What was the difficulty with that doctrine? In all that nothingness before creation where could that clock sit that should tell the universe when to come into being!

Today we have a little less difficulty with this point. We do not escape by saying that the universe goes through cycle after cycle of big bang and collapse, world without end. There is not the slightest warrant in general relativity for such a way of speaking. On the contrary, it provides no place whatsoever for a before before the big bang or an after after the big crunch. Quantum theory goes further. It tells us that however permissible it is to speak about space, it is not permissible to speak in other than approximate terms of spacetime. To do so would violate the uncertainty principle — as that principle applies to the dynamics of geometry. No, when it comes to small distances either in the here and the now or in the most extreme stages of gravitational collapse, spacetime loses all meaning, and time itself is not an ultimate category in the description of nature. No one who

19. G. Gamow, *My World Line* (New York: Viking, 1970), p. 44.

wrestles with the three gates of time, our greatest heritage of paradox — and of promise — from general relativity can escape the all-pervasive influence of the quantum.

Spinoza's influence on his thinking about cosmology Einstein could shake off — but not Spinoza's deterministic outlook. Proposition XXIX in *The Ethics* of Spinoza states: "Nothing in the universe is contingent, but all things are conditioned to exist and operate in a particular manner by the necessity of divine nature."[20] Einstein accepted determinism in his mind, his heart, his very bones.

Who then was first clearly to recognize that the real world, and the world of the quantum, is a world of chance and unpredictability? Einstein himself!

Why did Einstein, who in the beginning with Max Planck and Niels Bohr had done so much to give quantum physics to the world, in the end stand out so strongly and so lonesomely against the central point? What other explanation is there than this "set" he had received from Spinoza?

The early quantum work of Bohr and Einstein is almost a duet. Einstein, 1905: The energy of light is carried from place to place as quanta of energy, accidental in time and space in their arrival. Bohr, 1913: The atom is characterized by stationary states, and the difference in energy between one and another is given off in a light quantum. Einstein, 1916: The processes of light emission and light absorption are governed by the laws of chance, but satisfy the principle of detailed balance. Bohr, 1927: Complementarity prevents a detailed description in space and time of what goes on in the act of emission. Here Bohr and Einstein parted company. Einstein spoke against Einstein. The Einstein who in 1915 said there was no escape from the laws of chance was insisting by 1916, as he did all the rest of his life, against the evidence and against the views of his greatest colleagues, that "God does not [play] dice."[21]

If an army is being defeated it can still, by a sufficiently skillful rear-guard action, have an important influence on the outcome. No one who in all the great history of the quantum contested with Niels Bohr did more to sharpen and strengthen Bohr's position than Einstein. Never in recent centuries was there a dialogue between two

20. B. Spinoza, *Die Ethik*, Part One, Proposition XXIX (Hamburg: F. Meiner, 1955).

21. A. Einstein, *Albert Einstein und Max Born, Briefwechsel, 1916–1955, Kommentiert von Max Born* (München: Nymphenburg, 1969), pp. 129–30.

greater men over a longer period on a deeper issue at a higher level of colleagueship, nor a nobler theme for playwright, poet, or artist. From their earliest encounter, Einstein liked Bohr, writing him on May 2, 1920, "I am studying your great works — and when I get stuck anywhere — now have the pleasure of seeing your friendly young face before me smiling and explaining."[22] Bohr viewed Einstein with admiration and warm regard. Let him who will read Bohr's account of the famous dialogue, even today unsurpassed for its comprehensive articulation of the central issues. Who knows what the quantum means who does not know the friendly but deadly serious battles fought and won on the double-slit experiment, on the possibilities for weighing a photon, on the Einstein-Podolsky-Rosen experiment, and on the danger associated with unguarded use of the word "reality"? To help to clarify the issues brought up in the later years of the great dialogue, Bohr found himself forced to introduce the word "phenomenon"[23] to describe an elementary quantum process "brought to a close by an irreversible act of amplification."[24] Thanks to that word, brought in to withstand the criticism of Einstein, we have learned in our own time to state the central lesson of the quantum in a single simple sentence, "No elementary phenomenon is a phenomenon until it is an observed phenomenon."[25]

How could the correctness of quantum theory be by now so widely accepted, and its decisive point so well perceived, if there had been no great figure, no Einstein, to draw the embers of unease together in a single flame and thereby drive Bohr to that fuller formulation of the central lesson which he at last achieved?

If the quantum and the gates of time are the strongest features of this strange universe, and if they shall prove in time to come the doorways to that deeper view for which Einstein searched, mankind will forever remember with gratitude his absolutely decisive involvement with both.

No one who is a professor and receives his support from the

22. Letter to N. Bohr, 2 May 1920.

23. N. Bohr, "Discussion with Einstein on Epistemological Problems in Atomic Physics," in *Einstein: Philosopher-Scientist*, ed. P. A. Schilpp (Evanston, Ill.: Library of Living Philosophers, 1949), p. 238.

24. N. Bohr, *Atomic Physics and Human Knowledge* (New York: Wiley, 1958), pp. 73, 88.

25. J. A. Wheeler, "Frontiers of Time," in *Rendicotti della Scuola Internazionale di Fisica "Enrico Fermi,"* LXXII Corso, Problems in the Foundations of Physics, ed. N. Toraldo di Francia and Bas van Fraassen (Amsterdam: North-Holland, 1979), pp. 395–497.

larger community can rightly be unmindful of his obligations to it. He must speak to the higher values of all insofar as he is qualified and able to do so. Burden though it was for Einstein to take on this extra duty, he did it to the best of his ability. What he defended were no whims, no lightly held fancies, but goals he held and deeply desired for the world. If in this undertaking he had some of the character of an Old Testament prophet, he also had all of the eloquence. Statements from Einstein created an audience, and the audience created the pressure for more statements. What is long, Einstein felt, is lost. Pith and pungency were the points of his pronouncements. Who does not know the causes for which he stood! Whoever admires greatness, let him read Einstein's words about the goals and the greatness of recently departed colleagues, as well as heroes out of the deeper past. For social justice and social responsibility, Einstein spoke up time and again: "A hundred times every day I remind myself that my inner and outer life are based on the labors of other men, living and dead, and that I must exert myself in order to give in the same measure as I have received and am still receiving."[26] He stressed the necessity of a political system that does not rely on coercion if people are to contribute all that lies in them to achieve.

He expressed admiration for the system of social care, going back to Bismarck, that makes provision for the individual in case of illness or need. Living through the tragedy of two world wars, he protested many times about the wastefulness of war: lives lost, hatred engendered, and values perverted; but when it came to a choice between war or freedom and justice, he spoke for freedom and justice. He refused the invitation to become the first president of Israel, but he worked after that declination as effectively as before for the welfare of a unique community, remarking, "The pursuit of knowledge for its own sake, an almost fanatical love of justice and the desire for personal independence — these are the features of the Jewish tradition which make me thank my stars that I belong to it."[27]

Things did not go in the world as Einstein had hoped. Things did not go in physics as he had desired. Determinism stood in ruins. His search for a unified geometric theory of all the forces of nature came to nothing — though today, with a new and wider concept of what geometry is, in the sense of a so-called "gauge theory," mar-

26. A. Einstein, *Mein Weltbild,* trans. A. Harris, in *The World As I See It* (New York: Philosophical Library, 1949), p. 90.
27. *Ibid.,* p. 1.

velous new progress is now being made toward his dream of unification. He left us in general relativity with an ideal for a physical theory that has never been surpassed. He showed a unique talent for finding the central point in every subject to which his philosophical antecedents gave right of entry. He did as much as any man who ever lived to make us face up to the central mysteries of this strange world.

Einstein worked with all his force to the very end. In his last days he had a tired face. Everything that he had to give he had given for his causes, and among them that greatest of causes, the goal toward which he had climbed so high, that snowy peak whose light today shines brighter than ever: "A completely harmonious account of existence."[28]

As we look up at the distant intervening craggy slope, we are amazed suddenly to make out the faint sound of a high far-off violin. Then out of the valley behind and below us comes an answering burst of song, young voices all. They chorus of the loftiness of the peak, the danger of the climb, and the greatness of the climber, the man of peace with the white hair. He no longer belongs to any one country, any one group, any one age, we hear them singing, but to all friends of the future. Least of all, they tell us, does Einstein anymore belong to Einstein. He belongs to the world.

28. A. Einstein, *Essays in Science* (New York: Philosophical Library, 1934), p. 114.

# ALBERT EINSTEIN

■

The curious little manuscript from which the following excerpt is taken, composed at the behest of Paul Arthur Schilpp of Northwestern University, was the nearest that Albert Einstein (1879–1955) came to writing an autobiography. Befittingly for a man who viewed his chosen career as an escape "from the chains of the 'merely-personal' " and who insisted that "the essential in the being of a man of my type lies precisely in *what* he thinks and *how* he thinks, not in what he does or suffers," Einstein had little to say about his life but a great deal to say about his thought.

# *Autobiographical Notes*

HERE I SIT IN ORDER to write, at the age of 67, something like my own obituary. I am doing this not merely because Dr. [Paul] Schilpp has persuaded me to do it; but because I do, in fact, believe that it is a good thing to show those who are striving alongside of us, how one's own striving and searching appears to one in retrospect. After some reflection, I felt how insufficient any such attempt is bound to be. For, however brief and limited one's working life may be, and however predominant may be the ways of error, the exposition of that which is worthy of communication does nonetheless not come easy — today's person of 67 is by no means the same as was the one of 50, of 30, or of 20. Every reminiscence is colored by today's being what it is, and therefore by a deceptive point of view. This consideration could very well deter. Nevertheless much can be lifted out of one's own experience which is not open to another consciousness.

Even when I was a fairly precocious young man the nothingness of the hopes and strivings which chases most men restlessly through life came to my consciousness with considerable vitality. Moreover, I soon discovered the cruelty of that chase, which in those years was much more carefully covered up by hypocrisy and glittering words than is the case today. By the mere existence of his stomach everyone was condemned to participate in that chase. Moreover, it was possible to satisfy the stomach by such participation, but not man in so far as he is a thinking and feeling being. As the first way out there was religion, which is implanted into every child by way of the traditional education-machine. Thus I came — despite the fact that I was the son of entirely irreligious (Jewish) parents — to a deep religiosity, which, however, found an abrupt ending at the age of 12. Through the reading of popular scientific books I soon reached the conviction that much in the stories of the Bible could not be true. The conse- quence was a positively fanatic [orgy of] freethinking coupled with the impression that youth is intentionally being deceived by the state through lies; it was a crushing impression. Suspicion against every kind of authority grew out of this experience, a skeptical attitude towards the convictions which were alive in any specific social envi- ronment — an attitude which has never again left me, even though later on, because of a better insight into the causal connections, it lost some of its original poignancy.

It is quite clear to me that the religious paradise of youth, which was thus lost, was a first attempt to free myself from the chains of the "merely-personal," from an existence which is dominated by wishes, hopes and primitive feelings. Out yonder there was this huge world, which exists independently of us human beings and which stands before us like a great, eternal riddle, at least partially acces- sible to our inspection and thinking. The contemplation of this world beckoned like a liberation, and I soon noticed that many a man whom I had learned to esteem and to admire had found inner freedom and security in devoted occupation with it. The mental grasp of this extrapersonal world within the frame of the given pos- sibilities swam as highest aim half consciously and half unconsciously before my mind's eye. Similarly motivated men of the present and of the past, as well as the insights which they had achieved, were the friends which could not be lost. The road to this paradise was not as comfortable and alluring as the road to the religious paradise; but it has proved itself as trustworthy, and I have never regretted having chosen it.

What I have here said is true only within a certain sense, just as a drawing consisting of a few strokes can do justice to a compli- cated object, full of perplexing details, only in a very limited sense. If an individual enjoys well-ordered thoughts, it is quite possible that this side of his nature may grow more pronounced at the cost of other sides and thus may determine his mentality in increasing degree. In this case it is well possible that such an individual in retrospect sees a uniformly systematic development, whereas the actual experience takes place in kaleidoscopic particular situations. The manifoldness of the external situations and the narrowness of the momentary content of consciousness bring about a sort of atomizing of the life of every human being. In a man of my type the turning-point of the development lies in the fact that gradually the major interest disengages itself to a far-reaching degree from the momentary and the merely personal and turns towards the striving for a mental grasp of things. Looked at from this point of view the above schematic remarks contain as much truth as can be uttered in such brevity.

What, precisely, is "thinking"? When, at the reception of sense-impressions, memory-pictures emerge, this is not yet "thinking." And when such pictures form series, each member of which calls forth another, this too is not yet "thinking." When, however, a certain picture turns up in many such series, then — precisely through such return — it becomes an ordering element for such series, in that it connects series which in themselves are unconnected. Such an element becomes an instrument, a concept. I think that the transition from free association or "dreaming" to thinking is characterized by the more or less dominating rôle which the "concept" plays in it. It is by no means necessary that a concept must be connected with a sensorily cognizable and reproducible sign (word); but when this is the case thinking becomes by means of that fact communicable.

With what right — the reader will ask — does this man operate so carelessly and primitively with ideas in such a problematic realm without making even the least effort to prove anything? My defense: all our thinking is of this nature of a free play with concepts; the justification for this play lies in the measure of survey over the experience of the senses which we are able to achieve with its aid. The concept of "truth" cannot yet be applied to such a structure; to my thinking this concept can come in question only when a far-reaching agreement *(convention)* concerning the elements and rules of the game is already at hand.

For me it is not dubious that our thinking goes on for the most part without use of signs (words) and beyond that to a considerable degree unconsciously. For how, otherwise, should it happen that sometimes we "wonder" quite spontaneously about some experience? This "wondering" seems to occur when an experience comes into conflict with a world of concepts which is already sufficiently fixed in us. Whenever such a conflict is experienced hard and intensively it reacts back upon our thought world in a decisive way. The development of this thought world is in a certain sense a continuous flight from "wonder."

A wonder of such nature I experienced as a child of 4 or 5 years, when my father showed me a compass. That this needle behaved in such a determined way did not at all fit into the nature of events, which could find a place in the unconscious world of concepts (effect connected with direct "touch"). I can still remember — or at least believe I can remember — that this experience made a deep and lasting impression upon me. Something deeply hidden had to be behind things. What man sees before him from infancy causes no reaction of this kind; he is not surprised over the falling of bodies, concerning wind and rain, nor concerning the moon or about the fact that the moon does not fall down, nor concerning the differences between living and non-living matter.

At the age of 12 I experienced a second wonder of a totally different nature: in a little book dealing with Euclidian plane geometry, which came into my hands at the beginning of a schoolyear. Here were assertions, as for example the intersection of the three altitudes of a triangle in one point, which — though by no means evident — could nevertheless be proved with such certainty that any doubt appeared to be out of the question. This lucidity and certainty made an indescribable impression upon me. That the axiom had to be accepted unproved did not disturb me. In any case it was quite sufficient for me if I could peg proofs upon propositions the validity of which did not seem to me to be dubious. For example I remember that an uncle told me the Pythagorean theorem before the holy geometry booklet had come into my hands. After much effort I succeeded in "proving" this theorem on the basis of the similarity of triangles; in doing so it seemed to me "evident" that the relations of the sides of the right-angled triangles would have to be completely determined by one of the acute angles. Only something which did not in similar fashion seem to be "evident" appeared to me to be in need of any proof at all. Also, the objects with which geometry deals seemed to be of no different type than the objects of sensory perception, "which can be seen and touched." This primitive idea, which probably also lies at the bottom of the well known Kantian problematic concerning the possibility of "synthetic judgments *a priori*," rests obviously upon the fact that the relation of geometrical concepts to objects of direct experience (rigid rod, finite interval, etc.) was unconsciously present.

If thus it appeared that it was possible to get certain knowledge of the objects of experience by means of pure thinking, this "wonder" rested upon an error. Nevertheless, for anyone who experiences it for the first time, it is marvellous enough that man is capable at all to reach such a degree of certainty and purity in pure thinking as the Greeks showed us for the first time to be possible in geometry.

Now that I have allowed myself to be carried away sufficiently to interrupt my scantily begun obituary, I shall not hesitate to state here in a few sentences my epistemological credo, although in what precedes something has already incidentally been said about this. This credo actually evolved only much later and very slowly and does not correspond with the point of view I held in younger years.

I see on the one side the totality of sense-experiences, and, on the other, the totality of the concepts and propositions which are laid down in books. The relations between the concepts and propositions among themselves and each other are of a logical nature, and the business of logical thinking is strictly limited to the achievement of the connection between concepts and propositions among each other according to firmly laid down rules, which are the concern of logic. The concepts and propositions get "meaning," viz., "content," only through their connection with sense-experiences. The connection of the latter with the former is purely intuitive, not itself of a logical nature. The degree of certainty with which this relation, viz., intuitive connection, can be undertaken, and nothing else, differentiates empty phantasy from scientific "truth." The system of concepts is a creation of man together with the rules of syntax, which constitute the structure of the conceptual systems. Although the conceptual systems are logically entirely arbitrary, they are bound by the aim to permit the most nearly possible certain (intuitive) and complete co-ordination with the totality of sense-experiences; secondly they aim at greatest possible sparsity of their logically independent elements (basic concepts and axioms), i.e., undefined concepts and underived [postulated] propositions.

A proposition is correct if, within a logical system, it is deduced according to the accepted logical rules. A system has truth-content according to the certainty and completeness of its co-ordination-possibility to the totality of experience. A correct proposition borrows its "truth" from the truth-content of the system to which it belongs.

A remark to the historical development. Hume saw clearly that certain concepts, as for example that of causality, cannot be deduced from the material of experience by logical methods. Kant, thoroughly convinced of the indispensability of certain concepts, took them — just as they are selected — to be the necessary premises of every kind of thinking and differentiated them from concepts of empirical origin. I am convinced, however, that this differentiation is erroneous, i.e., that it does not do justice to the problem in a natural way. All concepts, even those which are closest to experience, are from the point of view of logic freely chosen conventions, just as is the case with the concept of causality, with which this problematic concerned itself in the first instance.

And now back to the obituary. At the age of 12–16 I familiarized myself with the elements of mathematics together with the principles of differential and integral calculus. In doing so I had the good fortune of hitting on books which were not too particular in their logical rigour, but which made up for this by permitting the main thoughts to stand out clearly and synoptically. This occupation was, on the whole, truly fascinating; climaxes were reached whose impression could easily compete with that of elementary geometry — the basic idea of analytical geometry, the infinite series, the concepts of differential and integral. I also had the good fortune of getting to know the essential results and methods of the entire field of the natural sciences in an excellent popular exposition, which limited itself almost throughout to qualitative aspects (Bernstein's *People's Books on Natural Science*, a work of 5 or 6 volumes), a work which I read with breathless attention. I had also already studied some theoretical physics when, at the age of 17, I entered the Polytechnic Institute of Zürich as a student of mathematics and physics.

There I had excellent teachers (for example, Hurwitz, Minkowski), so that I really could have gotten a sound mathematical education. However, I worked most of the time in the physical laboratory, fascinated by the direct contact with experience. The balance of the time I used in the main in order to study at home the works of Kirchhoff, Helmholtz, Hertz, etc. The fact that I neglected mathematics to a certain extent had its cause not merely in my stronger interest in the natural sciences than in mathematics but also in the following strange experience. I saw that mathematics was split up into numerous specialities, each of which could easily absorb the short lifetime granted to us. Consequently I saw myself in the position of Buridan's ass which was unable to decide upon any specific bundle of hay. This was obviously due to the fact that my intuition was not strong enough in the field of mathematics in order to differentiate clearly the fundamentally important, that which is really basic, from the rest of the more or less dispensable erudition. Beyond this, however, my interest in the knowledge of nature was also unqualifiedly stronger; and it was not clear to me as a student that the approach to a more profound knowledge of the basic principles of physics is tied up with the most intricate mathematical methods. This dawned upon me only gradually after years of independent scientific work. True enough, physics also was divided into separate fields, each of which was capable of devouring a short lifetime of work without having satisfied the hunger for deeper knowledge. The mass of insufficiently connected experimental data was overwhelming here also. In this field, however, I soon learned to scent out that which was able to lead to fundamentals and to turn aside from everything else, from the multitude of things which clutter up the mind and divert it from the essential. The hitch in this was, of course, the fact that one had to cram all this stuff into one's mind for the examinations, whether one liked it or not. This coercion had such a deterring effect [upon me] that, after I had passed the final examination, I found the consideration of any scientific problems distasteful to me for an entire year. In justice I must add, moreover, that in Switzerland we had to suffer far less under such coercion, which smothers every truly scientific impulse, than is the case in many another locality. There were altogether only two examinations; aside from these, one could just about do as one pleased. This was especially the case if one had a friend, as did I, who attended the lectures regularly and who worked over their content conscientiously. This gave one freedom in the choice of pursuits until a few months before the examination, a freedom which I enjoyed to a great extent and have gladly taken into the bargain the bad conscience connected with it as by far the lesser evil. It is, in fact, nothing short of a miracle that the modern methods of instruction have not yet entirely strangled the holy curiosity of inquiry; for this delicate little plant, aside from stimulation, stands mainly in need of freedom; without this it goes to wreck and ruin without fail. It is a very grave mistake to think that the enjoyment of seeing and searching can be promoted by means of coercion and a sense of duty. To the contrary, I believe that it would be possible to rob even a healthy beast of prey of its voraciousness, if it were possible, with the aid of a whip, to force the beast to devour continuously, even when not hungry, especially if the food, handed out under such coercion, were to be selected accordingly. . . .

Now to the field of physics as it presented itself at that time. In spite of all the fruitfulness in particulars, dogmatic rigidity prevailed in matters of principles: In the beginning (if there was such a thing) God created Newton's laws of motion together with the necessary masses and forces. This is all; everything beyond this follows from the development of appropriate mathematical methods by means of deduction. What the nineteenth century achieved on the strength of this basis, especially through the application of the partial differential equations, was bound to arouse the admiration of every receptive person. Newton was probably first to reveal, in his theory of sound-transmission, the efficacy of partial differential equations. Euler had already created the foundation of hydrodynamics. But the more precise development of the mechanics of discrete masses, as the basis of all physics, was the achievement of the 19th century. What made the greatest impression upon the student, however, was less the technical construction of mechanics or the solution of complicated problems than the achievements of mechanics in areas which apparently had nothing to do with mechanics: the mechanical theory of light, which conceived of light as the wave-motion of a quasi-rigid elastic ether, and above all the kinetic theory of gases: — the independence of the specific heat of monatomic gases of the atomic weight, the derivation of the equation of state of a gas and its relation to the specific heat, the kinetic theory of the dissociation of gases, and above all the quantitative connection of viscosity, heat-conduction and diffusion of gases, which also furnished the absolute magnitude of the atom. These results supported at the same time mechanics as the foundation of physics and of the atomic hypothesis, which latter was already firmly anchored in chemistry. However, in chemistry only the ratios of the atomic masses played any rôle, not their absolute magnitudes, so that atomic theory could be viewed more as a visualizing symbol than as knowledge concerning the factual construction of matter. Apart from this it was also of profound interest that the statistical theory of classical mechanics was able to deduce the basic laws of thermodynamics, something which was in essence already accomplished by Boltzmann.

We must not be surprised, therefore, that, so to speak, all physicists of the last century saw in classical mechanics a firm and final foundation for all physics, yes, indeed, for all natural science, and that they never grew tired in their attempts to base Maxwell's theory of electro-magnetism, which, in the meantime, was slowly beginning to win out, upon mechanics as well. Even Maxwell and H. Hertz, who in retrospect appear as those who demolished the faith in mechanics as the final basis of all physical thinking, in their conscious thinking adhered throughout to mechanics as the secured basis of physics. It was Ernst Mach who, in his *History of Mechanics,* shook this dogmatic faith; this book exercised a profound influence upon me in this regard while I was a student. I see Mach's greatness in his incorruptible skepticism and independence; in my younger years, however, Mach's epistemological position also influenced me very greatly, a position which today appears to me to be essentially untenable. For he did not place in the correct light the essentially constructive and speculative nature of thought and more especially of scientific thought; in consequence of which he condemned theory on precisely those points where its constructive-speculative character unconcealably comes to light, as for example in the kinetic atomic theory.

Before I enter upon a critique of mechanics as the foundation of physics, something of a broadly general nature will first have to be said concerning the points of view according to which it is possible to criticize physical theories at all. The first point of view is obvious: the theory must not contradict empirical facts. However evident this demand may in the first place appear, its application turns out to be quite delicate. For it is often, perhaps even always, possible to adhere to a general theoretical foundation by securing the adaptation of the theory to the facts by means of artificial additional assumptions. In any case, however, this first point of view is concerned with the confirmation of the theoretical foundation by the available empirical facts.

The second point of view is not concerned with the relation to the material of observation but with the premises of the theory itself, with what may briefly but vaguely be characterized as the "naturalness" or "logical simplicity" of the premises (of the basic concepts and of the relations between these which are taken as a basis). This point of view, an exact formulation of which meets with great difficulties, has played an important rôle in the selection and evaluation of theories since time immemorial. The problem here is not simply one of a kind of enumeration of the logically independent premises (if anything like this were at all unequivocally possible), but that of a kind of reciprocal weighing of incommensurable qualities. Furthermore, among theories of equally "simple" foundation that one is to be taken as superior which most sharply delimits the qualities of systems in the abstract (i.e., contains the most definite claims). Of the "realm" of theories I need not speak here, inasmuch as we are confining ourselves to such theories whose object is the *totality* of all physical appearances. The second point of view may briefly be characterized as concerning itself with the "inner perfection" of the theory, whereas the first point of view refers to the "external confirmation." The following I reckon as also belonging to the "inner perfection" of a theory: We prize a theory more highly if, from the logical standpoint, it is not the result of an arbitrary choice among theories which, among themselves, are of equal value and analogously constructed.

The meager precision of the assertions contained in the last two paragraphs I shall not attempt to excuse by lack of sufficient printing space at my disposal, but confess herewith that I am not, without more ado [immediately], and perhaps not at all, capable to replace these hints by more precise definitions. I believe, however, that a sharper formulation would be possible. In any case it turns out that among the "augurs" there usually is agreement in judging the "inner perfection" of the theories and even more so concerning the "degree" of "external confirmation."

And now to the critique of mechanics as the basis of physics.

From the first point of view (confirmation by experiment) the incorporation of wave-optics into the mechanical picture of the world was bound to arouse serious misgivings. If light was to be interpreted as undulatory motion in an elastic body (ether), this had to be a medium which permeates everything; because of the transversality of the lightwaves in the main similar to a solid body, yet incompressible, so that longitudinal waves did not exist. This ether had to lead a ghostly existence alongside the rest of matter, inasmuch as it seemed to offer no resistance whatever to the motion of "ponderable" bodies. In order to explain the refraction-indices of transparent bodies as well as the processes of emission and absorption of radiation, one would have had to assume complicated reciprocal actions between the two types of matter, something which was not even seriously tried, let alone achieved.

Furthermore, the electromagnetic forces necessitated the introduction of electric masses, which, although they had no noticeable inertia, yet interacted with each other, and whose interaction was, moreover, in contrast to the force of gravitation, of a polar type.

The factor which finally succeeded, after long hesitation, to bring the physicists slowly around to give up the faith in the possibility that all of physics could be founded upon Newton's mechanics, was the electrodynamics of Faraday and Maxwell. For this theory and its confirmation by Hertz's experiments showed that there are electromagnetic phenomena which by their very nature are detached from every ponderable matter — namely the waves in empty space which consist of electromagnetic "fields." If mechanics was to be maintained as the foundation of physics, Maxwell's equations had to be interpreted mechanically. This was zealously but fruitlessly attempted, while the equations were proving themselves fruitful in mounting degree. One got used to operating with these fields as independent substances without finding it necessary to give one's self an account of their mechanical nature; thus mechanics as the basis of physics was being abandoned, almost unnoticeably, because its adaptability to the facts presented itself finally as hopeless. Since then there exist two types of conceptual elements, on the one hand, material points with forces at a distance between them, and, on the other hand, the continuous field. It presents an intermediate state in physics without a uniform basis for the entirety, which — although unsatisfactory — is far from having been superseded. . . .

Now for a few remarks to the critique of mechanics as the foundation of physics from the second, the "interior," point of view. In today's state of science, i.e., after the departure from the mechanical foundation, such critique has only an interest in method left. But such a critique is well suited to show the type of argumentation which, in the choice of theories in the future, will have to play an all the greater rôle the more the basic concepts and axioms distance themselves from what is directly observable, so that the confrontation of the implications of theory by the facts becomes constantly more difficult and more drawn out. First in line to be mentioned is Mach's argument, which, however, had already been clearly recognized by Newton (bucket experiment). From the standpoint of purely geometrical description all "rigid" co-ordinate systems are among themselves logically equivalent. The equations of mechanics (for example this is already true of the law of inertia) claim validity only when referred to a specific class of such systems, i.e., the "inertial systems." In this the co-ordinate system as bodily object is without any significance. It is necessary, therefore, in order to justify the necessity of the specific choice, to look for something which lies outside of the objects (masses, distances) with which the theory is concerned. For this reason "absolute space" as originally determinative was quite explicitly introduced by Newton as the omnipresent active participant in all mechanical events; by "absolute" he obviously means uninfluenced by the masses and by their motion. What makes this state of affairs appear particularly offensive is the fact that there are supposed to be infinitely many inertial systems, relative to each other in uniform translation, which are supposed to be distinguished among all other rigid systems.

Mach conjectures that in a truly rational theory inertia would have to depend upon the interaction of the masses, precisely as was true for Newton's other forces, a conception which for a long time I considered as in principle the correct one. It presupposes implicitly, however, that the basic theory should be of the general type of Newton's mechanics: masses and their interaction as the original concepts. The attempt at such a solution does not fit into a consistent field theory, as will be immediately recognized.

How sound, however, Mach's critique is in essence can be seen particularly clearly from the following analogy. Let us imagine people construct a mechanics, who know only a very small part of the earth's surface and who also cannot see any stars. They will be inclined to ascribe special physical attributes to the vertical dimension of space (direction of the acceleration of falling bodies) and, on the ground of such a conceptual basis, will offer reasons that the earth is in most places horizontal. They might not permit themselves to be influenced by the argument that as concerns the geometrical properties space is isotrope and that it is therefore supposed to be unsatisfactory to postulate basic physical laws, according to which there is supposed to be a preferential direction; they will probably be inclined (analogously to Newton) to assert the absoluteness of the vertical, as proved by experience as something with which one simply would have to come to terms. The preference given to the vertical over all other spatial directions is precisely analogous to the preference given to inertial systems over other rigid co-ordination systems.

Now to [a consideration of] other arguments which also concern themselves with the inner simplicity, i.e., naturalness, of mechanics. If one puts up with the concepts of space (including geometry) and time without critical doubts, then there exists no reason to object to the idea of action-at-a-distance, even though such a concept is unsuited to the ideas which one forms on the basis of the raw experience of daily life. However, there is another consideration which causes mechanics, taken as the basis of physics, to appear as primitive. Essentially there exist two laws:

1. the law of motion
2. the expression for force or potential energy.

The law of motion is precise, although empty, as long as the expression for the forces is not given. In postulating the latter, however, there exists great latitude for arbitrary [choice], especially if one omits the demand, which is not very natural in any case, that the forces depend only on the co-ordinates (and, for example, not on their differential quotients with respect to time). Within the framework of theory alone it is entirely arbitrary that the forces of gravitation (and electricity), which come from one point, are governed by the potential function $(1/r)$. Additional remark: it has long been known that this function is the central-symmetrical solution of the simplest (rotation-invariant) differential equation $\Delta \phi = o$; it would therefore have been a suggestive idea to regard this as a sign that this function is to be regarded as determined by a law of space, a procedure by which the arbitrariness in the choice of the law of energy would have been removed. This is really the first insight which suggests a turning away from the theory of distant forces, a development which — prepared by Faraday, Maxwell and Hertz — really begins only later on under the external pressure of experimental data.

I would also like to mention, as one internal asymmetry of this theory, that the inert mass occurring in the law of motion also appears in the expression for the gravitational force, but not in the expression for the other forces. Finally I would like to point to the fact that the division of energy into two essentially different parts, kinetic and potential energy, must be felt as unnatural; H. Hertz felt this as so disturbing that, in his very last work, he attempted to free mechanics from the concept of potential energy (i.e., from the concept of force). . . .

Enough of this. Newton, forgive me; you found the only way which, in your age, was just about possible for a man of highest thought — and creative power. The concepts, which you created, are even today still guiding our thinking in physics, although we now know that they will have to be replaced by others farther removed from the sphere of immediate experience, if we aim at a profounder understanding of relationships.

"Is this supposed to be an obituary?" the astonished reader will likely ask. I would like to reply: essentially yes. For the essential in the being of a man of my type lies precisely in *what* he thinks and *how* he thinks, not in what he does or suffers. . . .

# ALBERT EINSTEIN

∎

Einstein wrote generous and perceptive tributes to many of the scientists he respected. This one was for Max Planck, the founder of quantum physics.

# A Tribute to Max Planck

... MANY KINDS OF MEN devote themselves to science, and not all for the sake of science herself. There are some who come into her temple because it offers them the opportunity to display their particular talents. To this class of men science is a kind of sport in the practice of which they exult, just as an athlete exults in the exercise of his muscular prowess. There is another class of men who come into the temple to make an offering of their brain pulp in the hope of securing a profitable return. These men are scientists only by the chance of some circumstance which offered itself when making a choice of career. If the attending circumstances had been different, they might have become politicians or captains of business. Should an angel of God descend and drive from the temple of science all those who belong to the categories I have mentioned, I fear the temple would be nearly emptied. But a few worshippers would still remain — some from former times and some from ours. To these latter belongs our Planck. And that is why we love him.

# C. P. SNOW

■

Ernest Rutherford (1871–1937) was one of the most forceful and colorful of the scientists who pieced together our modern conception of the atom. C. P. Snow (1905–1980) brings to this lively portrait of Rutherford his combined gifts as a physicist and a novelist.

# *Rutherford*

IN 1923, AT THE MEETING of the British Association for the Advancement of Science in Liverpool, Rutherford announced, at the top of his enormous voice: "We are living in the heroic age of physics." He went on saying the same thing, loudly and exuberantly, until he died, fourteen years later.

The curious thing was, all he said was absolutely true. There had never been such a time. The year 1932 was the most spectacular year in the history of science. Living in Cambridge, one could not help picking up the human, as well as the intellectual, excitement in the air. James Chadwick, grey-faced after a fortnight of work with three hours' sleep a night, telling the Kapitsa Club (to which any young man was so proud to belong) how he had discovered the neutron; P.M.S. Blackett, the most handsome of men, not quite so authoritative as usual, because it seemed too good to be true, showing plates which demonstrated the existence of the positive electron; John Cockcroft, normally about as much given to emotional display as the Duke of Wellington, skimming down King's Parade and saying to anyone whose face he recognized: "We've split the atom! We've split the atom!"

It meant an intellectual climate different in kind from anything else in England at the time. The tone of science was the tone of Rutherford: magniloquently boastful — boastful because the major discoveries were being made — creatively confident, generous, argumentative, lavish, and full of hope. The tone differed from the tone of literary England as much as Rutherford's personality differed from that of T. S. Eliot. During the twenties and thirties Cambridge was the metropolis of experimental physics for the entire world. Even in the late nineteenth century, during the professorships of Clerk Maxwell and J. J. Thomson, it had never quite been that. "You're always at the crest of the wave," someone said to Rutherford. "Well, after all, I made the wave, didn't I?" Rutherford replied.

I remember seeing him a good many times before I first spoke to him. I was working on the periphery of physics at the time, and so didn't come directly under him. I already knew that I wanted to write novels, and that was how I should finish, and this gave me a kind of ambivalent attitude to the scientific world; but, even so, I could not avoid feeling some sort of excitement, or enhancement of interest, whenever I saw Rutherford walking down Free School Lane.

He was a big, rather clumsy man, with a substantial bay-window that started in the middle of the chest. I should guess that he was less muscular than at first sight he looked. He had large staring blue eyes and a damp and pendulous lower lip. He didn't look in the least like an intellectual. Creative people of his abundant kind never do, of course, but all the talk of Rutherford looking like a farmer was unperceptive nonsense. His was really the kind of face and physique that often goes with great weight of character and gifts. It could easily have been the soma of a great writer. As he talked to his companions in the street, his voice was three times as loud as any of theirs, and his accent was bizarre. In fact, he came from the very poor: his father was an odd-job man in New Zealand and the son of a Scottish emigrant. But there was nothing Antipodean or Scottish about Rutherford's accent; it sounded more like a mixture of West Country and Cockney.

In my first actual meeting with him, perhaps I could be excused for not observing with precision. It was early in 1930; I had not yet been elected a Fellow of my own college, and so had put in for the Stokes studentship at Pembroke. One Saturday afternoon I was summoned to an interview. When I arrived at Pembroke, I found that the short list contained only two, Philip Dee and me. Dee was called in first; as he was being interviewed, I was reflecting without pleasure that he was one of the brightest of Rutherford's bright young men.

Then came my turn. As I went in, the first person I saw, sitting on the right hand of the Master, was Rutherford himself. While the

Master was taking me through my career, Rutherford drew at his pipe, not displaying any excessive interest in the proceedings. The Master came to the end of his questions, and said: "Professor Rutherford?"

Rutherford took out his pipe and turned on to me an eye which was blue, cold and bored. He was the most spontaneous of men; when he felt bored he showed it. That afternoon he felt distinctly bored. Wasn't his man, and a very good man, in for this job? What was this other fellow doing there? Why were we all wasting our time?

He asked me one or two indifferent questions in an irritated, impatient voice. What was my present piece of work? What could spectroscopy tell us anyway? Wasn't it just "putting things into boxes"?

I thought that was a bit rough. Perhaps I realized that I had nothing to lose. Anyway, as cheerfully as I could manage, I asked if he couldn't put up with a few of us not doing nuclear physics. I went on, putting a case for my kind of subject.

A note was brought round to my lodgings that evening. Dee had got the job. The electors wished to say that either candidate could properly have been elected. That sounded like a touch of Cambridge politeness, and I felt depressed. I cheered up a day or two later when I heard that Rutherford was trumpeting that I was a young man of spirit. Within a few months he backed me for another studentship. Incidentally, Dee was a far better scientist than I was or could have been, and neither Rutherford nor anyone else had been unjust.

From that time until he died, I had some opportunities of watching Rutherford at close quarters. Several of my friends knew him intimately, which I never did. It is a great pity that Tizard or Kapitsa, both acute psychological observers, did not write about him at length. But I belonged to a dining club which he attended, and I think I had serious conversations with him three times, the two of us alone together.

The difficulty is to separate the inner man from the Rutherfordiana, much of which is quite genuine. From behind a screen in a Cambridge tailor's, a friend and I heard a reverberating voice: "That shirt's too tight round the neck. Every day I grow in girth. *And* in mentality." Yet his physical make-up was more nervous than it seemed. In the same way, his temperament, which seemed exuberantly powerful, massively simple, rejoicing with childish satisfaction in creation and fame, was not quite so simple as all that. His was a personality of Johnsonian scale. As with Johnson, the façade was overbearing and unbroken. But there were fissures within.

No one could have enjoyed himself more, either in creative work or the honors it brought him. He worked hard, but with immense gusto; he got pleasure not only from the high moments, but also from the hours of what to others would be drudgery, sitting in the dark counting the alpha particle scintillations on the screen. His insight was direct, his intuition, with one curious exception, infallible. No scientist has made fewer mistakes. In the corpus of his published work, one of the largest in scientific history, there was nothing he had to correct afterwards. By thirty he had already set going the science of nuclear physics — single-handed, as a professor on five hundred pounds a year, in the isolation of late-Victorian Montreal. By forty, now in Manchester, he had found the structure of the atom — on which all modern nuclear physics depends.

It was an astonishing career, creatively active until the month he died. He was born very poor, as I have said. New Zealand was, in the 1880's, the most remote of provinces, but he managed to get a good education; enough of the old Scottish tradition had percolated there, and he won all the prizes. He was as original as Einstein, but unlike Einstein he did not revolt against formal instruction; he was top in classics as well as in everything else. He started research — on the subject of wireless waves — with equipment such as one might rustle up today in an African laboratory. That did not deter him: "I could do research at the North Pole," he once proclaimed, and it was true. Then he was awarded one of the 1851 overseas scholarships (which later brought to England Florey, Oliphant, Philip Bowden, a whole series of gifted Antipodeans). In fact, he got the scholarship only because another man, placed above him, chose to get married: with the curious humility that was interwoven with his boastfulness, he was grateful all of his life. There was a proposal, when he was Lord Rutherford, President of the Royal Society, the greatest of living experimental scientists, to cut down these scholarships. Rutherford was on the committee. He was too upset to speak: at last he blurted out:

"If it had not been for them, I shouldn't have been." That was nonsense. Nothing could have stopped him. He brought his wireless work to Cambridge, anticipated Marconi, and then dropped it because he saw a field — radioactivity — more scientifically interesting.

If he had pushed on with wireless, incidentally, he couldn't have avoided becoming rich. But for that he never had time to spare. He provided for his wife and daughter, they lived in comfortable middle-class houses, and that was all. His work led directly to the atomic energy industry spending, within ten years of his death, thousands of millions of pounds. He himself never earned, or wanted to earn, more than a professor's salary — about £1,600 a year at the Cavendish in the thirties. In his will he left precisely the value of his Nobel Prize, then worth £7,000. Of the people I am writing about, he died much the poorest (one has to leave Stalin out of this comparison): even G. H. Hardy, who by Rutherford's side looked so ascetic and unworldly, happened not to be above taking an interest in his investments.

As soon as Rutherford got on to radioactivity, he was set on his life's work. His ideas were simple, rugged, material: he kept them so. He thought of atoms as though they were tennis balls. He discovered particles smaller than atoms, and discovered how they moved or bounced. Sometimes the particles bounced the wrong way. Then he inspected the facts and made a new but always simple picture. In that way he moved, as certainly as a sleepwalker, from unstable radioactive atoms to the discovery of the nucleus and the structure of the atom.

In 1919 he made one of the significant discoveries of all time: he broke up a nucleus of nitrogen by a direct hit from an alpha particle. That is, man could get inside the atomic nucleus and play with it if he could find the right projectiles. These projectiles could either be provided by radioactive atoms or by ordinary atoms speeded up by electrical machines.

The rest of that story leads to the technical and military history of our time. Rutherford himself never built the great machines which have dominated modern particle physics, though some of his pupils, notably Cockcroft, started them. Rutherford himself worked with bizarrely simple apparatus: but in fact he carried the use of such apparatus as far as it would go. His researches remain the last supreme single-handed achievement in fundamental physics. No one else can ever work there again — in the old Cavendish phrase — with sealing wax and string.

It was not done without noise: it was done with anger and storms — but also with an overflow of creative energy, with abundance and generosity, as though research were the easiest and most natural avocation in the world. He had deep sympathy with the creative arts, particularly literature; he read more novels than most literary people manage to do. He had no use for critics of any kind. He felt both suspicion and dislike of the people who invested scientific research or any other branch of creation with an aura of difficulty, who used long, methodological words to explain things which he did perfectly by instinct. "Those fellows," he used to call them. "Those fellows" were the logicians, the critics, the metaphysicians. They were clever; they were usually more lucid than he was; in argument against them he often felt at a disadvantage. Yet somehow they never produced a serious piece of work, whereas he was the greatest experimental scientist of the age.

I have heard larger claims made for him. I remember one discussion in particular, a year or two after his death, by half-a-dozen men, all of whom had international reputations in science. Darwin was there; G. I. Taylor; Fowler and some others. Was Rutherford the greatest experimental scientist since Michael Faraday? Without any doubt. Greater than Faraday? Possibly so. And then — it is interesting, as it shows the anonymous Tolstoyan nature of organized science — how many years' difference would it have made if he had never lived? How much longer before the nucleus would have been understood as we now understand it? Perhaps ten years. More likely only five.

Rutherford's intellect was so strong that he would, in the long run, have accepted that judgment. But he would not have liked it. His estimate of his own powers was realistic, but if it erred at all, it did not err on the modest side. "There is no room for this particle in the atom as designed by *me*," I once heard him assure a large audience. It was part of his nature that, stupendous as his work was, he should consider it 10 per cent more so. It was also part of his nature that, quite without acting, he should behave constantly as though he were 10 per cent larger than life. Worldly success? He loved every minute of it: flattery, titles, the company of the high official world. He said in a speech: "As I was standing in the drawing-room at Trinity, a *clergyman* came in. And I said to him: 'I'm Lord Rutherford.' And he said to me: 'I'm the Archbishop of York.' And I don't suppose either of us believed the other."

He was a great man, a very great man, by any standards which we can apply. He was not subtle: but he was clever as well as creatively gifted, magnanimous (within the human limits) as well as hearty. He was also superbly and magnificently vain as well as wise — the combination is commoner than we think when we are young. He enjoyed a life of miraculous success. On the whole he enjoyed his own personality. But I am sure that, even quite late in his life, he felt stabs of a sickening insecurity.

Somewhere at the roots of that abundant and creative nature there was a painful, shrinking nerve. One has only to read his letters as a young man to discern it. There are passages of self-doubt which are not to be explained completely by a humble colonial childhood and youth. He was uncertain in secret, abnormally so for a young man of his gifts. He kept the secret as his personality flowered and hid it. But there was a mysterious diffidence behind it all. He hated the faintest suspicion of being patronized, even when he was a world figure. Archbishop Lang was once tactless enough to suggest that he supposed a famous scientist had no time for reading. Rutherford immediately felt that he was being regarded as an ignorant rough-neck. He produced a formidable list of his last month's reading. Then, half innocently, half malevolently: "And what do you manage to read, your Grice?" "I am afraid," said the Archbishop, somewhat out of his depth, "that a man in my position really doesn't have the leisure . . ." "Ah, yes, your Grice," said Rutherford in triumph, "it must be a dog's life! It must be a dog's life!"

Once I had an opportunity of seeing that diffidence face to face. In the autumn of 1934 I published my first novel, which was called *The Search* and the background of which was the scientific world. Not long after it came out, Rutherford met me in King's Parade. "What have you been doing to us, young man?" he asked vociferously. I began to describe the novel, but it was not necessary; he announced that he had read it with care. He went on to invite, or rather command, me to take a stroll with him round the Backs. Like most of my scientific friends, he was good-natured about the book, which has some descriptions of the scientific experience which are probably somewhere near the truth. He praised it. I was gratified. It was a sunny October afternoon. Suddenly he said: "I didn't like the erotic bits. I suppose it's because we belong to different generations."

The book, I thought, was reticent enough. I did not know how to reply.

In complete seriousness and simplicity, he made another suggestion. He hoped that I was not going to write all my novels about scientists. I assured him that I was not — certainly not another for a long time.

He nodded. He was looking gentler than usual, and thoughtful. "It's a small world, you know," he said. He meant the world of science. "Keep off us as much as you can. People are bound to think that you are getting at some of us. And I suppose we've all got things that we don't want anyone to see."

I mentioned that his intuitive foresight went wrong just once. As a rule, he was dead right about the practical applications of science, just as much as about the nucleus. But his single boss shot sounds ironic now. In 1933 he said, in another address to the British Association, "These transformations of the atom are of extraordinary interest to scientists, but we cannot control atomic energy to an extent which would be of any value commercially, and I believe we are not likely ever to be able to do so. A lot of nonsense has been talked about transmutations. Our interest in the matter is purely scientific."

That statement, which was made only nine years before the first pile worked, was not intended to be either optimistic or pessimistic. It was just a forecast, and it was wrong.

That judgment apart, people outside the scientific world often felt that Rutherford and his kind were optimistic — optimistic right against the current of the twentieth century literary-intellectual mood, offensively and brazenly optimistic. This feeling was not quite unjustified, but the difference between the scientists and the non-scientists was more complex than that. When the scientists talked of the individual human condition, they did not find it any more hopeful than the rest of us. Does anyone really imagine that Bertrand Russell, G. H. Hardy, Rutherford, Blackett and the rest were bemused by cheerfulness as they faced their own individual state? Very few of them had any of the consolations of religion: they believed, with the same certainty that they believed in Rutherford's atom, that they were going, after this mortal life, into annihilation. Several of them were men of deep introspective insight. They did not need teaching anything at all about the existential absurdity.

Nevertheless it is true that, of the kinds of people I have lived among, the scientists were much the happiest. Somehow scientists were buoyant at a time when other intellectuals could not keep away despair. The reasons for this are not simple. Partly, the nature of scientific activity, its complete success on its own terms, is itself a source of happiness; partly, people who are drawn to scientific activity tend to be happier in temperament than other clever men. By the nature of their vocation and also by the nature of their own temperament, the scientists did not think constantly of the individual human predicament. Since they could not alter it, they let it alone. When they thought about people, they thought most of what could be altered, not what couldn't. So they gave their minds not to the individual condition but to the social one.

There, science itself was the greatest single force for change. The scientists were themselves part of the deepest revolution in human affairs since the discovery of agriculture. They could accept what was happening, while other intellectuals shrank away. They not only accepted it, they rejoiced in it. It was difficult to find a scientist who did not believe that the scientific-technical-industrial revolution, accelerating under his eyes, was not doing incomparably more good than harm.

This was the characteristic optimism of scientists in the twenties and thirties. Is it still? In part, I think so. But there has been a change.

In the Hitler war, physicists became the most essential of military resources: radar, which occupied thousands of physicists on both sides, altered the shape of the war, and the nuclear bomb finished large scale "conventional" war for ever. To an extent, it had been foreseen by the mid-thirties that if it came to war (which a good many of us expected) physicists would be called on from the start. Tizard was a close friend of Rutherford's, and kept him informed about the prospects of RDF (as radar was then called). By 1938 a number of the Cavendish physicists had been secretly indoctrinated. But no one, no one at all, had a glimmering of how, for a generation afterwards, a high percentage of all physicists in the United States, the Soviet Union, this country, would remain soldiers-not-in-uniform. Mark Oliphant said sadly, when the first atomic bomb was dropped: "This has killed a beautiful subject." Intellectually that has turned out not to be true: but morally there is something in it. Secrecy, national demands, military influence, have sapped the moral nerve of physics. It will be a long time before the climate of Cambridge, Copenhagen, Göttingen in the twenties is restored: or before any single physicist can speak to all men with the calm authority of Einstein or Bohr. That kind of leadership has now passed to the biologists, who have so far not been so essential to governments. It will be they, I think, who are likely to throw up the great scientific spokesmen of the next decades. If someone now repeated Gorki's famous question, "Masters of culture, which side are you on?" it would probably be a biologist who spoke out for his fellow human beings.

In Rutherford's scientific world, the difficult choices had not yet formed themselves. The liberal decencies were taken for granted. It was a society singularly free from class or national or racial prejudice.

Rutherford called himself alternatively conservative or non-political, but the men he wanted to have jobs were those who could do physics. Niels Bohr, Otto Hahn, Georg von Hevesy, Hans Geiger were men and brothers, whether they were Jews, Germans, Hungarians — men and brothers whom he would much rather have near him than the Archbishop of Canterbury or one of "those fellows" or any damned English philosopher. It was Rutherford who, after 1933, took the lead in opening English academic life to Jewish refugees. In fact, scientific society was wide open, as it may not be again for many years. There was coming and going among laboratories all over the world, including Russia. Peter Kapitsa, Rutherford's favorite pupil, contrived to be in good grace with the Soviet authorities and at the same time a star of the Cavendish.

He had a touch of genius: in those days, before life sobered him, he had also a touch of the inspired Russian clown. He loved his own country, but he distinctly enjoyed backing both horses, working in Cambridge and taking his holidays in the Caucasus. He once asked a friend of mine if a foreigner could become an English peer; we strongly suspected that his ideal career would see him established simultaneously in the Soviet Academy of Sciences and as Rutherford's successor in the House of Lords.

At that time Kapitsa attracted a good deal of envy, partly because he could do anything with Rutherford. He called Rutherford "the Crocodile," explaining the crocodile means "father" in Russian, which it doesn't, quite: he had Eric Gill carve a crocodile on his new laboratory. He flattered Rutherford outrageously, and Rutherford loved it. Kapitsa could be as impertinent as a Dostoevskian comedian: but he had great daring and scientific insight. He established the club named after him (which again inspired some envy): it met every Tuesday night, in Kapitsa's rooms in Trinity, and was deliberately kept small, about thirty, apparently because Kapitsa wanted to irritate people doing physical subjects he disapproved of. We used to drink large cups of milky coffee immediately after hall (living was fairly simple, and surprisingly non-alcoholic, in scientific Cambridge), and someone gave a talk — often a dramatic one, like Chadwick's on the neutron. Several of the major discoveries of the thirties were first heard in confidence in that room. I don't think that the confidence was ever broken.

I myself enjoyed the one tiny scientific triumph of my life there. At the time Kapitsa barely tolerated me, since I did spectroscopy, a subject he thought fit only for bank clerks: in fact I had never dis-

covered why he let me join. One night I offered to give a paper outside my own subject, on nuclear spin, in which I had been getting
interested: I didn't know much about it, but I reckoned that most of
the Cavendish knew less. The offer was unenthusiastically accepted.
I duly gave the paper. Kapitsa looked at me with his large blue eyes,
with a somewhat unflattering astonishment, as at a person of low
intelligence who had contrived inadvertently to say something interesting. He turned to Chadwick, and said incredulously, "Jimmy, I
believe there *is* something in this."

It was a personal loss to Rutherford when Kapitsa, on one of
his holiday trips to Russia, was told by the Soviet bosses, politely but
unyieldingly, that he must stay: he was too valuable, they wanted his
services full-time. After a while Kapitsa made the best of it. He had
always been a patriotic Russian: though both he and his wife came
from the upper middle-class, if there was such a class in old Russia
(his father was a general in the Tsarist engineering corps), he took
a friendly attitude to the revolution. All that remained steady,
though I don't think he would mind my saying that his enthusiasm
for Stalin was not unqualified. Still, Kapitsa threw all his gifts into
his new work in the cause of Soviet science. It was only then that we,
who had known him in Cambridge, realized how strong a character
he was; how brave he was; and fundamentally what a good man. His
friendship with Cockcroft and others meant that the link between
Soviet and English science was never quite broken, even in the worst
days. Only great scientists like Lev Landau can say in full what he
has done for science in his own country. If he hadn't existed, the
world would have been worse: that is an epitaph that most of us
would like and don't deserve.

Between Leningrad and Cambridge, Kapitsa oscillated. Between
Copenhagen and Cambridge there was a stream of travellers, all
the nuclear physicists of the world. Copenhagen had become the second scientific metropolis on account of the personal influence of
one man, Niels Bohr, who was complementary to Rutherford as a
person — patient, reflective, any thought hedged with Proustian
qualifications — just as the theoretical quantum physics of which he
was the master was complementary to Rutherford's experimental
physics. He had been a pupil of Rutherford's, and they loved and
esteemed each other like father and son. (Rutherford was a pater-
familias born, and the death of his only daughter seems to have been
the greatest sorrow of his personal life. In his relations with Bohr
and Kapitsa and others, there was a strong vein of paternal emotion

diverted from the son he never had.) But, strong as Rutherford's liking for Bohr was, it was not strong enough to put up with Bohr's idea of a suitable length for a lecture. In the Cavendish lecture room, Bohr went past the hour; Rutherford began to stir. Bohr went past the hour and a half; Rutherford began plucking at his sleeve and muttering in a stage whisper about "another five minutes." Blandly, patiently, determined not to leave a qualification unsaid, as indefatigable as Henry James in his last period, Bohr went past the two hours; Rutherford was beginning to trumpet about "bringing the lecture to a close." Soon they were both on their feet at once.

Rutherford died suddenly when he was age sixty-six, still in full vigor. He died not only suddenly, but of something like a medical accident: he had a strangulated hernia. There was no discernible reason why he should not have lived into old age.

It was a sunny, tranquil October morning, the kind of day on which Cambridge looks so beautiful. I had just arrived at the crystallographic laboratory, one of the buildings in the old Cavendish muddle; why I was there I don't remember, nor whom I was talking to, except that it happened not to be Bernal. Someone put his head round the door and said: "The Professor's dead." I don't think anyone said much more. We were stupefied rather than miserable. It did not seem in the nature of things.

# JAGDISH MEHRA

■

Eccentric and brilliant, Paul Dirac (1902–1984) was the butt of many a fond anecdote recounted at a dinner held in his honor in Trieste, Italy, in 1972. When his friend and colleague Jagdish Mehra (b. 1937) told the following tale about Dirac, Dirac riposted with a story about the great German mathematician David Hilbert (1862–1943). Dirac's intention may have been to remind his colleagues that he was not the only oddball to have ornamented recent mathematical history.

# *Dirac*

MY FIRST MEETING with Paul Dirac took place in Cambridge in 1955. I had just returned to England after a couple of years with Heisenberg in Göttingen. A historian friend of mine in Cambridge, knowing of my great hero worship for Dirac, offered to take me with him to St. John's College, which was also his college, and to dine at the High Table. He thought we might see Dirac there. I went with him, and true to his word, he showed me that Professor Dirac was sitting there. We sat down. The weather outside was very bad, and since in England it is always quite respectable to start a conversation with the weather, I said to Dirac, "It is very windy, Professor." He said nothing at all, and a few seconds later he got up and left. I was mortified, as I thought that I had somehow offended him. He went to the door, opened it, looked out, came back, sat down, and said, "Yes."

## P A U L   A .   M .   D I R A C

■

# *Hilbert*

HILBERT . . . WAS PERHAPS the most absentminded man who ever lived. He was a great friend of the physicist James Franck. One day when Hilbert was walking in the street he met James Franck and he said, "James, is your wife as mean as mine?" Well, Franck was rather taken aback by this statement and didn't know quite what to say, and he said, "Well, what has your wife done?" And Hilbert said, "It was only this morning that I discovered quite by accident that my wife does not give me an egg for breakfast. Heaven knows how long that has been going on."

# JOHN TIERNEY

■

The Indian astrophysicist Subramanyan Chandrasekhar (b. 1910) brings to his work a freshness, vigor, and rigor that prompted a physicist thirty years his junior, black-hole theorist Kip S. Thorne, to call him "an inspiration to aging physicists like me." This profile of Chandrasekhar was written by the American science writer John Tierney.

# *Quest for Order*

SUBRAMANYAN CHANDRASEKHAR IS NOT sure what comes next. He survived a heart attack and open heart surgery during his latest project, an eight-year study of black holes. When he finished it in the spring of 1982 he was seventy-one, twice the age of practically everyone else in the field, a time of life when most scientists are either retired or enjoying an emeritus title — serving on committees, reminiscing at award dinners, directing graduate students, toying with a few leftover problems. That kind of scholarship would be impossible for Chandrasekhar. It would be the moral equivalent of going to the office without a plain dark suit, dark tie, and white shirt, the clothes that he has worn every working day for nearly half a century at the University of Chicago.

No, when he decides to work, he sits at a relentlessly neat desk searching for mathematical order for at least twelve hours a day, usually seven days a week, until after about a decade he has attained what he calls "a certain perspective" — which is to say, until some aspect of the universe has been completely reduced to a set of equations. Then, having written the definitive book on the subject, he puts all his files in the attic and looks for a totally different area of astrophysics to teach himself. Just talking about "Chandra's style" makes other astronomers tired. They can't understand how he regularly forces himself to abandon a subject and start over — how, in a discipline where a forty-year-old theoretician is considered way past his peak, a man of sixty-three could profitably *begin* analyzing what happens when things disappear down a black hole.

"He just batters his way through problems no one else could do," says his closest friend, Martin Schwarzschild, an astrophysicist at Princeton University. "Chandra's concentration is unbelievable — a mixture of sheer mathematical intelligence and phenomenal persistence. There is not one field that he's worked in where we are not now daily using some of his results."

Chandrasekhar, who is a great one for philosophizing about creativity and the aging scientific mind, turns uncomfortable when asked to explain his own professional longevity. But he allows that it probably has something to do with the meeting of the Royal Astronomical Society on January 11, 1935.

He arrived in London that Friday with great expectations for himself and mild suspicions concerning Sir Arthur Eddington. For months he and Eddington had been getting together, about twice a week after dinner, to discuss Chandrasekhar's latest calculations about the behavior of dying stars. They made an odd couple: the famed Eddington, eloquent, prepossessing, at fifty-two generally acknowledged as the world's finest astronomer, listening eagerly to a shy twenty-four-year-old from India who felt himself something of an outcast at Cambridge University. Chandrasekhar had been studying stellar structure for just a few years, ever since he won Eddington's classic book on the subject as a prize in a physics contest at Madras University in India. Now he was convinced that he had made a significant and startling discovery. That Friday afternoon he was to announce it.

But the day before, when a copy of the program for the meeting had arrived in Cambridge, Chandrasekhar had been amazed to discover that Eddington would also be speaking at the meeting. On the same subject. During all their discussions, while Chandrasekhar had been spewing out his figures, Eddington had never mentioned any work of his own in this area. It seemed an incredible breach of faith, yet Eddington offered no apology or explanation when the two men saw each other in the dining hall Thursday evening. His only remark was that he had used his influence to get Chandrasekhar extra time at the meeting — "so that you can present your work properly," as Chandrasekhar remembers him saying solicitously. Chandrasekhar was too deferential to mention Eddington's own paper, but the next day in London, at the tea before the meeting, another astronomer asked Eddington what he planned to say. Eddington wouldn't answer. He just turned to Chandrasekhar and smiled.

"That is a surprise for you," Eddington said.

Chandrasekhar's paper dealt with a fundamental question: what happens after a star has burned up all its fuel? According to the prevailing theory of the day, the cooling star would collapse under the force of its own gravity into a dense ball called a white dwarf. A star with the mass of the sun, for instance, would shrink to the size of the earth, at which point it would reach equilibrium. Chandrasekhar studied this collapse by considering what happens when a star's gas becomes so compressed that electrons move at nearly the speed of light — a state called relativistic degeneracy. He concluded that the enormous gravitational forces at work in a large star — any star more than 1.4 times as massive as the sun — would cause the star to go on collapsing beyond the white dwarf stage. The star would simply keep getting smaller and smaller and denser and denser until . . . well, that was an interesting question. Chandrasekhar delicately avoided it.

"A star of large mass cannot pass into the white dwarf stage," he concluded, "and one is left speculating on other possibilities."

Then it was Eddington's turn.

"I do not know whether I shall escape from this meeting alive, but the point of my paper is that there is no such thing as relativistic degeneracy," said Eddington, and proceeded to tear apart Chandrasekhar's paper. The speech was frequently interrupted by laughter. Eddington couldn't quarrel with Chandrasekhar's logic or calculations. But he claimed that the whole theory had to be wrong simply because it led to an inevitable and outlandish conclusion: "The star has to go on radiating and radiating and contracting and contracting until, I suppose, it gets down to a few kilometers radius, when gravity becomes strong enough to hold in the radiation, and the star can at last find peace."

Today, of course, such an object is called a black hole. That afternoon Eddington said it couldn't possibly exist.

"A *reductio ad absurdum*," he called it. "I think there should be a law of nature to prevent a star from behaving in this absurd way."

And there the matter rested, at least for the next few decades. Eventually, the theory would be vindicated, black holes would be accepted, and the dividing line mentioned in the paper (a stellar mass 1.4 times that of the sun) would go down in the textbooks as the

Chandrasekhar Limit. But not for a long time after Eddington's speech.

"At the end of the meeting," recalls Chandrasekhar, "everybody came up to me and said, 'Too bad, Chandra, too bad.' I had gone to the meeting thinking that I would be proclaimed as having found something very important. Instead, Eddington effectively made a fool of me. I was distraught. I didn't know whether to continue my career. I returned to Cambridge late that night, probably around one o'clock. I remember going into the common room, the place where the fellows would meet. Of course nobody was there. There was a fire still burning, and I remember standing there in front of it and repeating to myself, 'This is how the world ends, not with a bang but with a whimper.' "

Today he has a different perspective on that afternoon.

The argument with Eddington dragged on for years, ruined any chance of his getting a tenured position in England, and finally persuaded him to give up the subject altogether (although the two men, remarkably enough, remained friends throughout). He believed in his theory, but others didn't. So, shortly after arriving at the University of Chicago in 1937, he put the theory in a book and stopped worrying about it. He instead began studying the probability distributions of stars in galaxies and discovered the curious property called dynamical friction — the fact that any star hurtling through a galaxy tends to slow down because of the gravity of the stars surrounding it. Then he switched again and considered why the sky is blue. The simple answer to this problem — that the atmosphere's molecules scatter the short-wavelength blue light while allowing other colors to pass through — had been found last century by Britain's Lord Rayleigh. But Rayleigh and a succession of physicists had all failed to unravel the exact mathematics of how light is scattered. By the middle of the 1940s, Chandrasekhar had worked it all out. And he enjoyed it so much, this switching fields, that he decided to make a career of it. He went on to more topics: the behavior of hot fluids in magnetic fields, the stability of rotating objects, the general theory of relativity, and finally back to black holes (but from a completely different approach). He now thinks he was lucky to be driven out of his original specialty.

"Suppose Eddington had decided that there were black holes in nature," he says, pausing to consider this proposition as precisely as possible. He is impeccably formal, true to his Brahmin roots, determined to keep his conversation as structured as everything else in his life. He actually speaks in complete sentences and logical paragraphs. He does this with a soft-spoken, gentle charm, digressing occasionally with jokes and allusions to everything from Picasso to Mother Goose to Keats, but always sternly bringing himself back to the original question.

"It's very difficult to speculate. Eddington would have made the whole area a very spectacular one to investigate, and many of the properties of black holes might have been discovered twenty or thirty years ahead of time. I can easily imagine that theoretical astronomy would have been very different. It's not for me to judge whether that difference — well, the difference would have been salutary for astronomy, I think I would say that.

"But I do not think it would have been salutary for me. My position in science would have been radically altered as of that moment. Eddington's praise could make one very famous in astronomy. But I really do not know how I would have reacted to the temptation, to the glamour.

"How many young men after being successful and famous have survived for long periods of time? Not many. Even the very great men of the 1920s who made quantum mechanics — I mean Dirac, Heisenberg, Fowler — they never equaled themselves. Look at Maxwell. Look at Einstein."

Chandrasekhar hastily interrupts himself to say that he is not comparing himself to these scientists or trying to criticize them. "You must not confuse things large and small. Who am I to criticize Einstein?" It is the problem in the abstract, he insists, that interests him. He is struck by the fact that, at age forty-seven, Beethoven told a friend, "*Now* I know how to compose." Chandrasekhar doesn't think there has ever been a forty-seven-year-old scientist who announced, "*Now* I know how to do research."

"When you discuss the works of a great artist or writer, the assumption always is that there is a growth from the early period to the middle period to the mature work and the end. The artist's ability is refined. Clearly he's able to tackle difficult problems. It obviously required an enormous effort, an enormous emotional control, to be able to write a play like *King Lear*. Look at the contrast between that and an earlier play, *Romeo and Juliet*.

"Now why is a scientist *unable* to refine his mind? Einstein was one of the great scientific minds. He discovered special relativity and a number of things in 1905. He worked terribly hard and did the general theory of relativity in 1916, and then he did some fairly important work until the early 1920s. From that point on he detached himself from the progress of science, became a critic of quantum theory, and effectively did not add to science or to his own enlargement. There is nothing in Einstein's work after the age of forty which shows that he attained a greater intellectual perception than what he had before. Why?

"For lack of a better word, there seems to be a certain arrogance toward nature which people develop. These people have had great insights and made profound discoveries. They imagine afterwards that the fact that they succeeded so triumphantly in one area means they have a special way of looking at science which must therefore be right. But science doesn't permit that. Nature has shown over and over again that the kinds of truth which underlie nature transcend the most powerful minds.

"Take Eddington. He was a great man. He said that there must be a law of nature to prevent a star from becoming a black hole. Why should he say that? Just because he thought it was bad? Why does he assume that he has a way of deciding what the laws of nature should be? Similarly, this oft-quoted statement of Einstein's disapproving of the quantum theory: 'God does not play dice.' How does he know?"

The exception that Chandrasekhar likes to talk about is Lord Rayleigh, the nineteenth-century physicist who did the original study on the color of the sky. He remained steadily productive in a variety of fields for fifty years and turned out some of his best-known work (such as the discovery of argon gas) in the latter part of his career.

"You know, when Rayleigh was sixty-seven, his son asked him what he thought about the famous remark by Thomas Huxley — that a man of sixty in science does more harm than good. Rayleigh thought about it a great deal and said, 'Well, I don't see why that should be so, provided you do what you understand and do not contradict young people.' I don't think Einstein could have said that, or Dirac, or Heisenberg. Eddington wouldn't have said that. There is a certain modesty in that remark. Now on the other hand you could say, as Churchill said when somebody told him that Clement Attlee was a very modest man, 'He has much to be modest about.' The really great discoveries have been made by people who have had the arrogance to make judgments about nature. Certainly Rayleigh did not add any really great fundamental insights like Einstein or Maxwell. But his influence on science was enormous because he added to the great body of knowledge, constantly inventing many things that were not spectacular but were always important. I think one could say that a certain modesty toward understanding nature is a precondition to the continued pursuit of science."

He again insists that he is speaking in the abstract, not about himself. But he could just as easily have been describing his own career. Plunging into a new field every decade is guaranteed to produce modesty: how can you contradict young men when they've been in the field longer than you have? And like Rayleigh, he concerns himself with important but unspectacular work, with rigorous studies that enlarge a body of knowledge rather than overturn it. He doesn't go for quick hits, for the single blinding insight or the revolutionary discovery that wins a Nobel Prize. He has always insisted on a long and complete analysis of a whole field, no matter how useless it may seem to others.

In the 1960s, for instance, he wrote a book on tangerine-shaped geometric figures called ellipsoids, which at the time were guaranteed not to win anyone fame or fortune. His reason for writing this book, he said in the Introduction, was that all the previous research had left the subject "with many gaps and omissions and some plain errors and misconceptions. It seemed a pity that it should be allowed to remain in this destitute state." So he tidied it up by systematically analyzing the forces acting on a rotating ellipsoid — the gravity holding it together, the centrifugal force pulling it apart, and the point at which it becomes unstable. Other scientists thought he was wasting his time studying these idealized objects. Why study an abstraction that doesn't exist in the universe? Yet today, twenty years later, the book is being applied in ways that couldn't have been anticipated. It turns out, for example, that the properties of these imaginary objects are shared by many real galaxies, and scientists are using the book to understand what holds the Milky Way together as it spins.

"I think my motivations are different from many scientists'," he says. "James Watson wrote that when he was a young man, he wanted to solve a problem that would win him the Nobel Prize. So he went ahead and discovered DNA. Clearly that approach justified itself in his case. But my motive has not been to solve a single problem but instead to acquire a perspective of an entire area.

"I started studying black holes eight years ago, particularly the theory of how a rotating black hole reacts to external perturbations such as gravitational and electromagnetic waves. If you know that, you can determine what happens to a black hole when an object such as a star falls into it. Well, there are individual pieces of this work that have attracted attention, but to me what is important is the final point of view I have of the subject. That is why I wrote the book. I see it as a whole with a perspective. Obviously there are a number of problems in the area that I can still work on, but I don't feel inclined. If you make a sculpture, you finish it — you don't want to go on chipping it here and there."

In 1983 Chandrasekhar received the Nobel Prize in physics, in part for his work on black holes. So what next? This is a problem for a man in his seventies. The chief drawback to his style of research is that it requires enormous amounts of time and energy. It has meant starting work every day at 6:00 A.M. and continuing as late as midnight. One of his collaborators, a graduate student who had the misfortune of living in an apartment visible from Chandrasekhar's office, used to be surprised by late-night calls at home when Chandrasekhar needed help with a problem. This went on until Chandrasekhar happened to mention to the student's wife that he felt free to call when he could see a light in the couple's window. From then on the window was heavily shielded with a curtain.

"He hasn't really had time for other things, for travel and friends. He's always had an intense discipline about his work, an insistence that everything be neat and perfect," says Chandrasekhar's wife, Lalitha, who met him when they were both physics students at Madras University. She speaks uncomplainingly about his career and the sacrifices required of her — the long hours spent by herself, the abandoning of her career to follow him to the United States, the many years between visits to their families in India. But she also thinks that by now he has earned the right to relax, and so do his colleagues.

"Chandra's had to pay an enormous price, and it's steadily increased as he's gotten older," says Schwarzschild. "This last book was a *tour de force* — an example of willpower conquering exhaustion. I don't really know what he can do after that. His contributions have all come from this ability to keep pushing through problems that nobody else could push through. For Chandra, it would be completely out of character to drop down to something he knows a whole bunch of us could do."

Chandrasekhar tends to agree. "If I cannot pursue a subject in earnest," he says, "I would rather not make the effort at all." It is tempting for him to stop after this book if only for aesthetic reasons, to end with a study of the black holes that seemed so absurd at the beginning of his career. It would be a nice finale, especially since he regards it as perhaps the most difficult work he has ever undertaken. Yet he also talks about going into still another new field, maybe this time cosmology. "That's my habit of life, and it requires enormous discipline to change a habit of life. I haven't decided."

Although he doesn't consider himself a Hindu anymore — he classifies himself as an atheist — he sometimes wonders if he should follow the Hindu tradition of retirement: renouncing all worldly connections and going into the forest for solitary contemplation. To him, of course, this simply means giving up science, the ultimate change of field.

"One of the unfortunate facts about the pursuit of science the way I have done it is that it has distorted my personality. I had to sacrifice other interests in life — literature, music, traveling. I've devoted all my time, every living hour practically, to my work. I wanted to read all the plays of Shakespeare very carefully, line by line, word by word. I have never found the time to do it. I know I could have been a different person had I done this. I don't know if regret is the right term for what I feel. But sooner or later one has to reconcile these losses. One has to come to terms with oneself. One needs some time to get things in order."

P. R. HALMOS

■

The following is a sketch by one mathematician from Budapest about another. Paul Richard Halmos (b. 1916), once John von Neumann's assistant, went on to do distinguished work of his own. He has taught at four American universities and has written many books and articles on pure mathematics.

# *The Legend of John von Neumann*

JOHN VON NEUMANN WAS a brilliant mathematician who made important contributions to quantum physics, to logic, to meteorology, to war, to the theory and applications of high-speed computing machines, and, via the mathematical theory of games of strategy, to economics.

He was born December 28, 1903, in Budapest, Hungary. He was the eldest of three sons in a well-to-do Jewish family. His father was a banker who received a minor title of nobility from the Emperor Franz Josef; since the title was hereditary, von Neumann's full Hungarian name was Margittai Neumann János. (Hungarians put the family name first. Literally, but in reverse order, the name means John Neumann of Margitta. The "of," indicated by the final "i," is where the "von" comes from; the place name was dropped in the German translation. In ordinary social intercourse such titles were never used, and by the end of the first world war their use had gone out of fashion altogether. In Hungary von Neumann is and always was known as Neumann János and his works are alphabetized under N. Incidentally, his two brothers, when they settled in the U.S., solved the name problem differently. One of them reserves the title of nobility for ceremonial occasions only, but, in daily life, calls himself Neumann; the other makes it less conspicuous by amalgamating it with the family name and signs himself Vonneuman.)

Even in the city and in the time that produced Szilárd (1898), Wigner (1902), and Teller (1908), von Neumann's brilliance stood out, and the legends about him started accumulating in his childhood. Many of the legends tell about his memory. His love of history began early, and, since he remembered what he learned, he ultimately became an expert on Byzantine history, the details of the trial of Joan of Arc, and minute features of the battles of the American Civil War.

He could, it is said, memorize the names, addresses, and telephone numbers in a column of the telephone book on sight. Some of the later legends tell about his wit and his fondness for humor, including puns and off-color limericks. Speaking of the Manhattan telephone book he said once that he knew all the numbers in it — the only other thing he needed, to be able to dispense with the book altogether, was to know the names that the numbers belonged to.

Most of the legends, from childhood on, tell about his phenomenal speed in absorbing ideas and solving problems. At the age of 6 he could divide two eight-digit numbers in his head; by 8 he had mastered the calculus; by 12 he had read and understood Borel's *Théorie des Fonctions.*

These are some of the von Neumann stories in circulation. I'll report others, but I feel sure that I haven't heard them all. Many are undocumented and unverifiable, but I'll not insert a separate caveat for each one: let this do for them all. Even the purely fictional ones say something about him; the stories that men make up about a folk hero are, at the very least, a strong hint of what he was like.

In his early teens he had the guidance of an intelligent and dedicated high-school teacher, L. Rátz, and, not much later, he became a pupil of the young M. Fekete and the great L. Fejér, "the spiritual father of many Hungarian mathematicians." ("Fekete" means "Black," and "Fejér" is an archaic spelling, analogous to "Whyte.")

According to von Kármán, von Neumann's father asked him, when John von Neumann was 17, to dissuade the boy from becoming a mathematician, for financial reasons. As a compromise between father and son, the solution von Kármán proposed was chemistry. The compromise was adopted, and von Neumann studied chemistry in Berlin (1921–1923) and in Zürich (1923–1925). In 1926 he got both a Zürich diploma in chemical engineering and a Budapest Ph.D. in mathematics.

His definition of ordinal numbers (published when he was 20) is the one that is now universally adopted. His Ph.D. dissertation was about set theory too; his axiomatization has left a permanent mark on the subject. He kept up his interest in set theory and logic most of his life, even though he was shaken by K. Gödel's proof of the impossibility of proving that mathematics is consistent.

He admired Gödel and praised him in strong terms: "Kurt Gödel's achievement in modern logic is singular and monumental — indeed it is more than a monument, it is a landmark which will remain visible far in space and time. . . . The subject of logic has certainly completely changed its nature and possibilities with Gödel's achievement." In a talk entitled "The Mathematician," speaking, among other things, of Gödel's work, he said: "This happened in our lifetime, and I know myself how humiliatingly easily my own values regarding the absolute mathematical truth changed during this episode, and how they changed three times in succession!"

He was Privatdozent at Berlin (1926–1929) and at Hamburg (1929–1930). During this time he worked mainly on two subjects, far from set theory but near to one another: quantum physics and operator theory. It is almost not fair to call them two subjects: due in great part to von Neumann's own work, they can be viewed as two aspects of the same subject. He started the process of making precise mathematics out of quantum theory, and (it comes to the same thing really) he was inspired by the new physical concepts to make broader and deeper the purely mathematical study of infinite-dimensional spaces and operators on them. The basic insight was that the geometry of the vectors in a Hilbert space has the same formal properties as the structure of the states of a quantum-mechanical system. Once that is accepted, the difference between a quantum physicist and a mathematical operator-theorist becomes one of language and emphasis only. Von Neumann's book on quantum mechanics appeared (in German) in 1932. It has been translated into French (1947), Spanish (1949), and English (1955), and it is still one of the standard and one of the most inspiring treatments of the subject. Speaking of von Neumann's contributions to quantum mechanics, E. Wigner, a Nobel laureate, said that they alone "would have secured him a distinguished position in present day theoretical physics."

In 1930 von Neumann went to Princeton University for one term as visiting lecturer, and the following year he became professor there. In 1933, when the Institute for Advanced Study was founded, he was one of the original six professors of its School of Mathematics, and he kept that position for the rest of his life. . . .

In 1930 von Neumann married Marietta Kövesi; in 1935 their daughter Marina was born. (In 1956 Marina von Neumann graduated from Radcliffe *summa cum laude,* with the highest scholastic record in her class. In 1972 Marina von Neumann Whitman was appointed by President Nixon to the Council of Economic Advisers.) In the 1930's the stature of von Neumann, the mathematician, grew at the rate that his meteoric early rise had promised, and the legends about Johnny, the human being, grew along with it. He enjoyed life in America and lived it in an informal manner, very differently from the style of the conventional German professor. He was not a refugee and he didn't feel like one. He was a cosmopolite in attitude and a U.S. citizen by choice.

The parties at the von Neumanns' house were frequent, and famous, and long. Johnny was not a heavy drinker, but he was far from a teetotaller. In a roadside restaurant he once ordered a brandy with a hamburger chaser. The outing was in honor of his birthday and he was feeling fine that evening. One of his gifts was a toy, a short prepared tape attached to a cardboard box that acted as sounding board; when the tape was pulled briskly past a thumbnail, it would squawk "Happy birthday!" Johnny squawked it often. Another time, at a party at his house, there was one of those thermodynamic birds that dips his beak in a glass of water, straightens up, teeter-totters for a while, and then repeats the cycle. A temporary but firm house rule was quickly passed: everyone had to take a drink each time that the bird did.

He liked to drive, but he didn't do it well. There was a "von Neumann's corner" in Princeton, where, the story goes, his cars repeatedly had trouble. One often quoted explanation that he allegedly offered for one particular crack-up goes like this: "I was proceeding down the road. The trees on the right were passing me in orderly fashion at 60 miles an hour. Suddenly one of them stepped in my path. Boom!"

He once had a dog named "Inverse." He played poker, but only rarely, and he usually lost.

In 1937 the von Neumanns were divorced; in 1938 he married Klára Dán. She learned mathematics from him and became an expert programmer. Many years later, in an interview, she spoke about him. "He has a very weak idea of the geography of the house. . . . Once, in Princeton, I sent him to get me a glass of water;

he came back after a while wanting to know where the glasses were. We had been in the house only seventeen years. . . . He has never touched a hammer or a screwdriver; he does nothing around the house. Except for fixing zippers. He can fix a broken zipper with a touch."

Von Neumann was definitely not the caricatured college professor. He was a round, pudgy man, always neatly, formally dressed. There are, to be sure, one or two stories of his absentmindedness. Klára told one about the time when he left their Princeton house one morning to drive to a New York appointment, and then phoned her when he reached New Brunswick to ask: "Why am I going to New York?" It may not be strictly relevant, but I am reminded of the time I drove him to his house one afternoon. Since there was to be a party there later that night, and since I didn't trust myself to remember exactly how I got there, I asked how I'd be able to know his house when I came again. "That's easy," he said; "it's the one with that pigeon sitting by the curb."

Normally he was alert, good at rapid repartee. He could be blunt, but never stuffy, never pompous. Once the telephone interrupted us when we were working in his office. His end of the conversation was very short; all he said between "Hello" and "Goodbye" was "Fekete pestis!" which means "Black plague!" Remembering, after he hung up, that I understood Hungarian, he turned to me, half apologetic and half exasperated, and explained that he wasn't speaking of one of the horsemen of the Apocalypse, but merely of some unexpected and unwanted dinner guests that his wife just told him about.

On a train once, hungry, he asked the conductor to send the man with the sandwich tray to his seat. The busy and impatient conductor said "I will if I see him." Johnny's reply: "This train is linear, isn't it?"

The speed with which von Neumann could think was awe-inspiring. G. Pólya admitted that "Johnny was the only student I was ever afraid of. If in the course of a lecture I stated an unsolved problem, the chances were he'd come to me as soon as the lecture was over, with the complete solution in a few scribbles on a slip of paper." Abstract proofs or numerical calculations — he was equally quick with both, but he was especially pleased with and proud of his facility with numbers. When his electronic computer was ready for its first preliminary test, someone suggested a relatively simple problem involving powers of 2. (It was something of this kind: what is the smallest power of 2 with the property that its decimal digit fourth from the right is 7? This is a completely trivial problem for a present-day computer: it takes only a fraction of a second of machine time.) The machine and Johnny started at the same time, and Johnny finished first.

One famous story concerns a complicated expression that a young scientist at the Aberdeen Proving Grounds needed to evaluate. He spent ten minutes on the first special case; the second computation took an hour of paper and pencil work; for the third he had to resort to a desk calculator, and even so took half a day. When Johnny came to town, the young man showed him the formula and asked him what to do. Johnny was glad to tackle it. "Let's see what happens for the first few cases. If we put $n = 1$, we get . . ." — and he looked into space and mumbled for a minute. Knowing the answer, the young questioner put in "2.31?" Johnny gave him a funny look and said "Now if $n = 2$, . . ." and once again voiced some of his thoughts as he worked. The young man, prepared, could of course follow what Johnny was doing, and, a few seconds before Johnny finished, he interrupted again, in a hesitant tone of voice: "7.49?" This time Johnny frowned, and hurried on: "If $n = 3$, then . . ." The same thing happened as before — Johnny muttered for several minutes, the young man eavesdropped, and, just before Johnny finished, the young man exclaimed: "11.06!" That was too much for Johnny. It couldn't be! No unknown beginner could outdo him! He was upset and he sulked till the practical joker confessed.

Then there is the famous fly puzzle. Two bicyclists start twenty miles apart and head toward each other, each going at a steady rate of 10 m.p.h. At the same time a fly that travels at a steady 15 m.p.h. starts from the front wheel of the southbound bicycle and flies to the front wheel of the northbound one, then turns around and flies to the front wheel of the southbound one again, and continues in this manner till he is crushed between the two front wheels. Question: what total distance did the fly cover? The slow way to find the answer is to calculate what distance the fly covers on the first, northbound, leg of the trip, then on the second, southbound, leg, then on the third, etc., etc., and, finally, to sum the infinite series so obtained. The quick way is to observe that the bicycles meet exactly one hour after their start, so that the fly had just an hour for his travels; the answer must therefore be 15 miles. When the question was put to von Neumann, he solved it in an instant, and thereby disappointed the questioner: "Oh, you must have heard the trick before!" "What trick?" asked von Neumann; "all I did was sum the infinite series."

I remember one lecture in which von Neumann was talking about rings of operators. At an appropriate point he mentioned that they can be classified two ways: finite versus infinite, and discrete versus continuous. He went on to say: "This leads to a total of four possibilities, and, indeed, all four of them can occur. Or — let's see — can they?" Many of us in the audience had been learning this subject from him for some time, and it was no trouble to stop and mentally check off all four possibilities. No trouble — it took something like two seconds for each, and, allowing for some fumbling and shifting of gears, it took us perhaps 10 seconds in all. But after two seconds von Neumann had already said "Yes, they can," and he was two sentences into the next paragraph before, dazed, we could scramble aboard again. . . .

As a mathematical lecturer he was dazzling. He spoke rapidly but clearly; he spoke precisely, and he covered the ground completely. If, for instance, a subject has four possible axiomatic approaches, most teachers content themselves with developing one, or at most two, and merely mentioning the others. Von Neumann was fond of presenting the "complete graph" of the situation. He would, that is, describe the shortest path that leads from the first to the second, from the first to the third, and so on through all twelve possibilities.

His one irritating lecturing habit was the way he wielded an eraser. He would write on the board the crucial formula under discussion. When one of the symbols in it had been proved to be replaceable by something else, he made the replacement not by rewriting the whole formula, suitably modified, but by erasing the replaceable symbol and substituting the new one for it. This had the tendency of inducing symptoms of acute discouragement among note-takers, especially since, to maintain the flow of the argument, he would keep talking at the same time.

His style was so persuasive that one didn't have to be an expert to enjoy his lectures; everything seemed easy and natural. Afterward, however, the Chinese-dinner phenomenon was likely to occur. A couple of hours later the average memory could no longer support the delicate balance of mutually interlocking implications, and, puzzled, would feel hungry for more explanation.

As a writer of mathematics von Neumann was clear, but not

clean; he was powerful but not elegant. He seemed to love fussy detail, needless repetition, and notation so explicit as to be confusing. To maintain a logically valid but perfectly transparent and unimportant distinction, in one paper he introduced an extension of the usual functional notation: along with the standard $\phi(x)$ he dealt also with something denoted by $\phi((x))$. The hair that was split to get there had to be split again a little later, and there was $\phi(((x)))$, and, ultimately, $\phi((((x))))$. Equations such as

$$(\psi((((a)))))^2 = \phi((((a))))$$

have to be peeled before they can be digested; some irreverent students referred to this paper as von Neumann's onion.

Perhaps one reason for von Neumann's attention to detail was that he found it quicker to hack through the underbrush himself than to trace references and see what others had done. The result was that sometimes he appeared ignorant of the standard literature. If he needed facts, well-known facts, from Lebesgue integration theory, he waded in, defined the basic notions, and developed the theory to the point where he could use it. If, in a later paper, he needed integration theory again, he would go back to the beginning and do the same thing again.

He saw nothing wrong with long strings of suffixes, and subscripts on subscripts; his papers abound in avoidable algebraic computations. The reason, probably, is that he saw the large picture; the trees did not conceal the forest from him. He saw and he relished all parts of the mathematics he was thinking about. He never wrote "down" to an audience; he told it as he saw it. The practice caused no harm; the main result was that, quite a few times, it gave lesser men an opportunity to publish "improvements" of von Neumann. . . .

Von Neumann was not satisfied with seeing things quickly and clearly; he also worked very hard. His wife said "he had always done his writing at home during the night or at dawn. His capacity for work was practically unlimited." In addition to his work at home, he worked hard at his office. He arrived early, he stayed late, and he never wasted any time. He was systematic in both large things and small; he was, for instance, a meticulous proofreader. He would correct a manuscript, record on the first page the page numbers where he found errors, and, by appropriate tallies, record the number of errors that he had marked on each of those pages. Another example:

when requested to prepare an abstract of not more than 200 words, he would not be satisfied with a statistical check — there are roughly 20 lines with about 10 words each — but he would count every word.

When I was his assistant we wrote one paper jointly. After the thinking and the talking were finished, it became my job to do the writing. I did it, and I submitted to him a typescript of about 12 pages. He read it, criticized it mercilessly, crossed out half, and rewrote the rest; the result was about 18 pages. I removed some of the Germanisms, changed a few spellings, and compressed it into 16 pages. He was far from satisfied, and made basic changes again; the result was 20 pages. The almost divergent process continued (four innings on each side as I now recall it); the final outcome was about 30 typescript pages (which came to 19 in print).

Another notable and enviable trait of von Neumann's was his mathematical courage. If, in the middle of a search for a counter-example, an infinite series came up, with a lot of exponentials that had quadratic exponents, many mathematicians would start with a clean sheet of paper and look for another counterexample. Not Johnny! When that happened to him, he cheerfully said: "Oh, yes, a theta function . . . ," and plowed ahead with the mountainous computations. He wasn't afraid of anything.

He knew a lot of mathematics, but there were also gaps in his knowledge, most notably number theory and algebraic topology. Once when he saw some of us at a blackboard staring at a rectangle that had arrows marked on each of its sides, he wanted to know what that was. "Oh just the torus, you know — the usual identification convention." No, he didn't know. The subject is elementary, but some of it just never crossed his path, and even though most graduate students knew about it, he didn't.

Brains, speed, and hard work produced results. In von Neumann's *Collected Works* there is a list of over 150 papers. About 60 of them are on pure mathematics (set theory, logic, topological groups, measure theory, ergodic theory, operator theory, and continuous geometry), about 20 on physics, about 60 on applied mathematics (including statistics, game theory, and computer theory), and a small handful on some special mathematical subjects and general non-mathematical ones. A special number of the *Bulletin of the American Mathematical Society* was devoted to a discussion of his life and work (in May 1958).

Von Neumann's reputation as a mathematician was firmly established by the 1930's, based mainly on his work on set theory, quantum theory, and operator theory, but enough more for about three ordinary careers, in pure mathematics alone, was still to come. The first of these was the proof of the ergodic theorem. Various more or less precise statements had been formulated earlier in statistical mechanics and called the ergodic hypothesis. In 1931 B. O. Koopman published a penetrating remark whose main substance was that one of the contexts in which a precise statement of the ergodic hypothesis could be formulated is the theory of operators on Hilbert space — the very subject that von Neumann used earlier to make quantum mechanics precise and on which he had written several epoch-making papers. It is tempting to speculate on von Neumann's reaction to Koopman's paper. It could have been something like this: "By Koopman's remark the ergodic hypothesis becomes a theorem about Hilbert spaces — and if that's what it is I ought to be able to prove it. Let's see now . . ." Soon after the appearance of Koopman's paper, von Neumann formulated and proved the statement that is now known as the mean ergodic theorem for unitary operators. There was some temporary confusion, caused by publication dates, about who did what before whom, but by now it is universally recognized that von Neumann's theorem preceded and inspired G. D. Birkhoff's point ergodic theorem. In the course of the next few years von Neumann published several more first-rate papers on ergodic theory; and he made use of the techniques and results of that theory later, in his studies of rings of operators.

In 1900 D. Hilbert presented a famous list of 23 problems that summarized the state of mathematical knowledge at the time and showed where further work was needed. In 1933 A. Haar proved the existence of a suitable measure (which has come to be called Haar measure) in topological groups; his proof appears in the *Annals of Mathematics*. Von Neumann had access to Haar's result before it was published, and he quickly saw that that was exactly what was needed to solve an important special case (compact groups) of one of Hilbert's problems (the 5th). His solution appears in the same issue of the same journal, immediately after Haar's paper.

In the second half of the 1930's the main part of von Neumann's publications was a sequence of papers, partly in collaboration with F. J. Murray, on what he called rings of operators. (They are now called von Neumann algebras.) It is possible that this is the work for which von Neumann will be remembered the longest. It is a technically brilliant development of operator theory that makes contact with von Neumann's earlier work, generalizes many familiar facts about finite-dimensional algebra, and is currently one of the most powerful tools in the study of quantum physics.

A surprising outgrowth of the theory of rings of operators is what von Neumann called continuous geometry. Ordinary geometry deals with spaces of dimension 1, 2, 3, etc. In his work on rings of operators von Neumann saw that what really determines the dimension structure of a space is the group of rotations that it admits. The group of rotations associated with the ring of *all* operators yields the familiar dimensions. Other groups, associated with different rings, assign to spaces dimensions whose values can vary continuously; in that context it makes sense to speak of a space of dimension 3/4, say. Abstracting from the "concrete" case of rings of operators, von Neumann formulated the axioms that make these continuous-dimensional spaces possible. For several years he thought, wrote, and lectured about continuous geometries. In 1937 he was the Colloquium Lecturer of the American Mathematical Society and chose that subject for his topic.

The year 1940 was just about the half-way point of von Neumann's scientific life, and his publications show a discontinuous break then. Till then he was a topflight pure mathematician who understood physics; after that he was an applied mathematician who remembered his pure work. He became interested in partial differential equations, the principal classical tool of the applications of mathematics to the physical world. Whether the war made him into an applied mathematician or his interest in applied mathematics made him invaluable to the war effort, in either case he was much in demand as a consultant and advisor to the armed forces and to the civilian agencies concerned with the problems of war. His papers from this point on are mainly on statistics, shock waves, flow problems, hydrodynamics, aerodynamics, ballistics, problems of detonation, meteorology, and, last but not least, two non-classical, new aspects of the applicability of mathematics to the real world: games and computers.

Von Neumann's contributions to war were manifold. Most often mentioned is his proposal of the implosion method for bringing nuclear fuel to explosion (during World War II) and his espousal of the development of the hydrogen bomb (after the war). The citation that accompanied his honorary D.Sc. from Princeton in 1947 mentions (in one word) that he was a mathematician, but praises him for being a physicist, an engineer, an armorer, and a patriot.

His political and administrative decisions were rarely on the side that is described nowadays by the catchall term "liberal." He appeared at times to advocate preventive war with Russia. As early as 1946 atomic bomb tests were already receiving adverse criticism, but von Neumann thought that they were necessary and (in, for instance, a letter to the *New York Times*) defended them vigorously. He disagreed with J. R. Oppenheimer on the H-bomb crash program, and urged that the U.S. proceed with it before Russia could. He was, however, a "pro-Oppenheimer" witness at the Oppenheimer security hearings. He said that Oppenheimer opposed the program "in good faith" and was "very constructive" once the decision to go ahead with the super bomb was made. He insisted that Oppenheimer was loyal and was not a security risk.

As a member of the Atomic Energy Commission (appointed by President Eisenhower, he was sworn in on March 15, 1955), having to "think about the unthinkable," he urged a United Nations study of world-wide radiation effects. "We willingly pay 30,000–40,000 fatalities per year (2% of the total death rate)," he wrote, "for the advantages of individual transportation by automobile." He mentioned a fall-out accident in an early Pacific bomb test that resulted in one fatality and danger to 200 people, and he compared it with a Japanese ferry accident that "killed about 1,000 people, including 20 Americans — yet the . . . fall-out was what attracted almost world-wide attention." He asked: "Is the price in international popularity worth paying?" And he answered: "Yes: we have to accept it as part payment for our more advanced industrial position."

At about the same time that he began to apply his analytic talents to the problems of war, von Neumann found time and energy to apply his combinatorial insight to what he called the theory of games, whose major application was to economics. The mathematical cornerstone of the theory is one statement, the so-called minimax theorem, that von Neumann proved early (1928) in a short article (25 pages); its elaboration and applications are in the book he wrote jointly with O. Morgenstern in 1944. The minimax theorem says about a large class of two-person games that there is no point in playing them. If either play considers, for each possible strategy of play, the *maximum* loss that he can expect to sustain with that strategy, and then chooses the "optimal" strategy that *minimizes* the maximum loss, then he can be statistically sure of not losing more than that minimax value. Since (and this is the whole point of the theorem) that value is the negative of the one, similarly defined, that his opponent can guarantee for himself, the long-run outcome is completely determined by the rules.

Mathematical economics before von Neumann tried to achieve success by imitating the technique of classical mathematical physics. The mathematical tools used were those of analysis (specifically the calculus of variations), and the procedure relied on a not completely reliable analogy between economics and mechanics. The secret of the success of the von Neumann approach was the abandonment of the mechanical analogy and its replacement by a fresh point of view (games of strategy) and new tools (the ideas of combinatorics and convexity).

The role that game theory will play in the future of mathematics and economics is not easy to predict. As far as mathematics is concerned, it is tenable that the only thing that makes the Morgenstern–von Neumann book 600 pages longer than the original von Neumann paper is the development needed to apply the abstruse deductions of one subject to the concrete details of another. On the other hand, enthusiastic proponents of game theory can be found who go so far as to say that it may be "one of the major scientific contributions of the first half of the 20th century."

The last subject that contributed to von Neumann's fame was the theory of electronic computers and automata. He was interested in them from every point of view: he wanted to understand them, design them, build them, and use them. What are the logical components of the processes that a computer will be asked to perform? What is the best way of obtaining practically reliable answers from a machine with unreliable components? What does a machine need to "remember," and what is the best way to equip it with a "memory"? Can a machine be built that can not only save us the labor of computing but save us also the trouble of building a new machine — is it possible, in other words, to produce a self-reproducing automaton? (Answer: in principle, yes. A sufficiently complicated machine, embedded in a thick chowder of randomly distributed spare parts, its "food," would pick up one part after another till it found a usable one, put it in place, and continue to search and construct till its descendant was complete and operational.) Can a machine successfully imitate "randomness," so that when no formulae are available to solve a concrete physical problem (such as that of finding an optimal bombing pattern), the machine can perform a large number of probability experiments and yield an answer that is statistically accurate? (The last question belongs to the concept that is sometimes described as the Monte Carlo method.) These are some of the problems that von Neumann studied and to whose solutions he made basic contributions.

He had close contact with several computers — among them the MANIAC (Mathematical Analyzer, Numerical Integrator, Automatic Calculator), and the affectionately named JONIAC. He advocated their use for everything from the accumulation of heuristic data for the clarification of our intuition about partial differential equations to the accurate long-range prediction and, ultimately, control of the weather. One of the most striking ideas whose study he suggested was to dye the polar icecaps so as to decrease the amount of energy they would reflect — the result could warm the earth enough to make the climate of Iceland approximate that of Hawaii.

The last academic assignment that von Neumann accepted was to deliver and prepare for publication the Silliman lectures at Yale. He worked on that job in the hospital where he died, but he couldn't finish it. His notes for it were published, and even they make illuminating reading. They contain tantalizing capsule statements of insights, and throughout them there shines an attitude of faith in and dedication to knowledge. While physicists, engineers, meteorologists, statisticians, logicians, and computers all proudly claim von Neumann as one of theirs, the Silliman lectures prove, indirectly by their approach and explicitly in the author's words, that von Neumann was first, foremost, and always a mathematician. . . .

Von Neumann became ill in 1955. There was an operation, and the result was a diagnosis of cancer. He kept on working, and even travelling, as the disease progressed. Later he was confined to a wheelchair, but still thought, wrote, and attended meetings. In April 1956 he entered Walter Reed Hospital, and never left it. Of his last days his good friend Eugene Wigner wrote: "When von Neumann realized he was incurably ill, his logic forced him to realize that he would cease to exist, and hence cease to have thoughts. . . . It was heartbreaking to watch the frustration of his mind, when all hope was gone, in its struggle with the fate which appeared to him unavoidable but unacceptable."

Von Neumann was baptized a Roman Catholic (in the U.S.), but, after his divorce, he was not a practicing member of the church.

In the hospital he asked to see a priest — "one that will be intellectually compatible." Arrangements were made, he was given special instruction, and, in due course, he again received the sacraments. He died February 8, 1957.

The heroes of humanity are of two kinds: the ones who are just like all of us, but very much more so, and the ones who, apparently, have an extra-human spark. We can all run, and some of us can run the mile in less than 4 minutes; but there is nothing that most of us can do that compares with the creation of the Great G-minor Fugue. Von Neumann's greatness was the human kind. We can all think clearly, more or less, some of the time, but von Neumann's clarity of thought was orders of magnitude greater than that of most of us, all the time. Both Norbert Wiener and John von Neumann were great men, and their names will live after them, but for different reasons. Wiener saw things deeply but intuitively; von Neumann saw things clearly and logically.

What made von Neumann great? Was it the extraordinary rapidity with which he could understand and think and the unusual memory that retained everything he had once thought through? No. These qualities, however impressive they might have been, are ephemeral; they will have no more effect on the mathematics and the mathematicians of the future than the prowess of an athlete of a hundred years ago has on the sport of today.

The "axiomatic method" is sometimes mentioned as the secret of von Neumann's success. In his hands it was not pedantry but perception; he got to the root of the matter by concentrating on the basic properties (axioms) from which all else follows. The method, at the same time, revealed to him the steps to follow to get from the foundations to the applications. He knew his own strengths and he admired, perhaps envied, people who had the complementary qualities, the flashes of irrational intuition that sometimes change the direction of scientific progress. For von Neumann it seemed to be impossible to be unclear in thought or in expression. His insights were illuminating and his statements were precise.

# ANDREW HODGES

■

The logician Alan Turing, a towering figure in computer science whose development of the "Enigma machine" cracked Nazi secret codes and contributed to the Allied victory in World War II, died at age forty-one, evidently a suicide, after being prosecuted under England's antihomosexual statutes. His tragic demise is recounted by the mathematician Andrew Hodges (b. 1949) in this excerpt from his biography entitled *Alan Turing: The Enigma.*

# On the Beach

IT HAD NOT TAKEN the police long to detect Alan Turing's crime. It was almost inevitable once he had made the original report of the burglary, for the police had been able to identify Harry's fingerprints. He was already in custody on another charge in Manchester, and before long made a statement which referred to Arnold telling him of having "business" at Alan's home. The further information Alan had volunteered on the Sunday merely gave the police their opportunity to act with confidence.

Alan took them upstairs to where he was working with his desk calculator. The detectives, Mr. Wills and Mr. Rimmer, found themselves in an unfamiliar environment, the room littered with pieces of paper covered with mathematical symbols. They told Alan that they "knew all about it," leaving him unclear whether they were talking about the burglary or of something else. He later told Robin [Gandy, an old friend] that he had to admire their interrogation technique. They asked him to repeat the description he had given them on the Sunday morning, and Alan said, "He's about twenty-five years of age, five foot ten inches, with black hair." Imitation was not Alan Turing's strong point — perhaps an intelligent machine would have done better. This feeble attempt sank like a stone. Mr. Wills said, "We have reason to believe your description is false. Why are you lying?"

This was the moment for "I don't know what came over me," or the other phrases employed by more politically-minded persons, but once the detectives had shown their hand, Alan blurted out everything that they wanted to hear, in particular admitting that he had concealed the identity of the informant because he "had an affair with him." Mr. Wills asked "Would you care to tell us what kind of an affair you have had with him?", and this policemanlike question elicited from Alan a memorable phrase, detailing in semi-official language three of the activities that had taken place. "A very honourable man," the detectives thought him as they cautioned him in the usual way, and they were the more impressed when he volunteered a statement of five handwritten pages. Relieved of the usual necessity to translate human life into police language, they were most appreciative of what was "a lovely statement," written in "a flowing style, almost like prose," although "beyond them in some of its phraseology." They were particularly struck by his absence of shame. "He was a real convert . . . he really believed he was doing the right thing."

* * * * *

In more conservative circles, it was taken that the law only gave the final stamp of authority to the ostracism of society. King George V was supposed to have said, "I thought men like that shot themselves." Alan Turing, however, cared nothing for the opinion of society, and therefore was ahead of his time in laying bare the role of the state. For most gay men, the question of *who knew* would be of colossal significance, and life would be rigidly divided into two compartments, one for those who knew, and one for those who did not. Blackmail depended as much upon this fact as upon the legal penalties. The question was important to Alan too, but in a rather different way: it was because he did not wish to be accepted or respected as the person he was not. He was likely to drop a remark about an attractive young man, or something of the kind, on a third or fourth meeting with a generally friendly colleague. To be close to him, it was essential to accept him as a homosexual; it was one of the stringent conditions he imposed.

Exposure, therefore, held no intrinsic terror for him. But a criminal trial would involve not merely exposure as a homosexual, but all the concrete details. It would be one thing to be a martyr for an abstract cause, and quite another to have the sequence of events with Arnold rendered into an unflattering public form. It would expose him not only as a sexual outlaw, which at least carried with it a certain pride, but as a fool. In this respect his insouciance was amazing. But it was his all-or-nothing mentality at work. He had presumably decided long before that such things were part of the "large remnant of the random behaviour of infancy," and that it was absurd to be ashamed of anything harmlessly enjoyed, whether it be parlour games or bedroom pleasures. It meant that he had to take a stand not for an ideal, not for anything particularly rewarding or successful, but for that which was simply true. But he did not flinch. The detectives continued to be astonished when they visited him in connection with the case. He would take out his violin and play to them the Irish tune *Cockles and Mussels*, accompanied with glasses of wine.

\* \* \* \* \*

The case of *Regina v. Turing and Murray* was heard on 31 March 1952, at the Quarter Sessions held at the Cheshire town of Knutsford. The judge was Mr. J. Fraser Harrison. Alan was represented by Mr. G. Lind-Smith, and Arnold by Mr. Emlyn Hooson. Both were prosecuted by Mr. Robin David. The charges now amounted to twelve in number. With the Looking-Glass symmetry of symmetrical crimes, they began:

Alan Mathison Turing
1. On the 17th day of December, 1951, at Wilmslow, being a male person, committed an act of gross indecency with Arnold Murray, a male person.
2. On the 17th day of December, 1951, at Wilmslow, being a male person was party to the commission of an act of gross indecency with Arnold Murray, a male person.

and so forth, for each of the other two nights, and then Arnold was charged in exactly the same way so that the last accusation was that he:

12. On the 2nd day of February, 1952, at Wilmslow, being a male person, was party to the commission of an act of gross indecency with Alan Mathison Turing, a male person.

They both pleaded "guilty" to all the charges, although Alan was guilty of something for which he showed no guilt. The prosecuting counsel, in outlining the case, laid stress upon his unrepentant remarks.

There only lay his "character" to set against this admitted law-breaking. Normally, "good character" would be a disguised statement of class status, but in these circumstances his status told against him. The theme of the better public schools had been the balance of privilege and duty, and as one of the prefect class he was supposed to set an example, not to break the rules himself. Alan Turing, however, was little interested either in the privileges or the duties of his class. He never tried to pull rank on the detectives, who saw him as an "ordinary fellow," with his occasional visits to the local pub. Conversely, his crime was seen at least by an older generation as a betrayal of his class. Arnold likewise was made to feel by his family that his real crime had been that of dragging down a gentleman.

The OBE[1] was duly given a mention, and Hugh Alexander bore witness that Alan was "a national asset." Max Newman was asked whether he would receive such a man in his home, and replied that he had already done so, Alan being a personal friend of himself and his wife. He described Alan as "particularly honest and truthful." "He is completely absorbed in his work," he continued, "and is one of the most profound and original mathematical minds of his generation." Lind-Smith pleaded that he should not go to prison:

> He is entirely absorbed in his work, and it would be a loss if a man of his ability — which is no ordinary ability — were not able to carry on with it. The public would lose the benefit of the research work he is doing. There is treatment which could be given him. I ask you to think that the public interest would not be well served if this man is taken away from the very important work he is doing.

Mr. Hooson, however, defended Arnold as the innocent led astray by Alan's wiles:

> Murray is not a university Reader, he is a photo-printer. It was he who was approached by the other man. He has

---

1. The fact that he retained his OBE was itself an interesting detail of the case. The War Office would demand the return of medals from anyone guilty under the 1885 Act. The Foreign Office presumably took a different view.

not such tendencies as Turing, and if he had not met Turing he would not have indulged in that practice.

Max Newman and Hugh Alexander were amazed that Alan should go to the stake for Arnold, but Alexander was impressed by his "moral courage" and Newman by his "strong line." He answered back at the judge's remarks, and he did not recant, at an occasion whose very essence was the obtaining of a confession. Hilbert had written of Galileo that in his recantation "he was not an idiot. Only an idiot could believe that scientific truth needs martyrdom — that may be necessary in religion, but scientific results prove themselves in time." But this was not a trial of scientific truth.

The verdict quivered between the old and the new dispensations, and came down for the new. Bletchley Park scored a victory beyond its term. The state washed its hands, and handed Alan to the judgment of science. He was placed on probation, with the condition that he "submit for treatment by a duly qualified medical practitioner at Manchester Royal Infirmary."

The Wilmslow newspaper headline was:

UNIVERSITY READER PUT ON PROBATION
To have Organo-Therapic Treatment

Alan wrote to Philip Hall, two weeks later:

(postmarked 17 April 1952)
. . . I am *both* bound over for a year *and* obliged to take this organo-therapy for the same period. It is supposed to reduce sexual urge whilst it goes on, but one is supposed to return to normal when it is over. I hope they're right. The psychiatrists seemed to think it useless to try and do any psychotherapy.

The day of the trial was by no means disagreeable. Whilst in custody with the other criminals I had a very agreeable sense of irresponsibility, rather like being back at school. The warders rather like prefects. I was also quite glad to see my accomplice again, though I didn't trust him an inch.

Perhaps it was surprising that he chose the scientific alternative to prison. He was annoyed at having been circumcised, and at any editorial meddling with his writings — small interferences compared with this piece of doctoring. Neither did he care much for creature comforts, and a year in prison, even an English one, would not have been much more uncomfortable than Sherborne. On the other hand, to take that option would have impeded his work, and very likely would have forefeited his Manchester position and the computer. He had the choice between his body and feelings on the one hand, and his intellectual life on the other. It was a remarkable decision problem. He chose "thinking" and sacrificed "feeling."

There was no concept of a right to sexual expression in the Britain of 1952. People made jokes about bromides put in the servicemen's tea. Samuel Butler might well have laughed in his grave at the prophet of the intelligent machine being punished for being sick, and treated for committing a crime. But no one at the time perceived an irony in Alan Turing being on the receiving end of science. As for Jefferson, ranging himself with the humanists, or Polanyi, foe of the state's pretensions to order human life, and adherent of the Congress for Cultural Freedom — this was a private and distasteful medical matter, and did not gain the attention of the liberal intelligentsia of Manchester, discoursing on the folly and iniquity of treating minds like machines.

Harry the burglar was sent to a Borstal the same day, in another trial. Arnold, however, was conditionally discharged. He left the court in a daze, hardly knowing to what he had confessed, and then found himself pointed out in the street by his neighbours. After a few weeks he escaped back to London, found a job in the Lyons Corner House in the Strand, and rapidly made his way into anarchic Fitzrovia. Here in the coffee-bar world, meeting such people as Colin Wilson, he was accepted as an individual and learned to play the guitar.

For Alan, there were rather different consequences of the trial, because of the drug treatment.* He was rendered impotent, although scientific opinion was that the impotence was not permanent and potency would return when the medication was stopped. It had other physical effects, for

> To obtain the necessary effect mentally, it was necessary to maintain a moderate but not excessive degree of gynae-comastic response.

* The regimen that Turing underwent by court order consisted of regular injections of estrogen, the female sex hormone, in order to reduce the male sex drive; this was known in legal circles as "chemical castration." [editor's note]

Translating from the Greek, this meant that there could be no "reduction in libido" without the production of breasts. Again, according to the same authority,

> There is at least a possibility that oestrogen may have a direct pharmacological effect on the central nervous system. Zuckerman (1952) has demonstrated, through his experiments on rats, that learning can be influenced by sex hormones, and that oestrogen can act as a cerebral depressant in these rodents. While it has yet to be shown that a similar influence is exerted in humans, there are some indications clinically that performance may be impaired, though more investigation is needed before any conclusion is reached.

So perhaps "thinking" and "feeling" could not be so neatly separated after all.

There were some more minor consequences. The *News of the World* covered the case with a short article in its northern editions, headed ACCUSED HAD POWERFUL BRAIN. He remained under the auspices of the district probation officer. David Champernowne came to Manchester to do some work on the computer, and being invited to dinner at Hollymeade, found the probation officer another guest. Alan told a story about how the retired bishop of Liverpool had heard of the case and asked to see him — he had gone along, rather surprisingly for one who had written in 1936 that he would not tolerate bishops interfering in his private life. But nothing was private now. He had thought the bishop well-meaning but hopelessly old-fashioned. A further consequence, which for another person might have been major but which for Alan had little significance, was that, with a criminal record of "moral turpitude," he was henceforth automatically barred from the United States.

* * * * *

For Alan Turing there remained one freedom, not perhaps one that Tolstoy would have had in mind. It was that of exiled pleasure. The head master having taken action to prevent associations within the house, he had to fall back upon the possibilities offered by the boys in the other houses. For Alan would not let the system defeat him. On May Day 1952 there was a meeting of the Ratio Club at Cambridge, which he attended, and it was probably then that he saw

Norman Routledge at King's. Alan explained about the trial and the hormone treatment ("I'm growing *breasts!*") and Norman told him that he had heard that in Norway (of all places) there were dances "for men only."

In the summer of 1952 Alan went for a holiday in Norway, one which turned out to be a disappointment regarding the rumoured dances. But he met a number of Scandinavians, enough to have the addresses of five or six, and he was particularly struck by one young man called Kjell, whose photograph he showed to Robin on his return. Kjell had been somewhat coquettish, and little had transpired, but Alan had thereby demonstrated his unbroken will, which was perhaps the most important thing.

As for what the endocrinologists called intellectual "performance," Alan's work on the biological theory did continue, despite everything, to develop in range and scale. He was tackling the problems which he had outlined in the first paper. In particular, he was trying out on the computer the solution of the very difficult differential equations that arose when it was attempted to follow the chemical theory of morphogenesis beyond the moment of budding, taking into account the essential non-linearity. This was experimental work, in which he would be trying out many different initial conditions to see what happened. But it also required some rather sophisticated applied mathematics, which involved the use of "operators" rather as in quantum mechanics. Numerical analysis was also important, in deciding how to approximate the equations for the purpose of the calculation. In this it was like a private atomic bomb, the computer in both cases following the development of interacting fluid waves.

As a rather separate line of attack, he also developed a purely descriptive theory of leaf-arrangement, or "phyllotaxis" as it was called in biological Greek, in which he found ways of using matrices to represent the winding of spirals of leaves or seeds round a stem or flower-head. He brought into this theory a concept of "inverse lattices" somewhat like that used by crystallographers. It was also accompanied by a good deal of measurement-making of his own. The intention was that ultimately these two approaches would join up when he found a system of equations that would generate the Fibonacci patterns expressed by his matrices.

Although there was some correspondence with a number of biologists, this work was essentially done on his own. The Manchester botanist, C. W. Wardlaw, was particularly interested and wrote a paper describing, in biologists' terms, the significance of the first

Turing paper. This finally appeared in August 1952, and soon Alan had a letter from C. H. Waddington expressing interest but scepticism as to the correctness of the essential chemical hypothesis. But on the whole, Alan would tend to speak — to Lighthill, in particular — as though rather disappointed with the slow speed with which his ideas diffused, and the lack of reaction to them. There was, perhaps, an analogy with *Computable Numbers* in this respect, for the "confirmed solitary" whom Max Newman had diagnosed in 1936 still lacked the talent for patient, persistent pushing. Neither had his abilities as a lecturer improved. One side-line that developed was an interest in irreversible thermodynamics, and after giving his seminar in the chemistry department Alan had a meeting with W. Byers Brown to discuss the subject; but this soon faded out, Alan possibly being more interested in young Byers Brown than in this branch of physical chemistry. But one difference from the reaction to his earlier achievements was that this time no one had pre-empted his ideas. He was quite alone.

Robin had persuaded Alan to come with him to the 1952 British Mathematical Colloquium in the spring; it was held at the Royal Naval College, Greenwich, which meant that they had the excuse to take a jaunt on the Thames steamer. Alan found some interesting wild flowers on the Greenwich bombsites, and there was a nice moment at the lunchtime bar, when Alan suddenly disappeared through one doorway on spotting a particularly dull logician bear down upon him from another. By this time he was becoming quite famous as the author of *Computable Numbers*. He liked to hear references to Turing machines ("Pryce's buoy" in his story) but he did not like paying the price, that of being pinned down for shop-talk by those trying to make connections.

More to his taste was talking with Christopher Strachey, who had brought a fresh breeze from King's into the technical atmosphere of the Manchester computer laboratory. He had much the same attitudes as Alan, and the same sense of humour. His draughts program was much developed and played throughout the summer of 1952, this being the first time that the kind of automatic game-playing that Alan had so long talked about was seriously tried. But he and Alan also used the random number facility in a program to compose "love letters." One of these ran:

Darling Sweetheart,

You are my avid fellow feeling. My affection curiously clings to your passionate wish. My liking yearns to your heart. You are my wistful sympathy: my tender liking.

Yours beautifully, MUC

Those doing real men's jobs on the computer, concerned with optics or aerodynamics, thought this silly, but it was as good a way as any of investigating the nature of syntax, and it greatly amused Alan and Christopher Strachey — whose love lives, as it happened, were rather similar too.

Tony Brooker, meanwhile, had written a program system called FLOATCODE, which did the work of interpreting floating-point arithmetic, rather as Alan had envisaged in 1945, but had never bothered to do for Manchester. It was based on similar work he had done at Cambridge on the EDSAC. And Alick Glennie went further in 1952 with something called AUTOCODE which was, in effect, the first working high-level computer language in the world. Christopher Strachey was enthusiastic about it — AUTOCODE was in line with the ideas he had written about in 1951, of translating mathematical formulae into machine instructions. But Alan took little interest. Alick Glennie talked to him about it, but found that he was bored by mere translation, something that he had described as obvious in 1947 and had never chosen to take any further himself. He would have been interested in something that would actually *do* the algebra, rather than translate it.

The computer industry was now in a position to expand beyond the confines of a small trained élite, programming languages opening the universal machine to a much wider clientele. AUTOCODE did not in fact play this role, and was little known away from Manchester. But the American FORTRAN was not far behind, one of a chain of developments with which Alan Turing had quite parted company.

By 1952 the Manchester engineers not only had a Mark II machine in hand, but had begun the design of a small transistor-based prototype. No one could have guessed from his total lack of participation in these developments that Alan Turing had once been avid to keep abreast of the latest technological advances, and had broken unwritten rules in order to get his hands on them himself. All this had been dropped in 1949 when it finally became clear that the world saw such an interest only as a nuisance. There was no reflection of his attempts to make a practical contribution to computer development in a book called *Faster than Thought*, the definitive account of British computers as they stood in 1951–2. Here he appeared principally as the writer of part of a chapter on "Digital Computers applied to Games"; he wrote up his chess game with Alick Glennie, helped with some comments on the play by Hugh Alexander. Besides a one-line mention as the author of *Computable Numbers* and as one of Womersley's assistants, there was one telling entry in the glossary:

> *Türing Machine.* In 1936 Dr. Turing wrote a paper on the design and limitations of computing machines. For this reason they are sometimes known by his name. The umlaut is an unearned and undesirable addition, due, presumably, to an impression that anything so incomprehensible must be Teutonic.

The world of 1945 was now as remote, and effectively as secret, as that of 1942; a fact to which Alan seemed entirely resigned.

Robin kept him interested in the theory of types, and some work he had done stimulated Alan to dig out the paper he had written during the war, but had left unpublished, as his attempt to persuade mathematicians into a more careful use of "nouns" and "adjectives." Here again, the suggestions for "The Reform of Mathematical Notation," as his essay was entitled, were at a tangent to the development of post-war mathematics, in which the muddles to which Alan objected were being cured by other means.

Alan mentioned his "reform" proposals to Don Bayley when he paid a visit to him and his wife at Woburn Sands, near Bletchley, that summer. He gave Don some mathematical help, but the main point of the weekend was to make one last serious attempt to retrieve the silver bars. This time Don had got hold of a commercial metal detector, and they went out to the bridge near Shenley in his car. Alan said, "It looks a bit different," as he took off his socks and shoes and paddled in the mud. "Christ, do you know what's happened? They've rebuilt the bridge and concreted over the bed!" They tried for the other bar in the woods, finding that the pram in which he had wheeled the ingots in 1940 was still there, but without any more luck than before in locating the spot. They found nails and oddments, just as Alan had done on the earlier attempt with Donald Michie. Giving up both bars as lost for ever, they made their way to the Crown Inn at Shenley Brook End for some bread and cheese. The disappointment was not too great, and largely dissipated by the warm welcome accorded to him by Mrs. Ramshaw, his wartime land-lady.

When Don Bayley had met him at Bletchley station, he noticed Alan was carrying a Norwegian grammar. Alan explained that he had just had a holiday in Norway, and that the language interested him. Although at this point his knowledge was rudimentary, he made sufficient progress with Norwegian and Danish to read Hans Christian Andersen stories to his mother a year later. It did not occur to Don that the Norwegian holiday might have had a particular motive, even though Alan explained that now he would have to go abroad for pleasure. He had written to Don about the charge and the trial, as to his other friends, and on the visit spoke with his usual light-hearted bravado of the outcome. He also referred to a letter he had written to a titled lady politician, calling for a change in the law. This was no pleading, as Oscar Wilde had done, that it was no crime but sickness. He had drawn attention to the homosexuality of the politician's son. All he had received in reply was a brusque disclaimer from her secretary.

In October 1952, both Don Bayley and Robin went up to Wilmslow for a weekend, a re-creation of Hanslope. Don was there first and they waited together for Robin on the station, Alan showing Don the diffraction pattern that appeared when looking at the station lights through his handkerchief. On the summer visit Alan had taken his usual pleasure in being looked after for a weekend within the more conventional domesticity of the Bayley household, and Don was struck by the contrasting spartan, untidy arrangements of the Prof's home. Alan pointed to a stack of filing trays overflowing with letters from all over the world about logic, but said that he was not bothering to appear at the university now, and worked at home. He explained that he had an assistant who had taken over the organisation of the computer. Don advised him to watch out, or his assistant would take over. "Pooh!" said Alan, as if to say "see if I care!"

But if his computer days were over, this did not mean the end of his underlying interest in the human mind. October 1952 also saw Polanyi and the philosophy department at Manchester score something of a *coup* over the psychology department, by having the French psychologist Jean Piaget to give a course of lectures, which Alan attended. They concerned the child's learning of logical ideas, and connected symbolic logic with actual psychological observations. So perhaps for the first time, Alan found himself listening to arguments about learning and teaching that did not just come from his own experience, and which were touched by modern theories of education that no one at Sherborne would have known to exist. At about the same time he breached his self-sufficiency in another way. He began seeing a Jungian psychoanalyst, Franz Greenbaum.

There was an element of resistance in his attitude towards this step, because of the implication that there was something wrong with him, and in particular, that his homosexuality was something that should be changed. The 1950s were, indeed, witnessing a powerful come-back of psychoanalysis, and increasingly vocal claims to the effect that its techniques could eradicate homosexual desire. But Greenbaum did not take such a view; homosexuality was not "a problem" to him. He accepted Alan as a "natural homosexual," and as a Jungian, he did not consider human activities in terms of displaced or unconscious sexuality. Rather, as a German refugee of 1939 with a Jewish father and Catholic mother, it was the psychology of religion that interested him most. As with Jung himself, there was no devaluation of the intellect in Greenbaum's approach, and he was proud to know Alan as the inventor of the computer and as one who was working on the nature of life. His emphasis, as with Jung, was on the *integration* of "thinking" and "feeling." To apply intelligence to himself; to look at his own system from outside like Gödel, and break his own code — these were natural extensions of Alan's long-growing interest in psychology. A turning-point was indicated on 23 November 1952, when he wrote to Robin in connection with his now completed Ph.D. thesis, and added:

> Have decided to have another, and rather more co-operative go at the psychiatrist. If he can put me in a more resigned frame of mind it would be something.

Thereafter Franz Greenbaum had Alan write down all his dreams, and he filled three notebooks with them.[2] The relationship soon became more that of friendship than that of doctor and patient. But the professional status gave Alan the excuse to devote time and energy to all those things that he had pushed aside so long from the serious male business of "thinking." As with the war, he made the most of the situation in which he found himself.

2. Jung held that dreams had meanings, but did not believe that they could be deciphered according to some fixed scheme: "The interpretation of dreams and symbols demands intelligence. It cannot be turned into a mechanical system. . . . It demands . . . an increasing knowledge of the dreamer's individuality. . . ."

In the analysis of his dreams he was surprised to find that many concerned, or could be interpreted as relating, to his mother in hostile terms. In real life, his relationship with her had continued to grow warmer. The fact that she had taken the news of the trial as she had counted for a great deal. So in her seventieth year, Mrs. Turing found herself becoming one of Alan's few friends. By now she knew that he would never cease to be the "intellectual crank" she had been afraid of, and he knew that she would always concern herself with matters like fish-knives as though still arranging for dinner parties at Coonoor. A gentle bickering, with "*Really*, Alan!" answered by "Mother, don't be so *ridiculous*," characterised the occasional visits. But by this time he had perhaps come to appreciate some of *her* problems and frustrations, while she, in turn, had come a long way from being the muffled Dublin girl at Cheltenham Ladies College, and had perhaps come to realise that Alan's vivacity offered her a taste of the more artistic life that she had been denied. After looking so long for the higher and better in churches and institutions, ranks and titles, she found something of it in her son. For forty years she had been cross with him for doing everything the wrong way, but she found the capacity for change. Alan, too, became less totally dismissive of her preoccupations.

There was plenty of scope for unburying a forty-year-old resentment of a mother so unlike the sensuous, seductive figure of Freudian theory. Perhaps Alan also confronted the figure of his father, whose strength had somehow cancelled itself out, and who had not shown the marathon-runner's quality of his son. Perhaps too there was a hidden disappointment that his father had never even tried to penetrate his concerns in the way that his mother, however irritatingly, attempted. If Alan's friends heard him disparage his mother, they usually heard nothing of his father. But sorting out such complexities of inner feeling was one thing; coping with his situation in the real world of 1952 was quite another, and in this respect psychoanalysis was bound by the same limitation as his imitation game — it was the world of dreaming, not of doing. A private "free association" of ideas was allowed, but free association with male persons — that was the very thing that was forbidden. Franz Greenbaum could not supply the greenwood. Consistency and completeness of mind was not enough; something had to be done.

He had written to a politician about the state of the law, and there was little else that an individual could do — except in refusing to keep quiet. The problem did not lie at an individual level, where the only "solution" was that of being "resigned." He was not charged as one who had harmed another person, but as an enemy of social order. Alan Turing, however, had no interest in ordering other people, while he retained an almost untouched innocence of "why not?" in respect of sex. It was not an issue that could be resolved by rational argument, and not a problem that Dr. Greenbaum could solve.

\* \* \* \* \*

The probation period ended in April 1953. For the past three months they had put an implant of hormone into his thigh, instead of the dosage of pills. Suspecting, with some annoyance, that the effect would last more than three months, he had it taken out. Then he was free, the more so because his future at Manchester was secure. On 15 May 1953 the University Council formally voted to appoint him to a specially created Readership in the Theory of Computing when the five years of the old position ran out on 29 September. This he could reasonably expect to last for ten years, if he wanted it. In this respect, his insouciant "Pooh!" to Don Bayley had been justified: he had a small pay rise, and freedom to work exactly as he chose.

On 10 May Alan sent a letter to Maria Greenbaum, describing a complete solution to a solitaire puzzle, and ending:

> I hope you all have a very nice holiday in Italian Switzer-
> land. I shall not be very far away at Club Mediterrané,
> Ipsos-Corfu, Greece. Yours, Alan Turing.

He had already — most likely in 1951 — been to a Club Mediterrané on the French coast. In this summer of 1953, probably over the period of the coronation, Caliban escaped from the island for his brief ration of fun, to Paris for a short while, and then to Corfu. He would return with half a dozen Greek names and addresses, although from this point of view his exploration of the eastern Mediterranean proved disappointing. As at school, he made mistakes with the French, but still did better than with the Greek.

On the beach in Corfu, with the dark mountains of Albania on the horizon, he could study both the seaweed and the boys. Stalin was dead, and a watery sunshine was emerging over a new Europe. Even the cold shabbiness of British culture was not immune to change, and after more than ten years of ration books, a quite new mood, one that no one had planned for, was coming with the growth of the Fifties. Television, its development arrested in 1939, made its first mass impact with the coronation. In a far more complex and more affluent Britain, the boundaries of official and unofficial ideas would become less clear. An outsider, an intellectual beatnik like Alan Turing, might find more room to breathe.

Besides the general relaxation of manners, the diversification of life was most acute in questions of sex. As in the 1890s, the greater official consciousness of sexuality was matched by a greater outspokenness on the part of individuals — and most notably in America, where the process had begun earlier than in Britain. One particular example of this, the American novel *Finistère* which had appeared in 1951, was much admired by Alan. It described the relationship between a fifteen-year-old boy and his teacher, and like *The Cloven Pine* tried to see life through teenage eyes. It was, however, a relationship very different from the vague nuances of Fred Clayton's *cri de coeur*. In the old days Alan had often teased Fred, shocking him with rather over-simplified assertions about the prevalence of homosexual activity, and this was a book which caught up with the serious thread that had underlain that delight in gossip — a wish to defy the "social stigma" and discuss sex in the same way as one might discuss anything else. Meanwhile *Finistère* also did full justice to the reality of the "social taboo," and its plot followed a complex pattern of private and public disclosures. These the novelist made lead to a conclusion of hopeless doom, as though homosexual life were something inherently self-contradictory and fatal: "the strip of sand, the distinct footprints leading in one single trail into the black water."

In its tragic end, its suicide at a symbolic "end of the earth" — as also with its linking of the boy's longing for a man friend with the failure of his parents' marriage — *Finistère* took its place amidst the older genre of writing about homosexuality. It brought a post-war explicitness into an already dated form. By 1953 the point had been well made that gay men could muddle through like anyone else; thus the new English novel *The Heart in Exile* wended its way through the fading drama-ettes of upper middle class taboos, and the more modern obsession with psychological explanations, and rejected both for an ordinary, commonplace ending, tempered by the observation that "the battle must continue." Angus Wilson's 1952 *Hemlock and After*, with its bleak, black comedy of class and manners, was also close to the matter-of-fact modernity about sex that Alan liked to display. This was another book that he and Robin discussed — more evidence that officialdom and clinical management were not the only legacies of the Second World War. Yet Alan Turing could not share in this anarchic spirit as he might have wished. Less free than he appeared, he too was on the shore of life. A year later, on the evening of 7 June 1954, he killed himself.

Alan Turing's death came as a shock to those who knew him. It fell into no clear sequence of events. Nothing was explicit — there was no warning, no note of explanation. It seemed an isolated act of self-annihilation. That he was an unhappy, tense person; that he was consulting a psychiatrist and had suffered a blow that would have felled many people — all this was clear. But the trial was two years in the past, the hormone treatment had ended a year before, and he seemed to have risen above it all. There was no simple connection in the minds of those who had seen him in the previous two years. On the contrary, his reaction had been so different from the wilting, disgraced, fearful, hopeless figure expected by fiction and drama, that those who had seen it could hardly believe that he was dead. He was simply "not the type" for suicide. But those who resisted a stereotyped association of the trial in 1952 with the death in 1954 perhaps forgot that suicide did not have to be interpreted in terms of weakness or shame. As Alan had quoted Oscar Wilde in 1941, it could be the brave man that did it with a sword.

The inquest, on 10 June, established that it was suicide. The evidence was perfunctory, not for any irregular reason, but because it was so transparently clear a case. He had been found lying neatly in his bed by Mrs. C—— when she came in at five o'clock on Tuesday 8 June. (She would normally have been in on the Monday, but it was the Whitsun bank holiday, and she had had a day off.) There was froth round his mouth, and the pathologist who did the post-mortem that evening easily identified the cause of death as cyanide poisoning, and put the time of death as on the Monday night. In the house was a jar of potassium cyanide, and also a jam jar of a cyanide solution. By the side of his bed was half an apple, out of which several bites had been taken. They did not analyse the apple, and so it was never properly established that, as seemed perfectly obvious, the apple had been dipped in the cyanide. . . .

Anyone arguing that it was an accident would have had to admit that it was certainly one of suicidal folly. Alan Turing himself would have been fascinated by the difficulty of drawing a line between accident and suicide, a line defined only by a conception of free will.

Interested as he was by the idea of attaching a random element into a computer, a "roulette wheel," to give it the appearance of freedom, there might conceivably have been some Russian roulette aspect to his end. But even if this were so, his body was not one of a man fighting for life against the suffocation induced by cyanide poisoning. It was that of one resigned to death.

Like Snow White, he ate a poisoned apple, dipped in the witches' brew. But what were the ingredients of the brew? What would a less artificial inquest have made of his last years? It would depend upon the level of description, "not the will of man as such but our presentation of it." To ask what caused his death is like asking what caused the First World War: a pistol shot, the railway timetables, the armament race, or the logic of nationalism could all be held accountable. At one level the atoms were simply moving according to physical law; at other levels there was mystery; at another, a kind of inevitability. . . .

# JONATHAN M. BORWEIN
## AND PETER B. BORWEIN

■

Srinivasa Ramanujan (1887–1920) had little formal training, and his career as a full-time mathematician had lasted less than seven years when he died at the age of thirty-two, but his talent has been ranked with that of giants such as Gauss and Euler. The English mathematician G. H. Hardy used to recount, as an example of Ramanujan's intuitive powers, the story about a day he went to see Ramanujan in the Putney hospital where he lay dying. Hardy said, by way of starting the conversation, "The number of my taxicab was 1729. It seemed to me rather a dull number." Ramanujan instantly replied, "No, Hardy! No, Hardy! It is a very interesting number. It is the smallest number expressible as the sum of two cubes in two different ways." In the following article two Scottish mathematicians, Jonathan M. Borwein and his brother Peter, discuss today's wide application of just one of Ramanujan's discoveries, a way of finding the value of pi.

# *Ramanujan and Pi*

PI, THE RATIO OF ANY circle's circumference to its diameter, was computed in 1987 to an unprecedented level of accuracy: more than 100 million decimal places. Last year also marked the centenary of the birth of Srinivasa Ramanujan, an enigmatic Indian mathematical genius who spent much of his short life in isolation and poor health. The two events are in fact closely linked, because the basic approach underlying the most recent computations of pi was anticipated by Ramanujan, although its implementation had to await the formulation of efficient algorithms (by various workers including us), modern supercomputers and new ways to multiply numbers.

Aside from providing an arena in which to set records of a kind,

647

the quest to calculate the number to millions of decimal places may seem rather pointless. Thirty-nine places of pi suffice for computing the circumference of a circle girdling the known universe with an error no greater than the radius of a hydrogen atom. It is hard to imagine physical situations requiring more digits. Why are mathematicians and computer scientists not satisfied with, say, the first 50 digits of pi?

Several answers can be given. One is that the calculation of pi has become something of a benchmark computation: it serves as a measure of the sophistication and reliability of the computers that carry it out. In addition, the pursuit of ever more accurate values of pi leads mathematicians to intriguing and unexpected niches of number theory. Another and more ingenuous motivation is simply "because it's there." In fact, pi has been a fixture of mathematical culture for more than two and a half millenniums.

Furthermore, there is always the chance that such computations will shed light on some of the riddles surrounding pi, a universal constant that is not particularly well understood, in spite of its relatively elementary nature. For example, although it has been proved that pi cannot ever be exactly evaluated by subjecting positive integers to any combination of adding, subtracting, multiplying, dividing or extracting roots, no one has succeeded in proving that the digits of pi follow a random distribution (such that each number from 0 to 9 appears with equal frequency). It is possible, albeit highly unlikely, that after a while all the remaining digits of pi are 0's and 1's or exhibit some other regularity. Moreover, pi turns up in all kinds of unexpected places that have nothing to do with circles. If a number is picked at random from the set of integers, for instance, the probability that it will have no repeated prime divisors is six divided by the square of pi. No different from other eminent mathematicians, Ramanujan was prey to the fascinations of the number.

The ingredients of the recent approaches to calculating pi are among the mathematical treasures unearthed by renewed interest in Ramanujan's work. Much of what he did, however, is still inaccessible to investigators. The body of his work is contained in his "Notebooks," which are personal records written in his own nomenclature. To make matters more frustrating for mathematicians who have studied the "Notebooks," Ramanujan generally did not include formal proofs for his theorems. The task of deciphering and editing the

"Notebooks" is only now nearing completion, by Bruce C. Berndt of the University of Illinois at Urbana-Champaign.

To our knowledge no mathematical redaction of this scope or difficulty has ever been attempted. The effort is certainly worthwhile. Ramanujan's legacy in the "Notebooks" promises not only to enrich pure mathematics but also to find application in various fields of mathematical physics. Rodney J. Baxter of the Australian National University, for example, acknowledges that Ramanujan's findings helped him to solve such problems in statistical mechanics as the so-called hard-hexagon model, which considers the behavior of a system of interacting particles laid out on a honeycomblike grid. Similarly, Carlos J. Moreno of the City University of New York and Freeman J. Dyson of the Institute for Advanced Study have pointed out that Ramanujan's work is beginning to be applied by physicists in superstring theory.

Ramanujan's stature as a mathematician is all the more astonishing when one considers his limited formal education. He was born on December 22, 1887, into a somewhat impoverished family of the Brahmin caste in the town of Erode in southern India and grew up in Kumbakonam, where his father was an accountant to a clothier. His mathematical precocity was recognized early, and at the age of seven he was given a scholarship to the Kumbakonam Town High School. He is said to have recited mathematical formulas to his schoolmates — including the value of pi to many places.

When he was 12, Ramanujan mastered the contents of S. L. Loney's rather comprehensive *Plane Trigonometry*, including its discussion of the sum and products of infinite sequences, which later were to figure prominently in his work. (An infinite sequence is an unending string of terms, often generated by a simple formula. In this context the interesting sequences are those whose terms can be added or multiplied to yield an identifiable, finite value. If the terms are added, the resulting expression is called a series; if they are multiplied, it is called a product.) Three years later he borrowed the *Synopsis of Elementary Results in Pure Mathematics*, a listing of some 6,000 theorems (most of them given without proof) compiled by G. S. Carr, a tutor at the University of Cambridge. Those two books were the basis of Ramanujan's mathematical training.

In 1903 Ramanujan was admitted to a local government college. Yet total absorption in his own mathematical diversions at the expense of everything else caused him to fail his examinations, a pat-

tern repeated four years later at another college in Madras. Ramanujan did set his avocation aside — if only temporarily — to look for
a job after his marriage in 1909. Fortunately in 1910 R. Ramachan-
dra Rao, a well-to-do patron of mathematics, gave him a monthly
stipend largely on the strength of favorable recommendations from
various sympathetic Indian mathematicians and the findings he
already had jotted down in the "Notebooks."

In 1912, wanting more conventional work, he took a clerical
position in the Madras Port Trust, where the chairman was a British
engineer, Sir Francis Spring, and the manager was V. Ramaswami
Aiyar, the founder of the Indian Mathematical Society. They encouraged Ramanujan to communicate his results to three prominent
British mathematicians. Two apparently did not respond; the one
who did was G. H. Hardy of Cambridge, now regarded as the foremost British mathematician of the period.

Hardy, accustomed to receiving crank mail, was inclined to disregard
Ramanujan's letter at first glance the day it arrived, January 16, 1913.
But after dinner that night Hardy and a close colleague, John E.
Littlewood, sat down to puzzle through a list of 120 formulas and
theorems Ramanujan had appended to his letter. Some hours later
they had reached a verdict: they were seeing the work of a genius
and not a crackpot. (According to his own "pure-talent scale" of
mathematicians, Hardy was later to rate Ramanujan a 100, Littlewood a 30 and himself a 25. The German mathematician David Hilbert, the most influential figure of the time, merited only an 80.)
Hardy described the revelation and its consequences as the one
romantic incident in his life. He wrote that some of Ramanujan's
formulas defeated him completely, and yet "they must be true,
because if they were not true, no one would have had the imagination
to invent them."

Hardy immediately invited Ramanujan to come to Cambridge.
In spite of his mother's strong objections as well as his own reservations, Ramanujan set out for England in March of 1914. During
the next five years Hardy and Ramanujan worked together at Trinity
College. The blend of Hardy's technical expertise and Ramanujan's
raw brilliance produced an unequaled collaboration. They published
a series of seminal papers on the properties of various arithmetic
functions, laying the groundwork for the answer to such questions
as: How many prime divisors is a given number likely to have? How

many ways can one express a number as a sum of smaller positive integers?

In 1917 Ramanujan was made a Fellow of the Royal Society of London and a Fellow of Trinity College — the first Indian to be awarded either honor. Yet as his prominence grew his health deteriorated sharply, a decline perhaps accelerated by the difficulty of maintaining a strict vegetarian diet in war-rationed England. Although Ramanujan was in and out of sanatoriums, he continued to pour forth new results. In 1919, when peace made travel abroad safe again, Ramanujan returned to India. Already an icon for young Indian intellectuals, the 32-year-old Ramanujan died on April 26, 1920, of what was then diagnosed as tuberculosis but now is thought to have been a severe vitamin deficiency. True to mathematics until the end, Ramanujan did not slow down during his last, pain-racked months, producing the remarkable work recorded in his so-called "Lost Notebook."

Ramanujan's work on pi grew in large part out of his investigation of modular equations, perhaps the most thoroughly treated subject in the "Notebooks." Roughly speaking, a modular equation is an algebraic relation between a function expressed in terms of a variable $x$ — in mathematical notation, $f(x)$ — and the same function expressed in terms of $x$ raised to an integral power, for example $f(x^2)$, $f(x^3)$ or $f(x^4)$. The "order" of the modular equation is given by the integral power. The simplest modular equation is the second-order one:

$$f(x) = 2 \sqrt{f(x^2)}/[1 + f(x^2)].$$

Of course, not every function will satisfy a modular equation, but there is a class of functions, called modular functions, that do. These functions have various surprising symmetries that give them a special place in mathematics.

Ramanujan was unparalleled in his ability to come up with solutions to modular equations that also satisfy other conditions. Such solutions are called singular values. It turns out that solving for singular values in certain cases yields numbers whose natural logarithms coincide with pi (times a constant) to a surprising number of places. Applying this general approach with extraordinary virtuosity, Ramanujan produced many remarkable infinite series as well as single-term approximations for pi. Some of them are given in Ramanujan's one formal paper on the subject, *Modular Equations and Approxima-tions to* π, published in 1914.

Ramanujan's attempts to approximate pi are part of a venerable tradition. The earliest Indo-European civilizations were aware that the area of a circle is proportional to the square of its radius and that the circumference of a circle is directly proportional to its diameter. Less clear, however, is when it was first realized that the ratio of any circle's circumference to its diameter and the ratio of any circle's area to the square of its radius are in fact the same constant, which today is designated by the symbol π. (The symbol, which gives the constant its name, is a latecomer in the history of mathematics, having been introduced in 1706 by the English mathematical writer William Jones and popularized by the Swiss mathematician Leonhard Euler in the 18th century.)

Archimedes of Syracuse, the greatest mathematician of antiquity, rigorously established the equivalence of the two ratios in his treatise *Measurement of a Circle*. He also calculated a value for pi based on mathematical principles rather than on direct measurement of a circle's circumference, area and diameter. What Archimedes did was to inscribe and circumscribe regular polygons (polygons whose sides are all the same length) on a circle assumed to have a diameter of one unit and to consider the polygons' respective perimeters as lower and upper bounds for possible values of the circumference of the circle, which is numerically equal to pi.

This method of approaching a value for pi was not novel: inscribing polygons of ever more sides in a circle had been proposed earlier by Antiphon, and Antiphon's contemporary, Bryson of Her-aclea, had added circumscribed polygons to the procedure. What was novel was Archimedes' correct determination of the effect of dou-bling the number of sides on both the circumscribed and the inscribed polygons. He thereby developed a procedure that, when repeated enough times, enables one in principle to calculate pi to any number of digits. (It should be pointed out that the perimeter of a regular polygon can be readily calculated by means of simple trig-onometric functions: the sine, cosine and tangent functions. But in Archimedes' time, the third century B.C., such functions were only partly understood. Archimedes therefore had to rely mainly on geo-metric constructions, which made the calculations considerably more demanding than they might appear today.)

Archimedes began with inscribed and circumscribed hexagons, which yield the inequality $3 < \pi < 2\sqrt{3}$. By doubling the number of sides four times, to 96, he narrowed the range of pi to between $3^{10}/_{71}$ and $3^1/_7$, obtaining the estimate $\pi \approx 3.14$. There is some evidence that the extant text of *Measurement of a Circle* is only a fragment of a larger work in which Archimedes described how, starting with decagons and doubling them six times, he got a five-digit estimate: $\pi \approx 3.1416$.

Archimedes' method is conceptually simple, but in the absence of a ready way to calculate trigonometric functions it requires the extraction of roots, which is rather time-consuming when done by hand. Moreover, the estimates converge slowly to pi: their error decreases by about a factor of four per iteration. Nevertheless, all European attempts to calculate pi before the mid-17th century relied in one way or another on the method. The 16th-century Dutch mathematician Ludolph van Ceulen dedicated much of his career to a computation of pi. Near the end of his life he obtained a 32-digit estimate by calculating the perimeter of inscribed and circumscribed polygons having $2^{62}$ (some $10^{18}$) sides. His value for pi, called the Ludolphian number in parts of Europe, is said to have served as his epitaph.

The development of calculus, largely by Isaac Newton and Gottfried Wilhelm Leibniz, made it possible to calculate pi much more expeditiously. Calculus provides efficient techniques for computing a function's derivative (the rate of change in the function's value as its variables change) and its integral (the sum of the function's values over a range of variables). Applying the techniques, one can demonstrate that inverse trigonometric functions are given by integrals of quadratic functions that describe the curve of a circle. (The inverse of a trigonometric function gives the angle that corresponds to a particular value of the function. For example, the inverse tangent of 1 is 45 degrees or, equivalently, $\pi/4$ radians.)

(The underlying connection between trigonometric functions and algebraic expressions can be appreciated by considering a circle that has a radius of one unit and its center at the origin of a Cartesian $x$-$y$ plane. The equation for the circle — whose area is numerically equal to pi — is $x^2 + y^2 = 1$, which is a restatement of the Pythagorean theorem for a right triangle with a hypotenuse equal to 1. Moreover, the sine and cosine of the angle between the positive $x$ axis and any point on the circle are equal respectively to the point's coordinates, $y$ and $x$; the angle's tangent is simply $y/x$.)

Of more importance for the purposes of calculating pi, however, is the fact that an inverse trigonometric function can be "expanded" as a series, the terms of which are computable from the derivatives of the function. Newton himself calculated pi to 15 places by adding the first few terms of a series that can be derived as an expression for the inverse of the sine function. He later confessed to a colleague: "I am ashamed to tell you to how many figures I carried these calculations, having no other business at the time."

In 1674 Leibniz derived the formula $1 - 1/3 + 1/5 - 1/7 \cdots = \pi/4$, which is the inverse tangent of 1. (The general inverse-tangent series was originally discovered in 1671 by the Scottish mathematician James Gregory. Indeed, similar expressions appear to have been developed independently several centuries earlier in India.) The error of the approximation, defined as the difference between the sum of $n$ terms and the exact value of $\pi/4$, is roughly equal to the $n + 1$th terms in the series. Since the denominator of each successive term increases by only 2, one must add approximately 50 terms to get two-digit accuracy, 500 terms for three-digit accuracy and so on. Summing the terms of the series to calculate a value for pi more than a few digits long is clearly prohibitive.

An observation made by John Machin, however, made it practicable to calculate pi by means of a series expansion for the inverse-tangent function. He noted that pi divided by 4 is equal to 4 times the inverse tangent of 1/5 minus the inverse tangent of 1/239. Because the inverse-tangent series for a given value converges more quickly the smaller the value is, Machin's formula greatly simplified the calculation. Coupling his formula with the series expansion for the inverse tangent, Machin computed 100 digits of pi in 1706. Indeed, his technique proved to be so powerful that all extended calculations of pi from the beginning of the 18th century until recently relied on variants of the method.

Two 19th-century calculations deserve special mention. In 1844 Johann Dase computed 205 digits of pi in a matter of months by calculating the values of three inverse tangents in a Machin-like formula. Dase was a calculating prodigy who could multiply 100-digit numbers entirely in his head — a feat that took him roughly eight hours. (He was perhaps the closest precursor of the modern supercomputer, at least in terms of memory capacity.) In 1853 William

Shanks outdid Dase by publishing his computation of pi to 607 places, although the digits that followed the 527th place were wrong. Shanks' task took years and was a rather routine, albeit laborious, application of Machin's formula. (In what must itself be some kind of record, 92 years passed before Shanks' error was detected, in a comparison between his value and a 530-place approximation produced by D. F. Ferguson with the aid of a mechanical calculator.)

The advent of the digital computer saw a renewal of efforts to calculate ever more digits of pi, since the machine was ideally suited for lengthy, repetitive "number crunching." ENIAC, one of the first digital computers, was applied to the task in June, 1949, by John von Neumann and his colleagues. ENIAC produced 2,037 digits in 70 hours. In 1957 G. E. Felton attempted to compute 10,000 digits of pi, but owing to a machine error only the first 7,480 digits were correct. The 10,000-digit goal was reached by F. Genuys the following year on an IBM 704 computer. In 1961 Daniel Shanks and John W. Wrench, Jr., calculated 100,000 digits of pi in less than nine hours on an IBM 7090. The million-digit mark was passed in 1973 by Jean Guilloud and M. Bouyer, a feat that took just under a day of computation on a CDC 7600. (The computations done by Shanks and Wrench and by Guilloud and Bouyer were in fact carried out twice using different inverse-tangent identities for pi. Given the history of both human and machine error in these calculations, it is only after such verification that modern "digit hunters" consider a record officially set.)

Although an increase in the speed of computers was a major reason ever more accurate calculations for pi could be performed, it soon became clear that there were inescapable limits. Doubling the number of digits lengthens computing time by at least a factor of four, if one applies the traditional methods of performing arithmetic in computers. Hence even allowing for a hundredfold increase in computational speed, Guilloud and Bouyer's program would have required at least a quarter century to produce a billion-digit value for pi. From the perspective of the early 1970's such a computation did not seem realistically practicable.

Yet the task is now feasible, thanks not only to faster computers but also to new, efficient methods for multiplying large numbers in computers. A third development was also crucial: the advent of iterative algorithms that quickly converge to pi. (An iterative algorithm can be expressed as a computer program that repeatedly performs the same arithmetic operations, taking the output of one cycle as the input for the next.) These algorithms, some of which we constructed, were in many respects anticipated by Ramanujan, although he knew nothing of computer programming. Indeed, computers not only have made it possible to apply Ramanujan's work but also have helped to unravel it. Sophisticated algebraic-manipulation software has allowed further exploration of the road Ramanujan traveled alone and unaided 75 years ago.

One of the interesting lessons of theoretical computer science is that many familiar algorithms, such as the way children are taught to multiply in grade school, are far from optimal. Computer scientists gauge the efficiency of an algorithm by determining its bit complexity: the number of times individual digits are added or multiplied in carrying out an algorithm. By this measure, adding two $n$-digit numbers in the normal way has a bit complexity that increases in step with $n$; multiplying two $n$-digit numbers in the normal way has a bit complexity that increases as $n^2$. By traditional methods, multiplication is much "harder" than addition in that it is much more time-consuming.

Yet, as was shown in 1971 by A. Schönhage and V. Strassen, the multiplication of two numbers can in theory have a bit complexity only a little greater than addition. One way to achieve this potential reduction in bit complexity is to implement so-called fast Fourier transforms (FFT's). FFT-based multiplication of two large numbers allows the intermediary computations among individual digits to be carefully orchestrated so that redundancy is avoided. Because division and root extraction can be reduced to a sequence of multiplications, they too can have a bit complexity just slightly greater than that of addition. The result is a tremendous saving in bit complexity and hence in computation time. For this reason all recent efforts to calculate pi rely on some variation of the FFT technique for multiplication.

Yet for hundreds of millions of digits of pi to be calculated practically a beautiful formula known a century and a half earlier to Carl Friedrich Gauss had to be rediscovered. In the mid-1970's Richard P. Brent and Eugene Salamin independently noted that the formula produced an algorithm for pi that converged quadratically, that is, the number of digits doubled with each iteration. Between 1983 and the present Yasumasa Kanada and his colleagues at the University of Tokyo have employed this algorithm to set several world records for the number of digits of pi.

We wondered what underlies the remarkably fast convergence to pi of the Gauss-Brent-Salamin algorithm, and in studying it we developed general techniques for the construction of similar algorithms that rapidly converge to pi as well as to other quantities. Building on a theory outlined by the German mathematician Karl Gustav Jacob Jacobi in 1829, we realized we could in principle arrive at a value for pi by evaluating integrals of a class called elliptic integrals, which can serve to calculate the perimeter of an ellipse. (A circle, the geometric setting of previous efforts to approximate pi, is simply an ellipse with axes of equal length.)

Elliptic integrals cannot generally be evaluated as integrals, but they can be easily approximated through iterative procedures that rely on modular equations. We found that the Gauss-Brent-Salamin algorithm is actually a specific case of our more general technique relying on a second-order modular equation. Quicker convergence to the value of the integral, and thus a faster algorithm for pi, is possible if higher-order modular equations are used, and so we have also constructed various algorithms based on modular equations of third, fourth and higher orders.

In January, 1986, David H. Bailey of the National Aeronautics and Space Administration's Ames Research Center produced 29,360,000 decimal places of pi by iterating one of our algorithms 12 times on a Cray-2 supercomputer. Because the algorithm is based on a fourth-order modular equation, it converges on pi quartically, more than quadrupling the number of digits with each iteration. A year later Kanada and his colleagues carried out one more iteration to attain 134,217,000 places on an NEC SX-2 supercomputer and thereby verified a similar computation they had done earlier using the Gauss-Brent-Salamin algorithm. (Iterating our algorithm twice more — a feat entirely feasible if one could somehow monopolize a supercomputer for a few weeks — would yield more than two billion digits of pi.)

Iterative methods are best suited for calculating pi on a computer, and so it is not surprising that Ramanujan never bothered to pursue them. Yet the basic ingredients of the iterative algorithms for pi — modular equations in particular — are to be found in Ramanujan's work. Parts of his original derivation of infinite series and approximations for pi more than three-quarters of a century ago must have paralleled our own efforts to come up with algorithms for pi. Indeed, the formulas he lists in his paper on pi and in the "Notebooks" helped us greatly in the construction of some of our algorithms. For example, although we were able to prove that an 11th-order algorithm exists and knew its general formulation, it was not until we stumbled on Ramanujan's modular equations of the same order that we discovered its unexpectedly simple form.

Conversely, we were also able to derive all Ramanujan's series from the general formulas we had developed. The derivation of one, which converged to pi faster than any other series we knew at the time, came about with a little help from an unexpected source. We had justified all the quantities in the expression for the series except one: the coefficient 1,103, which appears in the numerator of the expression. We were convinced — as Ramanujan must have been — that 1,103 had to be correct. To prove it we had either to simplify a daunting equation containing variables raised to powers of several thousand or to delve considerably further into somewhat arcane number theory.

By coincidence R. William Gosper, Jr., of Symbolics, Inc., had decided in 1985 to exploit the same series of Ramanujan's for an extended-accuracy value for pi. When he carried out the calculation to more than 17 million digits (a record at the time), there was to his knowledge no proof that the sum of the series actually converged to pi. Of course, he knew that millions of digits of his value coincided with an earlier Gauss-Brent-Salamin calculation done by Kanada. Hence the possibility of error was vanishingly small.

As soon as Gosper had finished his calculation and verified it against Kanada's, however, we had what we needed to prove that 1,103 was the number needed to make the series true to within one part in $10^{10,000,000}$. In much the same way that a pair of integers differing by less than 1 must be equal, his result sufficed to specify the number: it is precisely 1,103. In effect, Gosper's computation became part of our proof. We knew that the series (and its associated algorithm) is so sensitive to slight inaccuracies that if Gosper had used any other value for the coefficient or, for that matter, if the computer had introduced a single-digit error during the calculation, he would have ended up with numerical nonsense instead of a value for pi.

Ramanujan-type algorithms for approximating pi can be shown to be very close to the best possible. If all the operations involved in the execution of the algorithms are totaled (assuming that the best techniques known for addition, multiplication and root extraction are applied), the bit complexity of computing $n$ digits of pi is only marginally greater than that of multiplying two $n$-digit numbers. But multiplying two $n$-digit numbers by means of an FFT-based technique is only marginally more complicated than summing two $n$-digit numbers, which is the simplest of the arithmetic operations possible on a computer.

Mathematics has probably not yet felt the full impact of Ramanujan's genius. There are many other wonderful formulas contained in the "Notebooks" that revolve around integrals, infinite series and continued fractions (a number plus a fraction, whose denominator can be expressed as a number plus a fraction, whose denominator can be expressed as a number plus a fraction, and so on). Unfortunately they are listed with little — if any — indication of the method by which Ramanujan proved them. Littlewood wrote: "If a significant piece of reasoning occurred somewhere, and the total mixture of evidence and intuition gave him certainty, he looked no further."

The herculean task of editing the "Notebooks," initiated 60 years ago by the British analysts G. N. Watson and B. N. Wilson and now being completed by Bruce Berndt, requires providing a proof, a source or an occasional correction for each of many thousands of asserted theorems and identities. A single line in the "Notebooks" can easily elicit many pages of commentary. The task is made all the more difficult by the nonstandard mathematical notation in which the formulas are written. Hence a great deal of Ramanujan's work will not become accessible to the mathematical community until Berndt's project is finished.

Ramanujan's unique capacity for working intuitively with complicated formulas enabled him to plant seeds in a mathematical garden (to borrow a metaphor from Freeman Dyson) that is only now coming into bloom. Along with many other mathematicians, we look forward to seeing which of the seeds will germinate in future years and further beautify the garden.

# LEE DEMBART

■

Freedom is perhaps the greatest gift proffered by a life of the mind — the freedom to think and do as one pleases, guided not by the commands of others but by the lamp of one's own reasoned judgment. Few thinkers have lived more freely than the Hungarian mathematician Paul Erdos (b. 1913), here profiled for the *Los Angeles Times* by science writer Lee Dembart (b. 1946), who has written many noteworthy newspaper reports on mathematics, philosophy, and science.

# *Paul Erdos: Mathematician*

PAUL ERDOS, ONE OF the world's foremost mathematicians, was giving a lecture at Washington University in St. Louis not long ago, and the university routinely asked for his Social Security number so they could pay him. He couldn't remember it.

That evening a university department chairman prevailed on the distinguished visitor to come to his home for dinner. He needn't have bothered. Other guests were present, mathematics was not discussed, and Erdos fell asleep at the table.

Wherever he goes, stories about Erdos abound, for in addition to being universally acclaimed for his spectacular contributions to mathematics, Erdos (pronounced ER-dish) leads an unusual if not eccentric life. He has no home, but travels around the world from university to university, living on honorariums and on the kindness of other mathematicians. All he owns he carries with him in a medium-sized suitcase. "I really have no feeling for property," he says. "I have only the bare minimum."

He has traveled continuously for the last 50 years, never spending longer than a month in one place. Wherever he goes, Erdos' main preoccupation is mathematics. Other scientists need equipment and laboratories to carry out experiments, but mathematicians need only a pencil and paper, and Erdos gets by with little more.

What mathematicians do is prove theorems, refining and adding to an elegant intellectual structure that traces back to the ancient Greeks. It is said that in every generation there is one great mathematician, and the rest don't do any harm.

Erdos, who may be this generation's great mathematician, turns 70 today. But unlike most mathematicians, whose careers are long since over by that age, he is still going strong. He published his first paper 52 years ago, and now has about 1,000 — an extraordinary total — to his credit. Each paper has contained at least one significant new result. He has several hundred co-authors, more than any mathematician in history.

"He will come to a place, and each person has saved up problems that he's been working on unsuccessfully," explained Solomon Golomb of USC. "They ask Erdos, how would you approach such and such a problem? Erdos will think about it for a minute or two, and then he will describe the solution. And the person who had been working on the problem will write up a joint paper. Erdos typically in such a collaboration is the one who supplied the solution."

Mark Kac of USC described his own collaboration with Erdos: "It was during a lecture I gave. I stated something which I believed to be true. He was peacefully asleep or dozing, but at the mention of number theory, he perked up and said, 'Would you repeat the question?' I repeated the question. Before the lecture was over, he had completed the proof."

To celebrate Erdos' birthday anniversary, international symposiums are being held in Cambridge, England, and in his native Budapest, where colleagues and collaborators will discuss his contributions to number theory, set theory, probability theory, combinatorics and much more.

"You can't say he is *the* greatest mathematician in the world today, but he's way, way, way up there," said Leon Bankoff, a Wilshire Boulevard dentist and amateur mathematician who has published papers with Erdos. "There's no way to ascertain who is *the* greatest.

"The stuff that Erdos fools with is so deep that many of the most advanced journals find difficulty in handling the convolutions of his brain," Bankoff said.

"He is as pure a mathematician as they come," said Ervin Y. Rodin, chairman of the systems science and mathematics department at Washington University. "His work is totally without use, yet his results and his theories are coming to be applied. Had it not been for his work, many of the things that are being done today couldn't have been done. In number theory and factorization, his results stimulate enormous interest."

"Very often his proofs are quite elegant," Golomb said. "Sometimes they involve just better and deeper techniques than some other people normally have at their fingertips."

A small, slight man whose English gives evidence of his native Hungary, Erdos' physical stamina is as untouched by age as his mental skills. He typically sleeps five hours or less a night. "Plenty of time to rest in the grave," is Erdos' standard response to people who tell him to slow down.

He maintains a whirlwind schedule. This month, before going to Europe, he spent time in Champaign, Ill.; St. Louis; Lincoln, Neb.; Houston; Memphis; Norman, Okla., and Atlanta.

Everywhere he meets with mathematics students and other mathematicians, asking them all the same questions: What are you working on? Is there anything interesting there? What results have you heard?

"When he comes to Los Angeles, he has many colleagues here and usually lines up his schedule so that he is more or less constantly working with somebody," said Bruce Rothschild of UCLA, who regularly puts Erdos up in Los Angeles.

"People come from the whole area, and he keeps up a phenomenal pace," Rothschild said. "He's always got lots of different problems. It's not uncommon for him to be in a room with several people and working on different things with each of them."

While his life style appears to require little money, Erdos is not without financial resources. The money he receives from his lectures is handled by Ronald Graham of Bell Laboratories, who puts it in a bank account and pays his bills.

And Erdos has been generous with his funds. To spur progress in mathematics, he offers cash prizes ranging from $100 to $10,000 for proofs of interesting theorems. He has hundreds of prize offers outstanding and so far has paid out several thousand dollars.

"People ask, what would happen to you if all of your theorems would be proved?" Erdos says. "I would be ruined. But what would happen to the strongest bank if all of the depositors suddenly asked for their money back? It would be wiped out. That is more likely than for all of my theorems to be proved."

When he's not talking mathematics face-to-face with someone, Erdos uses the telephone, usually dispensing with social niceties and getting right down to the subject at hand.

"You'll get a telephone call from him, and he'll immediately continue with the conversation you may have had some time ago," Rothschild said. "He'll say, 'Now suppose you have a set with $N$ elements. . . .' And you say, 'Now wait a minute, Paul, what problem are we talking about?'"

Erdos was born March 26, 1913, the only child of two mathematicians. He was a child prodigy who amazed people by multiplying three-digit numbers in his head. His parents were overly protective. By his own account, it was not until 1934, when he went to England at the age of 21, that he first buttered his own bread.

He was 18 when he published his first mathematical paper, a new and elegant proof of an old theorem that there is always a prime number between any two numbers, $N$ and $2N$. (A prime number is a number that has no divisors other than itself and 1.)

Perhaps his most famous — and most controversial — work came in the late 1940s when Erdos solved an old problem in mathematics by coming up with what is called an "elementary" proof of the prime number theorem. The theorem, first proposed in the 18th Century, says that the approximate number of primes up to some given number, $X$, is $X$ divided by the natural logarithm of $X$.

In 1896, the prime number theorem was proved using the theory of functions of a complex variable, but mathematicians considered this somehow inelegant. Erdos came up with a proof that did not use this technique.

Erdos, who has never married, traveled the world for many years with his mother, who died in 1971 in her 90s. Since then he has traveled by himself, cutting an unforgettable figure in rumpled suit, silk shirt (he has a bothersome skin condition that is easily irritated) and sandals. When he was in St. Louis recently, the weather was very warm, and Erdos removed his socks.

"You have to have your brain open," Erdos said of his style of work. "If an idea comes, you have to be able to drop everything and concentrate on this idea. Some people can only work if they withdraw quietly to their room. I could just take a paper and sit and think. In an airplane you can think very well. Also before going to sleep."

When Erdos comes to visit, his host knows that his time will not be his own. Erdos expects to be driven places, to have telephone numbers and airline schedules looked up and to have phone calls placed for him. Mathematicians call it "Uncle Paul sitting."

"Let's say he is an original," Kac said. "He has an original life style. He is a wonderful fellow. He has all the attributes of a saint, including being difficult to live with." He is known to be extremely generous, almost to a fault. He recently bought $1,000 worth of books from all over the world for young mathematicians in Hungary. He has established two prizes for students in Hungary and Israel. When a mathematician needs money, Erdos more than once has provided it, no questions asked.

Part of Erdos' originality involves a private language he has made up in which "bosses" are women, "slaves" are men, an "epsilon" is a child (epsilon is the mathematical symbol for a small amount) and "Sam and Joe" are the United States and the Soviet Union. "Poison, noise and bosses" means wine, women and song.

He is a well-known pessimist, having been complaining about aging since he was 30. "A 70th birthday should be mourned, not celebrated," Erdos said.

"As long as I've known him (more than 30 years), he has always complained of being old and sick and imminently close to death," Golomb said.

Erdos' absent-mindedness is as legendary as his mathematical skills. On a trip to Caltech in January, he lost his sweater twice in one day. The first time it was recovered; the second time it was not.

"It's the absent-mindedness that goes with single-mindedness," Golomb said. "If it doesn't involve mathematics, it's not something that he's going to devote much energy and attention to, and it will quite possibly slip his mind."

Erdos says he doesn't mind all of the stories about him, even if some of them are not true. "As long as the story is not malicious," he adds.

What motivates a man to organize his life this way? "This would be like if you asked Bach what pleasure does he get from composing," Erdos answers. "Probably when you suddenly see something which was hidden. Something esthetic."

Graham of Bell Laboratories explained, "Erdos needs to know, to understand. He's really driven to try to answer one more question, one more mountain to climb."

# NIGEL CALDER

■

Since winning a 1979 Nobel Prize for his work on unified theory, the Pakistani physicist Abdus Salam (b. 1926) has become something of a role model for aspiring scientists throughout the Islamic world. Nigel Calder (b. 1931), a British science writer known for his laudable BBC documentaries and for the entertaining and fact-filled books on which they are based, wrote this profile of Salam in 1967.

# A Man of Science — Abdus Salam

ONE SUMMER NOON IN 1940, Abdus Salam came cycling into Jhang, a country town in the Punjab region of British India. The townspeople had lined the streets to greet him because, at the age of 14, he had just made the highest marks ever recorded in the matriculation examination of Panjab University. The result was a national sensation, but nowhere more than in Jhang, which had so little tradition in schooling.

From that moment, Abdus Salam was public property. Scholarships were to relieve his family of the cost of his further education, first at Panjab University's Government College, Lahore, and later at St. John's College at Cambridge University in England.

Salam was to astonish the ablest men of his time and become a leader in theoretical physics. Today, at 41, he is international property. He directs the new International Centre for Theoretical Physics (ICTP) in Trieste, Italy, on leave of absence from Imperial College of Science and Technology, University of London, an institution similar to the Massachusetts Institute of Technology (MIT) in the United States. He is also chief scientific adviser to the president of Pakistan and one of the "wise men" entrusted by the United Nations (UN) with guiding the application of science and technology to the global war on poverty. But such public recognition says little about the man, or about his role in the world of physics.

Of course, Salam had been a child prodigy, but even his talents could have been stifled by neglect in his corner of the world. The boy had been lucky in his family circle, which has a long tradition of piety and learning. His father was a minor official in the farming community along a tributary of the mighty Indus River that gave India its name. Each day, when the boy came home from school, his father would question him closely on what he had studied. And if any other encouragement was needed, his maternal uncle, a former Moslem missionary to West Africa, supplied it.

As Salam's education proceeded, the traditions of Islam were complemented in his mind by Western studies. He read English literature as well as the *Koran*. . . . His prime subject was mathematics, but that would not have been sufficient to save him from the natural destiny of ambitious young men in his country — entry into the civil service. World War II had put a moratorium on new appointments, however, so, in 1946, he went to Cambridge University to continue his studies.

Cambridge captivated him, especially the flower gardens of St. John's. Later he was to turn down a fellowship at neighbouring Trinity College, considered the best college in Great Britain, for aesthetic reasons — the Trinity grounds were not as pleasing as those at St. John's. He became a *wrangler* (Cambridge's traditional term for a first-class mathematician) without much difficulty. Thereafter, Salam followed the advice of Fred Hoyle, the cosmologist, and took a course in advanced physics. "Otherwise," Hoyle told him, "you will never be able to look an experimental physicist in the eye."

Salam did more than take a course, he became a research student in experimental physics in the famous Cavendish Laboratory of Cambridge University. That move could have been a mistake. Salam was no good in the laboratory. He would get bizarre results in his experiments and explain them by inventing a new theory. He importuned the Cambridge theoreticians for something more to his taste. The rare self-confidence and fastidiousness of the young scholar demanded that he question the deepest qualities of nature.

To a Moslem mystic, Allah is to be sought in eternal beauty. And for Salam, beauty comes through finding new, subtle, yet simplifying patterns in the natural world. Anything that threatens to confuse the issue seems to him ugly, filling him with an almost physical revulsion and driving him to clean it away, much as one would remove mud from a shrine.

His first major piece of research, done at Cambridge, completed a vital cleaning operation to get rid of an absurdity in physics. In previous theory, there was nothing to stop an electron from having an infinite mass and an infinite electric charge. With great insight, physicists Julian Schwinger, Richard Feynman, and Freeman Dyson had indicated how the difficulty could be overcome, but the complete mathematical proof was lacking. That, Salam supplied.

During the period in which Salam has been active, since the late 1940s, physicists have shredded matter into smaller and smaller pieces, and have proposed new theories to explain them. In all the great advances, Salam has been in the thick of the action. Three of his contributions have been exceptionally important, and illustrate his quest for order.

The first had to do with parity — a theory of physics concerning the symmetry between an event and its mirror image. When a radioactive atom throws out an *electron* (beta particle) it also ejects that most elusive of particles, the neutrino. Both particles spin as they go, and the natural assumption was that the particles were just as likely to spin to the left as to the right. At a conference in Seattle, Wash., in 1956, the Chinese-born American physicists Tsung Dao Lee and Chen Ning Yang suggested that such parity of left and right might possibly not be the case.

The startling proposal, which challenged a 30-year-old law on the conservation of parity, nagged Salam as he flew back to England from the Seattle conference. If the ugly, irregular idea of "parity non-conservation" was to be tolerable at all, there had to be a beautiful explanation for it. He recalled that no one had satisfactorily explained why the neutrino had no mass. Any particle will tend to interact with its own field and thereby resist acceleration which is what we mean by mass. Salam saw that nature could dodge this outcome if the neutrino spun only in one direction — in other words, if parity was violated.

More precisely, parity violation had to balance parity conservation exactly. That would mean that of the electrons emerging with neutrinos from radioactive cobalt-60 atoms, an average of three electrons would spin one way to every one that spun the other. By the time his plane had landed in England, Salam had it all clear in his mind. Distinguished colleagues mocked the idea. In 1957, Chien-Shiung Wu of Columbia University in New York City performed the celebrated cobalt-60 experiment that proved the violation of parity — God was left-handed, as Austrian physicist Wolfgang Pauli put it. For every three electrons spinning to the left, one spun to the right, just as Salam had predicted.

But Salam, like many other physicists, was already after bigger game. Could the bewildering variety of particles be elementary? Or, was it, as Salam asked, that "some are more elementary than others?" The best thing was to seek family groupings, enabling one to say that if one particle exists, then others should exist with familial properties — similar but not identical.

The breakthrough began in 1960, when Yoshio Ohnuki of Nagoya University in Japan introduced the idea of "unitary symmetry" that might exist among particles. It started with the notion that most particles are made of the three entities which are themselves related to one another. Salam was the first non-Japanese physicist, perhaps in a sympathy of oriental minds, to accept the idea. Thus Imperial College, where Salam was professor of theoretical physics, became the centre of development of unitary symmetry.

Salam and John Ward, a visitor to Imperial College, used it in April, 1961, to predict an eightfold family of new particles having twice the spin of the proton, duly discovered some six months later. A research student working with Salam, Yuval Ne'eman from Israel, went on to show that the chief heavy particles, including the proton and neutron, also formed an eightfold family. About the same time, Murray Gell-Mann of the California Institute of Technology came to the same conclusion. He used the symmetry concept to predict a very strange particle — the omega minus — and when that, too, turned up early in 1964, the unitary ideas were established.

The next great advance came from American theorists who extended the unitary symmetry idea to link up separate families of heavy particles into a dynasty of 56 particles. But this theory left out the crucially important ideas of relativity, which omission launched Salam onto his third major contribution to science. This time, working with two associates, Robert Delbourgo and John Strathdee, Salam introduced Einstein's "four dimensions" (three dimensions of space, plus time) to arrive at a still higher pattern. "We are never going to be surprised by the discovery of a new particle again," Salam commented at the time. The earlier theory, leading up to the omega minus, had flaws, and these were carried over into the new theory, as Salam's fellow physicists were quick to point out. The fact remains that the valid portions of the theory represent the highest level of patternmaking in particle physics. As Salam puts it: "We have now run out of indexes."

According to Moslem colleagues, physics, for Salam, is a form of prayer. But he also treats physics as great fun. He holds onto the problem in his mind like a dog with a bone, yet he manages to remain relaxed. He pours out ideas in a continuous stream in discussions with his colleagues. Occasionally, Salam is right — and then his triumphant "I told you so!" might be irritating to anyone who recalled the 99 others, voiced with equal conviction, that were wrong.

The intensity of feeling and humor that goes into his theorizing were illustrated once when he was ill. "I'm sorry," he told a colleague, "I can't do physics now because I can't shout back at you." Generally, Salam talks quietly, thoughtfully, and fluently in a husky voice punctuated by laughter. But he always takes a positive attitude to ideas. "Some theorists are nihilists," he complains. "They are very good at showing where ideas are wrong, but they do not offer anything in their place. I prefer to build." He thinks more or less continuously about the patterns of nature and their mathematical representation, looking for order and beauty. "A broken symmetry breaks your heart," he says. He begins his day at 5 A.M. Like the wise man in the proverb, he goes to bed early, too.

That, then, is the story of the scholarly Punjabi boy who became an outstanding physicist. But there is another Salam: the man of the world in the most modern sense, a man concerned with the politics and organization of science, and with the terrible problems of poverty and backwardness in his homeland and in half the world.

In 1947, while Salam was finding his place in the unfamiliar world of Cambridge, the British dissolved their Indian empire and the new Moslem nation, Pakistan, came into existence. Four years later, at the age of 25, Salam went back to Lahore. He served as teacher of mathematics at his alma mater, Government College (1951 to 1954), and head of the mathematical department at Panjab University (1952 to 1954). He felt a duty to return home and work among and teach his own people. The move turned out to be unfortunate, although Salam did not give up easily. He spent three troubled years there before professional frustration drove him back to England. Reluctantly, he went down the "brain drain" which robs Asia of much of the talent that it so urgently needs. But he resolved to do all he could to save other young men from the "cruel choice" between homeland and profession.

At Lahore, the lack of facilities was the least of his worries — a theorist, after all, works with plain paper or a blackboard. But, the academic climate in Pakistan was wrong; science was ignored not only by the intellectual leaders of the new nation, but also by the brightest students. Salam, simply, was intellectually lonely. He dabbled fruitlessly in cosmology and the theory of superconductors. "You have to know what other physicists are thinking," he says, "and you have to talk to them. I feared that if I stayed in Lahore my work would deteriorate. Then what use would I be to my country?" Better to be a lecturer in Cambridge than a professor in Lahore.

Salam picked up the threads again with instant success. In 1955, he was asked to serve as a scientific secretary at the first Atoms for Peace Conference convened by the UN in Geneva, Switzerland. Like many others on that famous occasion, Salam was greatly moved and sensed the full strength of world science and its power to work great wonders for the benefit of all men. Two years later, he was chosen to found a department of theoretical physics at Imperial College. He was also elected the youngest fellow of Britain's most select association of scientists, the Royal Society.

Today, Salam is director of his International Centre for Theoretical Physics in Trieste. The possessive "his" is correct. Abdus Salam conceived the centre as a place where men from all countries could work alongside some of the most distinguished minds of physics. As delegate from Pakistan, he proposed its creation to the International Atomic Energy Agency (IAEA) in 1960, and he was himself appointed its first director in 1964. Advanced countries, such as France, Great Britain, the Soviet Union, and the United States, were cool to the idea at first, but they could not resist the enthusiastic support from developing countries that rallied behind Salam. The Italian government provided the greater share of the money for the centre's first four years, donated temporary premises and began work on a fine new building at the coastal resort of Miramare.

The advance that established the centre on the scientific stage and made it a magnet for the world's physicists was Salam's effort, along with Delbourgo and Strathdee, in carrying the unitary symmetry ideas forward. That work was announced within a few months after the centre had opened in October, 1964.

The centre, which Salam envisions as the first department of a UN university, provides a meeting place for leading theoreticians of East and West. In 1965, for instance, Salam arranged a year-long brain-storming session on the attempts to tame the H-bomb — to produce useful power from the hot, heavy gas. Out of the session, presided over by Marshall Rosenbluth, an American, and Raoul Sagdeev, a Russian, came something like an international policy for experimental work aimed at giving mankind access to an unlimited source of power.

Closest of all to Salam's heart is the centre's role in ending the loneliness of men working in the academically underdeveloped countries. Never again should any able theorist suffer the isolation that Salam himself felt when he went back to Lahore. From Africa, Asia and Latin America, professors and students come to spend a few weeks or months at Trieste, where they can "plug in" to the current excitement of physics, sample the latest ideas, and, most important of all, meet informally with the world leaders in the subject. One device pioneered by Salam is already being taken up in other institutions and has attracted special support from the Ford Foundation. It is the "associateship" plan whereby selected theorists in developing countries are given the privilege of coming to Trieste for three months a year, with the centre paying all the expenses.

The winter at Trieste is the time when many physicists come from the Southern Hemisphere during the summer vacation of their universities. It is for the scientists a time for renewal, an opportunity to communicate with kindred spirits. After teaching for four years at the University of Santiago, Chile, Igor Saavedra felt like "a squeezed lemon." He was tempted to accept a job in London, but the centre at Trieste opened in the nick of time, keeping Saavedra from joining the "brain drain." For East Europeans, Trieste is, above all other considerations, the only place in the world where effective collaboration is possible between the physicists of the East and the West. Salam is also gratified that, through the centre, important contributions to the subject by theoreticians from the developing nations in Africa have begun to appear.

Salam presides benevolently over the centre, aided by his deputy, Paolo Budini of Italy. Few of the visitors can know what battles Salam has fought and goes on fighting to make sure that the centre will survive. In February, 1967, for example, he took the night train to Vienna to talk the governors of IAEA into extending the centre's life indefinitely. He did not succeed, and he did not conceal his wrath. In former days, a Moslem warrior would draw his sword; Salam unleashes his words. He subscribes to the Islamic tradition that patience is a virtue only up to a certain point, that gentle persuasion can be tried only for so long if you are striving for higher goals.

Abdus Salam's name means, literally, Servant of Peace. The

ideal of human brotherhood cultivated in the abstruse mathematics and broken English of the Trieste centre finds a broader and plainer expression in Salam's work for the UN Advisory Committee on Science and Technology. Twice a year, he and 17 other men of learning spend 10 days together at one of the UN's centres — Geneva, Switzerland; New York City; Paris, France; and Rome, Italy. They try to specify ways in which scientific knowledge and technical skills can hasten the advancement of that half of the world now living in poverty.

The UN committee has produced a "World Plan of Action" for building up science and technology within the developing nations and for transferring technical knowledge to countries that desperately need it. The "wise men" have also named particular technologies, such as desalination and the elimination of disease-bearing insects, that need to be developed as fast as possible. Each member has his special pre-occupations and enthusiasms. Abdus Salam is particularly interested in involving leading scientists of advanced countries in the problems of world development.

On behalf of his own country, Pakistan, he did just that in a memorable fashion in 1962. The magnificent irrigation system built in the Indus Valley during the British era had deteriorated. Many years of seepage from the great irrigation canals had waterlogged huge areas of farmland, while evaporation from the soil had caused salts to accumulate. When Salam explained the problem, the U.S. government sent leading scientists, agriculturalists, and engineers to West Pakistan. After thorough studies, the team, led by Roger Revelle, then director of the Scripps Institution of Oceanography at La Jolla, Calif., and science adviser to the Secretary of the Interior, drew up a plan of wells and pumps for draining the land and washing out the salt. Several areas, each about a million acres in size, are already successfully under treatment west of Lahore. Over 30,000 farmers have adopted this procedure, greatly increasing agricultural production in West Pakistan.

President Ayub Khan of Pakistan appointed Salam as his personal scientific adviser in 1961, and a close and informal relationship has developed between them. Salam is frank about the human impediments in Pakistan, as in many developing countries, where scientists may proffer constructive suggestions, only to find them ignored by the administrators or dismissed because of the lack of resources to carry them out. Salam's most powerful colleague is Ishrat Usmani, chairman of the Pakistan Atomic Energy Commission. The commission has gone beyond its basic task of introducing nuclear power. It seeks to encourage general excellence among Pakistani scientists.

In the words of Usmani, "Most of the scientific effort in Pakistan is in a large measure due to Salam's imagination and the weight of his personality. Salam is a symbol of the pride and prestige of our nation in the world of science."

At the same time, Salam confesses that too little attention has been paid to food and agriculture and he is understandably prone to pessimism. In a forecast of the future, he has written: "Twenty years from now the less-developed countries will be as hungry, as relatively undeveloped, and as desperately poor, as today." Yet he recognizes slow progress in some directions. In Pakistan, the undue esteem given to the arts, at the expense of science, is being broken down. The president himself has come to share Salam's passionate interest in the publication of better science textbooks. More young people are studying science at the universities.

Since childhood, when he watched the apothecary at Jhang concoct aromatic sherbets from the ancient book of Avicenna, the Persian philosopher-physician, Salam has taken a proud interest in the former glories of Islamic science and literature. He likes to recall the days when Baghdad and Moorish Toledo were, for a time, the world's chief centres of learning. Even today, his vision of the future of Pakistan is not confined to the satisfaction of material needs. "Once a nation starts to think of higher things," he says, "scholars must find a role in that society." During his visits to Pakistan, it is not unusual to find him surrounded by a group of poets reading their verses to him and finding him an appreciative and critical listener.

In keeping with strong Islamic tradition, "Charity begins at home," no young Pakistani seeking help or guidance from Salam is left unaided. His Western students, too, find him generous to a fault in his support of them.

Salam is frequently on the move from continent to continent, yet unlike many of today's jet set scientists he refuses to let public business deflect him from his personal researches. Conversely, in his advisory work in Pakistan and for the UN, he does not allow his scientific sophistication to dampen the simple passion of a man born in a poor community and who knows that he is perhaps the luckiest of all his countrymen.

On the wall of the director's office in Trieste hangs an inscription of a 16th-century Persian prayer: "He cried: 'O Lord, work a miracle!' " Salam's strength is that he believes miracles are possible provided one goes out and helps them on their way.

# JULIAN SCHWINGER

∎

Dubbed "the Mozart of physics," the precocious Julian
Schwinger (b. 1918) was conversing on equal terms with giants
in the field by the time he completed his Ph.D. at Columbia
University at the age of nineteen. Here Schwinger eulogizes the
Japanese physicist Sin-itiro Tomonaga (1906–1979), with whom
he and Richard Feynman shared the 1965 Nobel Prize for
research in the quantum theory of electromagnetic fields.

# *Two Shakers of Physics: Memorial Lecture for Sin-itiro Tomonaga*

I AM DEEPLY HONORED to have the privilege of addressing you. It is
natural that I should do so, as the Nobel Prize partner whose work
on quantum electrodynamics was most akin in spirit to that of Sin-
itiro Tomonaga. But not until I began preparing this memorial did
I become completely aware of how much our scientific lives had in
common. I shall mention those aspects in due time. More immediately provocative is the curious similarity hidden in our names. The
Japanese character (the *kanji*) *shin* has, among other meanings, those
of "to wave" or "to shake." The beginning of my Germanic name,
Schwing, means "to swing" or "to shake." Hence my title: "Two
shakers of physics."

One cannot speak of Tomonaga without reference to Hideki
Yukawa and, of course, Yoshio Nishina. It is a remarkable coincidence that both Japanese Nobel Prize winners in physics were born
in Tokyo, both had their families move to Kyoto, both were sons of
professors at Kyoto University, both attended the Third High School
in Kyoto and both attended and graduated from Kyoto University

675

with degrees in physics. In their third and final year at the university, both learned the new quantum mechanics together. (Tomonaga would later remark, about this independent study, that he was happy not to be bothered by the professors.) Both graduated in 1929 into a world that seemed to have no place for them. (Yukawa later said that "the depression made scholars.") Accordingly, both stayed on as unpaid assistants to Professor Hidehiko Tamaki; Yukawa would eventually succeed him. In 1931, Nishina came on the scene. He gave a series of lectures at Kyoto University on quantum mechanics. Shoichi Sakata, then a student, later reported that Yukawa and Tomonaga asked the most questions afterward.

Nishina was a graduate in electrical engineering of Tokyo University. In 1917 he joined the recently founded Institute of Physical and Chemical Research, the Rikagaku Kenkyusho — Riken. A private institution, Riken was supported financially in various ways, including the holding of patents on the manufacture of sake. After several years at Riken, Nishina was sent abroad for further study, a pilgrimage that would last for eight years. He stopped at the Cavendish Laboratory in Cambridge, England, and at the University of Göttingen in Germany; then, finally, he went to Denmark and Niels Bohr in Copenhagen. He would stay there for six years, and out of that period came the famous Klein-Nishina formula. Nishina returned to Japan in December 1928 to begin building the Nishina group. It would, among other contributions, establish Japan in the forefront of research on nuclear and cosmic-ray physics — *soryu-shiron*.

There was a branch of Riken at Kyoto in 1931 when Nishina, the embodiment of the *Kopenhagener Geist*,* came to lecture and to be impressed by Tomonaga. The acceptance of Nishina's offer of a research position brought Tomonaga to Tokyo in 1932. (Three years earlier he had traveled to Tokyo to hear lectures at Riken given by Werner Heisenberg and Paul A. M. Dirac.) The year 1932 was a traumatic one for physics. The neutron was discovered; the positron was discovered. The first collaborative efforts of Nishina and Tomonaga dealt with the neutron and the problem of nuclear forces. Although there were no formal publications, this work was reported at the 1932 autumn and 1933 spring meetings that were regularly held by the Riken staff. Then, at the 1933 autumn meeting, the subject

---

*I.e., the Copenhagen school, dominated by Niels Bohr's view of quantum mechanics. [editor's note]

became the positron. It was the beginning of a joint research program that would see the publication of a number of papers concerned with various aspects of electron-positron pair creation and annihilation. Tomonaga's contributions to quantum electrodynamics had begun.

Although these papers were visible evidence of interest in quantum electrodynamics, we are indebted to Tomonaga for telling us, in his Nobel address, of an unseen but more important step: He read the 1932 paper of Dirac that attempted to find a new basis for electrodynamics. Dirac argued that "the role of the field is to provide a means for making observations of a system of particles," and therefore "we cannot suppose the field to be a dynamical system on the same footing as the particles and thus something to be observed in the same way as the particles." The attempt to demote the dynamical status of the electromagnetic field, or, in the more extreme later proposal of John Wheeler and Richard Feynman, to eliminate it entirely, was a false trail, contrary to the fundamental quantum duality between particle and wave, or field. Nevertheless, Dirac's paper was to be very influential. Tomonaga said:

> This paper of Dirac's attracted my interest because of the novelty of its philosophy and the beauty of its form. Nishina also showed a great interest in this paper and suggested that I investigate the possibility of predicting some new phenomena by this theory. Then I started computations to see whether the Klein-Nishina formula could be derived from this theory or whether any modification of the formula might result. I found out immediately, however, without performing the calculation through to the end, that it would yield the same answer as the previous theory. The new theory of Dirac's was in fact mathematically equivalent to the older Heisenberg-Pauli theory and I realized during the calculation that one could pass from one to the other by a unitary transformation. The equivalence of these two theories was also discovered by Rosenfeld and Dirac-Fock-Podolsky and was soon published in their papers.

I graduated from a high school that was named for Townsend Harris, the first American consul in Japan. Soon after, in 1934, I wrote but did not publish my first research paper. It was on quantum electrodynamics. Several years before, the Danish physicist Christian

677

Møller had proposed a relativistic interaction between two electrons, produced through the retarded intervention of the electromagnetic field. It had been known since 1927 that electrons could also be described by a field, one that had no classical macroscopic counterpart. And the dynamical description of this field was understood, when the electrons interacted instantaneously. I asked how things would be when the retarded interaction of Møller was introduced. To answer the question, I used the Dirac-Fock-Podolsky formulation. But because I was dealing entirely with fields, it was natural to introduce for the electron field, as well, the analogue of the unitary transformation that Tomonaga had already recognized as being applied to the electromagnetic field in Dirac's original version. Here was the first tentative use of what Tomonaga, in 1943, would correctly characterize as "a formal transformation which is almost self-evident" and I, years later, would call the interaction representation. No, neither of us, in the 1930s, had reached what would eventually be named the Tomonaga-Schwinger equation, but each of us held a piece that, in combination, would lead to that equation: Tomonaga appreciated the relativistic form of the theory, but was thinking in particle language; I used a field theory, but had not understood the need for a fully relativistic form. Had we met then, would history have been different?

The reports of the spring and autumn 1936 meetings of the Riken staff show something new: Tomonaga had resumed his interest in nuclear physics. In 1937 he went to Germany, to Heisenberg's Institute at Leipzig. He would stay for two years, working on nuclear physics and on the theory of mesons, to use the modern term. Tomonaga had come with a project in mind: treat Niels Bohr's liquid-drop model of the nucleus, and the way an impinging neutron heats it up, by using the macroscopic concepts of heat conduction and viscosity. This work was published in 1938. It was also the major part of the thesis submitted to Tokyo University in 1939 for the degree of Doctor of Science — *Rigakuhakushi*. Heisenberg's interest in cosmic rays then turned Tomonaga's attention to Yukawa's meson.

The not yet understood fact, that the meson of nuclear forces and the cosmic-ray meson observed at sea level are not the same particle, was beginning to thoroughly confuse matters at this time. Tomonaga wondered whether the problem of the meson lifetime could be overcome by including an indirect process in which the meson turns into a pair of nucleons (proton and neutron) that annihilate to produce the final electron and neutrino. The integral over all nucleon pairs, resulting from the perturbation calculation, was — infinite. Tomonaga kept a diary of his impressions during this German period. It poignantly records his emotional reactions to the difficulties he encountered. Here are some excerpts (translated by Fumiko Tanihara):

> It has been cold and drizzling since morning and I have devoted the whole day to physics in vain. As it got dark I went to the park. The sky was gray with a bit of the yellow of twilight in it. I could see the silhouetted white birch grove glowing vaguely in the dark. My view was partly obscured by my tired eyes; my nose prickled from the cold and upon returning home I had a nosebleed. After supper I took up my physics again, but at last I gave up. Ill-starred work indeed!

Then:

> Recently I have felt very sad without any reason, so I went to a film. . . . Returning home I read a book on physics. I don't understand it very well. Meanwhile I suffer. . . . Why isn't nature clearer and more directly comprehensible?

Again:

> As I went on with the calculation, I found the integral diverged — was infinite. After lunch I went for a walk. The air was astringently cold and the pond in Johanna Park was half frozen, with ducks swimming where there was no ice. I could see a flock of other birds. The flower beds were covered with chestnut leaves against the frost. . . . Walking in the park . . . I was no longer interested in the existence of neutron, neutrino. . . .

And, finally:

> I complained in emotional words to Professor Nishina about the slump in my work, whereupon I got his letter in reply this morning. After reading it my eyes were filled with tears. . . . He says: only fortune decides your progress in achievements. All of us stand on the dividing line from which the future is invisible. We need not be too anxious about the results, even though they may turn out quite dif-

ferent from what you expect. By-and-by you may meet a new chance for success. . . .

Toward the close of Tomonaga's stay in Leipzig, Heisenberg suggested a possible physical answer to the clear inapplicability of perturbation methods in meson physics. It involved the self-reaction of the strong meson field surrounding a nucleon. Heisenberg did a classical calculation, showing that the scattering of mesons by nucleons might thereby be strongly reduced, which would be more in conformity with the experimental results. About this idea, Tomonaga later remarked:

> Heisenberg, in this paper published in 1939, emphasized that the field reaction would be crucial in meson-nucleon scattering. Just at that time I was studying at Leipzig, and I still remember vividly how Heisenberg enthusiastically explained this idea to me and handed me galley proofs of his forthcoming paper. Influenced by Heisenberg, I came to believe that the problem of field reactions far from being meaningless was one which required a frontal attack.

Indeed, Tomonaga wanted to stay on for another year to work on the quantum mechanical version of Heisenberg's classical calculation. The growing clouds of war made this inadvisable, however, and Tomonaga returned to Japan by ship. As it happened, Yukawa, who had come to Europe to attend a Solvay Congress, which unfortunately was canceled, sailed on that same ship. When the ship docked at New York, Yukawa disembarked and, beginning at Columbia University, where I first met him, made his way across the United States, visiting various universities. But Tomonaga, after a day's sightseeing in New York that included the Japanese pavilion at the World's Fair, continued with the ship through the Panama Canal and on to Japan. About this, Tomonaga said: "When I was in Germany I had wanted to stay another year in Europe, but once I was aboard a Japanese ship I became eager to arrive in Japan." He also remarked about his one-day excursion in New York that "I found that I was speaking German rather than English, even though I had not spoken fluent German when I was in Germany."

Tomonaga had returned to Japan with some ideas concerning the quantum treatment of Heisenberg's proposal that attention to strong field reactions was decisive for understanding the meson-nucleon system. But soon after he began work, he became aware, through an abstract of a paper published in 1939, that Gregor Wentzel was also attacking this problem of strong coupling. Here is where the scientific orbits of Tomonaga and myself again crossed. At about the time that Tomonaga returned to Japan, I went to California to work with J. Robert Oppenheimer. Our first collaboration was a quantum electrodynamic calculation of the electron-positron pair emitted by an excited oxygen nucleus. And then we turned to meson physics. Heisenberg had suggested that meson-nucleon scattering would be strongly suppressed by field reaction effects. There also existed another proposal to the same end — that the nucleon possessed excited states, isobars, that would produce almost canceling contributions to the meson scattering process. We showed, classically, that the two explanations of suppressed scattering were one and the same: The effect of the strong field reaction, of the strong coupling, was to produce isobars, bound states of the meson about the nucleon. The problem of giving these ideas a correct quantum framework naturally arose. And then we became aware, through the published paper, of Wentzel's quantum considerations on a simple model of the strong coupling of meson and nucleon. I took on the quantum challenge myself. Not liking the way Wentzel had handled it, I redid his calculation in my own style and, in the process, found that Wentzel had made a mistake. In the short note that Oppenheimer and I eventually published, this work of mine is referred to as "to be published soon." And it was published, 29 years later, in a collection of essays dedicated to Wentzel. Recently, while surveying Tomonaga's papers, I came upon his delayed publication of what he had done along the same lines. I then scribbled a note: "It is as though I were looking at my own long unpublished paper." I believe that both Tomonaga and I gained from this episode added experience in using canonical (unitary) transformations to extract the physical consequences of a theory.

I must not leave the year 1939 without mentioning a work that would loom large in Tomonaga's later activities. But, to set the stage, I turn back to 1937. In that year, Felix Bloch and Arnold Nordsieck considered another kind of strong coupling — that between an electric charge and arbitrarily soft (extremely low frequency) light quanta. They recognized that in a collision, say between an electron and a nucleus, arbitrarily soft quanta will surely be emitted; a perfectly elastic collision cannot occur. Yet, if only soft photons, those of low energy, are considered, the whole scattering process goes on as though the electrodynamic interactions were ineffective. Once this was understood, it was clear that the real problem of electrodynamic field reaction begins when arbitrarily hard (unlimited high-energy) photons are reintroduced. In 1939 Sidney Dancoff performed such a relativistic scattering calculation both for electrons, which have spin 1/2, and for charged particles without spin. The spin-0 calculation gave a finite correction to the scattering, but for spin 1/2, the correction was infinite. This was confusing. And to explain why that was so, we must talk about electromagnetic mass.

It was already part of classical physics that the electric field surrounding an electrically charged body carries energy and contributes mass to the system. That mass varies inversely as a characteristic dimension of the body and therefore is infinite for a point charge. The magnetic field that accompanies a moving charge implies an additional momentum, an additional, electromagnetic, mass. It is very hard, at this level, to make those two masses coincide, as they must, in a relativistically invariant theory. The introduction of relativistic quantum mechanics, of quantum field theory, changes the situation completely. For the spin-1/2 electron-positron system, obeying Fermi-Dirac statistics, the electromagnetic mass, while still infinite, is only weakly, logarithmically, so. In contrast, the electromagnetic mass for a spin-0 particle, which obeys Bose-Einstein statistics, is more singular than the classical one. Thus, Dancoff's results were in contradiction to the expectation that spin 0 should exhibit more severe electromagnetic corrections.

Tomonaga's name had been absent from the Riken reports for the years from 1937 to 1939, when he was in Germany. It reappeared for the 1940 spring meeting under the title "On the Absorption and Decay of Slow Mesons." There the simple and important point was made that when cosmic-ray mesons are stopped in matter, the repulsion of the nuclear Coulomb field prevents positive mesons from being absorbed by the nucleus, whereas negative mesons will preferentially be absorbed before decaying. This was published as a *Physical Review* letter in 1940. Subsequent experiments showed that no such asymmetry existed in very light elements; the cosmic-ray meson does not interact strongly with nuclear particles. The Riken reports from autumn of 1940 to autumn of 1942 traced stages in the development of Tomonaga's strong- and intermediate-coupling meson theories. In particular, under the heading "Field Reaction and Multiple Production" there was discussed a coupled set of equations corresponding to various particle numbers that is the basis of an approximation scheme now generally called the Tamm-Dancoff approximation. This series of reports on meson theory was presented to the Meson symposium (*Chukanshi Toronkai*) that was initiated in September 1943, where also was heard the suggestion of Sakata's group that the cosmic-ray meson was not the meson responsible for nuclear forces.

But meanwhile there occurred the last of the Riken meetings held during the war, that of spring 1943. Tomonaga provided the following abstract with the title "Relativistically Invariant Formulation of Quantum Field Theory":

> In the present formulation of quantum fields as a generalization of ordinary quantum mechanics such non-relativistic concepts as probability amplitude, canonical commutation relations and Schrödinger equation are used. Namely these concepts are defined referring to a particular Lorentz frame in space-time. This unsatisfactory feature has been pointed out by many people and also Yukawa emphasized it recently. I made a relativistic generalization of these concepts in quantum mechanics such that they do not refer to any particular coordinate frame and reformulate the quantum theory of fields in a relativistically invariant manner.

In the previous year Yukawa had commented on the unsatisfactory nature of quantum field theory, pointing both to its lack of an explicit, manifestly covariant form and to the problem of divergences — infinities. He wished to solve both problems at the same time. To that end, he applied Dirac's decade-earlier suggestion of a generalized transformation function by proposing that the quantum field probability amplitude should refer to a closed surface in space-time. From the graphic presentation of such a surface as a circle, the proposal became known as the theory of *maru*. Tomonaga's reaction was to take one problem at a time, and he first proceeded to "reformulate the quantum theory of fields in a relativistically invariant manner." And in doing so he rejected Yukawa's more radical proposal in favor of retaining the customary concept of causality — the relation between cause and effect. What was Tomonaga's reformulation?

The abstract I have cited was that of a paper published in the *Bulletin of the Institute, Riken-Iho*. But its contents did not become known outside of Japan until it was translated into English to appear in the second issue, that of August–September, 1946, of the new journal *Progress of Theoretical Physics*. It would, however, be some time before this issue became generally available in the United States. Incidentally, in this 1946 paper Tomonaga gave his address as Physics Department, Tokyo Bunrika University. While retaining his connection with Riken, he had, in 1941, joined the faculty of this university that later, in 1949, became part of the Tokyo University of Education.

Tomonaga began his paper by pointing out that the standard commutation relations of quantum field theory, referring to two points of space at the same time, are not covariantly formulated; in a relatively moving frame of reference the two points will be assigned different times. This is equally true of the Schrödinger equation for time evolution, which uses a common time variable for different spatial points. He then remarked that there is no difficulty in exhibiting commutation relations for arbitrary space-time points when a noninteracting field is considered. The unitary transformation to which we have already referred, now applied to all the fields, provides them with the equations of motion of noninteracting fields, whereas in the transformed Schrödinger equation, only the interaction terms remain. About this, Tomonaga says: "In our formulation, the theory is divided into two sections. . . . One section gives the laws of behavior of the fields when they are left alone, and the other gives the laws determining the deviation from this behavior due to the interactions. This way of separating the theory can be carried out relativistically." Certainly commutation relations referring to arbitrary space-time points are four-dimensional in character. But what about the transformed Schrödinger equation, which still retains its single time variable? It demands generalization.

Tomonga was confident that he had the answer, for, as he put it later, "I was recalling Dirac's many-time theory which had enchanted me ten years before." In the theory of Dirac, and then of Dirac-Fock-Podolsky, each particle is assigned its own time variable. But, in a field theory, the role of the particle is played by the small volume elements of space. Therefore, assign to each spatial volume element an independent time coordinate. Thus the "super-many-time theory." Let me be more precise about that idea. At a common value of the time, distinct spatial volume elements constitute independent physical systems, for no physical influence is instantaneous. But more than that, no physical influence can travel faster than the speed of light. Therefore, any two space-time regions that cannot be connected, even by light signals, are physically independent; they are said to be in spacelike relationship. A three-dimensional domain such that any pair of points is in spacelike relationship constitutes a spacelike surface in the four-dimensional world. All of space at a common time is but a particular coordinate description of a plane spacelike surface. Therefore the Schrödinger equation, in which time advances by a common amount everywhere in space, should be regarded as describing the normal displacement of a plane spacelike surface. Its immediate generalization is to the change from one arbitrary spacelike surface to an infinitesimally neighboring one, which change can be localized in the neighborhood of a given space-time point. Such is the nature of the generalized Schrödinger equation that Tomonaga constructed in 1943, and to which I came toward the end of 1947.

By this time the dislocation produced by the war became dominant. Much later, Tomonaga recalled that "I myself temporarily stopped working on particle physics after 1943 and was involved in electronics research. Nevertheless the research on magnetrons and on ultrashortwave circuits was basically a continuation of quantum mechanics." Tatsuoki Miyazima had this remembrance:

> One day our boss Dr. Nishina took me to see several engineers at the Naval Technical Research Institute. They had been engaged in the research and development of powerful split anode magnetrons, and they seemed to have come to a concrete conclusion about the phenomena taking place in the electron cloud. . . . Since they were engineers their way of thinking was characteristic of engineers and it was quite natural that they spoke in an engineer's way, but unfortunately it was . . . completely foreign to me at the beginning. . . . Every time I met them, I used to report to Tomonaga how I could not understand them, but he must have understood something . . . because, after a month or so, he showed me his idea . . . [of] applying the idea of secular perturbation, well-known in celestial mechanics and quantum theory, to the motion of the electrons in the cloud. . . . I remember that the moment he told me I said, "This is it." Further investigation actually showed that the generation of electromagnetic oscillations in split anode magnetrons can be essentially understood by applying his idea.

When Tomonaga approached the problem of ultrashortwave

circuits, which is to say, the behavior of microwaves in waveguides and cavity resonators, he found the engineers still using the old language of impedance. He thought this artificial, because there no longer are unique definitions of current and voltage. Instead, being a physicist, Tomonaga began with the electromagnetic field equations of James C. Maxwell. But he quickly recognized that those equations contain much more information than is needed to describe a microwave circuit. One usually wants to know only a few things about a typical waveguide junction: If a wave of given amplitude moves into a particular arm, what are the amplitudes of the waves coming out of the various arms, including the initial one? The array of all such relations forms a matrix, even then familiar to physicists as the scattering matrix. I mention here the amusing episode of the German submarine that arrived bearing a dispatch stamped *Streng Geheim* (top secret). When delivered to Tomonaga, it turned out to be Heisenberg's paper on the scattering matrix. Copies of this top-secret document were soon circulating among the physicists. Tomonaga preferred to speak of the scattering matrix as the characteristic matrix, in this waveguide context. He derived properties of that matrix, such as its unitary character, and showed how various experimental arrangements could be described in terms of the characteristic matrix of the junction. In the paper published after the war, he remarked, concerning the utility of this approach, that "The final decision, however, whether or not the new concept is here preferable to impedance should of course be given not only by a theoretical physicist but also by general electro-engineers." But perhaps my experience is not irrelevant here.

During the war, I also worked on the electromagnetic problems of microwaves and waveguides. I also began with the physicist's approach, including the use of the scattering matrix. But long before this three-year episode was ended, I was speaking the language of the engineers. I should like to think that those years of distraction for Tomonaga and myself were not without their useful lessons. The waveguide investigations showed the utility of organizing a theory to isolate those inner structural aspects that are not probed under the given experimental circumstances. That lesson was soon applied in the effective-range description of nuclear forces. And it was this viewpoint that would lead to the quantum electrodynamic concept of self-consistent subtraction or renormalization.

Tomonaga already understood the importance of describing relativistic situations covariantly — without specialization to any particular coordinate system. At about this time I began to learn that lesson pragmatically, in the context of solving a physical problem. As the war in Europe approached its end, the American physicists responsible for creating a massive microwave technology began to dream of high-energy electron accelerators. One of the practical questions involved was posed by the strong radiation emitted by relativistic electrons swinging in circular orbits. In studying what is now called synchrotron radiation, I used the reaction of the field created by the electron's motion. One part of that reaction describes the energy and momentum lost by the electron to the radiation. The other part is an added inertial effect characterized by an electromagnetic mass. I have mentioned the relativistic difficulty that electromagnetic mass usually creates. But, in the covariant method I was using, based on action and proper time, a perfectly invariant form emerged. Moral: To end with an invariant result, use a covariant method and maintain covariance to the end of the calculation. And, in the appearance of an invariant electromagnetic mass that simply added to the mechanical mass to form the physical mass of the electron, neither piece being separately distinguishable under ordinary physical circumstances, I was seeing again the advantage of isolating unobservable structural aspects of the theory. Looking back at it, the basic ingredients of the coming quantum electrodynamic revolution were then in place. Lacking was an experimental impetus to combine them and take them seriously.

Suddenly the Pacific war was over. Amid total desolation, Tomonaga reestablished his seminar. But meanwhile, something had been brewing in Sakata's Nagoya group. It goes back to a theory of Møller and Leon Rosenfeld, who tried to overcome the nuclear force difficulties of meson theory by proposing a mixed field theory, with both pseudoscalar and vector mesons of equal mass. I like to think that my modification of this theory, in which the vector meson is more massive, was the prediction of the later discovered $\rho$ meson. Somewhat analogously, Sakata proposed that the massless vector photon is accompanied by a massive scalar meson called the cohesive or $C$ meson. About this, Tomonaga said:

> In 1946, Sakata proposed a promising method of eliminating the divergence of the electron mass by introducing the idea of a field of cohesive force. It was the idea that there exists an unknown field, of the type of the meson field, which interacts with the electron in addition to the

electromagnetic field. Sakata named this field the cohesive force field, because the apparent electromagnetic mass due to the interaction of this field and the electron, though infinite, is negative and therefore the existence of this field could stabilize the electron in some sense. Sakata pointed out the possibility that the electromagnetic mass and the negative new mass cancel each other and that the infinity could be eliminated by suitably choosing the coupling constant between this field and the electron. Thus the difficulty which had troubled people for a long time seemed to disappear insofar as the mass was concerned.

Let me break in here and remark that this solution of the mass divergence problem is, in fact, illusory. In 1950, Toichiro Kinoshita showed that the necessary relation between the two coupling constants will no longer cancel the divergences when the discussion is extended beyond the lowest order of approximation. Nevertheless, the $C$-meson hypothesis served usefully as one of the catalysts that led to the introduction of the self-consistent subtraction method. How that came about is described in Tomonaga's next sentence: "Then what concerned me most was whether the infinities appearing in the electron scattering process could also be removed by the idea of a plus-minus cancellation."

I have already referred to the 1939 calculation of Dancoff, on radiative corrections to electron scattering, that gave an infinite result. Tomonaga and his collaborators proceeded to calculate the additional effect of the cohesive force field. It encouragingly gave divergent results of the opposite sign, but they did not precisely cancel Dancoff's infinite terms. This conclusion was reported in a letter of November 1, 1947, submitted to *Progress of Theoretical Physics,* and also presented at a symposium on elementary particles held in Kyoto that same month. Meanwhile, parallel calculations of the electromagnetic effect were going on, repeating Dancoff's calculations, which had not been reported in detail. But then Tomonaga suggested a new and much more efficient method of calculation. It was to use the covariant formulation of quantum electrodynamics and subject it to a unitary transformation that immediately isolated the electromagnetic mass term. Tomonaga said:

Owing to this new, more lucid method, we noticed that among the various terms appearing in both Dancoff's and our previous calculation, one term had been overlooked.

There was only one missing term, but it was crucial to the final conclusion. Indeed, if we corrected this error, the infinities appearing in the scattering process of an electron due to the electromagnetic and cohesive force fields cancelled completely, except for the divergence of vacuum polarization type.

A letter of December 30, 1947, corrected the previous erroneous announcement.

But what is meant by "the divergence of vacuum polarization type"? From the beginning of Dirac's theory of positrons it had been recognized that, in a sense, the vacuum behaved as a polarizable medium; the presence of an electromagnetic field induced a charge distribution acting to oppose the inducing field. As a consequence, the charges of particles would appear to be reduced, although the actual calculation gave a divergent result. Nevertheless, the effect could be absorbed into a redefinition, a renormalization, of the charge. At this stage, then, Tomonaga had achieved a finite correction to the scattering of electrons by combining two distinct ideas: the renormalization of charge and the compensation mechanism of the $C$-meson field.

But meanwhile, another line of thought had been developing. In this connection, let me quote from a paper, published at about this time, by Mituo Taketani:

> The present state of theoretical physics is confronted with difficulties of extremely ambiguous nature. These difficulties can be glossed over but no one believes that a definite solution has been attained. The reason for this is that, on one hand, present theoretical physics itself has logical difficulties, while, on the other hand, there is no decisive experiment whereby to determine this theory uniquely.

In June of 1947 those decisive experiments were made known, in the United States.

For three days at the beginning of June, some 20 physicists gathered at Shelter Island, located in a bay near the tip of Long Island, New York. There we heard the details of the experiment by which Willis E. Lamb, Jr., and Robert Retherford had used the new microwave techniques to confirm the previously suspected upward displacement of the 2s level of hydrogen. Actually, rumors of this had already spread, and on the train to New York, Victor F. Weisskopf and I had agreed that electrodynamic effects were involved and that a relativistic calculation would give a finite prediction. But there was also a totally unexpected disclosure, by Isidor I. Rabi: The hyperfine structures in hydrogen and deuterium were larger than anticipated by a fraction of a percent. Here was another flaw in the Dirac electron theory, now referring to magnetic rather than electric properties.

Weisskopf and I had described at Shelter Island our idea that the relativistic electron-positron theory, then called the hole theory, would produce a finite electrodynamic energy shift. But it was Hans Bethe who quickly appreciated that a first estimate of this effect could be found without entering into the complications of a relativistic calculation. In a *Physical Review* article received on June 27, he said:

> Schwinger and Weisskopf, and Oppenheimer have suggested that a possible explanation might be the shift of energy levels by the interaction of the electron with the radiation field. This shift comes out infinite in all existing theories, and has therefore always been ignored. However, it is possible to identify the most strongly (linearly) divergent term in the level shift with an electromagnetic mass effect which must exist for a bound as well as a free electron. This effect should properly be regarded as already included in the observed *mass* of the electron, and we must therefore subtract from the theoretical expression, the corresponding expression for a free electron of the same average kinetic energy. The result then diverges only logarithmically (instead of linearly) in nonrelativistic theory: Accordingly, it may be expected that in the hole theory, in which the *main* term (self-energy of the electron) diverges only logarithmically, the result will be *convergent* after subtraction of the free electron expression. This would set an effective upper limit of the order of $mc^2$ to the frequencies of light which effectively contribute to the shift of the level of a bound electron. I have not carried out the relativistic calculations, but I shall assume that such an effective relativistic limit exists.

The outcome of Bethe's calculation agreed so well with the then not very accurately measured level shift that there could be no doubt of its electrodynamic nature. Nevertheless, the relativistic problem, of producing a finite and unique theoretical prediction, still remained.

The news of the Lamb-Retherford measurement and of Bethe's nonrelativistic calculation reached Japan in an unconventional way. Tomonaga said:

> The first information concerning the Lamb shift was obtained not through the *Physical Review,* but through the popular science column of a weekly U.S. magazine. This information about the Lamb shift prompted us to begin a calculation more exact than Bethe's tentative one.

He went on:

> In fact, the contact transformation method . . . could be applied to this case, clarifying Bethe's calculation and justifying his idea. Therefore the method of covariant contact transformations, by which we did Dancoff's calculation over again would also be useful for the problem of performing the relativistic calculation for the Lamb shift.

Incidentally, in speaking of contact transformations, Tomonaga was using another name for canonical or unitary transformations. Tomonaga announced his relativistic program at the previously mentioned Kyoto symposium of November 24 to 25, 1947. He gave it a name that appears in the title of a letter accompanying the one of December 30 that points out Dancoff's error. This title is "Application of the Self-Consistent Subtraction Method to the Elastic Scattering of an Electron." And so, at the end of 1947, Tomonaga was in full possession of the concepts of charge and mass renormalization.

Meanwhile, immediately following the Shelter Island conference, I found myself with a brand new wife, and for two months we wandered around the United States. Then it was time to go to work again. I also clarified for myself Bethe's nonrelativistic calculation by applying a unitary transformation that isolated the electromagnetic mass. This was the model for a relativistic calculation, based on the conventional hole-theory formulation of quantum electrodynamics. But here I held an unfair advantage over Tomonaga, for, owing to the communication problems of the time, I knew that there were two kinds of experimental effects to be explained: the electric one of Lamb and the magnetic one of Rabi. Accordingly, I carried out a calculation of the energy shift in a homogeneous magnetic field, which is the prediction of an additional magnetic moment of the electron, and also considered the Coulomb field of a nucleus in applications to scattering and to the energy shift of bound states. The results were described in a letter to the *Physical Review,* received on December 30, 1947, the very same date as Tomonaga's proposal of the self-consistent subtraction method. The predicted additional magnetic moment accounted for the hyperfine structure measurements and also for later, more accurate, atomic moment measurements. Concerning scattering, I said that "the finite radiative correction to the elastic scattering of electrons by a Coulomb field provides a satisfactory termination to a subject that has been beset with much confusion." Considering the absence of experimental data, this was perhaps all that needed to be said. But when it came to energy shifts, what I wrote was that "The values yielded by our theory differ only slightly from those conjectured by Bethe on the basis of a non-relativistic calculation, and are, thus, in good accord with experiment." Why did I not quote a precise number?

The answer to that was given in a lecture before the American Physical Society at the end of January 1948. Quite simply something was wrong. The coupling of the electron spin to the electric field was numerically different from what the additional magnetic moment would imply; relativistic invariance was violated in this noncovariant calculation. One could, of course, adjust that spin coupling to have the right value, and, in fact, the correct energy shift is obtained in this way. But there was no conviction in such a procedure. The need for a covariant formulation could no longer be ignored. At the time of this meeting, the covariant theory had already been constructed and applied to obtain an invariant expression for the electron electromagnetic mass. I mentioned this briefly. After the talk, Oppenheimer told me about Tomonaga's prior work.

A progress report on the covariant calculations, using the technique of invariant parameters, was presented at the Pocono Manor Inn conference held March 30 to April 1, 1948. At that very time, Tomonaga was writing a letter to Oppenheimer that would accompany a collection of manuscripts describing the work of his group. In response, Oppenheimer sent a telegram: "Grateful for your letter and papers. Found most interesting and valuable mostly paralleling much work done here. Strongly suggest you write a summary account of present state and views for prompt publication *Physical Review.* Glad to arrange. . . ." On May 28, 1948, Oppenheimer acknowledged the receipt of Tomonaga's letter entitled "On Infinite Field Reactions in Quantum Field Theory." He wrote:

> Your very good letter came two days ago and today your manuscript arrived. I have sent it on at once to the *Physical Review* with the request that they publish it as promptly as possible. . . . I also sent a brief note . . . which . . . may be of some interest to you in the prosecution of the higher order calculations. Particularly in the identification of light quantum self energies, it proves important to apply your relativistic methods throughout. We shall try to get an account of Schwinger's work on this and other subjects to you in the very near future.

He ended the letter expressing the "hope that before long you will spend some time with us at the Institute where we should all welcome you so warmly."

The point of Oppenheimer's added note is this: In examining the radiative correction to the Klein-Nishina formula, Tomonaga and his collaborators had encountered a divergence additional to those involved in mass and charge renormalization. It could be identified as a photon mass. But unlike the electromagnetic mass of the electron, which can be amalgamated, as Tomonaga put it, into an already existing mass, there is no photon mass in the Maxwell equations. Tomonaga noted the possibility of a compensation, a cancellation, analogous to the idea of Sakata. In response, Oppenheimer essentially quoted my observation that a gauge-invariant relativistic theory cannot have a photon mass and, further, that a sufficiently careful treatment would yield the required zero value. But Tomonaga was not convinced. In a paper submitted about this time, he spoke of the "somewhat quibbling way" in which it was argued that the photon mass must vanish. And he was right, for the real subtlety underlying the photon mass problem did not surface for another 10 years, in the eventual recognition of what others would call "Schwinger terms."

But even the concept of charge renormalization was troubling to some physicists. Abraham Pais, on April 13, 1948, wrote a letter to Tomonaga in which, after commenting on his own work parallel to that of Sakata, he remarked: "It seems one of the most puzzling problems how to 'renormalize' the charge of the electron and of the proton in such a way as to make the experimental values for these quantities equal to each other." Perhaps I was the first to fully appreciate that charge renormalization is a property of the electromagnetic field alone, which results in a renormalization, a fractional reduction of charge, that is the same for all. But while I'm congratulating myself, I must also mention a terrible mistake I made. Of course, I wasn't entirely alone — Feynman did it too. It occurred in the relativistic calculation of energy values for bound states. The effect of high-energy photons was treated covariantly; that of low-energy photons in the conventional way. These two parts had to be joined together, and a subtlety involved in relating the respective four- and three-dimensional treatments was overlooked for several months. But sometime around September 1948 it was straightened out, and apart from some uncertainty about the inclusion of vacuum polarization effects, all groups, Japanese and American, agreed on the answer. As I have mentioned, it was the result I had reached many months before by correcting the obvious relativistic error of my first noncovariant calculation.

In that same month, September 1948, Yukawa, accepting an invitation of Oppenheimer, went to the Institute for Advance Study at Princeton, New Jersey. The letters that he wrote back to Japan were circulated in a new informal journal called *Elementary Particle Physics Research — Soryushiron Kenkyu*. Volume 0 of that journal also contained the communications of Oppenheimer and Pais to which I have referred and a letter of Heisenberg to Tomonaga, inquiring whether Heisenberg's paper, sent during the war, had arrived. In writing to Tomonaga on October 15, 1948, Yukawa said, in part, "Yesterday I met Oppenheimer, who came back from the Solvay Conference. He thinks very highly of your work. Here, many people are interested in Schwinger's and your work and I think that this is the main reason why the demand for the *Progress of Theoretical Physics* is high. I am very happy about this."

During the period of intense activity in quantum electrodynamics, Tomonaga was also involved in cosmic-ray research. The results of a collaboration with Satio Hayakawa were published in 1949 under the title "Cosmic Ray(s) Underground." By now, the two mesons had been recognized and named: $\pi$ and $\mu$. This paper discussed the generation of, and the subsequent effects produced by, the deep-penetrating $\mu$ meson. Among other activities in that year of 1949, Tomonaga published a book on quantum mechanics that would be quite influential, and he accepted Oppenheimer's invitation to visit the Institute for Advanced Study. During the year he spent there, he turned in a new direction, one that would also interest me a number of years later. It is the quantum many-body problem. The resulting publication of 1950 is entitled "Remarks on Bloch's Method of Sound Waves Applied to Many-Fermion Problems." Five years later he would generalize this in a study of quantum collective motion.

But the years of enormous scientific productivity were coming to a close, owing to the mounting pressures of other obligations. In 1951, Nishina died, and Tomonaga accepted his administrative burdens. Tomonaga's attention turned toward improving the circumstances and facilities available to younger scientists, including the establishment of new institutes and laboratories. In 1956 he became president of the Tokyo University of Education, which post he held for six years. Then, for another six years, he was president of the Science Council of Japan, and also, in 1964, he assumed the presidency of the Nishina Memorial Foundation. I deeply regretted that he was unable to be with us in Stockholm on December 10, 1965, to accept his Nobel Prize. The lecture that I have often quoted here was delivered May 6, 1966.

Following his retirement in 1970, he began to write another volume of his book on quantum mechanics, which, unfortunately, was not completed. However, two other books, one left in an unfinished state, were published. To some extent, these books are directed to the general public rather than the professional scientist. And here, again, Tomonaga and I found a common path. I have recently completed a series of television programs that attempt to explain relativity to the general public. I very much hope that this series, which was expertly produced by the British Broadcasting Corporation, will eventually be shown in Japan.

On July 8, 1979, our story came to a close. But Sin-itiro Tomonaga lives on in the minds and hearts of the many people whose lives he touched, and graced.

# PHILIP J. HILTS

■

Anyone who stereotypes physicists as pasty-faced misfits should read this profile of Robert Wilson (b. 1914), a former cowboy who was about to become a sculptor when he agreed instead to build the Fermilab particle accelerator as a monument to both science and art. The author, Philip J. Hilts (b. 1947), has been a science reporter for the *Washington Post* and the *New York Times*.

# *Robert Wilson and the Building of Fermilab*

THE LEATHER WAS COLD and stiff as we pulled the saddles from their pegs and hauled them toward the stalls. It was dawn without the sun, thirty-seven degrees. Gray clouds pitched and reached in the wind. The snow had just melted in this part of the Illinois farmland, but the edge of winter had not yet gone. That morning, Dr. Robert Wilson, a particle physicist, wore a fedora that was black as jet and made of fur. He wore a quilted nylon jacket and a baggy pair of jeans. The jeans rode Western style, below his center of gravity, as low as was conceivable for a pair of trousers. The black boots he wore had pointed toes, and angled heels, and whorls of fancy stitching.

Wilson spoke to an attractive woman as he led his horse out of the barn. "This is how we used to mount a bucking bronco," he said with a grin. He hooked the rein with his elbow, and pulled it so the horse had to turn its head around to the saddle. "If her head's back this way, she can't start bucking so easily," he said, nodding toward the horse's hindquarters. He mounted with a single, easy arc of his foot.

His horse was a gray Arabian called Star. He walked her around the white fence of a pasture and then trotted her across the flat Illinois prairie, straight toward an odd geographic feature about a quarter of a mile away. There, the ground rose up in a twenty-foot welt. In this prairie, where the ground might not rise more than a few feet in a mile, the object ahead clearly was made by man. It was a long, earthen mound that stretched off to the right and left, curving out of sight. Larger than human scale, the whole mound cannot be discerned from horseback, but maps of the area show that it continues curving around for four miles. The curve meets itself directly opposite where we ride, completing the circle. The clearest sense of its size comes from photographs of northern Illinois taken from a satellite about three hundred miles up in space. They show the spreading towns and suburbs of the region as ragged blotches. Against this background appears one sharp and seemingly perfect circle. It is the only feature on earth of such size and regularity; it stands out against the Midwestern landscape like a dandelion in short grass.

Building the ring of tall earth was Robert Wilson's idea. It is a feature of the Fermi National Accelerator Laboratory, where he was the director from its inception in 1967 until 1978. Fermilab is among the largest of the world's laboratories, and all of its more than a hundred buildings, fifteen hundred workers, perhaps a million tons of equipment were created as support for a single great piece of machinery. The Fermilab machine is circular, a great four-mile loop, buried directly below the earthwork circle toward which we were riding. Physicists call the ring-shaped machine a particle accelerator; in newspaper stories it is called an atom smasher. There are only a few dozen of these machines in the world, and the one at Fermilab is the most powerful of its kind. Fermilab and its European twin, called CERN, are possibly the largest machines of any kind in the world.

It is unusual as machinery goes, not only for its great size, but because it has no practical purpose. It manufactures nothing and makes no profit. "Our work here is primarily spiritual," says Wilson. "We are concerned with the ultimate nature of matter." Congressmen and others asked him, frequently, what are the practical applications of the physics that go on in this machine? There are none, he says. No more, at least, than there are applications for literature, for theater, for poetry or painting. "Scientific understanding has inherent cultural value," he once explained to the Joint Committee on Atomic Energy. "It has great beauty. It adds to the satisfaction of our lives."

Since Wilson has some political as well as artistic sense, he usually adds as an aside the more practical answer of another scientist when put on the spot by a politician. When Faraday had found that electricity might be induced to flow in a trickling current, he hadn't any idea where this arcane fact might lead. But when Gladstone asked him, "What's the use of it?" Faraday replied boldly, "Sir, someday you will tax it."

We rode through the wheat-colored grass, which was still matted from the winter's weight of snow, toward the tall earthwork mound. Though the particle machine beneath it was on, there was no sound.

I had met Wilson in his role as laboratory director, and listened to Wilson in the role of physicist, in a comfortable office the day before. But there is a small revelation in watching him ride a horse. It is like watching the gears and levers of a machine, which seemed angled and awkward at rest, become all at once a thing of fine balance as it reaches speed. On the ground, Robert Wilson walks with an oddly rigid posture, which becomes exaggerated by his emphatic gestures. His hair is wild, like gray porcupine spikes. But once on horseback, everything begins to fit. The gestures and attitudes of body are not the habits of a physicist with too many years of lecturing, but natural parts of the deportment of a cowboy. Trotting, cantering, and running, he sits on his horse as comfortably as if he were relaxing against a fence. Walking or riding, his back is straight as a rod, and his limbs pivot from it with a floppy, cowhand's grace.

As the horses approached the twenty-foot-high hill on that cold March morning, they slowed to a walk. A few drops of freezing rain bit our faces as Wilson, with a turn of his wrist, reined his horse to walk her parallel to the long earth mound. The particle accelerator lies in a tunnel that is twenty feet below the bottom of the mound and forty feet below the fawn-colored tufts of grass that grow on the top. We walk the horses near what looks like a cement blockhouse sitting up against the mound. Here, and in several other places around the ring, there is a door and a deep stairwell leading down under the mound to the accelerator tunnel. If the machine were off and the doorway open, a visitor might climb down the stairway to the tunnel, which is ten feet wide, and constructed like a large cement sewer pipe. The actual accelerating machine runs on the inside curve of the cement tunnel, and consists mainly of rectangular iron boxes strung together like two-foot-high boxcars of a miniature freight train. There are about one thousand of the iron boxes in the four-mile length of tunnel, each hitched to its neighbor and the last completing the circle by connecting with the first. Each iron box, twenty feet long and weighing twenty thousand pounds, contains one huge, elongated magnet, and is surrounded by pipes feeding it water and electricity while retrieving from it electronic data. But the heart of this great ring-shaped machine is a thin pipe of stainless steel. It is only a few inches in diameter, it is all one piece, and it runs through the entire four-mile train of boxes just as a thread runs through the beads of a necklace. The stainless-steel pipe is vacuum-sealed and completely airtight. The pipe is like the inner tube of a tire four miles around. Protons enter the tube from a bulb of gas much as air enters an inner tube — through the stem of a valve. The stem of Fermilab's tube is one third of a mile long.

Protons are sent into the airless, frictionless space within the tube, and the resulting stream of particles is pulled by the magnets around the four-mile circumference about four hundred thousand times, gaining speed with each successive circling. The beam of particles has about the thickness of a flower stem or a piece of taut twine. As the particles circle, they reach 99.999 percent of the speed of light. The snake of protons is then extracted from the four-mile tube, guided down another tube, and aimed at a small aluminum rod. When the protons hit, a small portion of the rod is annihilated. The atoms of aluminum are destroyed, transformed into a smear of pure energy. The energy quickly congeals again, becoming a spray of matter. Among the array of hundreds of particles created this way, most of them do not exist freely on earth. They cannot be found in the rest of the universe either. The only other place, and time, in which such matter was formed was at the Creation and (in the billions of years since the Creation) during the sudden, immense stellar explosions which reprise the Creation. The particle accelerator thus reenacts, in the small chamber of a human machine, the beginning of the universe.

A simple scale of energy for physical events has been sketched by physicist Alfred K. Mann in a lecture he gave at Fermilab: Strike a match and the particles in that flare carry an energy of a few electron volts. Inside the furnace of the sun, each particle has an energy of millions of electron volts. And in the explosion of Creation, each particle raced apart from the others with an energy of billions of electron volts. The Fermilab's particles are accelerated to that same range, four hundred to five hundred billion electron volts.

For the past forty years, the foremost occupation of physics has

699

been to read, like a star map, the arcs and angles and trajectories which emerge from the flash of energy in particle accelerators. From this the elementary structure of matter may be deduced. The method is direct, and a little obscure, but it is the only way to see the skeleton of the universe, the particles which underlie and support the flesh of all appearance.

Alongside the great machine of physics, the physicists move in daily routines like priests repeating rituals of transubstantiation. If physics may be pictured as a single corpus of belief — as one work with no author and thousands of successive editors — then the bright flash at the accelerator target is perhaps the best symbol, the summary in a single action, of the whole achievement of physics.

The string of iron boxes, the tubes and pipes in the cement tunnel, and the circular mound above them, together make up a bundle of concentric rings. And there are more aboveground. There is a ring of road around the inside of the four-mile mound and a ring of ponds and fountains used for cooling the machine. The workers here refer to all the bands, above- and belowground, in a single phrase — "the main ring."

We rode out along the curve of the main ring for more than an hour — Wilson, myself, and Rich Orr, the owner of the horses and a physicist who was serving as the Fermilab's business manager. Under the violent clouds, the accelerator site looked more like fresh prairie than a federal laboratory facility. The landscape was a study in two plain colors, brown and gray. But present in a dozen contrasting shades, each color was animated by play against itself. The vertical stripes of the trees were carbon black, smoky gray, bluish gray, and silver. The grasses were white gold, fawn brown, reddish sienna, and dark umber brown.

Wilson has kept the Fermilab site as wild as possible. In fact, it is wilder than it was in 1967, when the land was owned by farmers. There are 1800 acres of the site devoted to corn, 650 acres left to pasture, 250 for hay, and some thousand still open, including the one and a quarter square miles in the center of the ring. There have been four-pound bass caught in Fermilab ponds and eight-pound catfish. During an Audubon bird count one year there were about forty different sorts of birds spotted at the lab, including about five hundred mallard ducks, a pair of the rarer canvasbacks seen floating in the cooling ponds, five rufous-sided towhees, about a score of hawks, a score of ring-necked pheasant, five purple finches, a dozen woodpeckers, and sixty-four common crows. There are deer, fox, raccoon, mink, and several lodges of beaver. When the accelerator suddenly overheated and shut down one afternoon some time ago, there was bewilderment in the control room. The cooling system had failed, but not for any apparent mechanical reason. Some time passed before it was found that muskrats, which like to make burrows in the main-ring bunker, had dug too close to the edge of a cooling pond. The side collapsed, the water emptied, and one segment of the machine quickly heated up.

Not long after Wilson completed the design and construction of Fermilab, an odd article about the lab appeared in *The New Republic* magazine. It was an art review of the place by critic Kenneth Evett, who began his review saying, "While advocates of the marriage of art and technology go marching down the aisle . . . whom should they meet coming in the opposite direction but Robert R. Wilson, designer and director of the Fermi National Laboratory at Batavia, Illinois, offering his creation — of the largest, most complex, and sophisticated instruments of scientific research on earth — as his version of that improbable consummation in which technology has become one with art. The inadvertent beauty of functional machines has long been widely noted, but Wilson, while responding to the rigorous necessities of efficiency and experimental precision, has, in addition, consciously designed his laboratory in every particular and on a gigantic scale to satisfy his own esthetic predilections. The rare combination of artistic and scientific aspiration . . . has produced a hybrid creation unprecedented in the history of art or science . . . the ensemble sits there, an island of intense Celebration and high Civilization amidst the suburban semirural expanses of northern Illinois." In addition to making the Fermilab artful, Wilson managed to build it within a tight budget, and give it double the experimental power planned for it.

In at least one respect, particle physics in the twentieth century is like astronomy in the seventeenth. In that time, just after the invention of telescopes, there was a race to create newer and bigger telescopes because each larger size practically guaranteed new discoveries through new depths of vision. Each telescope was superseded by another, and so astronomy advanced in leapfrog fashion. It has been the same with particle accelerators from the 1930s to today. There are some physicists who say that without Fermilab, and without Wilson's successful schedule and budget for the machine, accelerators around the world would not have reached such high experimental energies for another two or three decades. And when those energies finally came, they would not have been in American machines, but in European or Soviet ones. After Nobelist Ernest Lawrence, inventor of the circular particle accelerators, Robert Wilson has been the chief figure in physics always ready to build a newer, bigger machine — even when it seemed impossible, even at the expense of current experiments.

Wilson, who left the directorship of Fermilab in 1978, is still helping to build Fermilab's newest and most powerful accelerator, called the Tevatron because it is expected to produce beam energies of a trillion electron volts. Wilson, who is now sixty-eight years old, has also lived some length of time in Beijing, helping the Chinese to build their first large accelerator. He was pleased, he said, to find that the site chosen for the Chinese machine is an aesthetically fine one, within sight of the hallowed Ming Tombs.

As Wilson and I rode back toward the barn, a pair of Canadian geese flew up from the tall grass near one of the ponds; the birds bleated noisily as they joined a small squadron of geese flying over the main ring. We rode on, touring the borders of a thick stand of trees that ran right up to the edge of the main-ring bunker. During the building of the laboratory, after the site for the ring had been picked, Wilson noticed that the ring was going to cut through this stand of trees. Changing the plans at that point would have meant thousands of dollars in extra cost, and taking certain risks in placing the particle machine on the uncertain ground of a bog. Without hesitation Wilson moved the machine; the trees stood.

Since 1967, this land has reclaimed some of its past. The land by now would certainly be under several housing tracts, but it has been held open and is used as a public park. Some of the original Illinois prairie grass, which in the past two hundred years has been nearly choked off by strains of European grass, has been replanted, and large patches of it are seeded inside the main ring. At Fermilab there are also about thirty-five buffalo, which have not flourished here for about eight hundred years. At that time tribes living by the Fox River would send bands of hunters up to the site in the summer. They would make camp by the ponds, take their kill, and bring it back to the larger river settlement. The habit of hunting at this site goes back probably nine thousand years, perhaps more. Arrowheads, which not long ago littered the ground, have been collected and preserved.

The grounds of Fermilab are a patched quilt of sculpture, buildings, and equipment of Robert Wilson's design. So much is Wilson's that the display seems nearly indecent. Driving into Fermilab from the west, the first noticeable influence of Wilson's hand is at the front gate — or at the lack of a front gate. There is a token guardhouse by the road, but no guard, no barrier. Other similar federal facilities are so fierce about their privacy, one feels a flash of guilt passing through Fermilab's open entrance.

Wilson and assistant director Ned Goldwasser fought for the openness against constant demands by the government and Fermilab's own security forces. During the emotional storm of the 1960s, especially after a laboratory at the University of Wisconsin was bombed, the Atomic Energy Commission told Wilson to come up with a plan for handling activists should they appear at the National Accelerator Laboratory. Wilson wrote in response that he would face the students alone, "armed with the most potent student-stopper yet devised — a lecture on physics. I have found that when exposed to such strong verbal radiations strong men waver, brave men weep. A touch of Newton or a dose of quantum mechanics causes the eyes to glaze over, unconsciousness to set in. The fact is, I rather hesitate to use this weapon of overkill. . . ."

Far more than most physicists, Robert Wilson is aware of the aesthetic dimensions of physics. The giant particle accelerator might not have been built at all without Wilson, but it was designed and constructed with a sense of art as well as science. Before Wilson, the machines of physics were made with a chaos of wires and tubes and metal bars. They were the color of the parts, modified by dirt and grease. But the main body of the accelerator, the circular string of magnets resembling a four-mile-long freight train, is painted in bright colors. In this machine, as in others by Wilson, the magnets are bright blue, with every fourth one painted bright red. The magnets sit on yellow metal stands, which Wilson designed with decorative corners and scalloped edges. The wires and other paraphernalia he tucked neatly into steel tubes. Wilson's machines are gorgeous.

At one moment during our morning ride at the Fermilab, Wilson was trotting Star not far from the ring of tall earth under which the proton machine was working. He looked over and gauged the height of the mound, or "berm" as the physicists call it.

"What exactly is the purpose of the berm?" I asked him. Since the machine was already buried underground, I was uncertain just why this mound of dirt should also be piled on top of it. "Is it to absorb the extra radiation from the machine?"

"No, not really," he said. "Mostly we put it there to accentuate the ring. When we were building it, we looked out and saw that the ring just wasn't very visible. I was really disappointed. You could see where the cooling ponds were all right, but not the ring itself. So when we dug out the ponds, I had them pile the dirt up here. . . ." Wilson gestured toward the immense circle of dirt.

I looked at him, and thought of the view from three hundred miles up in space. "The berm is for aesthetic reasons?"

"Yes."

# STANISLAW M. ULAM

∎

When Stanislaw Ulam (1909–1984) first devoted himself to mathematics, he felt secure in the knowledge that he was embarking on an adventure in pure thought, one unfreighted with any possible practical consequences. In this chapter from his book *Adventures of a Mathematician*, Ulam wonderingly describes how he instead found himself recruited into one of the most grimly practical projects ever undertaken — the Manhattan Project.

# *Los Alamos*

DURING THE LATE SPRING of 1943, I wrote to [John] von Neumann about the possibility of war work. I knew he was involved, because his letters often came from Washington rather than from Princeton. I was not happy with teaching, although I did a lot of mathematics, wrote papers, organized colloquia, and taught war-related courses. Still it seemed a waste of my time; I felt I could do more for the war effort.

One day Johnny answered with an intimation that there was interesting work going on — he could not tell me where. From Princeton he said he was going west via Chicago, and suggested that I come to the Union Station there to talk to him since he had two hours between trains. That was in the early fall of 1943.

I went and, sure enough, Johnny appeared. What caught my attention were the two men escorting him, looking a bit like "gorillas." They were obviously guards, and that impressed me; he must be an important figure to rate this, I decided. One of the men went to do something about his railroad ticket, and we talked in the meantime.

Johnny said that there was some very exciting work going on in which I could possibly be of good use; he still could not tell me

705

where it was taking place, but he traveled rather often from Princeton to that location.

I don't know why — by pure chance or one of these incredible coincidences or prophetic insights? — but I answered jokingly, "Well, as you know, Johnny, I don't know much about engineering or experimental physics, in fact I don't even know how the toilet flusher works, except that it is a sort of autocatalytic effect." At this I saw him wince and his expression become quizzical. Only later did I discover that indeed the word autocatalytic was used in connection with schemes for the construction of an atomic bomb.

Then another coincidence occurred. I said, "Recently I have been looking at some work on branching processes." There was a paper by a Swedish mathematician about processes in which particles multiply quite like bacteria, for example. It was prewar work, and an elegant theory of probabilistic processes. That, too, could have had something to do with the mathematics of neutron multiplication. And again he looked at me almost with suspicion or wonder and smiled wanly.

The Wisconsin astronomer Joel Stebbins, whom I saw occasionally, had told me about some work going on with uranium and about the release of energy from very heavy elements. I wondered if, subconsciously, this prompted my remarks.

During this meeting at the station, Johnny and I also discussed what seemed a general lack of imagination in the scientific community's planning of work useful for the war effort — especially in computations for hydrodynamics and aerodynamics. I pointed out my doubts about the age of some of the main participants (people over forty-five seemed to me at that time old). Johnny agreed there were obvious elements of senility. As usual, we tried to lighten our sadness with jocular comments, observing that someone should establish a "gerontological" society, whose members would be scientists interested in war work and afflicted with premature or "galloping" senility.

Since Johnny could not or would not tell me where he was going except that it was to the Southwest, I remembered an old Jewish story about two Jews on a train in Russia. One asks the other, "Where are you going?" and the second replies, "To Kiev." Whereupon the first says, "You liar, you tell me you are going to Kiev so I would think you are going to Odessa. But I know you are going to Kiev, so why do you lie?" And I told Johnny, "I know you can't tell me, but you say you are going Southwest in order that I should think that you are going Northeast. But I know you are going Southwest, so why do you lie?" He laughed. We talked a while longer about the war situation, politics, and the world; then his two companions reappeared and he left.

I saw him once more, I think, in Chicago, before I received an official invitation to join an unidentified project that was doing important work, the physics having something to do with the interior of stars. The letter inviting me was signed by the famous physicist Hans Bethe. It came together with a letter from the personnel department with details of the appointment, salary, clearance procedures, indications on how to get there, and so forth. I accepted immediately with excitement and eagerness.

The pay was slightly above my university salary, but on a twelve-month basis — around $5,000, if I remember correctly. The professional physicists like Bethe who were there already received little more than their university salaries. I learned later that a chemist from Harvard, George Kistiakowski, had the allegedly astronomical salary of $9,000 or $10,000.

I informed my university of this opportunity to join an obviously important war project and secured a leave of absence for the duration.

A student of mine, Joan Hinton, had left for an unknown destination a few weeks before. Joan was taking a course I gave in classical mechanics. One day she appeared in my office in North Hall to ask if I could give her an examination three or four weeks before the end of the term so that she could start some war work. She produced a letter from Professor Ingraham, the chairman, authorizing me to do that. She was a good student, a rather eccentric girl, blonde, sturdy, good-looking. Her uncle was G. I. Taylor, the English physicist. She was also a great-granddaughter of George Boole, the famous nineteenth-century logician. I wrote a number of questions on the back of an envelope; Joan took some sheets of paper, sat down on the floor with her notebook, wrote out her exam, passed, and disappeared from Madison.

Soon after, other people I knew well began to vanish one after the other, without saying where — cafeteria acquaintances, young physics professors and graduate students like David Frisch, and his wife Rose, who was a graduate student in my calculus class, Joseph McKibben, Dick Taschek, and others.

Finally I learned that we were going to New Mexico, to a place not far from Santa Fe. Never having heard about New Mexico, I went to the library and borrowed the Federal Writers' Project Guide to New Mexico. At the back of the book, on the slip of paper on which borrowers signed their names, I read the names of Joan Hinton, David Frisch, Joseph McKibben, and all the other people who had been mysteriously disappearing to hush-hush war jobs without saying where. I had uncovered their destination in a simple and unexpected fashion. It is next to impossible to maintain absolute secrecy and security in war time.

This reminds me of another story. Since I knew Stebbins well, about a month after arriving at Los Alamos, I wrote to him. I did not say where I was but mentioned that in January or February I had seen the star Canopus on the horizon. Later it occurred to me that as an astronomer he could easily have deduced my latitude since this star of the Southern skies is not visible about the 38th parallel.

I shall pass over our problems in getting train reservations. Even with the priorities I had, our departure was delayed by about a month. On the train I had to offer a gratuity to the conductor to obtain a berth for Françoise, who was two months pregnant at the time. This was the first — and I think last — time in my life that I "bribed" anyone.

We arrived at a remote, lonely, unimpressive little whistle-stop — Lamy, New Mexico. To my infinite surprise, there to meet us was Jack Calkin, a mathematician I knew well. I had met him several years before at the University of Chicago and had seen him a number of times since. Calkin had been Johnny's assistant and had gone with him to London to discuss probability problems in aerial-bombing patterns and methods. Just a few weeks before, he had joined the Manhattan Project. He was a tall, pleasant-looking man, with more savoir-faire than most mathematicians. Having heard that I was coming, he borrowed a car from the Army motor pool and drove to meet us at the train.

The sun shone brilliantly, the air was crisp and heady, and it was warm even though there was a lot of snow on the ground — a lovely contrast to the rigors of winter in Madison. Calkin drove us into Santa Fe, and we stopped for lunch at the Hotel La Fonda, where we sat at the low Spanish-style tables in the bar. After an interesting New Mexico–style meal, we walked to a small doorway in a one-story building on a little street that bordered the central Plaza. In a modest suite of rooms, a smiling middle-aged lady invited me to fill out a few forms, turned a crank on a primitive desk machine, and produced the sheets of paper that were our passes to the Los

Alamos Project. This inconspicuous little office was the entrance to the gigantic Los Alamos complex. The scene, very much like a British cloak-and-dagger mystery, brought memories of my boyhood fascination with such tales.

The project site was about forty miles northwest of Santa Fe. The ride was hair-raising, Jack having elected to show us the countryside by taking us on a short cut — a muddy track through sparse Mexican and Indian villages — until we came to the Rio Grande, which we crossed on a narrow wooden bridge.

The setting was romantic. We were going up and up into a strange, mysterious landscape of mesas, cliffs, piñon trees and brush. This became a forest of pines as we gained elevation. At a military gate in a barbed-wire fence, we showed our passes and drove on to a helter-skelter collection of one- and two-story wooden structures built along muddy, unpaved, narrow streets and paths.

We were assigned a small cottage by a pond (with the promise of larger quarters as soon as they were built). I then followed Jack to my first visit to the technical area.

We entered an office, where I was surprised to find Johnny deep in conversation with a man of middle stature, bushy eyebrows, an intense expression. He limped slightly as he paced back and forth in front of a blackboard. This was Edward Teller, to whom Johnny introduced me.

They were talking about things which I only vaguely understood. There were tremendously long formulae on the blackboard, which scared me. Seeing all these complications of analysis, I was dumbfounded, fearing I would never be able to contribute anything. However, when day after day the same equations remained and were not changed every few hours as I had expected, I regained my confidence and some hope of being able to add something to the theoretical work.

I understood snatches of their conversation, and an hour later, Johnny took me aside and explained to me formally and clearly the nature of the project and its status at the moment. The work in Los Alamos had started in earnest only about two or three months before. Von Neumann seemed very certain of its importance and radiated confidence about the ultimate success of the enterprise whose objective was the construction of an atomic bomb. He told me of all the possibilities which had been considered, of the problems relating to the assembling of fissionable materials, about plutonium (which did not yet physically exist even in the most microscopic quantities at Los Alamos). I remember very well, when a couple of months later I saw Robert Oppenheimer running excitedly down a corridor holding a small vial in his hand, with Victor Weisskopf trailing after him. He was showing some mysterious drops of something at the bottom of the vial. Doors opened, people were summoned, whispered conversations ensued, there was great excitement. The first quantity of plutonium had just arrived at the lab.

Needless to say, I soon ran across most of the Wisconsinites who had so mysteriously disappeared from Madison before us. I met Hans Bethe on the first day. I knew more about him than about Teller. I gradually met the entire group of theoretical and experimental physicists. I had known many mathematicians in Europe and in this country, but not as many physicists.

I had some knowledge of theoretical physics, despite the joke I had told Johnny about my not understanding even the autocatalytic action of a toilet. Astronomy, of course, had been my first interest, and then physics and mathematics. I had even given a course in classical mechanics at Harvard, but it is one thing to know about physics abstractly, and quite another to have a practical encounter with problems directly connected with experimental data, such as the very novel technology which was to come from Los Alamos.

I found out that the main ability to have was a visual, and also an almost tactile, way to imagine the physical situations, rather than a merely logical picture of the problems.

The feeling for problems in physics is quite different from purely theoretical mathematical thinking. It is hard to describe the kind of imagination that enables one to guess at or gauge the behavior of physical phenomena. Very few mathematicians seem to possess it to any great degree. Johnny, for example, did not have to any extent the intuitive common sense and "gut" feeling or penchant for guessing what happens in given physical situations. His memory was mainly auditory, rather than visual.

Another thing that seems necessary is the knowledge of a dozen or so physical constants, not merely of their numerical value, but a real feeling for their relative orders of magnitude and interrelations, and, so to speak, an instinctive ability to "estimate."

I knew, of course, the values of constants like the velocity of light and maybe three or four other fundamental constants — the Planck constant $h$, a gas constant $R$, etc. Very soon I discovered that if one gets a feeling for no more than a dozen other radiation and nuclear constants, one can imagine the subatomic world almost tangibly, and manipulate the picture dimensionally and qualitatively, before calculating more precise relationships.

Most of the physics at Los Alamos could be reduced to the study of assemblies of particles interacting with each other, hitting each other, scattering, sometimes giving rise to new particles. Strangely enough, the actual working problems did not involve much of the mathematical apparatus of quantum theory although it lay at the base of the phenomena, but rather dynamics of a more classical kind — kinematics, statistical mechanics, large-scale motion problems, hydrodynamics, behavior of radiation, and the like. In fact, compared to quantum theory the project work was like applied mathematics as compared with abstract mathematics. If one is good at solving differential equations or using asymptotic series, one need not necessarily know the foundations of function space language. It is needed for a more fundamental understanding, of course. In the same way, quantum theory is necessary in many instances to explain the data and to explain the values of cross sections. But it was not crucial, once one understood the ideas and then the facts of events involving neutrons reacting with other nuclei.

Teller, in whose group I was supposed to work, talked to me on that first day about a problem in mathematical physics that was part of the necessary theoretical work in preparation for developing the idea of a "super" bomb, as the proposed thermonuclear hydrogen bomb was then called. The idea of thermonuclear reactions that would release enormous amounts of energy was, of course, older. Their role in the reactions in the interior of stars was discussed in theoretical papers in the 1930s by the physicists Geoffrey S. Atkinson and Fritz Houtermans. The idea of using a uranium fission explosion to trigger a thermonuclear reaction can be credited to Teller, Bethe, Konopinski, I believe, and perhaps some others.

Teller's problem concerned the interaction of an electron gas with radiation, and it had more to do with the thermonuclear possibilities than with the assembly of the fission bomb, which was the main problem and work of Los Alamos. He guessed a formula for energy transfers connected with the so-called Compton effect about the rate of energy transfer. This formula, based on dimensional grounds and his intuition alone, was quite simple; he asked me to try to derive it more rigorously. As it was presented, there was no numerical factor in front. This seemed curious to me. I asked him explicitly about this a day or two later, and he said "Oh, the factor should be 1."

This was the first technical problem in theoretical physics I had ever tackled in my life, and I approached it from a very elementary point of view. I read papers on statistical mechanics, on properties of the radiation field, and started working with rather naive and common-sense kinematic pictures. I tried some arithmetic and obtained a formula much like Teller's, but with a numerical factor of about four in front as the rate of transfer. It was a messy little job. Edward was not satisfied with my rather elementary derivations.

Shortly after I had discussed this work with Teller, a young, more professional mathematical physicist, Henry Hurwitz, Jr., joined Teller's group and with his much better mathematical techniques and experience in the special functions that were used in this type of problem, he obtained a formula, much more scholarly than mine, involving Bessel functions. Indeed, the exact numerical factor was not very different from four. If I remember correctly, it was a root of a certain Bessel function.

The idea was to have some thermonuclear material — deuterium — next to the fission bomb, and to let it ignite after the uranium bomb had exploded. How to do it in detail was the big problem, and it was by no means easy to see how such an arrangement would ignite and not just sputter and fizzle out. There was also, theoretically at least, the hazard of getting more of an explosion than intended and of having the whole atmosphere of the earth ignite! The well-known physicist Gregory Breit was involved in calculating the chances of the ignition of the atmosphere. These, of course, had to be zero before one could even think of tampering with thermonuclear reactions on earth.

I think it was Bethe, with Emil Konopinski, a well-known theoretical physicist, who suggested tritium instead of deuterium as a material easier to ignite, given the temperature of the fission bomb. Such an engineering suggestion from theoretical work came from his superb knowledge of theoretical nuclear physics.

Bethe was the head of the theoretical division, as it was called. Actually it was his and Robert F. Bacher's papers in *Reviews of Modern Physics* which were used as the "bible" of the Los Alamos scientists, for they contained the bulk of the theoretical ideas and experimental facts known at the time. Bethe, now a Nobel Prize winner for his earlier discovery of the mechanism of energy generation in the sun and other stars (the so-called carbon cycle), is, among other things, a virtuoso in the techniques of mathematical physics. As Feynman once put it, at Los Alamos he was, with his rigorous and definitive work, like a battleship moving steadily forward, surrounded by a flotilla of smaller vessels, the younger theoretical workers of the laboratory. He is one of the few persons for whom I merely had respect in the beginning but over the years have continuously developed liking and admiration.

When I first met Teller, he appeared youthful, always intense, visibly ambitious, and harboring a smoldering passion for achievement in physics. He was a warm person and clearly desired friendship with other physicists. Possessing a very critical mind, he also showed quickness, sense, and great determination and persistence. However, I think he also showed less feeling for true simplicity in the more fundamental levels of theoretical physics. To exaggerate a bit, I would say his talents were more in the direction of engineering, construction, and the surveying of existing methods. But undoubtedly he also had great ingenuity.

Teller was well known for his work on molecules, but he may have considered this as a sort of secondary field. I think it was the ease with which Gamow had new ideas without any technical arsenal at his disposal that pushed Teller into trying to emulate him and to attempt more fundamental work.

After he got into personal difficulties with Teller on the organizational features of the hydrogen work, Gamow later told me that before the war Teller was, in his view, a different person — helpful, willing, and able to work on other people's ideas without insisting on everything having to be his own. According to Gamow, something changed in him after he joined the Los Alamos Project.

Of course, many physicists who were almost congenitally ivory-tower types got their heads turned with the sudden realization of not only the practical but worldwide historical importance of their work — not to mention the more trivial but obvious matter of the enormous sums of money and physical facilities that surpassed anything in their previous experience. Perhaps this played a role in the personality change of some principals; with Oppenheimer, the director, it may have had a bearing on his subsequent activities, career, ideas, and role as a universal sage. Like Teller, Oppenheimer may have had a feeling of inadequacy as compared with the creators of great new physical theories. He was equal to or even more brilliant and quick than Teller, but perhaps lacked the ultimate creative spark of originality. With his fantastic intelligence, he must have realized this himself. In speed of understanding and in critical ability, he probably surpassed Bethe or Fermi.

Teller wanted to have his own stamp on much of the essential work of Los Alamos, at first via his own approach to the fission bomb. He was pushing for milder explosions, dilution of material, etc. In spite of calculations by Konopinski and others that gave a poor outlook for some of these plans, he was trying by every means to have his own adopted. Collaboration with Bethe, who was head of the theoretical division, became increasingly difficult.

As disagreements between Teller and Bethe became more frequent and acute, Teller threatened to leave. Oppenheimer, who did not want to lose such a brilliant scientist, agreed to let him and his group work in a more future-oriented field, independent of the project's main line. This is how Teller began to concentrate and organize the theoretical work for the "Super." Konopinski, Weisskopf, Serber, Richard Feynman, William Rarita, and many others all had special contributions to make, but it was really Teller who kept the thing together and moving forward during the war.

After Fermi's arrival, Teller's group became a part of Fermi's division. Fermi took great interest in the theoretical work on the thermonuclear reactions and H-bomb possibilities; at the end of the war he gave a series of lectures magnificently summarizing the work done until then — thanks mostly to the investigations of Teller and his group.

But even then, before the success of the atomic bomb itself, some of Teller's actions gave clues to what led to much of the unpleasantness and waste of time in the so-called H-Bomb controversy.

Teller's group was composed of a number of very interesting young physicists, younger even than Teller, Konopinski, or myself. It included Nick Metropolis, a Greek-American with a wonderful personality; Harold and Mary Argo, a husband-and-wife team, eager and talented; Jane Roeberg, a young woman who gave the impression of being competent; and a few others whose names I do not now remember.

There was, of course, much contact with other groups of physicists who were on the borderline of problems concerning the possibilities of a "super"; discussions with them were frequent, pleasant, and concerned many different branches of physics. One could hear about the pros and cons of the idea of implosion, which was new and vigorously debated in many of the offices. These discussions were completely open. Nothing was concealed from anybody who was a scientist.

The more formal way of letting people know what was going on was the weekly colloquia, which were held in a big hangar that also served as the movie theater. These talks covered progress of the work of the whole laboratory and the specific problems which the project encountered. They were run by Oppenheimer himself.

As for myself, after this first work on Edward's problem, I spread out my interests to other related questions, one being the problem of statistics of neutron multiplication. This was more tangible for me from the purely mathematical side. I discussed such problems of branching and multiplying patterns with David Hawkins. We wrote a report on multiplicative branching processes, which had some practical application and relevance to the problem of the initial detonation of the bomb by a few neutrons. This problem was also studied by Stan Fraenkel and by Feynman, in a more technical and classical way. Our paper could be considered the beginning of what would come to be known in mathematics as branching processes theory, a sub-field of probability theory.

I also talked a lot with von Neumann and Calkin about problems of hydrodynamics, especially those concerning the process of implosion. Somewhat to my surprise I found my purely abstract intellectual habits as a mathematician immediately useful in the work with these more practical, special, and tangible problems. I have never felt the "gap" between the mode of thinking in pure mathematics and the thinking in physics, on which many mathematicians place so much stress. Anything amenable to mental analysis was congenial for me. I do not mean the distinction between rigorous thinking and more vague "imaginings"; even in mathematics itself, all is not a question of rigor, but rather, at the start, of reasoned intuition and imagination, and, also, repeated guessing. After all, most thinking is a synthesis or juxtaposition of advances along a line of syllogisms — perhaps in a continuous and persistent "forward" movement, with searching, so to speak "sideways," in directions which are not necessarily present from the very beginning and which I describe as "sending out exploratory patrols" and trying alternative routes. It is all a multicolored thing, not very easy to describe in a way that a reader can appreciate. But I hope this kind of personal analysis of thinking in science is one of the possible interests of this book.

A discussion with von Neumann which I remember from early 1944 took several hours, and concerned ways to calculate the course of an implosion more realistically than the first attempts outlined by him and his collaborators. The hydrodynamical problem was simply stated, but very difficult to calculate — not only in detail, but even in order of magnitude.

In particular, the questions concerned values of certain numbers relating to compression versus pressure, and such. These had to be known, let us say within ten per cent or better, but the simplifications made in the outline of the calculations were of such a nature that they could not guarantee accuracy within a factor of two or three. All the ingenious shortcuts and theoretical simplifications which von Neumann and other mathematical physicists suggested, and which he tried to execute with the help of Calkin, seemed inadequate to me. In this discussion I stressed pure pragmatism and the necessity for attempting to get a heuristic survey of the general problem by simpleminded brute force — that is, more realistic, massive numerical work. At that time, in 1944, with the available computing facilities, the accuracy of the necessary numerical work could not be satisfactory. This was one of the first reasons for pressing for the development of electronic computers.

One of the charms and great attractions of life in Los Alamos in those days was the lunches at the Lodge, in the midst of friends. I was very surprised to find there and gradually to meet so many famous persons I had heard about.

Los Alamos was a very young place. At thirty-four, I was already one of the older people. What impressed me most was the very great competence of the younger people and the variety of their fields of specialization. It was almost like having an encyclopedia to look at, something that I so much like to do. I had the same feeling when talking to the young scientists around the laboratory. It is not the right expression perhaps but, roughly speaking, they were more accomplished in depth than in breadth. The older men, many of whom were European-born, had a more general knowledge. Yet science had become so ramified, specialization had proceeded so far, that it was quite difficult to retain knowledge of all the details and the overall view at the same time.

The younger scientists showed a lot of common sense in their own fields, but in general a great hesitation to engage in speculation outside their areas. Perhaps this stemmed from a fear of not being "absolutely right." Many displayed a certain anti-philosophical spirit — not anti-intellectual, but anti-philosophical. This was perhaps because of the pragmatic nature of American attitudes.

I was also struck by the well-known American talent for cooperation, the team spirit, and how it contrasted with what I had known in continental Europe. I remembered how Jules Verne had anticipated this when he wrote about the collective effort needed for the organization of his "Voyage to the Moon." People here were willing to assume minor roles for the sake of contributing to a common enterprise. This spirit of team work must have been characteristic of life in the nineteenth century and was what made the great industrial empires possible. One of its humorous side effects in Los Alamos was a fascination with organizational charts. At meetings, theoretical talks were interesting enough to the audience, but whenever an organizational chart was displayed, I could feel the whole audience come to life with pleasure at seeing something concrete and definite ("Who is responsible to whom," etc.). Organization was and perhaps still is a great American talent, although this is written at a time when the so-called energy crisis appears to me to be more a crisis of momentum than of energy (a crisis of enterprise, solidarity, common spirit, determination, and cooperation for the common good).

It is difficult to describe for the general reader the intellectual flavor, the feeling, of a scientific "atmosphere." There is no specific English word for this impression. Odor and smell have unpleasant connotations; perfume is artificial; aura is suggestive of mystery, of the supernatural. The younger scientists did not have much of an aura, they were bright young men, not geniuses. Perhaps only Feynman among the young ones had a certain aura.

Six or seven years younger than I, he was brilliant, witty, eccentric, original. I remember one day Bethe's laughter shook the corridor walls, making me rush out of my office to see what was so funny. Three doors down in Bethe's office, Feynman was standing — talking and gesticulating. He was telling the story of how he had failed his draft physical examination, re-enacting his now famous gesture: when a doctor asked him to show his hands, he chose to stretch them in front of him one with palm up, the other palm down. The doctor said, "The other side!" and he reversed both hands. This and other incidents of his physical examination had caused an explosion of laughter on the entire floor. It was on the first or second day in Los Alamos that I met Feynman, and remarked to him about my surprise that $E = mc^2$ — which I of course believed in theoretically but somehow did not really "feel" — was, in fact, the basis of the whole thing and would bring about a bomb. What the whole Project was working on depended on those few little signs on paper. Einstein himself, when he was first told before the war about radioactive phenomena showing the equivalence of mass and energy, allegedly replied, *"Ist das wirklich so? ist das wirklich so?"* (Is that really so?)

Jokingly I told Feynman, "One day people will discover that a cubic centimeter of vacuum is really worth ten thousand dollars — it is equivalent to so much energy." He immediately agreed and added, "Yes, but of course it will have to be *pure* vacuum!" Indeed, people now know about the polarization of vacuum. The force between two electrons or two protons is not $e^2/r^2$, but an infinite series of which this is the first term. It works on itself, like two almost-parallel mirrors, which show a reflection of a reflection of a reflection, ad infinitum.

Writing this reminds me of a feeling I once had when I visited the cyclotron in Chicago with Fermi. He took me around and made me walk through an incredibly heavy door, which he said "would flatten you into a piece of paper were it closed on you." We walked between the poles of the magnet, and I reached into my pocket for the penknife with which I sometimes play. Suddenly it was jerked out of my hand when I touched it. The power of the vacuum! This made me physically conscious of the reality of "empty" space.

Feynman was also interested in many purely mathematical re-creations not related to physics. I remember how he once gave an amusing talk about triangular numbers and managed to entertain everybody with his humor. At the same time, he was doing mathematics, and showing the foolishness of excessive cleverness and the irrationality of such strange interests.

One day he recited to me the following:

> I wonder why I wonder,
> I wonder why I wonder why I wonder.
> I wonder why I wonder why I wonder why,

and so on.

It all depends on where you put the intonation, conveying a different meaning in every case. He did it marvelously in five or six styles, each with different stresses, as it were, most humorously.

Physically, Los Alamos consisted of a collection of two- and four-apartment buildings, temporary Army structures which turned out to be sturdy enough to survive for many years after the end of the war. To his everlasting credit, Oppenheimer insisted that they be laid out along the contours of the land, retaining as many trees as possible, instead of in the monotonous rectangular pattern of army camps and company towns. Still they were rather primitive, equipped with coal furnaces and coal stoves in the kitchens. People griped about the inadequacies of the housing situation, and wives had all sorts of complaints. But I found Los Alamos on the whole quite comfortable. The climate of New Mexico — Los Alamos in particular, at an elevation of seventy-two hundred feet — was one of the best I have ever lived in.

Placzek, a physicist who joined the project after the war, felt that east of the Rocky Mountains the United States was on the whole climatically uninhabitable, "*unbewohnbar*." This is true especially for Europeans who are not accustomed to hot and muggy summers or to penetrating winter cold. In Cambridge, I used to tell my friends that the United States was like the little child in a fairy tale, at whose birth all the good fairies came bearing gifts, and only one failed to come. It was the one bringing the climate.

Soon after my arrival in Los Alamos I met David Hawkins, a young philosopher from Berkeley, one of the people Oppenheimer had brought with him to staff the administration of the Laboratory. We hit it off intellectually right away.

Hawkins is a tallish, blue-eyed, blond descendant of early New Mexico settlers. His father, Judge Hawkins, was a famous figure at the turn of the century. He was a lawyer and an official of the Territory, important in the Santa Fe Railroad operations. David was brought up in the small community of La Luz, in the southern part of the State. I mention this because later, when the bomb was exploded in the Jornada del Muerte desert near Alamagordo, David worried that blinding flashes or the heat and shock phenomena might be dangerous for people living in La Luz, some thirty or forty miles away, where his sister had her home.

Hawkins is a man of wide interests, with great breadth of knowledge, very good education, and a very logical mind. He regards scientific problems not as a narrow specialist, but from a general epistemological and philosophical point of view. To top it off, he is the most talented amateur mathematician I know. He told me that at Stanford he took some courses from Ouspenski, the Russian émigré specialist in probability and number theory, but he has not had any extensive training in mathematics. He has a very great natural feeling for it and a talent for manipulation. He is the most impressive of the non-professional mathematicians or physicists I have met anywhere in the world.

We discussed problems of neutron chain reactions and the probability problems of branching processes, or multiplicative processes, as we called them in 1944.

I was interested in the purely stylized problem of a branching tree of progeny from one neutron which may multiply, into zero (that is, the death of a neutron by absorption), or one (that just continues itself), or two or three or four (that is, causes the emergence of new neutrons), each possibility with a given probability. The problem is to follow the future course and the chain of possibilities through many generations.

Very early Hawkins and I detected a fundamental trick to help study such branching chains mathematically. The so-called characteristic function, a device invented by Laplace and useful for normal "addition" of random variables, turned out to be just the thing to study "multiplicative" processes. Later we found that observations to this effect had been made before us by the statistician Lotka, but the real theory of such processes, based on the operation of iteration of a function or of operators allied to the function (a more general process), was begun by us in Los Alamos, starting with a short report. This work was strongly generalized and broadened in 1947, after the war, by Everett and myself after he joined me in Los Alamos. Some time later, Eugene Wigner brought up a question of priorities. He was eager to note that we did this work quite a bit before the celebrated mathematician Andrei N. Kolmogoroff and other Russians and some Czechs had laid claim to having obtained similar results.

I liked Hawkins's general curiosity, his almost unique knowledge of the fundamentals of several scientific theories — not only in the conceptual elements of physics but in biology and even economics. I liked his interest and genuinely original work in what was to become known as "information theory" after it became formalized by Wiener and especially by Claude Shannon. David applied to economic problems the mathematical ideas of von Neumann and Morgenstern in game theory.

Hawkins has since written several interesting papers and an excellent book on the philosophy of science, or rather on the philosophy of rational thinking, called *The Language of Nature*.

Hawkins's position in Los Alamos at first was as a liaison between Oppenheimer's office and the military. Some years later he wrote two volumes, since declassified, about the organization and the scientific history from the early days of Los Alamos until the end of the war. I did not know it at the time (and it was not obvious from conversations with him) that in the nineteen thirties he had been involved on the West Coast with communist sympathizer groups. That caused him great trouble before and during the McCarthy era, including hearings in Washington. He came out of it completely vindicated.

His wife, Frances, an extremely interesting person, became friendly with Françoise, and we saw each other a great deal. At the time of my illness in California in 1946, the Hawkinses were immensely helpful to us in caring for our daughter Claire, who was then an eighteen-month-old infant.

Hawkins left Los Alamos after the end of the war to take the post of professor of philosophy at the University of Colorado in Boulder, where he is today.

The Los Alamos community was completely different from any where I had ever lived and worked. Even Lwów, which had a dense concentration of people and where the mathematicians and university people were in daily contact and spent much time together in restaurants and coffee houses, did not have the degree of togetherness of Los Alamos. It was even more pronounced there because of the isolation and the smallness of the town, and the proximity of all the buildings. People visited each other constantly at all hours after work. What was novel to me was that these were not mathematicians (except von Neumann and two or three younger persons), but physicists, chemists, and engineers — psychologically quite different from my more inward-oriented mathematical colleagues. The variety and richness of the physicists was interesting and delightful to observe. On the whole, theoreticians and experimentalists differed in temperament.

It has been said that at lunch in Fuller Lodge one could see as many as eight or ten Nobel Prize winners eating at the same time (Rabi, Lawrence, Fermi, Bloch, Bohr, Chadwick, and others). Their interests were wide because physics has more definite and obvious central problems than mathematics, which splits into many almost independent domains of thought. They considered not only the main problem — the construction of an atomic bomb and related physical questions about phenomena that would attend the explosion — the strictly project work — but also general questions about the nature of physics, the future of physics, the impact of nuclear experiments on the technology of the future, and contrastingly its influence on the future development of theory. Beyond this, I remember very many after-dinner discussions about the philosophy of science, and of course on the world situation, from daily progress on the war fronts to prospects of victory in the months to come.

The intellectual quality of so many interesting persons and their being constantly together was unique. In the entire history of science there had never been anything even remotely approaching such a concentration. The radar project in Cambridge, Massachusetts, proceeding at the same time, had some of these characteristics, but without the same intensity. It was more technological perhaps, and did not touch as many fundamental questions of physics.

Who were some of the luminaries of this fantastic assembly? Von Neumann, Fermi, Bethe, Bohr, Feynman, Teller, Oppenheimer, O. R. Frisch, Weisskopf, Segré, and many more. I have already tried to sketch the personalities of some of them and can describe a few more.

I first met Fermi when he arrived at Los Alamos, a few months after us, after the Chicago pile had been successfully completed. I remember sitting at lunch in Fuller Lodge before his arrival with six or seven people, including von Neumann and Teller. Teller said, "It is quite certain now that Enrico will arrive next week." I had learned earlier that Fermi was referred to as "the pope" because of the infallibility of his pronouncements. So immediately I intoned: "*Annuncio vobis gaudium maximum, papam habemus,*" which is the classical way cardinals announce the election of a pope on the balcony overlooking St. Peter's Square, after the white smoke comes out of the chimney in the Vatican. Johnny, who understood, explained this reference, and the allusion was applauded by the entire table.

Fermi was short, sturdily built, strong in arms and legs, and rather fast moving. His eyes, darting at times, would be fixed reflectively when he was considering some question. His fingers often nervously played with a pencil or a slide rule. He usually appeared in good humor, with a smile almost perpetually playing around his lips.

He would look at a questioner in an inquiring way. His conversation included many questions rather than expressions of opinion. His questions were formulated in such a way, however, that it was clear which way Fermi's beliefs or guesses went. He would try to elucidate other persons' thoughts by asking questions in a Socratic manner, yet more concretely than in Plato's succession of problems.

Sublimated common sense characterized his thoughts. He had will power and control; and not obstinacy but persistence in following a line, all the while looking very carefully at possible ramifications.

He would not neglect the opportunities that presented themselves, often by chance, from random observations in scientific work.

Once when we discussed another physicist, he characterized him as too systematically obstinate. Yet he also told me that he liked to work very systematically in an orderly fashion in order to keep everything under control. At the same time he had decided in his youth to spend at least one hour a day thinking in a speculative way. I liked this paradox of a systematic way of thinking unsystematically. Fermi had a whole arsenal of mental pictures, illustrations, as it were, of important laws or effects, and he had a great mathematical technique, which he used only when necessary. Actually it was more than mere technique; it was a method for dissecting a problem and attacking each part in turn. With our limited knowledge of introspection this cannot be explained at the present time. It is still an "art" rather than a "science." I would say that Fermi was overwhelmingly rational. Let me explain what I mean: the special theory of relativity was strange, irrational, seen against the background of what was known before. There was no simple way to develop it through analogies with previous ideas. Fermi probably would not have tried to develop such a revolution.

I think he had a supreme sense of the important. He did not disdain work on the so-called smaller problems; at the same time, he kept in mind the order of importance of things in physics. This quality is more vital in physics than in mathematics, which is not so uniquely tied to "reality." Strangely enough, he started as a mathematician. Some of his first papers with very elegant results were devoted to the problem of ergodic motion. When he wanted to, he could do all kinds of mathematics. To my surprise, once on a walk he discussed a mathematical question arising from statistical mechanics which John Oxtoby and I had solved in 1941.

Fermi's will power was obvious, even to the extent of controlling his impulsive gestures. In my opinion, he deliberately avoided volatile Latin mannerisms, and perhaps by a conscious decision controlled gesticulations and avoided exclamations. But Enrico smiled and laughed very readily.

In all activities, scientific or otherwise, he had a mixture of semi-logical whimsical humor about common-sense points of view. When he played tennis, for instance, if he lost four games to six, he would say: "It does not count because the difference is less than the square root of the sum of the number of games." (This is a measure of purely random fluctuations in statistics.)

He loved political discussions, and he loved trying — not too seriously — to foresee the future. He would ask people in a group to write down what they thought would happen, and put it in a sealed envelope to be opened a couple of months later. On the whole he was very pessimistic about the long-range outlook politically, concluding that humanity is still foolish and would destroy itself one day.

He could be also quite a tease. I remember his Italian inflections when he would taunt Teller with statements like: "Edward-a how com-a the Hungarians have not-a invented anything?" Once Segré, who was very fond of fishing on weekends in the streams of the Los Alamos mountains, was expounding on the subtleties of the art, saying that it was not easy to catch trout. Enrico, who was not a fisherman, said with a smile, "Oh, I see, Emilio, it is a battle of wits."

In conversations with friends about the personalities of others, he tried to be entirely detached and objective, allowing little of personal or subjective opinions or feelings to surface. About himself, he had tremendous self-assurance. He knew that he had the touch as well as luck on top of his supreme common sense, enormous mathematical technique, and knowledge of physics.

Enrico was fond of walking; several times we walked all the way from Los Alamos down the walls of a canyon and along a stream to the Bandelier National Monument. It was a walk of seven or eight miles during which we had to cross the stream more than thirty times. The walk lasted several hours, and we discussed many subjects.

I should mention here one of my own peculiarities: I do not like walking uphill. I don't really know why. Some people tell me that I tend to go too fast from impatience and get winded for that reason. I do not mind walking on level ground and I actually enjoy walking down hill. Years ago I bought a German travel guidebook called "One Hundred Downhill Walks in the Alps." Certainly a humorous title.

After the war, on one of these downhill excursions in Frijoles Canyon, I told Fermi how in my last year of high school I was reading popular accounts of the work of Heisenberg, Schrödinger, and De Broglie on the new quantum theory. I learned that the solution of the Schrödinger equation gives levels of hydrogen atoms with a precision of six decimals. I wondered how such an artificially abstracted equation could work to better than one part in a million. A partial differential equation pulled out of thin air, it seemed to me, despite the appearances of derivation by analogies. I was relating this to

Fermi, and at once he replied: "It [the Schrödinger equation] has no business being that good, you know, Stan."

He went on to say that in the fall he intended to give a really logical introduction and derivation of quantum theory in his course at the University of Chicago. He apparently worked at it, but told me the next summer, when he returned to Los Alamos, "No, I didn't succeed to my satisfaction in giving a really rational introduction to quantum theory." It is not just a question of axioms as some naïve purist might think. The question is why such and no other axiom? Any working algorithm can be axiomatized. How to introduce, justify, tie up or simplify the axioms, historically or conceptually, and how to base them on experiments — that is the problem.

Von Neumann and Fermi were really quite different in personality. Johnny was perhaps broader in his interests than Enrico. He had more specifically expressed interests in other fields, certainly, for example, in ancient history. Fermi did not show any great interest in or liking for the arts. I never remember him discussing music, painting, or literature. Current affairs, politics, yes; history, no. Von Neumann was interested in both. Fermi did not indulge in quotations or allusions, Latin or otherwise, although he liked epigrammatic formulations occasionally. But he did not display a gymnasium or lycée type of education or the resultant mental habits. His overwhelming characteristic was his Latin clarity. Von Neumann did not consciously insist on simplicity; on the contrary, he liked to show clever complications on occasion.

In their lectures to students or scientific gatherings, they demonstrated their different approaches. Johnny did not mind showing off brilliancy or special ingenuity; Fermi, on the contrary, always strived for the utmost simplicity, and when he talked everything appeared in a most natural, direct, bright, clear light. After students had gone home, they were often unable to reconstruct Fermi's dazzlingly simple explanation of some phenomenon or his deceptively simple-looking idea on how to treat a physical problem mathematically. In contrast, von Neumann showed the effects of his sojourns at German universities. He was absolutely devoid of pomposity, but in his language structure he could be complicated, though perfect logic always gave a unique interpretation to his words.

They held high opinions of each other. I remember a discussion of some hydrodynamical problem Fermi had been thinking about. Von Neumann showed a way to consider it, using a formal mathematical technique. Fermi told me later with admiration, "He is really a professional, isn't he!" As for von Neumann, he always took external evidences of success seriously; he was quite impressed by Fermi's Nobel Prize. He also appreciated wistfully other people's ability to get results by intuition or seemingly pure luck, especially by the apparent effortlessness of Fermi's fundamental physics discoveries. After all, Fermi was perhaps the last all-around physicist in the sense that he knew the theory, did original work in many branches, and knew what experiments to suggest and even do himself; he was the last to be great both in theory and as an experimenter.

Niels Bohr, the discoverer of the quantized electron orbits in the atom and a great pioneer of quantum theory, was in Los Alamos for several months. He was not very old. To me at thirty-five, he seemed ancient, even though Bohr in his late fifties was very active and energetic, physically as well as mentally. He walked, skied, and hiked in the Los Alamos mountains. Somehow he seemed the embodiment of wisdom. (Wisdom, perhaps not genius in the sense of Newton or Einstein.) He knew what not to attempt and how much could be done without mathematics, which he left to others. This enormous wisdom is what I liked about him.

Departing from his usual caution about expressing opinions about other people, Fermi remarked once that when Bohr talked he sometimes gave the impression of a Catholic priest celebrating mass. It was an iconoclastic statement, since so many physicists are still under the spell of Bohr.

He had his own kind of genius that made him a great physicist but, to my mind, some of his students were almost benighted by his complementarity philosophy of "one can say this, but on the other hand one can . . ." or "one cannot say sharply what this means." People without his great sense and intuitive wisdom were led astray and lost the precision and sharpness of their intellectual or scientific approach, in my opinion. But he still has many admirers. Victor F. Weisskopf is one.

It seems to me that as a philosophical guideline, complementarity is essentially negative. It can only console. Whether it can be positively useful other than in philosophical consolations is a question which troubles me.

Bohr's speech was very difficult to understand, and anecdotes about him abound. Most of the time it was impossible to get his exact words. One day a young physicist, Ruby Scherr, was called on the public address system. Here I should explain that every day at periodic intervals the halls of the laboratory resounded with announcements and requests, the most frequent of which was a call for J. J. Gutierrez, who was a supply factotum and jack of all trades. Other calls were requests for the return of such and such an instrument, or even the Sears, Roebuck catalogue. One day among other announcements came one asking Ruby Scherr to please go to Nicholas Baker's office. (Nicholas Baker was the pseudonym of Bohr for security reasons; Fermi's was Farmer.) As Ruby Scherr tells the story, he went to the office, saw several physicists sitting around and obviously listening to a presentation by Bohr. Bohr stopped, mumbled a few incomprehensible sentences in the direction of Scherr, and suddenly ended with a crystal-clear three words: "Guess how much?" Scherr, who had not understood a word of the question, blushed with embarrassment, shook his head shyly and remained silent. After a moment Bohr again in a clear voice said "$10^{41}$." Whereupon everybody laughed. To this day, Scherr does not know what it was all about.

Another Bohr story illustrates the absentmindedness of scientists: it was well known throughout Los Alamos that Nicholas Baker was Niels Bohr; nevertheless, his true name was never supposed to be mentioned in public. At one colloquium, Weisskopf referred to "the well-known Bohr principle." "Oh excuse me," he fumbled, "the Nicholas Baker principle!" General laughter greeted this security breach.

Not all of us were unduly security conscious. Every scientist, old or young, had in his office a safe where secret documents had to be kept. Indeed, the Project must have had more safes than all the banks in New York. Once in Bohr's office I watched him struggle to open his safe. The safes could be opened with a rather simple combination of three two-digit numbers. He tried and tried for a long time, finally succeeding. He pulled out the drawer and exclaimed delightedly: "I believe I have done enough for the day." The story that Dick Feynman could open safes whose combinations had been forgotten by their owners is true. He apparently listened to the clicks of the tumblers and sometimes he guessed which combinations of digits of numbers like $\pi$ or $e$ in mathematics, or $c$, the velocity of light, or $h$, Planck's constant, had been selected by the owners for the combinations.

One thing that relieved the repetition and alternation of work, intellectual discussions, evening gatherings, social family visits and dinner parties, was when a group of us would play poker about once a week. The group included Metropolis, Davis, Calkin, Flanders, Langer, Long, Konopinski, von Neumann (when he was in town), Kistiakowski sometimes, Teller, and others. We played for small stakes; the naïveté of the game and the frivolous discussions laced with earthy exclamations and rough language provided a bath of refreshing foolishness from the very serious and important business that was the raison d'être of Los Alamos.

In playing such a game, unless you are vitally interested in the game itself, and not merely in its relaxing qualities, you will not do well. Von Neumann, Teller, and I would think about completely unrelated subjects during the bidding or betting; consequently, more often than not we were the losers. Metropolis once described what a triumph it was to win ten dollars from John von Neumann, author of a famous treatise on game theory. He then bought his book for five dollars and pasted the other five inside the cover as a symbol of his victory. It may not be clear to non-scientists or non-mathematicians that one can do theoretical work in one's head and pursue it quite intensely while literally carrying on some other more prosaic activity.

The Trinity test, Hiroshima, V-J Day, and the story of Los Alamos exploded over the world almost simultaneously with the A-Bomb. Publicity over the secret wartime Project filled the newspapers and its administrative heads were thrown in the limelight. In one newspaper interview out of many published the day after Hiroshima, E. O. Lawrence "modestly admitted," according to the interviewer, "that he more than anyone else was responsible for the atomic bomb." Similar statements by and about others filled the media. Oppenheimer was reported to have described his feelings after the unearthly light of the initial flash of the Trinity experiment by quoting from the Hindu epic, the Bhagavad Gita: "It flashed to my mind that I had become the Prince of Darkness, the destroyer of Universes."

What is true is that as I was reading this item in a newspaper, something else flashed through my mind, a story of a "pension" in Berlin before the war. I told it immediately to Johnny, who was eating dinner in our house. The Berlin boarders were sitting around a table for dinner and dishes were passed for each person to help himself to his share. One man was taking most of the asparagus that was on the platter. Whereupon another man stood up shyly and said: "Excuse me, Mr. Goldberg, we also like asparagus!" And the expression "asparagus" became a code word in our private conversations for trying to obtain an unduly large share of credit for scientific work or any other accomplishment of a joint or group character. Johnny loved this story so much that in our humorous conversations we played on developing the theme. We would plan to write a twenty-volume treatise on "Asparagetics through the Ages." Johnny would do "*Die Asparagetics im Altertum*" and I the final volume "*Rückblick und Ausblick*" in the manner of heavy German scholarship. Later, Carson Mark put his own stamp on these jokes by composing a song, "Oh, How I Love Asparagus," to the tune of a current popular song.

But levities like these could hardly alleviate the general feeling of foreboding upon entering into the era of history that would be called the Atomic Age. The war was over, the world and the nation had to reorganize themselves. Life would never be the same.

# LUIS W. ALVAREZ

■

The versatile experimental physicist Luis Alvarez (1911–1988) tried his hand at everything from inventing an instrumented landing system for commercial aircraft to x-raying the Great Pyramid of Cheops. He won the Nobel Prize in 1968 for his studies of subatomic particles. In this chapter of an autobiography, *Alvarez: Adventures of a Physicist,* he recounts his baptism of fire as a young postdoctoral assistant at Berkeley in the 1930s.

# *Coming Into My Own*

MY TRANSITION FROM APPRENTICE to professional was not without incident. I fitted the Radiation Laboratory's style comfortably, and the staff regarded me as hardworking and useful, but like my fellows I was sometimes consigned with good reason to the laboratory doghouse.

We usually disgraced ourselves by cutting corners, trying to work faster than prudent operation warranted. . . .

I mistakenly buggered an interlock. (Those familiar only with the vulgar meaning of this verb should know that *Webster's* defines "bugger up" as "put into disorder.") Interlocks are devices that protect machines and people from accidents. An interlock disconnects the electrical power from a television set, for example, when the back of the cabinet is removed, to guard the unsuspecting from contact with the 30,000-volt circuit that powers the color tube. A television repairman has to bugger the interlock and stay away from the high voltage to fix the set. Buggering interlocks is accepted practice, then, but only if you are certain you know what you're doing and only if you unbugger the interlock when you're through.

On cyclotron crew one day I went through the routine of opening a valve that admitted cooling oil into the upper coil tank. A turning of the valve handle simultaneously lifted a weight attached to a length of clothesline. An electrical relay normally kept the weight from falling when the valve was fully open. I had never examined the plumbing for this part of the cyclotron, because it was one of the few systems that hadn't broken down in my many months of crew duty.

The relay refused to work. The weight dropped and closed the valve. Since I couldn't operate the cyclotron with the valve closed, I buggered the interlock by putting a clamp on the clothesline just above the pulley. Then I turned on the cyclotron and ran it until my shift ended. After dinner I returned to the laboratory with Gerry to find an extremely angry evening crew chief. In front of my bride he thoroughly chewed me out, after which he led me to the scene of the crime in the basement below the cyclotron, where I had often spent hours working in the two-foot crawl space between the dirt subfloor and the wooden floor beams. We found the subfloor covered now with transformer oil; the oil valve had been interlocked to prevent oil from leaking from the top tank and overflowing the bottom tank. Positive oil pressure normally kept the valve open; the oil pump had apparently stopped working. Because of my tampering all the oil in the top tank had drained out, exposing the upper coils, and it was only good fortune that spared a burnout. If the coils had burned out, the magnet would have been shut down for repairs for weeks. I was saved from such absolute disgrace because the cyclotron had been operating only marginally at the time.

That was a close call; it taught me to be much more careful in making changes on collective equipment. I had been a lone wolf in Chicago. If I damaged my apparatus there, I hurt only myself. Now my calculated risks could jeopardize my friends as well. I learned my lesson and thereafter buggered no interlocks unless I understood their function.

Group operation under the conditions of that era exposed us to other hazards as well. My first published work at Berkeley concerned removing the ion beam of the cyclotron from the magnetic field. It was possible to direct the beam of deuterons as it came out of the dees through a window of thin platinum foil into the open air, where it made a luminescent blue glow several inches long. Ernest Lawrence's love affair with the deuteron beam was legendary. He led every laboratory visitor into the cyclotron room to see the luminescent beam. As the beam current increased with improved equipment and larger cyclotrons, the blue glow became visible even in a fully lighted room.

731

But we seldom kept adequate records of cyclotron modifications. Some months after I arrived in Berkeley, the chamber of the 27-inch cyclotron was removed for major improvements. Rather than idle the instrument, we temporarily reinstalled the previous chamber. It didn't permit the beam to leave the vacuum chamber, but it had a reentrant viewing cylinder that could be operated either evacuated or filled with air. With air the glowing ion beam was visible; without air there was nothing to see, but the beam impacting invisibly on its beryllium target made a flood of neutrons.

Alvin Weinberg, a friend from graduate-student days in Chicago, dropped in unexpectedly one afternoon to see the laboratory. It fascinated him as much as it had fascinated me when I first toured it. Before he dropped in, I had watched the beam streaking blue across the cylinder; now I invited Al to watch with me. I asked the crew chief to turn on the cyclotron, and then Al and I knelt leaning on the lower coil tank with our heads about a foot from the viewing window. After a time I shouted to the crew chief that he didn't have a beam. He shouted back that he did. I said it wasn't enough to see, and the crew chief said it ought to be more than enough. Out of the corner of my eye I then noticed the beryllium target glowing bright red. I grabbed Al and pulled him away from the cyclotron and at the same time shouted to the operator to turn off the machine. In the hour since I had watched the beam alone, someone had evacuated the air from the viewing cylinder; what I saw from a distance of one foot was the beryllium target heated by a strong current of 5 MeV deuterons. Al Weinberg and I probably received the largest dose of fast neutrons anyone had ever experienced up to that time. Luckily we escaped without injury and both went on to full scientific careers.

* * * * *

Since the discovery of artificial radioactivity, in France in 1934, the cyclotron had been used primarily to induce radioactivity in most of the known chemical elements. The Radiation Laboratory used the cyclotron as a radioactivity factory first of all because great numbers of new radioisotopes could be discovered that way with very little effort. The gold rush of isotopes lasted four years, and only after the Second World War was there time and adequate instrumentation for the researchers to follow through with detailed analyses of all the radiations the hundreds of radioisotopes emitted.

Contingency also affected the choice; the cyclotron was available only a small fraction of the time. We spent many hours finding leaks, adjusting equipment, repairing oscillators, and developing cyclotron technology. Lawrence's job as director also required him to consider the necessities of public relations. He wanted radioisotopes used in all branches of science, and this led him to missionary work with the other Berkeley science departments. We recognized the importance of these activities, but after spending days in cyclotron repairs we grumbled when a physiologist or a biologist turned up to claim the fruits of the first bombardment. We grumbled among ourselves, that is; we knew the strength of Ernest's convictions and were much too loyal to allow outsiders to discover our ambivalence.

Once Lawrence came close to violating our constitutional guarantees against cruel and unusual punishment. For most of his long, fruitful career Ernest Rutherford, Lord Rutherford of Nelson, had expressed his preference for experiments managed on modest equipment assembled with legendary Cavendish string and sealing wax. He had quarreled about the matter with James Chadwick, the discoverer of the neutron and a Rutherford protégé, when Chadwick had wanted to build a cyclotron — quarreled unhappily enough that, after sixteen years at the Cavendish, Chadwick had gone off to the University of Liverpool and there built England's first cyclotron. Rutherford had now finally seen the light and had sent John Cockcroft to Berkeley to learn about cyclotrons. The money became available when Peter Kapitsa, a Russian physicist working at the Cavendish, had been detained in his homeland by the Soviet government while attending a conference and forbidden to return to England. Kapitsa had raised funds in England and despite Rutherford's prejudices, which he alone seems to have been able to circumvent, had built and equipped in the Cavendish courtyard a massively instrumented new laboratory devoted to research into powerful magnetic fields. Rutherford appealed to the Soviets to allow Kapitsa to return to England and continue his scientific work, to which they responded that they would be equally pleased to have Rutherford in Russia. They agreed, however, to buy all of Kapitsa's apparatus. The value of the equipment came to some £30,000. Part of that fund was earmarked for a cyclotron.

Few occasions can have given Lawrence more pleasure than his receiving an emissary from the great Ernest Rutherford himself. Lawrence had known John Cockcroft since Cockcroft's visit to Berkeley in 1933. They had found they had much in common and

733

had immediately become friends. Ernest personally showed John the laboratory, taught him how to operate the cyclotron, and told him what he could expect to see when the cyclotron tank was pulled for the inevitable repairs.

As Cockcroft's departure approached, however, the cyclotron for the first time in memory stubbornly refused to break down. Since Ernest had promised John a look inside the chamber, I was prepared to be ordered to shut down and pull the vacuum tank for inspection. Instead Ernest concocted a truly diabolical subterfuge. He turned off the compressed air that cooled the glass insulators that supported the dee stems and their cantilevered dees. While Ernest operated the cyclotron and John looked over his shoulder, I watched with mounting horror as the radio-frequency fields heated the glass insulators to the softening point. One of them collapsed. Air rushed into the vacuum chamber. We all dashed to the rescue, just as we would have done had the "accident" been accidental. Ernest pointed out the site of the failure to his guest and blithely informed us that they would return after dinner to watch the reassembly. We were speechless.

Bethe's Bible and a long discussion with my father during one of his visits to Berkeley led me to another experimental interest. Dad and I talked about the identifying of problems that are really worth working on. Looking back on his career, he said, he realized he should have taken time out more often to sit down and think about what he was doing. He had always felt pressed to follow a mental list and had missed important leads. He had once thought hard about pernicious anemia, he told me, in those days a fatal disease. He had fed ground liver to a few anemic dogs with no certain results and then moved on to other pressing projects. He was the more unhappy a year later when his immediate superior at the Hooper Foundation, Dr. George Whipple, demonstrated that massive doses of raw liver restored the health of anemic dogs. Whipple's work established a standard treatment for human pernicious anemia that won him and two colleagues the 1934 Nobel Prize in medicine. Dad never claimed he might have won that Nobel Prize, but he did say more than once that he would have been a better researcher if he had occasionally let his mind wander over the full range of his work. He advised me to sit every few months in my reading chair for an entire evening, close my eyes, and try to think of new problems to solve. I took his advice very seriously and have been glad ever since that I did.

By the spring of 1937 my literature survey was far enough

along to allow me to put my father's suggestion into practice. Two gallbladder attacks offered unwanted assistance. Doctors think of gallbladder patients as female, fair, fat, and forty. I was male, fair, thin, and young, an uncomfortable exception to the rule. My second attack caused me intense pain, which massive doses of morphine barely attenuated. In bed for some time recovering from jaundice, I had the leisure to follow my father's advice.

Niels Bohr had delivered the prestigious Hitchcock Lectures at Berkeley that spring. The audience at the first lecture overflowed the auditorium but quickly dwindled to a hard core of physicists and chemists thereafter. Bohr whispered his lectures, *dropping* his voice to emphasize a point. Wandering from the podium microphone to the blackboard, he was barely audible to those in the first few rows. I was fortunate to find a seat in those rows for all five lectures. His work on the liquid-drop model of the nucleus impressed me deeply.

It led me to think about the problem of slow-neutron capture, which happens with greater frequency at certain neutron energies than at others, a phenomenon known as resonance. Bohr drew extensively on our meager experimental knowledge of neutron capture resonances. All we could establish at that time was the rank order in resonant energy between one element and another. If optical spectroscopy were as primitive as neutron spectroscopy, Ed McMillan had quipped in a lecture the preceding year, instead of using prisms and diffraction gratings we'd be using colored shards of broken beer bottles.

I meditated at length on the problem and decided to look into the possibility of measuring neutron energies by measuring their velocities. As far as I know, I was the first person to think of the method, now commonplace, of measuring time of flight. The problem with neutrons was sorting out fast neutrons from slow. Under most circumstances they arrive at a detector jumbled together. Cadmium is a powerful absorber of very slow ("thermal") neutrons; by taking a series of measurements of a neutron beam with and without cadmium sheets interposed, investigators had been able to identify the effects of slow neutrons as the difference between the two conditions. But slow-neutron absorption experiments had been compromised by the ever-present and apparently unavoidable fast-neutron background.

Soon after the discovery of slow neutrons by Fermi's group in Rome in 1934, John Dunning, Emilio Segré, and their colleagues at Columbia showed, by using rotating shutters, that thermal neutrons

735

really moved as slowly as hydrogen atoms at room temperature did. But they couldn't get rid of their fast-neutron background, so their important experiment didn't lead anywhere. But the fact that slow neutrons take an appreciable time to travel distances on the order of meters — 1/240 of a second for eight meters, in the case of a thermal neutron — was the basis for the method I devised.

I had hoped to explore the "resonance region," where neutrons with energies of a few electron volts are preferentially captured. After many attempts to turn the cyclotron beam off fast enough to work in this region, I had to give up; the ion source we were using didn't permit it. So I backed off and used the slower "modulation times" that let me work with what was the first beam of pure thermal neutrons — in fact, thermal neutrons with a wide range of effective temperatures, down to very low values.

I arranged to turn the cyclotron beam on and off 120 times per second. That generated fast neutrons in the beryllium target in pulses about half the time. Near the target I placed a block of paraffin wax. Some of the fast neutrons passing through the paraffin would collide with its hydrogen atoms and slow down. The paraffin block faced one end of a long cadmium tube; at the other end, some eight meters away, I fitted a boron trifluoride ionization chamber that could register the arriving neutrons and display them through a linear amplifier on an oscilloscope screen. The linear amplifier I arranged to be sensitive only when the cyclotron beam was off.

Since fast neutrons have a very short life span — a mean of less than $10^{-5}$ second — I could turn on the cyclotron beam for a few milliseconds sixty times a second and detect only room-temperature neutrons. The faster neutrons would have passed the detector before it was sensitive, and the slower ones would reach it after it had been deactivated. Operationally speaking, I now had a beam of pure thermal neutrons. People who had worked with neutrons were amazed to see that when I interposed a thin sheet of cadmium between the cyclotron target and my neutron detector, the counting rate dropped to zero, as if the cadmium were a sheet of black paper intercepting ordinary light and as if fast neutrons didn't exist.

In 1937 no one thought it was possible to make a beam of pure thermal neutrons. The only materials anyone used to slow down neutrons were those containing hydrogen. The materials unfortunately also absorbed thermal neutrons more effectively than they did fast neutrons. So in hydrogenous materials fast neutrons were always present among the slow. Later, when Fermi and Leo Szilard in the

United States and Frédéric Joliot and his colleagues in France were trying independently to invent a nuclear reactor, they realized that other materials that absorbed fewer slow neutrons would serve their purposes better than water or paraffin. Necessity was the mother of invention; the U.S. tried beryllium and then carbon and found that neutrons could be slowed to thermal velocities at which they could travel long distances without appreciable loss. It only took a lot more collisions to do the job. The French physicists hoped to use heavy water, but before they could accumulate enough of it the Second World War intervened. Szilard and Fermi concluded that graphite was best suited for the job, and the rest is history.

My "pure" thermal beam achieved its effects differently from the graphite columns Fermi later devised to produce true beams of pure thermal neutrons, but operationally it was the first such machine. It had strange manifestations. If I arranged to delay my measurements by ten milliseconds instead of five milliseconds, a guest observer would be startled to find that the apparent temperature of my neutron beam had dropped from room temperature, 300 degrees Kelvin, to 75 degrees Kelvin, which is below the temperature of liquid air. These cold neutrons took twice as long to reach the detector. With half the velocity of their predecessors, they were only one-fourth as energetic and therefore that much cooler.

To pulse the flow of neutrons, I had to turn the cyclotron oscillator on and off. And to make sure I was really stopping and starting the deuteron beam, I made a test in which I ran the deuterons into a thin ionization chamber that just happened to have lithium fluoride smeared on one of its electrodes. The beam current dropped to zero about two milliseconds after the oscillator voltage peaked, as expected, but a background of heavily ionizing particles appeared uniformly across the oscilloscope screen. These particles were obviously related to the deuteron beam because when the cyclotron was turned off, the background decreased with about a one-second half-life.

It took me a few minutes to realize that I had independently rediscovered artificial radioactivity, and by a method that should have led to its discovery at Berkeley before Frédéric Joliot and Irène Joliot-Curie discovered it in Paris in 1934. (That discovery had been denied to the Berkeley group because they thought they couldn't make good Geiger counters — because, as we now know, their whole laboratory was artificially radioactive. But they were skilled in the use of thin ionization chambers of the kind that I was now using and that were insensitive to beta rays but very sensitive to the alpha particles I was now seeing. And lithium is almost unique in giving delayed alpha particles.) Bombarding the lithium in my detector made it radioactive; it responded by emitting alpha particles following beta decay. I had just made a really extraordinary discovery. If I hadn't been three years too late, I would certainly have told someone about it.

I did look into why Berkeley missed it. To make accelerated deuteron beams, G. N. Lewis, Ernest Lawrence, Stan Livingston, and Malcolm Henderson had used equipment almost identical to the equipment I was now using; they had described this work at the meeting of the American Physical Society I had attended during the Chicago exposition in 1933. Lewis, the dean of American chemists, had concentrated the heavy water from which the deuterium was electrolyzed, so his name was first on the paper even though he wasn't very familiar with the cyclotron or the detection equipment. Lawrence and Livingston were undoubtedly busy keeping the cyclotron running. So the Berkeley person who missed discovering artificial radioactivity, for which the Joliot-Curies won the 1935 Nobel Prize in chemistry, was Malcolm Henderson.

I couldn't understand why Malcolm missed seeing what I had just observed. Like me, he had used a thin ionization chamber with deuterons hitting a lithium target. Like mine, his cyclotron switched on and off sixty times a second (to save money in those days, Berkeley used "raw AC" to power its oscillator). I went back over Malcolm's earlier papers to familiarize myself with his techniques. If he had displayed his detector signal on an oscilloscope, he would have seen the effect, just as I had. I doubt that he could have tuned up his apparatus without an oscilloscope. He may have missed so important a discovery because he simply didn't expect it and therefore didn't watch his oscilloscope as closely as I had. Or he may have turned it off to increase its useful life; oscilloscopes were expensive in those days.

I had now made two major discoveries in nuclear physics: K-electron capture and artificial radioactivity. I haven't ever mentioned the second of the two before; I do so here only to show that I had now learned my trade and noticed things I hadn't expected to see, which is, of course, essential to scientific discovery. . . .

John Lawrence and Joe Hamilton were the two medical doctors who spent most of their time in the laboratory. I certainly owe my life to John; had he not come to Berkeley in 1935, I would probably have died of radiation sickness long ago. John's first experiment was designed to determine the degree of radiation hazard associated with neutron bombardment; the neutron fluxes at the Berkeley cyclotron were enormously more intense than those available at any other physics laboratory in the world. John had a brass cylinder made just large enough to hold a mouse. Rubber tubes carried a stream of air to the mouse when the cylinder was placed close to the beryllium target in the 27-inch cyclotron. After a fifteen-minute exposure to the target neutrons, the mouse was removed. It was dead. When I arrived on the scene several months later, people were still talking about the shocked silence that greeted that observation. By then John's secret had leaked out: the mouse died of asphyxiation; someone had forgotten to turn on the compressed air. John subsequently showed that fast neutrons were several times more dangerous per unit of radiation dose than the gamma rays from radium. As a result, the cyclotron control table had been moved as far from the target as possible. As Ernest continued his relentless drive to raise the beam current in the cyclotron, John decided that the cyclotron operators were getting too high a dose of neutrons. He led a campaign then to surround the entire machine with water tanks. I had been a cyclotron operator for several months before the tanks were put in place, so my debt to John Lawrence is large, and frequently remembered.

Joe Hamilton became more deeply involved in the techniques of nuclear physics than John, but they both did important experiments in medical physics. In 1937 Joe asked me if I would like to collaborate on an experiment he had been thinking about to use radioactive sodium, which has a half-life of fifteen hours, to measure the absorption of sodium ions from the stomach wall into the bloodstream. He proposed that each of us would drink a millicurie of sodium 24 while holding a Geiger counter shielded inside a lead-lined box and watch the counting rate increase with time. I told him that I would calculate the dose that night; if it looked all right to me, he could count me in. Radiation doses increase as the inverse square of the distance, so they shoot up fast as the radioactive source approaches the body. But what happens if the source actually enters the body? That's the kind of problem Newton invented the calculus to solve; I was soon able to conclude that Joe was proposing a reasonable experiment (it wouldn't be permitted today, of course). As medical tradition dictates, Joe did the experiment first on himself. A few hours later I became the second guinea pig. As far as I know, it was the first experiment of its kind.

When Ernest gave his Faculty Research Lecture a few months after Joe's experiment, he demonstrated it onstage — using a tenth of a millicurie of sodium 24 and with Robert Oppenheimer as the guinea pig.

In May 1986, after the Soviet reactor fire at Chernobyl, U.S. newspapers carried daily front-page stories about the radioactivity deposited on our soil by its fallout. The typical amount of radioactivity measured was ten picocuries per liter of rainwater. The one millicurie Joe and I each considered safe to drink was a hundred million times stronger than the ten picocuries that apparently scared the American public in 1986 sufficiently to cause many citizens to cancel long-held European travel reservations. I have been outraged for years by the media's exaggeration of the dangers of trivial amounts of radioactivity. But the public reaction may be the good news about Chernobyl. Both our leaders and those of the Soviet Union saw then how their citizens react to the presence of even tiny amounts of radioactivity. I can't believe that either leadership would ever attempt a preemptive nuclear strike, which would certainly be answered in kind, when the resulting levels of radioactivity would be measured in kilocuries or megacuries, a million or a billion times more dangerous than the one millicurie Joe and I each drank, which was in turn a hundred million times more than the amount that frightened Americans in 1986. . . .

# C. P. SNOW

∎

When people talk of "the two cultures" of science and literature they are harkening back to the following essay by C. P. Snow. Originally published in *The New Statesman and Nation* in 1956, it touched off a debate about educational and cultural values that continues to this day. Much of its piquancy derives from the fact that Snow, having made a name for himself as both a novelist and a scientist, was uniquely well qualified to comment on the status of the two camps.

# *The Two Cultures*

"IT'S RATHER ODD," said G. H. Hardy, one afternoon in the early Thirties, "but when we hear about 'intellectuals' nowadays, it doesn't include people like me and J. J. Thomson and Rutherford." Hardy was the first mathematician of his generation, J. J. Thomson the first physicist of his; as for Rutherford, he was one of the greatest scientists who have ever lived. Some bright young literary person (I forget the exact context) putting them outside the enclosure reserved for intellectuals seemed to Hardy the best joke for some time. It does not seem quite such a good joke now. The separation between the two cultures has been getting deeper under our eyes; there is now precious little communication between them, little but different kinds of incomprehension and dislike.

The traditional culture, which is, of course, mainly literary, is behaving like a state whose power is rapidly declining — standing on its precarious dignity, spending far too much energy on Alexandrian intricacies, occasionally letting fly in fits of aggressive pique quite beyond its means, too much on the defensive to show any generous imagination to the forces which must inevitably reshape it. Whereas the scientific culture is expansive, not restrictive, confident at the roots, the more confident after its bout of Oppenheimerian self-criticism, certain that history is on its side, impatient, intolerant, creative rather than critical, good-natured and brash. Neither culture knows the virtues of the other; often it seems they deliberately do not want to know. The resentment which the traditional culture feels for the scientific is shaded with fear; from the other side, the resentment is not shaded so much as brimming with irritation. When scientists are faced with an expression of the traditional culture, it tends (to borrow Mr. William Cooper's eloquent phrase) to make their feet ache.

It does not need saying that generalisations of this kind are bound to look silly at the edges. There are a good many scientists indistinguishable from literary persons, and vice versa. Even the stereotype generalisations about scientists are misleading without some sort of detail — *e.g.*, the generalisation that scientists as a group stand on the political Left. This is only partly true. A very high proportion of engineers is almost as conservative as doctors; of pure scientists, the same would apply to chemists. It is only among physicists and biologists that one finds the Left in strength. If one compared the whole body of scientists with their opposite numbers of the traditional culture (writers, academics, and so on), the total result might be a few per cent more towards the Left wing, but not more than that. Nevertheless, as a first approximation, the scientific culture is real enough, and so is its difference from the traditional. For anyone like myself, by education a scientist, by calling a writer, at one time moving between groups of scientists and writers in the same evening, the difference has seemed dramatic.

The first thing, impossible to miss, is that scientists are on the up and up; they have the strength of a social force behind them. If they are English, they share the experience common to us all — of being in a country sliding economically downhill — but in addition (and to many of them it seems psychologically more important) they belong to something more than a profession, to something more like a directing class of a new society. In a sense oddly divorced from politics, they are the new men. Even the staidest and most politically conservative of scientific veterans, lurking in dignity in their colleges, have some kind of link with the world to come. They do not hate it as their colleagues do; part of their mind is open to it; almost against their will, there is a residual glimmer of kinship there. The young English scientists may and do curse their luck; increasingly they fret about the rigidities of their universities, about the ossification of the traditional culture which, to the scientists, makes the universities cold and dead; they violently envy their Russian counterparts who have money and equipment without discernible limit, who have the whole field wide open. But still they stay pretty resilient: they are swept on by the same social force. Harwell and Winscale have just as much spirit as Los Alamos and Chalk River: the neat petty bourgeois houses, the tough and clever young, the crowds of children: they are symbols, frontier towns.

There is a touch of the frontier qualities, in fact, about the whole scientific culture. Its tone is, for example, steadily heterosexual. The difference in social manners between Harwell and Hampstead, or as far as that goes between Los Alamos and Greenwich Village, would make an anthropologist blink. About the whole scientific culture, there is an absence — surprising to outsiders — of the feline and oblique. Sometimes it seems that scientists relish speaking the truth, especially when it is unpleasant. The climate of personal relations is singularly bracing, not to say harsh: it strikes bleakly on those unused to it, who suddenly find that the scientists' way of deciding on action is by a full-dress argument, with no regard for sensibilities and no holds barred. No body of people ever believed more in dialectic as the primary method of attaining sense; and if you want a picture of scientists in their off-moments it could be just one of a knock-about argument. Under the argument there glitter egotisms as rapacious as any of ours: but, unlike ours, the egotisms are driven by a common purpose.

How much of the traditional culture gets through to them? The answer is not simple. A good many scientists, including some of the most gifted, have the tastes of literary persons, read the same things, and read as much. Broadly, though, the infiltration is much less. History gets across to a certain extent, in particular social history: the sheer mechanics of living, how men ate, built, travelled, worked, touches a good many scientific imaginations, and so they have fastened on such works as Trevelyan's *Social History,* and Professor Gordon Childe's books. Philosophy, the scientific culture views with indifference, especially metaphysics. As Rutherford said cheerfully to Samuel Alexander: "When you think of all the years you've been talking about those things, Alexander, and what does it all add up to? *Hot air,* nothing but *hot air.*" A bit less exuberantly, that is what contemporary scientists would say. They regard it as a major intellectual virtue, to know what not to think about. They might touch their hats to linguistic analysis, as a relatively honourable way of wasting time; not so to existentialism.

743

The arts? The only one which is cultivated among scientists is music. It goes both wide and deep; there may possibly be a greater density of musical appreciation than in the traditional culture. In comparison, the graphic arts (except architecture) score little, and poetry not at all. Some novels work their way through, but not as a rule the novels which literary persons set most value on. The two cultures have so few points of contact that the diffusion of novels shows the same sort of delay, and exhibits the same oddities, as though they were getting into translation in a foreign country. It is only fairly recently, for instance, that Graham Greene and Evelyn Waugh have become more than names. And, just as it is rather startling to find that in Italy Bruce Marshall is by a long shot the best-known British novelist, so it jolts one to hear scientists talking with attention of the works of Nevil Shute. In fact, there is a good reason for that: Mr. Shute was himself a high-class engineer, and a book like *No Highway* is packed with technical stuff that is not only accurate but often original. Incidentally, there are benefits to be gained from listening to intelligent men, utterly removed from the literary scene and unconcerned as to who's in and who's out. One can pick up such a comment as a scientist once made, that it looked to him as though the current preoccupations of the New Criticism, the extreme concentration on a tiny passage, had made us curiously insensitive to the total flavour of a work, to its cumulative effects, to the epic qualities in literature. But, on the other side of the coin, one is just as likely to listen to three of the most massive intellects in Europe happily discussing the merits of *The Wallet of Kai-Lung.*

When you meet the younger rank-and-file of scientists, it often seems that they do not read at all. The prestige of the traditional culture is high enough for some of them to make a gallant shot at it. Oddly enough, the novelist whose name to them has become a token of esoteric literary excellence is that difficult highbrow Dickens. They approach him in a grim and dutiful spirit as though tackling *Finnegans Wake,* and feel a sense of achievement if they manage to read a book through. But most young technicians do not fly so high. When you ask them what they read — "As a married man," one says, "I prefer the garden." Another says: "I always like just to use my books as tools." (Difficult to resist speculating what kind of tool a book would make. A sort of hammer? A crude digging instrument?)

That, or something like it, is a measure of the incommunicability of the two cultures. On their side the scientists are losing a great deal. Some of that loss is inevitable: it must and would happen in any society at our technical level. But in this country we make it quite unnecessarily worse by our educational patterns. On the other side, how much does the traditional culture lose by the separation?

I am inclined to think, even more. Not only practically — we are familiar with those arguments by now — but also intellectually and morally. The intellectual loss is a little difficult to appraise. Most scientists would claim that you cannot comprehend the world unless you know the structure of science, in particular of physical science. In a sense, and a perfectly genuine sense, that is true. Not to have read *War and Peace* and *La Cousine Bette* and *La Chartreuse de Parme* is not to be educated; but so is not to have a glimmer of the Second Law of Thermodynamics. Yet that case ought not to be pressed too far. It is more justifiable to say that those without any scientific understanding miss a whole body of experience: they are rather like the tone deaf, from whom all musical experience is cut off and who have to get on without it. The intellectual invasions of science are, however, penetrating deeper. Psycho-analysis once looked like a deep invasion, but that was a false alarm; cybernetics may turn out to be the real thing, driving down into the problems of will and cause and motive. If so, those who do not understand the method will not understand the depths of their own cultures.

But the greatest enrichment the scientific culture could give us is — though it does not originate like that — a moral one. Among scientists, deep-natured men know, as starkly as any men have known, that the individual human condition is tragic; for all its triumphs and joys, the essence of it is loneliness and the end death. But what they will not admit is that, because the individual condition is tragic, therefore the social condition must be tragic, too. Because a man must die, that is no excuse for his dying before his time and after a servile life. The impulse behind the scientists drives them to limit the area of tragedy, to take nothing as tragic that can conceivably lie within men's will. They have nothing but contempt for those representatives of the traditional culture who use a deep insight into man's fate to obscure the social truth — or to do something pettier than obscure the truth, just to hang on to a few perks. Dostoevski sucking up to the Chancellor Pobedonostsev, who thought the only thing wrong with slavery was that there was not enough of it; the political decadence of the *avant garde* of 1914, with Ezra Pound finishing up broadcasting for the Fascists; Claudel agreeing sanctimoniously with the Marshal about the virtue in others' suffering;

Faulkner giving sentimental reasons for treating Negroes as a different species. They are all symptoms of the deepest temptation of the clerks — which is to say: "Because man's condition is tragic, everyone ought to stay in their place, with mine as it happens somewhere near the top." From that particular temptation, made up of defeat, self-indulgence, and moral vanity, the scientific culture is almost totally immune. It is that kind of moral health of the scientists which, in the last few years, the rest of us have needed most; and of which, because the two cultures scarcely touch, we have been most deprived.

# VIVIAN GORNICK

∎

The paucity of women in physics, astronomy, and mathematics prompts concern over why these vital sciences have not yet availed themselves of the brainpower of half the human species. The journalist and literary critic Vivian Gornick (b. 1935) wrote this profile of physicist Alma Norovsky (b. 1934) for her book *Women in Science*, published in 1983.

# *Women in Science*

FORTY-NINE YEARS OLD; strong-featured face, brooding eyes, a mass of sexy dark hair she tosses about like a forties movie vamp, the walk seductive and knowing, the mouth sullen and grievance-collecting in repose, then surprisingly girlish in laughter when the eyes fill with a sudden shimmering light. Alma Norovsky is a theoretical physicist at a university renowned for its devotion to the life of the mind. Of her colleagues Alma says drily: "They're very theoretical. People are always asking me how women are treated here. 'Women?' I answer. 'They're a theoretical concept.' "

Divorced four years from the physicist husband she married in graduate school, on her own for the first time in her life, in love with her new independence and happy to be working here, Alma nevertheless sighs. "How do you work in physics, or live among academic liberal men, and not explode all day long every day? Once in a while I'm able to control myself. . . . Last year at a conference I was standing with a group of physicists, all men, and I was introduced to a new member of the group. He said, 'You're the first good-looking physicist I've ever met.' I casually indicated the man standing beside me and said, 'Oh, that's not true. You know Richard here. He's good-looking, and he's a physicist.' They all looked startled, and then some of them nodded their heads appreciatively. I was proud of myself then, but usually it's awful. Still. Always. At every dinner table, in the office, the constant little indications that you don't really exist. You've got to remind them that you're a thinking, working being just like themselves all the time. It's wearing."

She was a pretty girl, bold and flirtatious, loved exercising her power to attract, she'd be damned if she'd give that up ("Why? What for? It was so much fun"). So nobody took her seriously although she was a fine physics student from high school on. Her father, a frustrated scientist, adored and encouraged her, but he floundered. She went to a small women's college where the science courses were bad and the teachers worse, but sexual success made her stubborn and determined on her own seriousness. She bulled her way through into an Ivy League education.

In her second year in graduate school she met and married Lawrence Norovsky, a strongly ambitious fellow student in physics. She says of this time: "I enjoyed being one of the few women in physics, but I certainly did not enjoy it when I realized women in physics were considered ugly, undesirable, clumsy eccentrics. I wanted to be sexually lovely and desirable, and still be a fine physicist." What she doesn't say — although it's apparent — is that she also wanted to attract the attention of a man of power, the kind of man whose interest in her would always be mixed: compelled and antagonistic, attracted by her brains but enjoying her subordinate status as his wife, his relationship to her over the years intermittently eroticized by her professional intelligence, but his sense of her life as equal to his own never maturing.

When Lawrence accepted a job at a university in northern Massachusetts, Alma had had a baby and had not yet finished her degree. She remembers that everyone seemed to be patting her on the head, as though indulging a child's whimsical insistence, when she said she would finish her degree in Massachusetts. But "There were four women in science at the university in Massachusetts. When I think back on it, how those women influenced my life! They arrived one day, sat down in my living room and began giving me full instructions in how to organize myself, my time, my babysitter problem, my shopping and laundry problems. They never for a moment assumed that I wouldn't finish the degree. They had come as compatriots, in a situation they knew nobody but themselves understood, to give me the benefit of their experience. And I did as they said, and finally I finished. But not with any help from my husband, I can assure you.

"When I was just beginning to write the dissertation, and it was in its earliest stages, Lawrence suddenly decided to go to the Physics

748

Institute in Paris for two years. He said to me, 'You can finish in two years. What's the difference?' Then I put my foot down. I don't know why it was, but I suddenly said no, absolutely no, this I will not do. He sent all his friends to see me, to persuade me to go with him. They said I was ruining his career, how could I be so selfish, so I'd finish in a year or two, what was the big deal? I didn't answer any of them, but I stayed where I was. He went to Paris, I sent the children to my parents, and I worked furiously at finishing my degree. Which I did in one year. Then I collected the children and followed him to Paris. Where I was immediately given an office, lab space, treated like a scientist. And the first thing my husband announced was that I was not welcome to join his group for lunch, or anything else for that matter. That was the first time he told me that under no circumstances would we ever work together, that physics was competitive, and he wouldn't have his marriage disfigured by a competitive relationship forming between us. Hah! His idea of avoiding competition between us was for me never to become anything.

"We came back home. Lawrence went to work at Brookhaven. I had three children. I didn't really want to work as a full-time physicist. I got an associateship at Stony Brook. I never was an equal of my husband's physics friends, but I never thought there was anything wrong with that. After all, they worked hard and long, and what was I doing? Why should they treat me as an equal?

"In 1971 I lost my job at Stony Brook, and I couldn't find another one. Suddenly, I was walking around in a daze. What had happened? I'd been a good physicist. I had wanted to do important work. What was I? Nothing. How had I ended up here? What had gone wrong with my career?

"That summer I talked with Nina Braverman at Aspen. We compared notes. It was astonishing how similar the pattern of our lives had been! For the first time it hit me that my life had developed as it had because I was a woman, and I'd made women's choices, and ended up where women in science end up. It hit me like a ton of bricks. All at once, I saw everything. From that moment on I became a rabid feminist. And I mean rabid. Shortly after that my marriage fell apart.

"I remember thinking back to one incident. After I'd lost my job at Stony Brook I was going crazy sitting home, not doing any science at all. One day I was visiting my husband at the lab. Another scientist there, a friend of ours, suddenly said to me, 'Alma, I've got an extra desk in my office. Why don't you come and use it as a guest

749

worker?' That meant no pay but the privilege of working in the lab, using its facilities, and talking with the scientists. I responded gratefully and went to work there. Afterward, I thought, Why is it Lawrence never thought of this for me? How come a friend saved me but my husband never thought of doing so?

"Lawrence was insecure and that insecurity always made him more aware of protecting himself than of seeing his behavior toward me as unjust. And then there was the related, deeper truth that he never took me seriously as a physicist. Why should he when it was so much more convenient not to? He wasn't like Leon Braverman, and if I didn't do it for myself he certainly wasn't going to do it for me.

"I'm sure I'm being unfair to him, and if he were here he'd put a whole different construction on these same events. But that's the reality, isn't it? It's not a matter of fair, it's what we all did to each other, and no amount of looking at it fairly will make me feel any better about my marriage or my husband or the lost working years of my life."

# PRIMO LEVI

■

The industrial chemist Primo Levi was a prisoner at Auschwitz. His books — among them *Survival in Auschwitz, Moments of Reprieve,* and *The Damned and the Saved* — invoke unflinching memories of Nazi horrors. "Iron" is a chapter from his book *The Periodic Table,* published in the United States in 1984. Born in Turin, Italy, in 1919, Levi committed suicide in 1987. The Speaker of the Italian Chamber of Deputies wrote to his widow, "We must consider the tragic death of Primo Levi a sign of the endlessness of that episode against man and civilization that was the Nazi genocide."

# *Iron*

NIGHT LAY BEYOND THE WALLS of the Chemical Institute, the night of Europe: Chamberlain had returned from Munich duped, Hitler had marched into Prague without firing a shot, Franco had subdued Barcelona and was ensconced in Madrid. Fascist Italy, the small-time pirate, had occupied Albania, and the premonition of imminent catastrophe condensed like grumous dew in the houses and streets, in wary conversations and dozing consciences.

But the night did not penetrate those thick walls; Fascist censorship itself, the regime's masterwork, kept us shut off from the world, in a white, anesthetized limbo. About thirty of us had managed to surmount the harsh barrier of the first exams and had been admitted to the second year's Qualitative Analysis laboratory. We had entered that enormous, dark, smoky hall like someone who, coming into the House of the Lord, reflects on each of his steps. The previous lab, where I had tackled zinc, seemed an infantile exercise to us now, similar to when as children we had played at cooking: something, by hook or crook, in one way or another, always came of it, perhaps too little, perhaps not very pure, but you really had to be a hopeless case or pigheaded not to get magnesium sulfate from magnesite, or potassium bromide from bromine.

Not here: here the affair had turned serious, the confrontation with Mother-Matter, our hostile mother, was tougher and closer. At two in the afternoon, Professor D., with his ascetic and distracted air, handed each of us precisely one gram of a certain powder: by the next day we had to complete the qualitative analysis, that is, report what metals and non-metals it contained. Report in writing, like a police report, only yes and no, because doubts and hesitations were not admissible: it was each time a choice, a deliberation, a mature and responsible undertaking, for which Fascism had not prepared us, and from which emanated a good smell, dry and clean.

Some elements, such as iron and copper, were easy and direct, incapable of concealment; others, such as bismuth and cadmium, were deceptive and elusive. There was a method, a toilsome, age-old plan for systematic research, a kind of combined steamroller and fine-toothed comb which nothing (in theory) could escape, but I preferred to invent each time a new road, with swift, extemporaneous forays, as in a war of movement, instead of the deadly grind of a war of position. Sublimate mercury into droplets, transform sodium into chloride, and identify it as trough-shaped chips under my microscope. One way or another, here the relationship with Matter changed, became dialectical: it was fencing, a face-to-face match. Two unequal opponents: on one side, putting the questions, the unfledged, unarmed chemist, at his elbow the textbook by Autenrieth as his sole ally (because D., often called to help out in difficult cases, maintained a scrupulous neutrality, refused to give an opinion: a wise attitude, since whoever opens his mouth can put his foot in it, and professors are not supposed to do that); on the other side, responding with enigmas, stood Matter, with her sly passivity, ancient as the All and portentously rich in deceptions, as solemn and subtle as the Sphinx. I was just beginning to read German words and was enchanted by the word *Urstoff* (which means "element": literally, "primal substance") and by the prefix *Ur* which appeared in it and which in fact expresses ancient origin, remote distance in space and time.

In this place, too, nobody wasted many words teaching us how to protect ourselves from acids, caustics, fires, and explosions; it appeared that the Institute's rough and ready morality counted on the process of natural selection to pick out those among us most qualified for physical and professional survival. There were few ventilation hoods; each student, following his text's prescriptions, in the course of systematic analysis, conscientiously let loose into the air a good dose of hydrochloric acid and ammonia, so that a dense, hoary mist of ammonium chloride stagnated permanently in the lab, depositing minute scintillating crystals on the windowpanes. Into the hydrogen sulfide room with its murderous atmosphere withdrew couples seeking privacy and a few lone wolves to eat their snacks.

Through the murk and in the busy silence, we heard a Piedmontese voice say: *"Nuntio vobis gaudium magnum. Habemus ferrum."* "I announce to you a great joy. We have iron." It was March 1939, and a few days earlier an almost identical solemn announcement (*"Habemus Papam"*) had closed the conclave that had raised to Peter's Throne Cardinal Eugenio Pacelli, in whom many put their hopes, since one must after all put one's hope in someone or something. The blasphemous announcement came from Sandro, the quiet one.

In our midst, Sandro was a loner. He was a boy of medium height, thin but muscular, who never wore an overcoat, even on the coldest days. He came to class in worn corduroy knickers, knee socks made of homespun wool and sometimes a short black cape which made me think of the Tuscan poet Renato Fucini. He had large, calloused hands, a bony, rugged profile, a face baked by the sun, a low forehead beneath the line of his hair, which he wore very short and cut in a brush. He walked with the peasant's long, slow stride.

A few months before, the racial laws against the Jews had been proclaimed, and I too was becoming a loner. My Christian classmates were civil people; none of them, nor any of the teachers, had directed at me a hostile word or gesture, but I could feel them withdraw and, following an ancient pattern, I withdrew as well: every look exchanged between me and them was accompanied by a minuscule but perceptible flash of mistrust and suspicion. What do you think of me? What am I for you? The same as six months ago, your equal who does not go to Mass, or the Jew who, as Dante put it, "in your midst laughs at you"?

I had noticed with amazement and delight that something was happening between Sandro and me. It was not at all a friendship born from affinity; on the contrary, the difference in our origins made us rich in "exchangeable goods," like two merchants who meet after coming from remote and mutually unknown regions. Nor was it the normal, portentous intimacy of twenty-year-olds: with Sandro I never reached this point. I soon realized that he was generous, subtle, tenacious, and brave, even with a touch of insolence, but he had an elusive, untamed quality; so that, although we were at the age when one always has the need, instinct, and immodesty of inflicting on one another everything that swarms in one's head and elsewhere (and this is an age that can last long, but ends with the first compromise), nothing had gotten through his carapace of reserve, nothing of his inner world, which nevertheless one felt was dense and fertile — nothing save a few occasional, dramatically truncated hints. He had the nature of a cat with whom one can live for decades without ever being permitted to penetrate its sacred pelt.

We had many concessions to make to each other. I told him we were like cation and anion, but Sandro did not seem to acknowledge the comparison. He was born in Serra d'Ivrea, a beautiful but niggardly region. He was the son of a mason and spent his summers working as a shepherd. Not a shepherd of souls: a shepherd of sheep, and not because of Arcadian rhetoric or eccentricity, but happily, out of love for the earth and grass and an abundance of heart. He had a curious mimetic talent, and when he talked about cows, chickens, sheep, and dogs he was transformed, imitating their way of looking, their movements and voices, becoming very gay and seeming to turn into an animal himself, like a shaman. He taught me about plants and animals, but said very little about his family. His father had died when he was a child; they were simple, poor people, and since the boy was bright, they had decided to make him study so that he would bring money home: he had accepted this with Piedmontese seriousness but without enthusiasm. He had traveled the long route of high school — *liceo* — aiming at the highest marks with the least effort. He was not interested in Catullus and Descartes, he was interested in being promoted, and spending Sunday on his skis and climbing the rocks. He had chosen chemistry because he had thought it better than other studies; it was a trade that dealt with things one can see and touch, a way to earn one's bread less tiring than working as a carpenter or a peasant.

We began studying physics together, and Sandro was surprised when I tried to explain to him some of the ideas that at the time I was confusedly cultivating. That the nobility of Man, acquired in a hundred centuries of trial and error, lay in making himself the conqueror of matter, and that I had enrolled in chemistry because I wanted to remain faithful to this nobility. That conquering matter is to understand it, and understanding matter is necessary to understanding the universe and ourselves: and that therefore Mendeleev's Periodic Table, which just during those weeks we were laboriously learning to unravel, was poetry, loftier and more solemn than all the poetry we had swallowed down in liceo; and come to think of it, it even rhymed! That if one looked for the bridge, the missing link, between the world of words and the world of things, one did not have to look far: it was there, in our Autenrieth, in our smoke-filled labs, and in our future trade.

And finally, and fundamentally, an honest and open boy, did he not smell the stench of Fascist truths which tainted the sky? Did he not perceive it as an ignominy that a thinking man should be asked to believe without thinking? Was he not filled with disgust at all the dogmas, all the unproved affirmations, all the imperatives? He did feel it; so then, how could he not feel a new dignity and majesty in our study, how could he ignore the fact that the chemistry and physics on which we fed, besides being in themselves nourishments vital in themselves, were the antidote to Fascism which he and I were seeking, because they were clear and distinct and verifiable at every step, and not a tissue of lies and emptiness, like the radio and newspapers?

Sandro listened to me with ironical attention, always ready to deflate me with a couple of civil and terse words when I trespassed into rhetoric. But something was ripening in him (certainly not all my doing; those were months heavy with fateful events), something that troubled him because it was at once new and ancient. He, who until then had read only Salgari, Jack London, and Kipling, overnight became a furious reader: he digested and remembered everything, and everything in him spontaneously fell into place as a way of life; together with this, he began to study, and his average shot up from C to A. At the same time, out of unconscious gratitude, and perhaps also out of a desire to get even, he in turn took an interest in my education and made it clear to me that it had gaps. I might even be right: it might be that Matter is our teacher and perhaps also, for lack of something better, our political school; but he had another form of matter to lead me to, another teacher: not the powders of the Analytical Lab but the true, authentic, timeless *Urstoff*, the rocks and ice of the nearby mountains. He proved to me without too much difficulty that I didn't have the proper credentials to talk about matter. What commerce, what intimacy had I had, until then, with Empedocles' four elements? Did I know how to light a stove? Wade across a torrent? Was I familiar with a storm high up in the mountains? The sprouting of seeds? No. So he too had something vital to teach me.

A comradeship was born, and there began for me a feverish season. Sandro seemed to be made of iron, and he was bound to iron by an ancient kinship: his father's fathers, he told me, had been tinkers (*magnín*) and blacksmiths in the Canavese valleys: they made nails on the charcoal forges, sheathed wagon wheels with red-hot hoops, pounded iron plates until deafened by the noise; and he himself when he saw the red vein of iron in the rock felt he was meeting a friend. In the winter when it suddenly hit him, he would tie his skis on his rusty bike, leaving early in the morning and pedaling away until he reached the snow, without a cent, an artichoke in one pocket and the other full of lettuce; then he came back in the evening or even the next day, sleeping in haylofts, and the more storms and hunger he suffered the happier and healthier he was.

In the summer, when he went off by himself, he often took along a dog to keep him company. This was a small yellow mongrel with a downcast expression; in fact, as Sandro had told me, acting out in his way the animal episode, as a puppy he had had a mishap with a cat. He had come too close to a litter of newborn kittens, the mother cat was miffed and became enraged, and had begun to hiss, getting all puffed up; but the puppy had not yet learned the meaning of those signals and remained there like a fool. The cat had attacked him, chased him, caught him, and scratched his nose; the dog had been permanently traumatized. He felt dishonored, and so Sandro had made him a cloth ball, explained to him that it was the cat, and every morning presented it to him so that he could take his revenge on it for the insult and regain his canine honor. For the same therapeutic motive, Sandro took him to the mountains, so he could have some fun: he tied him to one end of a rope, tied himself to the other, set the dog firmly on a rock ledge, and then climbed up; when the rope ended, he pulled it up slowly, and the dog had learned to walk up with his muzzle pointed skywards and his four paws against the nearly vertical wall of rock, moaning softly as though he were dreaming.

Sandro climbed the rocks more by instinct than technique, trusting to the strength of his hands and saluting ironically, in the projecting rock to which he clung, the silicon, calcium, and magnesium he had learned to recognize in the course on mineralogy. He seemed to feel that he had wasted a day if he had not in some way gotten to the bottom of his reserve of energy, and then even his eyes became brighter and he explained to me that, with a sedentary life, a deposit of fat forms behind the eyes, which is not healthy; by working hard the fat is consumed and the eyes sink back into their sockets and become keener.

He spoke grudgingly about his exploits. He did not belong to that species of persons who do things in order to talk about them (like me); he did not like high-sounding words, indeed words. It appeared that in speaking, as in mountain climbing, he had never received lessons; he spoke as no one speaks, saying only the core of things.

If necessary he carried a thirty-kilo pack, but usually he traveled without it; his pockets were sufficient, and in them he put some vegetables, as I have said, a chunk of bread, a pocketknife, sometimes the dog-eared Alpine Club guide, and a skein of wire for emergency repairs. In fact he did not carry the guide because he believed in it, but for the opposite reason. He rejected it because he felt that it shackled him; not only that, he also saw it as a bastard creature, a detestable hybrid of snow and rock mixed up with paper. He took it into the mountains to vilify. Happy if he could catch it in an error, even if it was at his and his climbing companion's expense. He could walk for two days without eating, or eat three meals all together and then leave. For him, all seasons were good. In the winter he skied, but not at the well-equipped, fashionable slopes, which he shunned with laconic scorn: too poor to buy ourselves the sealskin strips for the ascents, he showed me how you sew on rough hemp cloths, Spartan devices which absorb the water and then freeze like codfish, and must be tied around your waist when you ski downhill. He dragged me along on exhausting treks through the fresh snow, far from any sign of human life, following routes that he seemed to intuit like a savage. In the summer, from shelter to shelter, inebriating ourselves with the sun, the effort, and the wind, and scraping the skin of our fingertips on rocks never before touched by human hands: but not on the famous peaks, nor in quest of memorable feats; such things did not matter to him at all. What mattered was to know his limitations, to test and improve himself; more obscurely, he felt the need to prepare himself (and to prepare me) for an iron future, drawing closer month by month.

To see Sandro in the mountains reconciled you to the world and made you forget the nightmare weighing on Europe. This was his place, what he had been made for, like the marmots whose whistle and snout he imitated: in the mountains he became happy, with a silent, infectious happiness, like a light that is switched on. He aroused a new communion with the earth and sky, into which flowed my need for freedom, the plenitude of my strength, and a hunger to understand the things he had pushed me toward. We would come out at dawn, rubbing our eyes, through the small door of the Martinotti bivouac, and there, all around us, barely touched by the sun, stood the white and brown mountains, new as if created during the night that had just ended and at the same time innumerably ancient. They were an island, an elsewhere.

In any event, it was not always necessary to go high and far. In the in-between seasons Sandro's kingdom was the rock gymnasiums. There are several, two or three hours by bike from Turin, and I would be curious to know whether they are still frequented: the Straw Stack Pinnacles with the Wolkmann Tower, the Teeth of Cumiana, Patanüa Rock (which means Bare Rock), the Plô, the Sbarüa, and others, with their homely, modest names. The last, the Sbarüa, I think was discovered by Sandro himself and a mythical brother of his whom Sandro never let me see but who, from his few scanty hints, must have stood in the same relationship to him as he stood to the run of humanity. Sbarüa is a noun from the verb sbarüé, which means "to terrify"; the Sbarüa is a prism of granite that towers about a hundred meters above a modest hill bristling with brambles and a brushwood coppice; like Dante's Veglio di Creta — the Old Man of Crete — it is split from base to summit by a fissure that gets narrower as it rises, finally forcing the climber to come out on the rock face itself, where, precisely, he is terrified, and where at that time there was just a single piton, charitably left behind by Sandro's brother.

Those were curious places, frequented by a few dozen amateurs of our stamp, all of whom Sandro knew either by name or sight: we climbed up, not without technical problems, and surrounded by the irritating buzz of enormous bluebottle flies attracted by our sweat, crawling up good solid rock walls interrupted by grassy ledges where ferns and strawberries grew or, in the fall, blackberries: often enough we used as holds the trunks of puny little trees, rooted in the cracks, and after a few hours we reached the peak, which was not a peak at all but mostly placid pastureland where cows stared at us with indifferent eyes. Then we descended at breakneck speed, in a few minutes, along paths strewn with old and recent cow dung, to recover our bikes.

At other times our exploits were more demanding; never any quiet jaunts, since Sandro said that we would have plenty of time when we were forty to look at the scenery. "Let's go, shall we?" he said to me one day in February — which in his language meant that, since the weather was good, we should leave in the afternoon for the winter climb of the Tooth of M., which for some weeks had been one of our projects. We slept in an inn and left the next day, not too early, at some undetermined hour (Sandro did not like watches: he felt their quiet continuous admonishment to be an arbitrary intrusion). We plunged boldly into the fog and came out of it about one o'clock, in gleaming sunlight and on the big crest of a peak which was not the right one.

I then said that we should be able to go down about a hundred meters, cross over halfway up the mountain, and go up along the next ridge: or, better yet, since we were already there, continue climbing and be satisfied with the wrong peak, which in any case was only forty meters lower than the right one. But Sandro, with splendid bad faith, said in a few dense syllables that my last proposal was fine, but from there "by way of the easy northwest ridge" (this was a sarcastic quotation from the abovementioned Alpine Club guide) we could also reach the Tooth of M. in half an hour; and what was the point of being twenty if you couldn't permit yourself the luxury of taking the wrong route.

The easy ridge must really have been easy, indeed elementary in the summer; but we found it in a very discomforting state. The rock was wet on the side facing the sun and covered with a black layer of ice in the shade; between one large outcrop of rock and another lay pockets of melting snow into which we sank to our waists. We reached the top at five; I dragged myself along so pitifully that it was painful, while Sandro was seized by a sinister hilarity that I found very annoying.

"And how do we get down?"

"As for getting down, we shall see," he replied, and added mysteriously: "The worst that can happen is to have to taste bear meat." Well, we tasted bear meat in the course of that night, which seemed very, very long. We got down in two hours, helped badly by the rope, which was frozen; it had become a malignant, rigid tangle that snagged on each projection and rang against the rock face like the cable of a funicular. At seven we were on the bank of a frozen pond and it was dark. We ate the little that was left, built a useless dry stone wall facing the wind, and lay down on the ground to sleep, pressed to each other. It was as though time itself had frozen; every so often we got to our feet to reactivate our circulation, and it was always the same time: the wind never stopped blowing, there was always the same ghost of a moon, always at the same point in the sky, and in

759

front of the moon passed a fantastic cavalcade of tattered clouds, always the same. We had taken off our shoes, as described in Lammer's books, so dear to Sandro, and we kept our feet in our packs; at the first funereal light, which seemed to seep from the snow and not the sky, we rose with our limbs benumbed and our eyes glittering from lack of sleep, hunger, and the hardness of our bed. And we found our shoes so frozen that they rang like bells, and to get them on we had to hatch them out like brood hens.

But we went back down to the valley under our own steam; and to the innkeeper who asked us, with a snicker, how things had gone, and meanwhile was staring at our wild, exalted faces, we answered flippantly that we had had an excellent outing, then paid the bill and departed with dignity. This was it — the bear meat; and now that many years have passed, I regret that I ate so little of it, for nothing has had, even distantly, the taste of that meat, which is the taste of being strong and free, free also to make mistakes and be the master of one's destiny. That is why I am grateful to Sandro for having led me consciously into trouble, on that trip and other undertakings which were only apparently foolish, and I am certain that they helped me later on.

They didn't help Sandro, or not for long. Sandro was Sandro Delmastro, the first man to be killed fighting in the Resistance with the Action Party's Piedmontese Military Command. After a few months of extreme tension, in April of 1944 he was captured by the Fascists, did not surrender, and tried to escape from the Fascist Party house in Cuneo. He was killed with a tommygun burst in the back of the neck by a monstrous child-executioner, one of those wretched murderers of fifteen whom Mussolini's Republic of Salò recruited in the reformatories. His body was abandoned in the road for a long time, because the Fascists had forbidden the population to bury him.

Today I know that it is a hopeless task to try to dress a man in words, make him live again on the printed page, especially a man like Sandro. He was not the sort of person you can tell stories about, nor to whom one erects monuments — he who laughed at all monuments: he lived completely in his deeds, and when they were over nothing of him remains — nothing but words, precisely.

# The Poetry of Science

Science (or the "natural philosophy" from which science evolved) has long provided poets with raw material, inspiring some to praise scientific ideas and others to react against them. Lucretius wrote paeans to the atomic theory of the Epicureans in the first century B.C. Milton interviewed Galileo about cosmology. Blake excoriated Newton; Wordsworth celebrated him. Goethe found room for both attitudes in his rangy mind. He wrote more scientific papers than poems, and esteemed his scientific work more highly than his poetry, but was quick to use his poet's pen to ridicule the metaphysical pretensions of science. By the twentieth century a great many poets were studying scientific concepts — some (Marianne Moore, the physician Miroslav Holub) to applaud them, others (T. S. Eliot, Robinson Jeffers, Robert Frost) to poke fun at them.

The skepticism that many poets display toward science reflects, and to some extent perpetuates, the myth that science is cold and inhuman, poetry warm and romantic. Yet science is more romantic than is generally realized, poetry less so, and the scientists and the poets ultimately are allies. Both are creative and unpredictable (and therefore dangerous). Neither can tolerate authoritarianism, blind obedience, or cant. And both, to do their best work, must draw on aesthetic as well as intellectual resources; a logical but ugly mathematical theorem is as unsatisfactory as a pretty but silly sonnet.

This is not to say that scientists should try to emulate poets, or that poets should turn proselytes for science. Poetry and science are both too powerful to benefit from so bland and bourgeois a marriage, and their relationship is likely to remain stormy so long as each remains vital. But they need each other, and the world needs them both.

WALT WHITMAN
(1819—1892)

■

# *When I Heard the Learn'd Astronomer*

When I heard the learn'd astronomer,
When the proofs, the figures, were ranged in columns
    before
    me,
When I was shown the charts and diagrams, to add, divide,
    and measure them,
When I sitting heard the astronomer where he lectured
    with much applause in the lecture-room,
How soon unaccountable I became tired and sick,
Till rising and gliding out I wander'd off by myself,
In the mystical moist night air, and from time to time,
Look'd up in perfect silence at the stars.

GERARD MANLEY HOPKINS
(1844—1889)

■

# *"I am Like a Slip of Comet . . ."*

    — I am like a slip of comet,
Scarce worth discovery, in some corner seen
Bridging the slender difference of two stars,
Come out of space, or suddenly engender'd
By heady elements, for no man knows;
But when she sights the sun she grows and sizes
And spins her skirts out, while her central star
Shakes its cocooning mists; and so she comes
To fields of light; millions of travelling rays

Pierce her; she hangs upon the flame-cased sun,
And sucks the light as full as Gideon's fleece:
But then her tether calls her; she falls off,
And as she dwindles shreds her smock of gold
Between the sistering planets, till she comes
To single Saturn, last and solitary;
And then she goes out into the cavernous dark.
So I go out: my little sweet is done:
I have drawn heat from this contagious sun:
To not ungentle death now forth I run.

# EMILY DICKINSON
## (1830—1886)

∎

# *Arcturus*

"Arcturus" is his other name –
I'd rather call him "Star."
It's very mean of Science
To go and interfere!

I slew a worm the other day –
A "Savant" passing by
Murmured "Resurgam" – "Centipede"!
"Oh Lord – how frail are we"!

I pull a flower from the woods –
A monster with a glass
Computes the stamens in a breath –
And has her in a "class"!

Whereas I took the Butterfly
Aforetime in my hat –
He sits erect in "Cabinets" –
The Clover bells forgot.

What once was "Heaven"
Is "*Zenith*" now –
Where I proposed to go

When Time's brief masquerade was done
Is mapped and charted too.
What if the poles should frisk about
And stand upon their heads!
I hope I'm ready for "the worst" –
Whatever prank betides!

Perhaps the "Kingdom of Heaven's" changed –
I hope the "Children" there
Won't be "new fashioned" when I come –
And laugh at me – and stare –

I hope the Father in the skies
Will lift his little girl –
Old fashioned – naughty – everything –
Over the stile of "Pearl."

ROBINSON JEFFERS
(1887–1962)

■

# *Star-Swirls*

The polar ice-caps are melting, the mountain glaciers
Drip into rivers; all feed the ocean;
Tides ebb and flow, but every year a little bit higher.
They will drown New York, they will drown London.
And this place, where I have planted tree and built a stone
  house,
Will be under sea. The poor trees will perish,
And little fish will flicker in and out the windows. I built it well,
Thick walls and Portland cement and gray granite,
The tower at least will hold against the sea's buffeting; it will
  become
Geological, fossil and permanent.
What a pleasure it is to mix one's mind with geological
Time, or with astronomical relax it.
There is nothing like astronomy to pull the stuff out of man.
His stupid dreams and red-rooster importance: let him count the
  star-swirls.

RICHARD RYAN
(b. 1946)

■

# *Galaxy*

faint
  in deep space,
    immense as a brain

down
  through the thought-
    shaft it drifts, a wale

of light to
  which the retina
    opens and is entered

time and
  space dis-
    appearing as the mind

recedes
  to a soundless
    flickering somewhere

deeper
  than consciousness
    where, permanent as

change
  a whorl of light
    rides, wheeling in darkness

## JAMES CLERK MAXWELL
(1831—1879)

■

# *Molecular Evolution*
### Belfast, 1874

At quite uncertain times and places,
  The atoms left their heavenly path,
And by fortuitous embraces,
  Engendered all that being hath.
And though they seem to cling together,
  And form "associations" here,
Yet, soon or late, they burst their tether,
  And through the depths of space career.

So we who sat, oppressed with science,
  As British asses, wise and grave,
Are now transformed to wild Red Lions,
  As round our prey we ramp and rave.
Thus, by a swift metamorphösis,
  Wisdom turns wit, and science joke,
Nonsense is incense to our noses,
  For when Red Lions speak, they smoke.

Hail, Nonsense! dry nurse of Red Lions,
  From thee the wise their wisdom learn,
From thee they cull those truths of science,
  Which into thee again they turn.
What combinations of ideas,
  Nonsense alone can wisely form!
What sage has half the power that she has,
  To take the towers of Truth by storm?

A . M . S U L L I V A N
(1896—1980)

■

# *Atomic Architecture*

Take Carbon for example then
What shapely towers it constructs
To house the hopes of men!
What symbols it creates
For power and beauty in the world
Of patterned ring and hexagon —
Building ten thousand things
Of earth and air and water!
Pride searches in the flues of earth
For the diamond and its furious sun,
Love holds its palms before the glow
Of anthracite and purrs.
Five senses take their fill
Of raiment, rainbows and perfumes,
Of sweetness and of monstrous pain.

If life begins in carbon's dancing atoms
Moving in quadrilles of light
To the music of pure numbers,
Death is the stately measure
Of Time made plausible
By carbon's slow procession
Out of the shifting structure
Of crumbling flesh and bone.

HOWARD NEMEROV
(b. 1920)

■

## *Seeing Things*

Close as I ever came to seeing things
The way the physicists say things really are
Was out on Sudbury Marsh one summer eve
When a silhouetted tree against the sun
Seemed at my sudden glance to be afire:
A black and boiling smoke made all its shape.

Binoculars resolved the enciphered sight
To make it clear the smoke was a cloud of gnats,
Their millions doing such a steady dance
As by the motion of the many made the one
Shape constant and kept it so in both the forms
I'd thought to see, the fire and the tree.

Strike through the mask? you find another mask,
Mirroring mirrors by analogy
Make visible. I watched till the greater smoke
Of night engulfed the other, standing out
On the marsh amid a hundred hidden streams
Meandering down from Concord to the sea.

JOHN UPDIKE
(b. 1932)

■

## *Cosmic Gall*

Neutrinos, they are very small.
They have no charge and have no mass

768

And do not interact at all.
The earth is just a silly ball
    To them, through which they simply pass.
Like dustmaids down a drafty hall
    Or photons through a sheet of glass.
    They snub the most exquisite gas,
Ignore the most substantial wall,
    Cold-shoulder steel and sounding brass,
Insult the stallion in his stall,
    And, scorning barriers of class,
Infiltrate you and me! Like tall
And painless guillotines, they fall
    Down through our heads into the grass.
At night, they enter at Nepal
    And pierce the lover and his lass
From underneath the bed — you call
    It wonderful; I call it crass.


# ANTHONY PICCIONE
(b. 1939)

■

# *Nomad*

The particle scientist
is more or less
happy. He has no home.

All his ladders
go straight down
and claim the nameless.

JOHN HAINES
(b. 1924)

■

# *Little Cosmic Dust Poem*

Out of the debris of dying stars,
this rain of particles
that waters the waste with brightness;

the sea-wave of atoms hurrying home,
collapse of the giant,
unstable guest who cannot stay;

the sun's heart reddens and expands,
his mighty aspiration is lasting,
as the shell of his substance
one day will be white with frost.

In the radiant field of Orion
great hordes of stars are forming,
just as we see every night,
fiery and faithful to the end.

Out of the cold and fleeing dust
that is never and always,
the silence and waste to come —

this arm, this hand,
my voice, your face, this love.

WALLACE STEVENS
(1879—1955)

■

# *Connoisseur of Chaos*

## I

A.  A violent order is disorder; and
B.  A great disorder is an order. These
Two things are one. (Pages of illustrations.)

## II

If all the green of spring was blue, and it is;
If the flowers of South Africa were bright
On the tables of Connecticut, and they are;
If Englishmen lived without tea in Ceylon, and they do;
And if it all went on in an orderly way,
And it does; a law of inherent opposites,
Of essential unity, is as pleasant as port,
As pleasant as the brush-strokes of a bough,
An upper, particular bough in, say, Marchand.

## III

After all the pretty contrast of life and death
Proves that these opposite things partake of one,
At least that was the theory, when bishops' books
Resolved the world. We cannot go back to that.
The squirming facts exceed the squamous mind,
If one may say so. And yet relation appears,
A small relation expanding like the shade
Of a cloud on sand, a shape on the side of a hill.

## IV

A.  Well, an old order is a violent one.
This proves nothing. Just one more truth, one more
Element in the immense disorder of truths.
B.  It is April as I write. The wind
Is blowing after days of constant rain.

All this, of course, will come to summer soon.
But suppose the disorder of truths should ever come
To an order, most Plantagenet, most fixed . . .
A great disorder is an order. Now, A
And B are not like statuary, posed
For a vista in the Louvre. They are things chalked
On the sidewalk so that the pensive man may see.

V

The pensive man . . . He sees that eagle float
For which the intricate Alps are a single nest.


RAINER MARIA RILKE
(1875—1926)
(translated by Robert Bly)

■

# "*The Kings of the World . . .*"


The kings of the world are growing old,
and they shall have no inheritors.
Their sons died while they were boys,
and their neurasthenic daughters abandoned
the sick crown to the mob.

The mob breaks it into tiny bits of gold.
The Lord of the World, master of the age,
melts them in fire into machines,
which do his orders with low growls:
but luck is not on their side.

The ore feels homesick. It wants to abandon
the minting houses and the wheels
that offer it such a meager life.
And out of factories and payroll boxes
it wants to go back into the veins
of the thrown-open mountain,
which will close again behind it.

# ANNIE DILLARD
(b. 1945)

■

# *The Windy Planet*

pole
polar      easterlies      pola
r easterlies polar easterlies polar easterli
es polar easterlies polar easterlies polar easter

roaring forties roaring forties roaring forties roarin
forties westerlies roaring forties roaring forties roar
ing forties roaring forties roaring forties roaring forties

horse latitudes horse latitudes horse latitudes horse latitudes

the trades the trades the trades the trades the trades the trades the
trades the northeast trades the trades the trades the trades the trades
the trades the trades the trades the trades the trades the trades the tr
ades the trades the trades the northeast trades the trades the trades t
doldrums doldrums doldrums doldrums doldrums doldrums doldrums doldr

the trades the trades the trades the trades the trades the trades the t
rades the southeast trades the trades the trades the trades the trades
the trades the trades the trades the trades the trades the trades the
trades the trades the trades the southeast trades the trades the trade

horse latitudes horse latitudes horse latitudes horse latitudes horse latitud

roaring forties roaring forties roaring forties roaring fo
rties westerlies roaring forties roaring forties roaring forties roari
ng forties roaring forties roaring forties roaring f

polar easterlies polar easterlies polar easterli
es polar easterlies polar easterlies pol
ar easterlies polar easterlies pol
pole

# DIANE ACKERMAN
(b. 1948)

■

# *Space Shuttle*

By all-star orchestra, they dine in space
in a long steel muscle so fast it floats,
in a light waltz they lie still as amber
watching Earth stir in her sleep beneath them.

They have brought along a plague
of small winged creatures, whose brains are tiny
as computer chips. Flight is the puzzle,
the shortest point between two times.

In zero gravity, their hearts will be light,
not three pounds of blood, dream and gristle.
When they were young, the sky was a tree
whose cool branches they climbed,
sweaty in August, and now they are the sky
children imagine as invisible limbs.

On the console, a light summons them
to the moment, and they must choose
between the open-mouthed delirium in their cells,
the awe ballooning beyond the jetstream,
or husband all that is safe and tried.

They are good providers. Their eyes do not wander.
Their fingers do not pause at the prick
of a switch. Their mouths open for sounds
no words rush into. Answer the question
put at half-garble. Say again
how the cramped world turns, say again.

# RICHARD WILBUR
(b. 1921)

■

# *Epistemology*

## I

Kick at the rock, Sam Johnson, break your bones:
But cloudy, cloudy is the stuff of stones.

## II

We milk the cow of the world, and as we do
We whisper in her ear, "You are not true."

# WILLIAM WATSON
(1858—1935)

■

# *What Science Says to Truth*

As is the mainland to the sea,
Thou art to me;
Thou standest stable, while against thy feet
I beat, I beat!

Yet from thy cliffs so sheer, so tall,
Sands crumble and fall;
And golden grains of thee my tides each day
Carry away.

775

# ALEXANDER POPE
## (1688—1744)

■

# From *An Essay on Man*

All Nature is but Art, unknown to thee;
All Chance, Direction, which thou canst not see;
All Discord, Harmony, not understood;
All partial Evil, universal Good. . . .

# JOHANN WOLFGANG VON GOETHE
## (1749—1832)
### (translated by Michael Hamburger)

■

# *True Enough: To the Physicist*

"Into the core of Nature" —
O Philistine —
"No earthly mind can enter."
The maxim is fine;
But have the grace
To spare the dissenter,
Me and my kind.
We think: in every place
We're at the center.
"Happy the mortal creature
To whom she shows no more
Than the outer rind,"
For sixty years I've heard your sort announce.
It makes me swear, though quietly;
To myself a thousand times I say:

All things she grants, gladly and lavishly;
Nature has neither core
Nor outer rind,
Being all things at once.
It's yourself you should scrutinize to see
Whether you're center or periphery.

## GEORGE BRADLEY
(b. 1953)

■

# *About Planck Time*

Once upon a time, way back in the infinitesimal
First fraction of a second attending our creation,
A tiny drop containing all of it, all energy
And all its guises, burst upon the scene,
Exploding out of nothing into everything
Virtually instantaneously, the way our thoughts
Leap eagerly to occupy the abhorrent void.
Once, say ten or twenty billion years ago,
In Planck time, in no time at all, the veil
Available to our perceptions was flung out
Over space at such a rate the mere imagination
Cannot keep up, so rapidly the speed of light
Lags miraculously behind, producing a series
Of incongruities that has led our curiosity,
Like Ariadne's thread, through the dim labyrinth
Of our conclusions to the place of our beginning.
In Planck time, everything that is was spread so thin
That all distance is enormous, between each star,
Between subatomic particles, so that we are composed
Almost entirely of emptiness, so that what separates
This world, bright ball floating in its midnight blue,
From the irrefutable logic of no world at all
Has no more substance than the traveller's dream,
So that nothing can be said for certain except

That sometime, call it Planck time, it will all just
Disappear, a parlor trick, a rabbit back in its hat,
Will all go up in a flash of light, abracadabra,
An idea that isn't being had anymore.

JOHN FREDERICK NIMS
(b. 1913)

■

# *The Observatory Ode*
### Harvard, June 1978

I

The Universe:
We'd like to understand,
But any piece, in the palm, gets out of hand,
Any stick, any stone,
— How mica burns! — or worse,
Any star we catch in pans of glass,
Sift to a twinkle the vast nuclear zone,
Lava-red, polar-blue,
Apple-gold (noon our childhood knew),
Colors that through the prism, like dawn through Gothic, pass,
Or in foundries sulk among grots and gnomes, in glare of zinc or brass.
Would Palomar's flashy cannon say? Would you,
Old hourglass, galaxy of sand,
You, the black hole where Newton likes to stand?

II

Once on this day,
Our Victorian renaissance-man,
Percival Lowell — having done Japan,
And soon to be seen
Doing over all heaven his way —
Spoke poems here. (These cheeks, a mite
Primped by the laurel leaves' symbolical green,
Should glow like the flustered beet

To scuff, in his mighty shoes, these feet.)
He walked high ground, each long cold Arizona night,
Grandeurs he'd jot: put folk on Mars, but guessed a planet right,
Scribbling dark sums and ciphers at white heat
For his Pluto, lost. Till — there it swam!
Swank, with his own P L for monogram.

### III

Just down the way
The Observatory. And girls
Attending, with lint of starlight in their curls,
To lens, 'scope, rule.
Sewing bee, you could say:
They stitch high heaven together here,
Save scraps of the midnight sky. Compile, poll, pool.
One, matching star with star,
Learns that *how bright* can mean *how far*.
That widens the galaxies! Each spiraling chandelier
In three-dimensional glamour hangs; old flat nights disappear.
Desk-bound, they explore the immensities. Who are
These women that, dazed at dusk, arise?
— No Helen with so much heaven in her eyes.

### IV

With what good night
Did the strange women leave?
What did the feverish planet-man achieve?
A myth for the sky:
All black. Then a haze of light,
A will-o'-the-wisp, hints *time* and *place*.
Whirling, the haze turned fireball, and let fly
Streamers of bright debris,
The makings of our land and sea.
Great rafts of matter crash, their turbulence a base
For furnaces of nuclear fire that blast out slag in space.
Primal pollution, dust and soot, hurl free
Lead, gold — all that. Heaven's gaudy trash.
This world — with our joy in June — is a drift of ash.

V

That fire in the sky
On the Glorious Fourth, come dark,
Acts "Birth of the Universe" out, in Playland Park.
Then a trace of ash
In the moon. Suppose we try
— Now only suppose — to catch in a jar
That palmful of dust, on bunsens burn till it flash,
Could we, from that gas aglow,
Construct the eventful world we know,
Or a toy of it, in the palm? Yet our world came so: we are
Debris of a curdled turbulence, and dust of a dying star
— The children of nuclear fall-out long ago.
No wonder if late world news agree
With Eve there's a creepy varmint in the Tree.

VI

The Universe:
. . . *Such stuff as dreams are made on* . . .
Yet stuff to thump, to call a spade a spade on.
No myth — Bantu,
Kurd, Urdu, Finnish, Erse —
Had for the heaven such hankering
As ours, that made new eyes for seeing true.
For seeing what we are:
Sun-bathers of a nuclear star,
Scuffing through curly quarks — mere fact a merry thing!
Then let's, with the girls and good P.L., sing carols in a ring!
Caution: combustible myth, though. Near and far
The core's aglow. No heat like this,
No heat like science and poetry when they kiss.

# Philosophy and Science

## ISAAC ASIMOV

■

When science first burst upon the world in full force, a scant three hundred years or so ago, society reacted by treating it as something apart from the rest of the world — as an alien activity practiced with a cold and almost inhuman objectivity. This may have been a normal reaction to so powerful and novel an innovation, the equivalent of a living organism's rejection of a biological invader. In truth, however, science is as human an endeavor as, well, fishing — as Isaac Asimov explains.

# The Nature of Science

A NUMBER OF YEARS AGO, when I was a freshly appointed instructor, I met, for the first time, a certain eminent historian of science. At the time I could only regard him with tolerant condescension.

I was sorry for a man who, it seemed to me, was forced to hover about the edges of science. He was compelled to shiver endlessly in the outskirts, getting only feeble warmth from the distant sun of science-in-progress; while I, just beginning my research, was bathed in the heady liquid heat at the very center of the glow.

In a lifetime of being wrong at many a point, I was never more wrong. It was I, not he, who was wandering in the periphery. It was he, not I, who lived in the blaze.

I had fallen victim to the fallacy of the "growing edge"; the belief that only the very frontier of scientific advance counted; that everything that had been left behind by that advance was faded and dead.

But is that true? Because a tree in spring buds and comes greenly into leaf, are those leaves therefore the tree? If the newborn twigs and their leaves were all that existed, they would form a vague halo of green suspended in mid-air, but surely that is not the tree. The leaves, by themselves, are no more than trivial fluttering decoration. It is the trunk and limbs that give the tree its grandeur and the leaves themselves their meaning.

There is not a discovery in science, however revolutionary, however sparkling with insight, that does not arise out of what went before. "If I have seen further than other men," said Isaac Newton, "it is because I have stood on the shoulders of giants."

And to learn that which goes before does not detract from the beauty of a scientific discovery but, rather, adds to it; just as the gradual unfolding of a flower, as seen by time-lapse photography, is more wonderful than the mature flower itself, caught in stasis.

In fact, an overly exclusive concern with the growing edge can kill the best of science, for it is not on the growing edge itself that growth can best be seen. If the growing edge only is studied, science begins to seem a revelation without a history of development. It is Athena, emerging adult and armed from the forehead of Zeus, shouting her fearful war cry with her first breath.

How dare one aspire to add to such a science? How can one ward off bitter disillusion when part of the structure turns out to be wrong. The perfection of the growing edge is meretricious while it exists, hideous when it cracks.

But add a dimension!

Take the halo of leaves and draw it together with branches that run into limbs that join to form a trunk that firmly enters the ground. It is the tree of science that you will then see, an object that is a living, growing, and permanent thing; not a flutter of leaves at the growing edge, insubstantial, untouchable, and dying with the frosts of fall.

Science gains reality when it is viewed not as an abstraction, but as the concrete sum of work of scientists, past and present, living and dead. Not a statement in science, not an observation, not a thought exists in itself. Each was ground out of the harsh effort of some man, and unless you know the man and the world in which he worked; the assumptions he accepted as truths; the concepts he considered untenable; you cannot fully understand the statement or observation or thought.

Consider some of what the history of science teaches.

First, since science originated as the product of men and not as a revelation, it may develop further as the continuing product of men. If a scientific law is not an eternal truth but merely a generalization which, to some man or group of men, conveniently described a set of observations, then to some other man or group of men, another generalization might seem even more convenient. Once it is grasped that scientific truth is limited and not absolute, scientific truth becomes capable of further refinement. Until that is understood, scientific research has no meaning.

Second, it reveals some important truths about the humanity of scientists. Of all the stereotypes that have plagued men of science, surely one above all has wrought harm. Scientists can be pictured as "evil," "mad," "cold," "self-centered," "absentminded," even "square" and yet survive easily. Unfortunately, they are usually pictured as "right" and that can distort the picture of science past redemption.

Scientists share with all human beings the great and inalienable privilege of being, on occasion, wrong; of being egregiously wrong sometimes, even monumentally wrong. What is worse still, they are sometimes perversely and persistently wrongheaded. And since that it true, science itself can be wrong in this aspect or that.

With the possible wrongness of science firmly in mind, the student of science today is protected against disaster. When an individual theory collapses, it need not carry with it one's faith and hope and innocent joy. Once we learn to expect theories to collapse and to be supplanted by more useful generalizations, the collapsing theory becomes not the gray remnant of a broken today, but the herald of a new and brighter tomorrow.

Third, by following the development of certain themes in science, we can experience the joy and excitement of the grand battle against the unknown. The wrong turnings, the false clues, the elusive truth nearly captured half a century before its time, the unsung prophet, the false authority, the hidden assumption and cardboard syllogism, all add to the suspense of the struggle and make what we slowly gain through the study of the history of science worth more than what we might quickly gain by a narrow glance at the growing edge alone.

To be sure, the practical thought might arise: But would it not be better if we learned the truth at once? Would we not save time and effort?

Yes, we might, but it is not as important to save time and effort as to enjoy the time and effort spent. Why else should a man rise before dawn and go out in the damp to fish, waiting happily all day for the occasional twitch of his line when, without getting out of bed, he might have telephoned the market and ordered all the fish he wanted?

# HORACE FREELAND JUDSON

∎

The American science writer Horace Freeland Judson (b. 1931) has worked harder than most to document the process by which working scientists make their way into the unknown. This excerpt is from Judson's book *The Search for Solutions*, first published in 1980.

# *The Art of Discovery*

SCIENCE IS OUR CENTURY'S ART. Nearly four hundred years ago, when modern science was just beginning, Francis Bacon wrote that *knowledge is power*. Yet Bacon was not a scientist. He wrote as a bureaucrat in retirement. His slogan was actually the first clear statement of the promise by which, ever since, bureaucrats justify to each other and to king or taxpayer the spending of money on science. Knowledge is power; today we would say, less grandly, that science is essential to technology. Bacon's promise has been fulfilled abundantly, magnificently. The rage to know has been matched by the rage to make. Therefore — with the proviso, abundantly demonstrated, that it's rarely possible to predict which program of fundamental research will produce just what technology and when — the promise has brought scientists in the Western world unprecedented freedom of inquiry. Nonetheless, Bacon's promise hardly penetrates to the thing that moves most scientists. Science has several rewards, but the greatest is that it is the most interesting, difficult, pitiless, exciting, and beautiful pursuit that we have yet found. Science is our century's art.

The takeover can be dated more precisely than the beginning of most eras: Friday, June 30, 1905, will do, when Albert Einstein, a clerk in the Swiss patent office in Bern, submitted a thirty-one-page paper, "On the Electrodynamics of Moving Bodies," to the journal *Annalen der Physik*. No poem, no play, no piece of music written since then comes near the theory of relativity in its power, as one strains to apprehend it, to make the mind tremble with delight. Whereas fifty years ago it was often said that hardly twoscore people understood the theory of relativity, today its essential vision, as Einstein himself said, is within reach of any reasonably bright high school student — and that, too, is characteristic of the speed of assimilation of the new in the arts.

Consider also the molecular structure of that stuff of the gene, the celebrated double helix of deoxyribonucleic acid. This is two repetitive strands, one winding up, the other down, but hooked together, across the tube of space between them, by a sequence of pairs of chemical entities — just four sorts of these entities, making just two kinds of pairs, with exactly ten pairs to a full turn of the helix. It's a piece of sculpture. But observe how form and function are one. That sequence possesses a unique duality: one way, it allows the strands to part and each to assemble on itself, by the pairing rules, a duplicate of the complementary strand; the other way, the sequence enciphers, in a four-letter alphabet, the entire specification for the substance of the organism. The structure thus encompasses both heredity and embryological growth, the passing-on of potential and its expression. The structure's elucidation, in March of 1953, was an event of such surpassing explanatory power that it will reverberate through whatever time mankind has remaining. The structure is also perfectly economical and splendidly elegant. There is no sculpture made in this century that is so entrancing.

If to compare science to art seems — in the last quarter of this century — to undervalue what science does, that must be, at least partly, because we now expect art to do so little. Before our century, everyone of course supposed that the artist imitates nature. Aristotle had said so; the idea was obvious, it had flourished and evolved for two thousand years; those who thought about it added that the artist imitated not just nature as it accidentally happens, but by penetrating to nature as it has to be. Yet today that describes the scientist. "Scientific reasoning," Medawar also said, "is a constant interplay or interaction between hypotheses and the logical expectations they give rise to: there is a restless to-and-fro motion of thought, the formulation and reformulation of hypotheses, until we arrive at a hypothesis which, to the best of our prevailing knowledge, will satisfactorily meet the case." Thus far, change only the term "hypothesis" and Medawar described well the experience the painter or the poet has of his own work. "Scientific reasoning is a kind of dialogue between

785

the possible and the actual, between what might be and what is in fact the case," he went on — and there the difference lies. The scientist enjoys the harsher discipline of what is and is not the case. It is he, rather than the painter or the poet in this century, who pursues in its stringent form the imitation of nature.

Many scientists — mathematicians and physicists especially — hold that beauty in a theory is itself almost a form of proof. They speak, for example, of "elegance." Paul Dirac predicted the existence of antimatter (what would science fiction be without him?) several years before any form of it was observed. He won a share in the Nobel Prize in physics in 1933 for the work that included that prediction. "It is more important to have beauty in one's equations than to have them fit experiment," Dirac wrote many years later. "It seems that if one is working from the point of view of getting beauty in one's equations, and if one has really a sound insight, one is on a sure line of progress."

Here the scientist parts company with the artist. The insight must be sound. The dialogue is between what might be and what is in fact the case. The scientist is trying to get the thing right. The world is there.

And so are other scientists. The social system of science begins with the apprenticeship of the graduate student with a group of his peers and elders in the laboratory of a senior scientist; it continues to collaboration at the bench or the blackboard, and on to formal publication — which is a formal invitation to criticism. The most fundamental function of the social system of science is to enlarge the interplay between imagination and judgment from a private into a public activity. The oceanic feeling of well-being, the true touchstone of the artist, is for the scientist, even the most fortunate and gifted, only the midpoint of the process of doing science.

# THOMAS S. KUHN

∎

Thomas Kuhn (b. 1922) has taught at Berkeley, Harvard, and the Institute for Advanced Study at Princeton. His *Structure of Scientific Revolutions* is one of the most influential modern studies of the philosophy of science. Here, in the book's introduction, Kuhn outlines the book's main premise — that major changes in scientific thought have less to do with building new facts upon old than with changing the very paradigm through which scientists and others view the world.

# *The Structure of Scientific Revolutions*

HISTORY, IF VIEWED AS A repository for more than anecdote or chronology, could produce a decisive transformation in the image of science by which we are now possessed. That image has previously been drawn, even by scientists themselves, mainly from the study of finished scientific achievements as these are recorded in the classics and, more recently, in the textbooks from which each new scientific generation learns to practice its trade. Inevitably, however, the aim of such books is persuasive and pedagogic; a concept of science drawn from them is no more likely to fit the enterprise that produced them than an image of a national culture drawn from a tourist brochure or a language text. This essay attempts to show that we have been misled by them in fundamental ways. Its aim is a sketch of the quite different concept of science that can emerge from the historical record of the research activity itself.

Even from history, however, that new concept will not be forthcoming if historical data continue to be sought and scrutinized mainly to answer questions posed by the unhistorical stereotype drawn from science texts. Those texts have, for example, often seemed to imply that the content of science is uniquely exemplified by the observations, laws, and theories described in their pages. Almost as regularly, the same books have been read as saying that scientific methods are simply the ones illustrated by the manipulative techniques used in gathering textbook data, together with the logical operations employed when relating those data to the textbook's theoretical generalizations. The result has been a concept of science with profound implications about its nature and development.

If science is the constellation of facts, theories, and methods collected in current texts, then scientists are the men who, successfully or not, have striven to contribute one or another element to that particular constellation. Scientific development becomes the piecemeal process by which these items have been added, singly and in combination, to the ever growing stockpile that constitutes scientific technique and knowledge. And history of science becomes the discipline that chronicles both these successive increments and the obstacles that have inhibited their accumulation. Concerned with scientific development, the historian then appears to have two main tasks. On the one hand, he must determine by what man and at what point in time each contemporary scientific fact, law, and theory was discovered or invented. On the other, he must describe and explain the congeries of error, myth, and superstition that have inhibited the more rapid accumulation of the constituents of the modern science text. Much research has been directed to these ends, and some still is.

In recent years, however, a few historians of science have been finding it more and more difficult to fulfill the functions that the concept of development-by-accumulation assigns to them. As chroniclers of an incremental process, they discover that additional research makes it harder, not easier, to answer questions like: When was oxygen discovered? Who first conceived of energy conservation? Increasingly, a few of them suspect that these are simply the wrong sorts of questions to ask. Perhaps science does not develop by the accumulation of individual discoveries and inventions. Simultaneously, these same historians confront growing difficulties in distinguishing the "scientific" component of past observation and belief from what their predecessors had readily labeled "error" and "superstition." The more carefully they study, say, Aristotelian dynamics, phlogistic chemistry, or caloric thermodynamics, the more certain they feel that those once current views of nature were, as a whole, neither less scientific nor more the product of human idiosyncrasy than those current today. If these out-of-date beliefs are to be called myths, then myths can be produced by the same sorts of methods and held for the same sorts of reasons that now lead to scientific knowledge. If, on the other hand, they are to be called science, then science has included bodies of belief quite incompatible with the ones we hold today. Given these alternatives, the historian must choose the latter. Out-of-date theories are not in principle unscientific because they have been discarded. That choice, however, makes it difficult to see scientific development as a process of accretion. The same historical research that displays the difficulties in isolating individual inventions and discoveries gives ground for profound doubts about the cumulative process through which these individual contributions to science were thought to have been compounded.

The result of all these doubts and difficulties is a historiographic revolution in the study of science, though one that is still in its early stages. Gradually, and often without entirely realizing they are doing so, historians of science have begun to ask new sorts of questions and to trace different, and often less than cumulative, developmental lines for the sciences. Rather than seeking the permanent contributions of an older science to our present vantage, they attempt to display the historical integrity of that science in its own time. They ask, for example, not about the relation of Galileo's views to those of modern science, but rather about the relationship between his views and those of his group, i.e., his teachers, contemporaries, and immediate successors in the sciences. Furthermore, they insist upon studying the opinions of that group and other similar ones from the viewpoint — usually very different from that of modern science — that gives those opinions the maximum internal coherence and the closest possible fit to nature. Seen through the works that result, works perhaps best exemplified in the writings of Alexandre Koyré, science does not seem altogether the same enterprise as the one discussed by writers in the older historiographic tradition. By implication, at least, these historical studies suggest the possibility of a new image of science. This essay aims to delineate that image by making explicit some of the new historiography's implications.

What aspects of science will emerge to prominence in the course of this effort? First, at least in order of presentation, is the insufficiency of methodological directives, by themselves, to dictate a unique substantive conclusion to many sorts of scientific questions. Instructed to examine electrical or chemical phenomena, the man who is ignorant of these fields but who knows what it is to be scientific may legitimately reach any one of a number of incompatible conclusions. Among those legitimate possibilities, the particular conclusions he does arrive at are probably determined by his prior experience in other fields, by the accidents of his investigation, and by his own individual makeup. What beliefs about the stars, for example, does he bring to the study of chemistry or electricity? Which of the many conceivable experiments relevant to the new field does he elect to perform first? And what aspects of the complex phenomenon that then results strike him as particularly relevant to an elucidation of the nature of chemical change or of electrical affinity? For the individual, at least, and sometimes for the scientific community as well, answers to questions like these are often essential determinants of scientific development. We shall note, for example, . . . that the early developmental stages of most sciences have been characterized by continual competition between a number of distinct views of nature, each partially derived from, and all roughly compatible with, the dictates of scientific observation and method. What differentiated these various schools was not one or another failure of method — they were all "scientific" — but what we shall come to call their incommensurable ways of seeing the world and of practicing science in it. Observation and experience can and must drastically restrict the range of admissible scientific belief, else there would be no science. But they cannot alone determine a particular body of such belief. An apparently arbitrary element, compounded of personal and historical accident, is always a formative ingredient of the beliefs espoused by a given scientific community at a given time.

That element of arbitrariness does not, however, indicate that any scientific group could practice its trade without some set of received beliefs. Nor does it make less consequential the particular constellation to which the group, at a given time, is in fact committed. Effective research scarcely begins before a scientific community thinks it has acquired firm answers to questions like the following: What are the fundamental entities of which the universe is composed? How do these interact with each other and with the senses? What questions may legitimately be asked about such entities and what techniques employed in seeking solutions? At least in the mature sciences, answers (or full substitutes for answers) to questions like these are firmly embedded in the educational initiation that prepares and licenses the student for professional practice. Because that education is both rigorous and rigid, these answers come to exert a deep hold on the scientific mind. That they can do so does much to account both for the peculiar efficiency of the normal research activity and for the direction in which it proceeds at any given time. When examining normal science . . . we shall want finally to describe that research as a strenuous and devoted attempt to force nature into the conceptual boxes supplied by professional education. Simultaneously, we shall wonder whether research could proceed without such boxes, whatever the element of arbitrariness in their historic origins and, occasionally, in their subsequent development.

Yet that element of arbitrariness is present, and it too has an important effect on scientific development. . . . Normal science, the activity in which most scientists inevitably spend almost all their time, is predicated on the assumption that the scientific community knows what the world is like. Much of the success of the enterprise derives from the community's willingness to defend that assumption, if necessary at considerable cost. Normal science, for example, often suppresses fundamental novelties because they are necessarily subversive of its basic commitments. Nevertheless, so long as those commitments retain an element of the arbitrary, the very nature of normal research ensures that novelty shall not be suppressed for very long. Sometimes a normal problem, one that ought to be solvable by known rules and procedures, resists the reiterated onslaught of the ablest members of the group within whose competence it falls. On other occasions a piece of equipment designed and constructed for the purpose of normal research fails to perform in the anticipated manner, revealing an anomaly that cannot, despite repeated effort, be aligned with professional expectation. In these and other ways besides, normal science repeatedly goes astray. And when it does — when, that is, the profession can no longer evade anomalies that subvert the existing tradition of scientific practice — then begin the extraordinary investigations that lead the profession at last to a new set of commitments, a new basis for the practice of science. The extraordinary episodes in which that shift of professional commitments occurs are the ones known in this essay as scientific revolutions. They are the tradition-shattering complements to the tradition-bound activity of normal science.

The most obvious examples of scientific revolutions are those famous episodes in scientific development that have often been labeled revolutions before . . . [such as] the major turning points in scientific development associated with the names of Copernicus, Newton, Lavoisier, and Einstein. More clearly than most other episodes in the history of at least the physical sciences, these display what all scientific revolutions are about. Each of them necessitated the community's rejection of one time-honored scientific theory in favor of another incompatible with it. Each produced a consequent shift in the problems available for scientific scrutiny and in the standards by which the profession determined what should count as an admissible problem or as a legitimate problem-solution. And each transformed the scientific imagination in ways that we shall ultimately need to describe as a transformation of the world within which scientific work was done. Such changes, together with the controversies that almost always accompany them, are the defining characteristics of scientific revolutions.

These characteristics emerge with particular clarity from a study of, say, the Newtonian or the chemical revolution. It is, however, a fundamental thesis of this essay that they can also be retrieved from the study of many other episodes that were not so obviously revolutionary. For the far smaller professional group affected by them, Maxwell's equations were as revolutionary as Einstein's, and they were resisted accordingly. The invention of other new theories regularly, and appropriately, evokes the same response from some of the specialists on whose area of special competence they impinge. For these men the new theory implies a change in the rules governing the prior practice of normal science. Inevitably, therefore, it reflects upon much scientific work they have already successfully completed. That is why a new theory, however special its range of application, is seldom or never just an increment to what is already known. Its assimilation requires the reconstruction of prior theory and the re-evaluation of prior fact, an intrinsically revolutionary process that is seldom completed by a single man and never overnight. No wonder historians have had difficulty in dating precisely this extended process that their vocabulary impels them to view as an isolated event.

Nor are new inventions of theory the only scientific events that have revolutionary impact upon the specialists in whose domain they occur. The commitments that govern normal science specify not only what sorts of entities the universe does contain, but also, by implication, those that it does not. It follows, though the point will require extended discussion, that a discovery like that of oxygen or X-rays does not simply add one more item to the population of the scientist's world. Ultimately it has that effect, but not until the professional community has re-evaluated traditional experimental procedures, altered its conception of entities with which it has long been familiar, and, in the process, shifted the network of theory through which it deals with the world. Scientific fact and theory are not categorically separable, except perhaps within a single tradition of normal-scientific practice. That is why the unexpected discovery is not simply factual in its import and why the scientist's world is qualitatively transformed as well as quantitatively enriched by fundamental novelties of either fact or theory. . . .

This extended conception of the nature of scientific revolutions . . . admittedly . . . strains customary usage. Nevertheless, I shall continue to speak even of discoveries as revolutionary, because it is just the possibility of relating their structure to that of, say, the Copernican revolution that makes the extended conception seem to me so important. . . .

Undoubtedly, some readers will already have wondered whether historical study can possibly effect the sort of conceptual transformation aimed at here. An entire arsenal of dichotomies is available to suggest that it cannot properly do so. History, we too often say, is a purely descriptive discipline. The theses suggested above are, however, often interpretive and sometimes normative. Again, many of my generalizations are about the sociology or social psychology of scientists; yet at least a few of my conclusions belong traditionally to logic or epistemology. In the preceeding paragraph I may even seem to have violated the very influential contemporary distinction between "the context of discovery" and "the context of justification." Can anything more than profound confusion be indicated by this admixture of diverse fields and concerns?

Having been weaned intellectually on these distinctions and others like them, I could scarcely be more aware of their import and force. For many years I took them to be about the nature of knowledge, and I still suppose that, appropriately recast, they have something important to tell us. Yet my attempts to apply them, even *grosso modo*, to the actual situations in which knowledge is gained, accepted, and assimilated have made them seem extraordinarily problematic. Rather than being elementary logical or methodological distinctions, which would thus be prior to the analysis of scientific knowledge, they now seem integral parts of a traditional set of substantive answers to the very questions upon which they have been deployed. That circularity does not at all invalidate them. But it does make them parts of a theory and, by doing so, subjects them to the same

793

scrutiny regularly applied to theories in other fields. If they are to have more than pure abstraction as their content, then that content must be discovered by observing them in application to the data they are meant to elucidate. How could history of science fail to be a source of phenomena to which theories about knowledge may legitimately be asked to apply?

# KARL POPPER

■

The Viennese philosopher Karl Popper (b. 1902) recollects that when he first expressed the hypothesis for which he was to become world-famous — that "what we call 'scientific knowledge' [is] hypothetical, and often not true, let alone certainly or probably true" — it met with a rude response. Popper made his remarks, in halting English, to a meeting of the Aristotelian Society at Oxford in 1936. He was in dead earnest, but, he recalls, "the audience took this for a joke, or a paradox, and they laughed and clapped." Today most scientists would agree with Popper that theories are never really confirmed by experiment, but can at best survive from one test to the next, remaining hostage to possible disproof tomorrow.

# The Logic of Scientific Discovery

ONE MAY DISCERN something like a general direction in the evolution of physics — a direction from theories of a lower level of universality to theories of a higher level. This is usually called the "inductive" direction; and it might be thought that the fact that physics advances in this "inductive" direction could be used as an argument in favour of the inductive method.

Yet an advance in the inductive direction does not necessarily consist of a sequence of inductive inferences. Indeed we have shown that it may be explained in quite different terms — in terms of degree of testability and corroborability. For a theory which has been well corroborated can only be superseded by one of a higher level of universality; that is, by a theory which is better testable and which, in addition, *contains* the old, well corroborated theory — or at least a good approximation to it. It may be better, therefore, to describe that trend — the advance towards theories of an ever higher level of universality — as "quasi-inductive."

The quasi-inductive process should be envisaged as follows. Theories of some level of universality are proposed, and deductively tested; after that, theories of a higher level of universality are proposed, and in their turn tested with the help of those of the previous levels of universality, and so on. The methods of testing are invariably based on deductive inferences from the higher to the lower level;[1] on the other hand, the levels of universality are reached, in the order of time, by proceeding from lower to higher levels.

The question may be raised: "Why not invent theories of the highest level of universality straight away? Why wait for this quasi-inductive evolution? Is it not perhaps because there is after all an inductive element contained in it?" I do not think so. Again and again suggestions are put forward — conjectures, or theories — of all possible levels of universality. Those theories which are on too high a level of universality, as it were (that is, too far removed from the level reached by the testable science of the day), give rise, perhaps, to a "metaphysical system." In this case, even if from this system statements should be deducible (or only semi-deducible, as for example in the case of Spinoza's system), which belong to the prevailing scientific system, there will be no *new* testable statement among them; which means that no crucial experiment can be designed to test the system in question.[2] If, on the other hand, a crucial experiment can be designed for it, then the system will contain, as a first approximation, some well corroborated theory, and at the same time also something new — and something that can be tested. Thus the system will not, of course, be "metaphysical." In this case, the system in question may be looked upon as a new advance in the quasi-inductive evolution of science. This explains why a link with the science of the day is as a rule established only by those theories which are proposed in an attempt to meet the current problem situation; that is, the current difficulties, contradictions, and falsifications. In proposing a solution to these difficulties, these theories may point the way to a crucial experiment.

To obtain a picture or model of this quasi-inductive evolution

---

1. The "deductive inferences from the higher to the lower level" are, of course, *explanations* . . . ; thus the hypotheses on the higher level are *explanatory* with respect to those on the lower level.

2. It should be noted that I mean by a crucial experiment one that is designed to refute a theory (if possible) and more especially one which is designed to bring about a decision between two competing theories by refuting (at least) one of them — without, of course, proving the other. . . .

of science, the various ideas and hypotheses might be visualized as particles suspended in a fluid. Testable science is the precipitation of these particles at the bottom of the vessel: they settle down in layers (of universality). The thickness of the deposit grows with the number of these layers, every new layer corresponding to a theory more universal than those beneath it. As the result of this process ideas previously floating in higher metaphysical regions may sometimes be reached by the growth of science, and thus make contact with it, and settle. Examples of such ideas are atomism; the idea of a single physical "principle" or ultimate element (from which the others derive); the theory of terrestrial motion (opposed by Bacon as fictitious); the age-old corpuscular theory of light; the fluid-theory of electricity (revived as the electron-gas hypothesis of metallic conduction). All these metaphysical concepts and ideas may have helped, even in their early forms, to bring order into man's picture of the world, and in some cases they may even have led to successful predictions. Yet an idea of this kind acquires scientific status only when it is presented in falsifiable form; that is to say, only when it has become possible to decide empirically between it and some rival theory.

My investigation has traced the various consequences of the decisions and conventions — in particular of the criterion of demarcation — adopted at the beginning of this book. Looking back, we may now try to get a last comprehensive glimpse of the picture of science and of scientific discovery which has emerged. (What I have here in mind is not a picture of science as a biological phenomenon, as an instrument of adaptation, or as a roundabout method of production: I have in mind its epistemological aspects.)

Science is not a system of certain, or well-established, statements; nor is it a system which steadily advances towards a state of finality. Our science is not knowledge (*epistēmē*): it can never claim to have attained truth, or even a substitute for it, such as probability.

Yet science has more than mere biological survival value. It is not only a useful instrument. Although it can attain neither truth nor probability, the striving for knowledge and the search for truth are still the strongest motives of scientific discovery.

*We do not know: we can only guess.* And our guesses are guided by the unscientific, the metaphysical (though biologically explicable) faith in laws, in regularities which we can uncover — discover. Like Bacon, we might describe our own contemporary science — "the method of reasoning which men now ordinarily apply to nature" —

as consisting of "anticipations, rash and premature," and of "prejudices."[3]

But these marvellously imaginative and bold conjectures or "anticipations" of ours are carefully and soberly controlled by systematic tests. Once put forward, none of our "anticipations" are dogmatically upheld. Our method of research is not to defend them, in order to prove how right we were. On the contrary, we try to overthrow them. Using all the weapons of our logical, mathematical, and technical armoury, we try to prove that our anticipations were false — in order to put forward, in their stead, new unjustified and unjustifiable anticipations, new "rash and premature prejudices," as Bacon derisively called them.[4]

It is possible to interpret the ways of science more prosaically. One might say that progress can ". . . come about only in two ways: by gathering new perceptual experiences, and by better organizing those which are available already."[5] But this description of scientific progress, although not actually wrong, seems to miss the point. It is too reminiscent of Bacon's induction: too suggestive of his industrious gathering of the "countless grapes, ripe and in season,"[6] from which he expected the wine of science to flow: of his myth of a sci-

3. Bacon, *Novum Organum* I, 26.
4. Bacon's "anticipation" (*"anticipatio"*; *Novum Organum* I, 26) means almost the same as "hypothesis" (in my usage). Bacon held that, to prepare the mind for the intuition of the true *essence* or *nature* of a thing, it has to be meticulously cleansed of all anticipations, prejudices, and idols. For the source of all error is the impurity of our own minds; Nature itself does not lie. The main function of eliminative induction is (as with Aristotle) to assist the purification of the mind. (*See also* my *Open Society*, chapter 24; note 59 to chapter 10; note 33 to chapter 11, where Aristotle's theory of induction is briefly described.) Purging the mind of prejudices is conceived as a kind of ritual, prescribed for the scientist who wishes to prepare his mind for the interpretation (the unbiassed reading) of the Book of Nature; just as the mystic purifies his soul to prepare it for the vision of God. (*Cf.* the Introduction to my *Conjectures and Refutations* [1963] 1965.)
5. P. Frank, *Das Kausalgesetz und seine Grenzen*, 1932. The view that the progress of science is due to the accumulation of perceptual experiences is still widely held (*cf.* my second Preface, 1958). My denial of this view is closely connected with the rejection of the doctrine that science or knowledge is *bound* to advance since our experiences are *bound* to accumulate. As against this, I believe that the advance of science depends upon the free competition of thought, and thus upon freedom, and that it must come to an end if freedom is destroyed (though it may well continue for some time in some fields, especially in technology). This view is more fully expounded in my *Poverty of Historicism* (section 32). I also argue there (in the Preface) that the growth of our knowledge is unpredictable by scientific means, and that, as a consequence, the future course of our history is also unpredictable.
6. Bacon, *Novum Organum* I, 123.

entific method that starts from observation and experiment and then proceeds to theories. (This legendary method, by the way, still inspires some of the newer sciences which try to practice it because of the prevalent belief that it is the method of experimental physics.)

The advance of science is not due to the fact that more and more perceptual experiences accumulate in the course of time. Nor is it due to the fact that we are making ever better use of our senses. Out of uninterpreted sense-experiences science cannot be distilled, no matter how industriously we gather and sort them. Bold ideas, unjustified anticipations, and speculative thought are our only means for interpreting nature: our only organon, our only instrument, for grasping her. And we must hazard them to win our prize. Those among us who are unwilling to expose their ideas to the hazard of refutation do not take part in the scientific game.

Even the careful and sober testing of our ideas by experience is in its turn inspired by ideas: experiment is planned action in which every step is guided by theory. We do not stumble upon our experiences, nor do we let them flow over us like a stream. Rather, we have to be active: we have to *"make"* our experiences. It is we who always formulate the questions to be put to nature; it is we who try again and again to put these questions so as to elicit a clear-cut "yes" or "no" (for nature does not give an answer unless pressed for it). And in the end, it is again we who give the answer; it is we ourselves who, after severe scrutiny, decide upon the answer to the question which we put to nature — after protracted and earnest attempts to elicit from her an unequivocal "no." "Once and for all," says Weyl,[7] with whom I fully agree, "I wish to record my unbounded admiration for the work of the experimenter in his struggle to wrest *interpretable facts* from an unyielding Nature who knows so well how to meet our theories with a decisive *No* — or with an inaudible *Yes*."

The old scientific ideal of *epistēmē* — of absolutely certain, demonstrable knowledge — has proved to be an idol. The demand for scientific objectivity makes it inevitable that every scientific statement must remain *tentative for ever*. It may indeed be corroborated, but every corroboration is relative to other statements which, again, are tentative. Only in our subjective experiences of conviction, in our subjective faith, can we be "absolutely certain."

With the idol of certainty (including that of degrees of imper-

---

7. Weyl, *Gruppentheorie und Quantenmechanik*, 1931, p. 2. English translation by H. P. Robertson: *The Theory of Groups and Quantum Mechanics*, 1931, p. xx.

fect certainty or probability) there falls one of the defences of obscurantism which bar the way of scientific advance. For the worship of this idol hampers not only the boldness of our questions, but also the rigour and the integrity of our tests. The wrong view of science betrays itself in the craving to be right; for it is not his *possession* of knowledge, of irrefutable truth, that makes the man of science, but his persistent and recklessly critical *quest* for truth.

Has our attitude, then, to be one of resignation? Have we to say that science can fulfil only its biological task; that it can, at best, merely prove its mettle in practical applications which may corroborate it? Are its intellectual problems insoluble? I do not think so. Science never pursues the illusory aim of making its answers final, or even probable. Its advance is, rather, towards an infinite yet attainable aim: that of ever discovering new, deeper, and more general problems, and of subjecting our ever tentative answers to ever renewed and ever more rigorous tests.

# NIELS BOHR

■

The Danish quantum physicist Niels Bohr (1885–1962) was a protean thinker who took the philosophical implications of his work to heart. Bohr's theory of "complementarity," here discussed in an essay he wrote in 1958, has become the most widely accepted interpretation of quantum theory.

# Causality and Complementarity

THE SIGNIFICANCE OF PHYSICAL science for philosophy does not merely lie in the steady increase of our experience of inanimate matter, but above all in the opportunity of testing the foundation and scope of some of our most elementary concepts. Notwithstanding refinements of terminology due to accumulation of experimental evidence and developments of theoretical conceptions, all account of physical experience is, of course, ultimately based on common language, adapted to orientation in our surroundings and to tracing relationships between cause and effect. Indeed, Galileo's programme — to base the description of physical phenomena on measurable quantities — has afforded a solid foundation for the ordering of an ever larger field of experience.

In Newtonian mechanics, where the state of a system of material bodies is defined by their instantaneous positions and velocities, it proved possible, by the well-known simple principles, to derive, solely from the knowledge of the state of the system at a given time and of the forces acting upon the bodies, the state of the system at any other time. A description of this kind, which evidently represents an ideal form of causal relationships, expressed by the notion of *determinism*, was found to have still wider scope. Thus, in the account of electromagnetic phenomena, in which we have to consider a propagation of forces with finite velocities, a deterministic description could be upheld by including in the definition of the state not only the positions and velocities of the charged bodies, but also the direction and intensity of the electric and magnetic forces at every point of space at a given time.

The situation in such respects was not essentially changed by the recognition, embodied in the notion of *relativity,* of the extent to which the description of physical phenomena depends on the reference frame chosen by the observer. We are here concerned with a most fruitful development which has made it possible to formulate physical laws common to all observers and to link phenomena which hitherto appeared uncorrelated. Although in this formulation use is made of mathematical abstractions such as a four-dimensional non-Euclidean metric, the physical interpretation for each observer rests on the usual separation between space and time, and maintains the deterministic character of the description. Since, moreover, as stressed by Einstein, the space-time coordination of different observers never implies reversal of what may be termed the causal sequence of events, relativity theory has not only widened the scope but also strengthened the foundation of the deterministic account, characteristic of the imposing edifice generally referred to as classical physics.

A new epoch in physical science was inaugurated, however, by Planck's discovery of the *elementary quantum of action,* which revealed a feature of *wholeness* inherent in atomic processes, going far beyond the ancient idea of the limited divisibility of matter. Indeed, it became clear that the pictorial description of classical physical theories represents an idealization valid only for phenomena in the analysis of which all actions involved are sufficiently large to permit the neglect of the quantum. While this condition is amply fulfilled in phenomena on the ordinary scale, we meet in experimental evidence concerning atomic particles with regularities of a novel type, incompatible with deterministic analysis. These quantal laws determine the peculiar stability and reactions of atomic systems, and are thus ultimately responsible for the properties of matter on which our means of observation depend.

The problem with which physicists were confronted was therefore to develop a rational generalization of classical physics, which would permit the harmonious incorporation of the quantum of action. After a preliminary exploration of the experimental evidence by more primitive methods, this difficult task was eventually accomplished by the introduction of appropriate mathematical abstractions. Thus, in the quantal formalism, the quantities by which the state of a physical system is ordinarily defined are replaced by symbolic operators subjected to a non-commutative algorism involving Planck's constant. This procedure prevents a fixation of such quantities to the extent which would be required for the deterministic description of classical physics, but allows us to determine their spectral distribution as revealed by evidence about atomic processes. In conformity with the non-pictorial character of the formalism, its physical interpretation finds expression in laws, of an essentially statistical type, pertaining to observations obtained under given experimental conditions.

Notwithstanding the power of quantum mechanics as a means of ordering an immense amount of evidence regarding atomic phenomena, its departure from accustomed demands of causal explanation has naturally given rise to the question whether we are here concerned with an exhaustive description of experience. The answer to this question evidently calls for a closer examination of the conditions for the unambiguous use of the concepts of classical physics in the analysis of atomic phenomena. The decisive point is to recognize that the description of the experimental arrangement and the recording of observations must be given in plain language, suitably refined by the usual physical terminology. This is a simple logical demand, since by the word "experiment" we can only mean a procedure regarding which we are able to communicate to others what we have done and what we have learnt.

In actual experimental arrangements, the fulfilment of such requirements is secured by the use, as measuring instruments, of rigid bodies sufficiently heavy to allow a completely classical account of their relative positions and velocities. In this connection, it is also essential to remember that all unambiguous information concerning atomic objects is derived from the permanent marks — such as a spot on a photographic plate, caused by the impact of an electron — left on the bodies which define the experimental conditions. Far from involving any special intricacy, the irreversible amplification effects on which the recording of the presence of atomic objects rests rather remind us of the essential irreversibility inherent in the very concept of observation. The description of atomic phenomena has in these respects a perfectly objective character, in the sense that no explicit reference is made to any individual observer and that therefore, with proper regard to relativistic exigencies, no ambiguity is involved in the communication of information.

As regards all such points, the observation problem of quantum physics in no way differs from the classical physical approach. The essentially new feature in the analysis of quantum phenomena is, however, the introduction of a *fundamental distinction between the measuring apparatus and the objects under investigation.* This is a direct consequence of the necessity of accounting for the functions of the measuring instruments in purely classical terms, excluding in principle any regard to the quantum of action. On their side, the quantal features of the phenomenon are revealed in the information about the atomic objects derived from the observations. While, within the scope of classical physics, the interaction between object and apparatus can be neglected or, if necessary, compensated for, in quantum physics this interaction thus forms an inseparable part of the phenomenon. Accordingly, the unambiguous account of proper quantum phenomena must, in principle, include a description of all relevant features of the experimental arrangement.

The very fact that repetition of the same experiment, defined on the lines described, in general yields different recordings pertaining to the object, immediately implies that a comprehensive account of experience in this field must be expressed by statistical laws. It need hardly be stressed that we are not concerned here with an analogy to the familiar recourse to statistics in the description of physical systems of too complicated a structure to make practicable the complete definition of their state necessary for a deterministic account. In the case of quantum phenomena, the unlimited divisibility of events implied in such an account is, in principle, excluded by the requirement to specify the experimental conditions. Indeed, the feature of wholeness typical of proper quantum phenomena finds its logical expression in the circumstance that any attempt at a well-defined subdivision would demand a change in the experimental arrangement incompatible with the definition of the phenomena under investigation.

Within the scope of classical physics, all characteristic properties of a given object can in principle be ascertained by a single experimental arrangement, although in practice various arrangements are often convenient for the study of different aspects of the phenomena. In fact, data obtained in such a way simply supplement each other and can be combined into a consistent picture of the behaviour of the object under investigation. In quantum physics, however, evidence about atomic objects obtained by different experimental arrangements exhibits a novel kind of complementary relationship. Indeed, it must be recognized that such evidence which appears contradictory when combination into a single picture is attempted, exhausts all conceivable knowledge about the object. Far from restricting our efforts to put questions to nature in the form of experiments, the notion of *complementarity* simply characterizes the answers we can receive by such inquiry, whenever the interaction between the measuring instruments and the objects forms an integral part of the phenomena.

Although, of course, the classical description of the experimental arrangement and the irreversibility of the recordings concerning the atomic objects ensure a sequence of cause and effect conforming with elementary demands of causality, the irrevocable abandonment of the ideal of determinism finds striking expression in the complementary relationship governing the unambiguous use of the fundamental concepts on whose unrestricted combination the classical physical description rests. Indeed, the ascertaining of the presence of an atomic particle in a limited space-time domain demands an experimental arrangement involving a transfer of momentum and energy to bodies such as fixed scales and synchronized clocks, which cannot be included in the description of their functioning, if these bodies are to fulfil the role of defining the reference frame. Conversely, any strict application of the laws of conservation of momentum and energy to atomic processes implies, in principle, a renunciation of detailed space-time coordination of the particles.

These circumstances find quantitative expression in Heisenberg's indeterminacy relations which specify the reciprocal latitude for the fixation, in quantum mechanics, of kinematical and dynamical variables required for the definition of the state of a system in classical mechanics. In fact, the limited commutability of the symbols by which such variables are represented in the quantal formalism corresponds to the mutual exclusion of the experimental arrangements required for their unambiguous definition. In this context, we are of course not concerned with a restriction as to the accuracy of measurements, but with a limitation of the well-defined application of space-time concepts and dynamical conservation laws, entailed by the necessary distinction between measuring instruments and atomic objects.

In the treatment of atomic problems, actual calculations are most conveniently carried out with the help of a Schrödinger state function, from which the statistical laws governing observations obtainable under specified conditions can be deduced by definite mathematical operations. It must be recognized, however, that we are here dealing with a purely symbolic procedure, the unambiguous physical interpretation of which in the last resort requires a reference to a complete experimental arrangement. Disregard of this point has sometimes led to confusion, and in particular the use of phrases like "disturbance of phenomena by observation" or "creation of physical attributes of objects by measurements" is hardly compatible with common language and practical definition.

In this connection, the question has even been raised whether recourse to multivalued logics is needed for a more appropriate representation of the situation. From the preceding argumentation it will appear, however, that all departures from common language and ordinary logic are entirely avoided by reserving the word "phenomenon" solely for reference to unambiguously communicable information, in the account of which the word "measurement" is used in its plain meaning of standardized comparison. Such caution in the choice of terminology is especially important in the exploration of a new field of experience, where information cannot be comprehended in the familiar frame which in classical physics found such unrestricted applicability.

It is against this background that quantum mechanics may be seen to fulfil all demands on rational explanation with respect to consistency and completeness. Thus, the emphasis on permanent recordings under well-defined experimental conditions as the basis for a consistent interpretation of the quantal formalism corresponds to the presupposition, implicit in the classical physical account, that every step of the causal sequence of events in principle allows of verification. Moreover, a completeness of description like that aimed at in classical physics is provided by the possibility of taking every conceivable experimental arrangement into account.

Such argumentation does of course not imply that, in atomic physics, we have no more to learn as regards experimental evidence and the mathematical tools appropriate to its comprehension. In fact, it seems likely that the introduction of still further abstractions into the formalism will be required to account for the novel features revealed by the exploration of atomic processes of very high energy. The decisive point, however, is that in this connection there is no question of reverting to a mode of description which fulfils to a higher degree the accustomed demands regarding pictorial representation of the relationship between cause and effect.

The very fact that quantum regularities exclude analysis on classical lines necessitates, as we have seen, in the account of experience a logical distinction between measuring instruments and atomic objects, which in principle prevents comprehensive deterministic description. Summarizing, it may be stressed that, far from involving any arbitrary renunciation of the ideal of causality, the wider frame of complementarity directly expresses our position as regards the account of fundamental properties of matter presupposed in classical physical description, but outside its scope.

Notwithstanding all difference in the typical situations to which the notions of relativity and complementarity apply, they present in epistemological respects far-reaching similarities. Indeed, in both cases we are concerned with the exploration of harmonies which cannot be comprehended in the pictorial conceptions adapted to the account of more limited fields of physical experience. Still, the decisive point is that in neither case does the appropriate widening of our conceptual framework imply any appeal to the observing subject, which would hinder unambiguous communication of experience. In relativistic argumentation, such objectivity is secured by due regard to the dependence of the phenomena on the reference frame of the observer, while in complementary description all subjectivity is avoided by proper attention to the circumstances required for the well-defined use of elementary physical concepts.

In general philosophical perspective, it is significant that, as regards analysis and synthesis in other fields of knowledge, we are confronted with situations reminding us of the situation in quantum physics. Thus, the integrity of living organisms and the characteristics of conscious individuals and human cultures present features of wholeness, the account of which implies a typical complementary mode of description. Owing to the diversified use of the rich vocabulary available for communication of experience in those wider fields, and above all to the varying interpretations, in philosophical literature, of the concept of causality, the aim of such comparisons has sometimes been misunderstood. However, the gradual development of an appropriate terminology for the description of the simpler situation in physical science indicates that we are not dealing with more or less vague analogies, but with clear examples of logical relations which, in different contexts, are met with in wider fields.

# ALBERT EINSTEIN

■

Albert Einstein could not bring himself to accept the view pro-
mulgated by Niels Bohr (and by most other physicists) that
quantum uncertainty imposes a permanent limitation on our
knowledge of events. "God does not play dice," he said. With
no one did he argue this point longer, or in better spirits, than
with his good friend the German physicist Max Born (1882–
1970). Their letters, from which these excerpts are taken, remain
a model of intellectual debate at its loftiest. The two contested
the matter for years; neither convinced the other, nor did either
ever deviate in the slightest from the attitude of mutual affection
and respect with which they had begun.

# From *Letters to Max Born*

[*4 December 1926*]

Quantum mechanics is certainly imposing. But an inner voice tells
me that it is not yet the real thing. The theory says a lot, but does
not really bring us any closer to the secret of the "Old One." I, at
any rate, am convinced that *He* is not playing at dice.

[*29 April 1924*]

I should not want to be forced into abandoning strict causality
without defending it more strongly than I have so far. I find the idea
quite intolerable that an electron exposed to radiation should choose
*of its own free will,* not only its moment to jump off, but also its direction. In that case, I would rather be a cobbler, or even an employee
in a gaming-house, than a physicist. Certainly my attempts to give
tangible form to the quanta have foundered again and again, but I
am far from giving up hope. And even if it never works there is
always that consolation that this lack of success is entirely mine.

[*1 September 1919*]

The causal way of looking at things . . . always answers only the question "Why?", but never the question "To what end?" No utility principle and no natural selection will make us get over that. However, if someone asks "To what purpose should we help one another, make life easier for each other, make beautiful music or have inspired thoughts?", he would have to be told: "If you don't feel it, no one can explain it to you." Without this primary feeling we are nothing and had better not live at all.

[*9 September 1920*]

Don't be too hard on me. Everyone has to sacrifice at the altar of stupidity from time to time, to please the Deity and the human race.

[*7 September 1944*]

We have become Antipodean in our scientific expectations. You believe in the God who plays dice, and I in complete law and order in a world which objectively exists, and which I, in a wildly speculative way, am trying to capture. . . . Even the great initial success of the quantum theory does not make me believe in the fundamental dice-game, although I am well aware that our younger colleagues interpret this as a consequence of senility. No doubt the day will come when we will see whose instinctive attitude was the correct one.

[*18 March 1948*]

I really understand very well why you consider me an impenitent old sinner. But I feel sure that you do not understand how I came by my lonely ways; it would certainly amuse you, even if there is not the slightest chance of your approving of my attitude. I would enjoy picking your positivistic philosophical attitude to pieces myself. But this is hardly likely to happen during our lifetime.

[*Undated*]

Our respective hobby-horses have irretrievably run off in different directions — yours, however, enjoys far greater popularity as a result of its remarkable practical success, while mine, on the other hand, smacks of quixotism, and even I myself cannot adhere to it with absolute confidence. But at least mine does not represent a blind-man's buff with the idea of reality.

# JACOB BRONOWSKI

■

The origins of science are intertwined with the pursuit of occult subjects the contemplation of which would make many a modern scientist shudder with distaste. When John Maynard Keynes bought a trunk full of Isaac Newton's papers and inspected them, he was startled to find that Newton spent as much time studying alchemy and numerology as he did formulating his laws of motion. Newton, he declared, "was the last of the magicians." But not all magic is the same, as the scientist-philosopher Jacob Bronowski (1908–1974) explains.

# Black Magic and White Magic

... THE FORM OF MAGIC that I shall discuss is the notion that there is a way of having power over nature which simply depends on hitting the right key. If you say "open sesame" then nature will open for you; if you are an expert then nature will open for you; if you are a specialist of some kind or if you are remote, if you are esoteric, if you are an initiate there is some way of getting into nature which is not accessible to other people.

Now this was the dominant theme of all those centuries up to the fifteenth. And all primitive forms of magic — sympathetic magic, the kind of magic you read about in Lévi-Strauss for instance, magic that structuralists talk about — all come back to this notion: there is a way of having power which is esoteric and does not depend on generally accessible knowledge. Now I think that is fundamentally false and I also think, of course, that it is terribly dangerous, because it recurs in every generation. But let me say something about it in this highly specific context of magic up to the fifteenth century.

One of the things that must have struck you if you have read any book about magic is that there is a tendency for the rituals of magic to turn nature upside-down. For instance, if you have ever seen an illustration of a witch riding a broomstick, she does not ride the broomstick sitting forward, she rides the broomstick sitting backward. Now it may seem a childish thing for an eminent intellectual historian to be discussing which way witches sit on broomsticks. But the fact is that intellectual history is made of exactly such points. Why did people think that satanic rituals had to be set backwards? Why did people celebrate the black mass by going through the mass in reverse? Because the concept of that conquest of nature was that whatever the laws of nature were, the magic consisted of turning them back. What Joshua said was, "Sun stand thou still"; he didn't say anything about ellipses and inverse square laws. He said "let us stop the laws of nature and turn them back in their tracks." And really, one could say, if I may put this terribly crudely, that until the year 1500 any attempt to get power from nature had inherent in it the idea that you could only do this if you forced nature to provide it against her will. Nature had to be subjugated, and magic was a form of words, actions, and pictures which forced nature to do something which she wouldn't of herself do.

Let me note here that science does exactly the opposite. But it is important to realize that the subjugation of nature is the theme of all magical practice. We must get her to do something for us which she wouldn't do for everybody else — which means we must make her disobey her own laws. Of course, people before 1500 didn't really have much of an idea of what a law of nature was. But insofar as they conceived of nature following a natural course, magic was something which reversed it.

What I am saying is my view, although it is not the view of all those who have written about magic. Lynn Thorndike, for example, was an eminent writer on the subject. His eight-volume work on the subject is entitled *The History of Magic and Experimental Science* (New York: Columbia University Press, 1923–58). It would be impertinent of me not to state that he thought differently. However, the very title *The History of Magic and Experimental Science* implies a view of science which is different from mine. What Lynn Thorndike said was that there are in magical and particularly alchemical practices many techniques which later formed an important part of technology and experimental science. Now that's undoubtedly true. But alas, in my view, this has nothing to do with the case. Of course there were people of all kinds practicing all kinds of alchemy right up to the days of Newton, whose alchemical writings are so voluminous that they were never published. Nevertheless, my main interest is in their attitude toward how the world works and how you make it obey, and not at all in their discoveries of how you smelt this or how you make that process in metallurgy work. It was the view of Lynn Thorndike, and it has been the view of some other eminent historians of science, that there is a continuity running from even before the Middle Ages into modern science. This is what Pierre Duhem was anxious to show, and in a way this is what George Sarton said also. And of course, there is some truth in that. There is not the slightest doubt that any particular piece of science that you have today can be traced to some fantasy in the Middle Ages. . . .

But it is my view that those continuities give a false perspective of the great threshold from which the burst of modern science comes. And I would put this quite simply: I don't know whether science was born before 1500 or not (though I don't believe it was) but I do know that, mysteriously, magic in fact died after 1500.

I ought also just to pay a small obeisance to those historians who think that we ought not to look at the history of the Middle Ages or the Renaissance as if in some way it were the forerunner of today. I wrote a book of intellectual history and was amused to find that one of my kinder critics said that it was all very fine, but why did I think that the present age was any better than the fifteenth century? Well, I don't know whether it is better, but it seems to me terribly interesting that the fifteenth century has led to the present age and that the present age has not led to the fifteenth century.

My view of history is essentially an evolutionary one. I think it is right that we should look at history with hindsight, because I think, for two reasons, that the most important species-specific thing which man possesses and which started him off on his evolutionary career is exactly hindsight. If you make any plans, only hindsight will tell you whether they were any good. Secondly, we know from work on memory that it is only from hindsight, only from memory, that imagination and foresight develops. So I make no apologies for the fact that I shall discuss the history of the past as if the most exciting thing about it is that it has led us to the present. . . .

A point which must be made very forcibly about science is that it took an irreversible step in the cultural evolution of man. I have noted that we did not lay enough stress on the fact not only that science has made our lives different but also that it was a threshold of this kind. Holding that view I am bound to say that the dates of the scientific revolution between 1500 and 1700 do represent a major threshold in the development of science.

Now that seems strange to people who have tried to trace the history of science back beyond that time, because they point out that, after all, there was a school of people — who read Aristotle, who were Averroists, who continued to talk about scientific truth and distinguish it from spiritual truth — in a number of universities, such as Paris and Padua, through the thirteenth, fourteenth, and fifteenth centuries.

The school that looks for a continuity in the development of science looks for it there. Now I think that view is mistaken because I think that in about 1500 something very remarkable happened in all intellectual history, of which science is a part and a crucial part. It's not terribly fashionable to talk about the Renaissance now, because everyone is very busy explaining how it all really started much earlier. And it's not very fashionable even to talk about humanism, because very eminent scholars, including Professor Kristeller of Columbia University, have pointed out that humanism was a special kind of academic syllabus that led to the elaboration of rhetoric and theories of language at the expense of theology and other practices. In itself, humanism did not make a new way of life, and of course it appears to have had no influence on science. All that seems to me to be quite right. And yet it is absolutely true that Florence in 1500 was a different city (I cite Kristeller again) from Florence in 1400. Something had happened in Italy which made a great inroad in established, authoritarian, and traditional views of life.

When we come to revalue the Renaissance over the next twenty or thirty years of scholarship, the view that we are sure to come up with is that the most important thing was not that people in Florence started reading Plato instead of Aristotle, or that people from Padua argued about this or that, or that Ficino wrote this and Pomponazzi wrote that, but that in some way a dissolution of tradition took place and there developed an interest in new things in which the particular character of the new was not nearly so important as the shaking up of the old. And that characteristic was crucial to the development of science at that particular time. To my mind, the most extraordinary thing is that about 1500 the incursion of neo-Platonic and mystical ideas gave that impulse to the human mind, made that intellectual revaluation from which science and the arts took off together. The view that I am putting forward is that this revolution worked as much in the sciences as in the arts and that it is impossible to understand the really radical change that the Renaissance made unless we see

813

science not as an afterthought but as an integral part of that humanism — rhetoric and linguistics and all.

Now to some small, interesting, specific examples: Between 1450 and 1465 Cosimo de'Medici began to collect a library of Greek manuscripts. They were being brought to the West by scholars and he sent his own merchants out to buy them up. They brought back the *Dialogues* of Plato, which had still not been translated from Greek, and they brought back also an incomplete manuscript of the *Corpus Hermeticum*, the fabulous book of magic of the Middle Ages of which, again, only a small part had been translated into Latin. His secretary was a man called Marsilio Ficino, and in 1463 he was translating the dialogues of Plato when Cosimo de'Medici told him to translate the *Corpus Hermeticum* first. In fact, Cosimo died the next year and he obviously felt that this esoteric knowledge, this magic, he had to know. Now the *Corpus Hermeticum* is an extraordinary book which remains in our language simply because we still use the phrase "hermetically sealed" to mean sealed by secret alchemical formulae. Although the book is called the *Corpus Hermeticum* because it was supposed to be a book about Hermes Trismegistus (Hermes is the Three Times Great who was supposed to have been a mythological Greek-Hebrew character), Ficino thought he might have been Moses himself. I need not tell you that the books are fakes, but that was not discovered until 150 years later — and anyway, as I have already explained, fakery was no crime at that time.

Ficino was an extraordinary character because he came at a moment when the old black magic, the witches' Sabbath, and so on, was not done anymore. Ficino never got up and said "I won't go to a coven of witches." He was too polite, too much of a gentleman, to do all this romping about in the nude in damp fields with satanic imagery and goats, and he was from a new kind of upper class society that was taking an interest in magic. They were sophisticated and gentlemanly, and this was not the kind of black magic in which they could be interested. And yet, Ficino really did sing hymns; he really did believe that he was conjuring down the influences of the planets and that in some way the world was opening up, that Orpheus and Pythagoras and all those planetary influences were one.

This is really the central point of neo-Platonism as Ficino introduced it. The world is a great harmony, and harmony is the crucial word. It literally meant both music and mathematics and incidentally also poetry. All these things were different aspects of the universal spirit, the *anima mundi,* the thing that Plato and Plotinus said was like a great organic creature of the world. Indeed Giorgio said that all this was simply a description, that the universe was the face of God and that all its aspects — music, poetry, and mathematics — were different expressions of the fact that it was a harmonious whole. Music and mathematics go together because Pythagoras and the Greeks, three thousand years ago, had discovered that in order to make an octave you have to make a musical string twice as long and in order to make the other main notes you have to have whole-number separations. And this extraordinary notion that the length of the vibrating string also gives pleasure to the ear and fills your soul with harmony had come down from the Greeks. For instance, Pythagoras invented the phrase "the music of the spheres." Shortly after the time I am describing, Kepler tried to fit the five Platonic solids into the orbits of the solar system because he naturally felt that all these things must go together — mathematics, music, harmony were one. Harmony is indeed the word to hold on to.

The exciting thing about these neo-Platonists is, first of all, that they made people interested in mathematics. It was from that moment onward that Greek mathematics was rediscovered, became exciting to people again; they started arguing about Playfair's axiom and all kinds of things that were unclear in Euclid. That led to natural knowledge through mathematics of which Newton really set the keystone. Secondly, people like Ficino had this marvelous sense that the world was at once both intelligible and beautiful. The phrase (which does not come from Ficino) about its being the face of God is the crux here. We have the sense that suddenly the Middle Ages were over. That rather heavy view of God sitting on the world with man quietly padding around making sure that he doesn't give offense had ended. There was suddenly a rainbow in the sky, the world was beautiful. We have the transcending sense of the beauty of nature, but above all of the beauty of the creation.

We see this outlook in Copernicus when in the next century, in 1543, he published the book that he had been working on for nearly thirty years about the revolution of the planets. He talked about the sun, about how marvelous it is. Of course, the textbooks just tell us that in effect he said, "well, it works simpler if we put the sun in the center of the universe." But that's not what he said. He said the sun was fit to be the center of the universe. The sun was marvelous. And he took that straight out of Ficino, who in fact wrote a book called *Of the Sun.* It has only recently been discovered that when Giordano Bruno came to Oxford in the next century and lectured about the

Copernican system, the Oxford Dons treated him with grave suspicion, and particularly because they all spotted the quotations from Ficino that he thought they wouldn't know.

This sense that man and the universe are one, that the presence of God in the universe is a different kind of presence, is what makes the neo-Platonic revolution crucial in the science of the Renaissance. I have called it antiauthoritarian, but I ought really to have said anti-traditional. Now I do not mean by this that people would suddenly go around saying God is dead, which would have been inconceivable then. What was happening was something quite different: there had been a hierarchy of God, man, nature. And that in that hierarchy God and man had moved into one position. Man was still dominating nature, but there was no longer the sense that he was under any higher authority. Everything that God had expressed was expressed in man.

We see this best in a follower of Ficino, Pico della Mirandola, who in 1487 proposed to dispute a famous series of theses, which have since come down to us under the title *Of the Dignity of Man*. Now, the dispute really was to a considerable extent about the dignity of man, provided that you understood this equivalence of man and God. Pico della Mirandola was saying above all that man was a unique animal because he was the only animal that made himself, that had no species-specific properties. Well, that is something of an exaggeration, but you know it is not quite as bad animal behavior as one would think, because it is certainly true that the most important part of the human equipment is its enormously greater flexibility and adaptability than that of any other animal. In biology we generally express this by saying that whereas every other animal fits into an evolutionary niche, man essentially is busy hewing his evolutionary niche out of nature for himself. . . .

Pico della Mirandola was also very much against astrology. And he says so in the oration, which is full of all kinds of stuff out of the Cabala — all kinds of things that nobody reads anymore except in corners. And yet, he says in 1487, astrology is wrong. Why? — because it outrages the dignity of man that we should be subject to the influence of the planets out there traveling on their immutable, dreary predestined courses. That cannot be consonant with the dignity of man. It is a gorgeous thought; naturally we would put it rather differently now. We would say it's not consonant with the dignity of a *planet*.

Both Ficino and Pico as well as a number of other people about

1500 were practitioners of magic, and yet their magic had a different quality. They were no longer trying to force nature into a different mode. In some way, they were trying to exploit a preordained harmony in nature. Ficino says this quite firmly: "When I sing a song to the sun it is not because I expect the sun to change its course but [because] I expect to put myself into a different cast of mind in relation to the sun." Now this is a very important concept that developed between 1500 and 1550 — the notion that yes, there is a magic, but it is a *natural* magic, a *white* magic. No one knows quite how it works, but it attempts to extract out of the universe its own harmonies for our good. And here we are on the way to science as we understand it. If one had to put a date to this, one would say that roughly speaking between 1500 and the publication of Porta's book in 1558, which was called *Natural Magic,* the turning point took place. Of course, I am not trying to make the sort of point that on the 27th of February it all changed. History doesn't work like that. But what did happen was that highly intelligent people were bothered about demons and angels and all the oppositions in the old magic; they were convinced that the universe was harmonious, that man could be in contact with it, and they asked themselves how this could be done. There is a fascinating series of writings to be explored still between 1500 and 1550 (complicated of course by the occurrence of the new humanism) — those of people like Erasmus, Luther (Protestantism) which keep on saying — "well, how could magic work?"

Now we come to several very interesting trends of thought. There are some people who say, "well, you see it's all psychological" and you get this. They say, "the faithful or the superstitious are in a special frame of mind where they really see these apparitions and feel these influences." All right — now nobody is much bothered about this, because psychology was the one thing they really understood. And they understood about the art of memory and about human imagination and about the power of imagination, but by 1500 they were asking themselves a very crucial question — "can this be transferred to the outside world? We are all convinced that a man can hold the audience spellbound, but will this spell work on the chairs? Will this spell work on dead nature?" And this question — "is white magic transitive — can it be transferred to dead objects?" This is what was now engaging everybody's attention. There was wonderful excitement about how different people treat this, but they all came back to the same thing. Now they took different lines.

Some of them said, "well, yes, but in very special circumstances"; or "well, yes, but it isn't exactly that you can make nature obey your will, but if you choose to do it at a moment when nature is ready, then you can just slightly distort her," and so on. Some of them, of course, began to say about this time that it's just not true. There just isn't any magic at all. For several hundred years everybody had been saying that it is well known that menstruating women must not look in mirrors because they tarnish the mirror. And then, around 1550, people said, "Have I been looking in mirrors recently? I haven't noticed any mirrors tarnishing." And of course, at that moment, all those delicious old wives' tales about the influence of man on the environment began to disappear. To summarize, I quote from Pomponazzi, who, in a book called *Of Incantations,* says quite firmly,

> It is possible to justify any experience by natural causes and natural causes only. There is no reason that could ever compel us to make any perception depend on demonic powers. There is no point in introducing supernatural agents. It is ridiculous as well as frivolous to abandon the evidence of natural reason and to search for things that are neither probable nor rational.

Well, of course, that's a very wild voice by this time. And Pomponazzi was an Aristotelian from Padua to whom these things came, as it were, from the outside; but he did mark a great turning point in this period, the time when black magic was at an end; everyone had gone through the white magic period. In black magic, the belief was that you would make nature run against her will. In white magic, you began to say, "Well you know, let's make nature work with us. There is a harmony; we could exploit it." Finally came the concept of natural law itself. And that was represented, in a most spectacular way, for the first time in the writing of Francis Bacon between 1600 and 1620. It was Francis Bacon, whom I was quoting, who was the first person to say "knowledge is power." It was Francis Bacon who said in the *Novum Organum* "we cannot command nature except by obeying her." At this point, the scientific revolution was really complete. This is an important issue because there has been a good deal of argument about who Francis Bacon was — whether he wrote Shakespeare, for example. It is particularly important to determine how he fits into all this. And it's really only since the publication of Paolo Rossi's book, *Francis Bacon from Magic to Science* (University of Chicago Press, 1968), that it has really become clear to us that he went through all this; he understood all this Italian stuff. And then at the end, he came out with this simple notion: it wouldn't work. That's a very English thing to do. One could have an entirely separate chapter on that Puritan frame of mind which made him say, "all this stuff about the face of God and the harmony of the spheres and the number, mythology, and the love of God — how does it really work?" At any rate, it's clear that he was outraged by many of the fancies of the sixteenth-century writers on memory and magic, and that he came to this crucial conclusion, "we cannot command nature except by obeying her." There are laws of nature, and what you really do is not to turn them back but to exploit them.

If I might give you one spectacular example, who would have thought in 1569 — when they were already well on the way to this concept — that if you really wanted to make the biggest bang that you ever made on earth, you would not in any way call up the sun, call up the volcanos, call up the mystic power; you would just take ordinary atoms of uranium and you would put the $U_{238}$ atoms in one box and the $U_{235}$ atoms in another box and that this simple rearrangement of nature by her own laws would blow up 120,000 people in Japan.

I've made a passing reference to a shift to England at this time and it would not be fair if I didn't draw your attention to the importance of Protestantism and Puritanism. There is a very curious history about magic I believe to be true. It has always been a puzzle why, certainly from 1640 onward and probably even before, the Protestant countries began to take the lead in science. Obviously the trial of Galileo in 1633 had a tremendous influence, but there must be something in the background of the period 1500–1600 which began to shift the center of gravity. Now I believe that this has a great deal to do with magic, for a very curious reason. By 1600 it was open to anybody to say, "you can't persuade people of one thing when another is true simply by using words." But unfortunately, Thomas Aquinas had committed himself to the statement that the words which are used in the elevation of the host have absolute power to change the bread into the body of Christ and the wine into the blood of Christ. And the statement that Thomas Aquinas made back in the 1250s was so absolute that it was really impossible to get around it. The words, "this is my body," the words "this is my blood," would, if uttered, make a difference even if they were made by a priest in bad faith, by a priest in unworthy circumstances, or by a priest who was not thinking about the subject. And if he did not utter those words, then no transubstantiation would take place. That was a very big issue throughout the sixteenth century because you could not get round the authority of Aquinas on this, and yet here was something which in some way had to be explained away, and it was very adequately explained away. One could write about Duns Scotus's view, the Thomist view, how all this was dealt with in the sixteenth century. But the fact of the matter was that it created an attitude, in my view, about the nature of science and the existence of magical powers which was different in Roman Catholic countries and in the new Protestant countries. And we know this, because the Protestant writers were busy attacking what they called the superstition of the church. And of course, in Puritan England this was especially true.

I have made this long historical excursion because I wanted to demonstrate what I think Ficino did when he suddenly opened up the world and made the rainbow full of color and said nature and man are in harmony. I said . . . that I couldn't think of any way of being a human being other than by being an intellectual. To me, being an intellectual doesn't mean knowing about intellectual issues; it means taking pleasure in them. And that to my mind is exactly what happened — exactly what transformed the attitude to science about the year 1500. The sudden sense of an opening universe — you get it in Copernicus, you get it in Galileo. If you read Galileo's *Dialogues* and all those corny jokes and all that leg pulling, here is a man who is in love with his subject and who is no longer practicing Faustian demonic magic and swearing to the devil. He is out in the open; he just thinks it's marvelous. I think, of course, that science is wonderful in that way. And I end it with Francis Bacon for that very reason — that the Elizabethan Age, to us an age of literature, was exactly that age when all this science and literature together came to fruition in England.

I quoted the *Novum Organum* of 1620. It was as late as 1620 that "knowledge is power" was written for the first time. Twenty-five years later, on Christmas day in 1645, Isaac Newton was born; in another forty years, Isaac Newton published the *Principia,* and quite suddenly the world was transformed into something which is both rational and beautiful in just the way that the neo-Platonists believed with all their Averroist and their Aristotelian tradition. . . . At one moment in history, science and the arts rose together, because of the simple sense of man's pleasure in his own gifts.

# WERNER HEISENBERG

■

For the founders of quantum physics, philosophical questions about causation and probability were pressing, practical concerns and not merely the stuff of academic treatises. Werner Heisenberg here recounts how, years after the dust seemed to have settled, Niels Bohr was still probing such issues with Socratic exuberance.

# *Positivism, Metaphysics, and Religion*

THE RESUMPTION OF international contacts once again brought together old friends. Thus, in the early summer of 1952, atomic physicists assembled in Copenhagen to discuss the construction of a European accelerator. I was most interested in this project because I was hoping that a large accelerator would help us to determine whether or not the high-energy collision of two elementary particles could lead to the production of a host of further particles, as I had assumed; whether, indeed, we were entitled to assume the existence of many new particles and, if so, whether, like the stationary states of atoms or molecules, they differed only in their symmetries, masses, and lifetimes. The main topic of the meeting was thus a matter of great personal concern, and if I do not report it here, it is simply because I must relate a conversation with Niels [Bohr] and Wolfgang [Pauli] on that occasion. Wolfgang had come over from Zurich, and the three of us were sitting in the small conservatory that ran from Bohr's official residence down to the park. We were discussing the old theme, namely, whether our interpretation of quantum theory in this very spot, twenty-five years ago, had been correct, and whether or not our ideas had since become part of the intellectual stock-in-trade of all physicists. Niels had this to say:

"Some time ago there was a meeting of philosophers, most of them positivists, here in Copenhagen, during which members of the Vienna Circle played a prominent part. I was asked to address them on the interpretation of quantum theory. After my lecture, no one raised any objections or asked any embarrassing questions, but I must say this very fact proved a terrible disappointment to me. For those who are not shocked when they first come across quantum theory cannot possibly have understood it. Probably I spoke so badly that no one knew what I was talking about."

Wolfgang objected: "The fault need not necessarily have been yours. It is part and parcel of the positivist creed that facts must be taken for granted, sight unseen, so to speak. As far as I remember, Wittgenstein says: 'The world is everything that is the case.' 'The world is the totality of facts, not of things.' Now if you start from that premise, you are bound to welcome any theory representative of the 'case.' The positivists have gathered that quantum mechanics describes atomic phenomena correctly, and so they have no cause for complaint. What else we have had to add — complementarity, interference of probabilities, uncertainty relations, separation of subject and object, etc. — strikes them as just so many embellishments, mere relapses into prescientific thought, bits of idle chatter that do not have to be taken seriously. Perhaps this attitude is logically defensible, but, if it is, I for one can no longer tell what we mean when we say we have understood nature."

Niels [commented]: "For my part, I can readily agree with the positivists about the things they want, but not about the things they reject. All the positivists are trying to do is to provide the procedures of modern science with a philosophical basis, or, if you like, a justification. They point out that the notions of the earlier philosophies lack the precision of scientific concepts, and they think that many of the questions posed and discussed by conventional philosophers have no meaning at all, that they are pseudo problems and, as such, best ignored. Positivist insistence on conceptual clarity is, of course, something I fully endorse, but their prohibition of any discussion of the wider issues, simply because we lack clear-cut enough concepts in this realm, does not seem very useful to me — this same ban would prevent our understanding of quantum theory."

"Positivists," I tried to point out, "are extraordinarily prickly about all problems having what they call a prescientific character. I remember a book by Philipp Frank on causality, in which he dismisses a whole series of problems and formulations on the ground that all of them are relics of the old metaphysics, vestiges from the period of prescientific or animistic thought. For instance, he rejects the biological concepts of 'wholeness' and 'entelechy' as prescientific ideas and tries to prove that all statements in which these concepts are commonly used have no verifiable meaning. To him 'metaphysics' is a synonym for 'loose thinking,' and hence a term of abuse."

"This sort of restriction of language doesn't seem very useful to me either," Niels said. "You all know Schiller's poem, 'The Sentences of Confucius,' which contains these memorable lines: 'The full mind is alone the clear, and truth dwells in the deeps.' The full mind, in our case, is not only an abundance of experience but also an abundance of concepts by means of which we can speak about our problems and about phenomena in general. Only by using a whole variety of concepts when discussing the strange relationship between the formal laws of quantum theory and the observed phenomena, by lighting this relationship up from all sides and bringing out its apparent contradictions, can we hope to effect that change in our thought processes which is a *sine qua non* of any true understanding of quantum theory.

"You mentioned Philipp Frank's book on causality. Philipp Frank was one of the philosophers to attend the congress in Copenhagen, and he gave a lecture in which he used the term 'metaphysics' simply as a swearword or, at best, as a euphemism for unscientific thought. After he had finished, I had to explain my own position, and this I did roughly as follows:

"I began by pointing out that I could see no reason why the prefix 'meta' should be reserved for logic and mathematics — Frank had spoken of metalogic and metamathematics — and why it was anathema in physics. The prefix, after all, merely suggests that we are asking further questions, i.e., questions bearing on the fundamental concepts of a particular discipline, and why ever should we not be able to ask such questions in physics? But I should start from the opposite end. Take the question 'What is an expert?' Many people will tell you that an expert is someone who knows a great deal about his subject. To this I would object that no one can ever know very much about any subject. I would much prefer the following definition: an expert is someone who knows some of the worst mistakes that can be made in his subject, and how to avoid them. Hence Philipp Frank ought to be called an expert on metaphysics, one who knows how to avoid some of its worst mistakes — I was not quite sure whether Frank was very happy about my praise, though I was certainly not offering it tongue-in-cheek. In all such discussions what matters most to me is that we do not simply talk the 'deeps in which the truth dwells' out of existence. That would mean taking a very superficial view."

That same evening Wolfgang and I continued the discussion alone. It was the season of the long nights. The air was balmy, twilight lasted until almost midnight, and as the sun traveled just beneath the horizon, it bathed the city in a subdued, bluish light. And so we decided to walk along the Langelinie, a beautiful harbor promenade, with freighters discharging their cargo on either side. In the south, the Langelinie begins roughly where Hans Christian Andersen's Little Mermaid rests on a rock beside the beach; in the north, it is continued by a jetty that swings out into the basin and marks the entrance to Frihavn with a small beacon. After we had been looking at the toing and froing of the ships in the twilight for quite a while, Wolfgang asked me:

"Were you quite satisfied with Niels' remarks about the positivists? I gained the impression that you are even more critical of them than Niels himself, or rather that your criterion of truth differs radically from theirs."

"I should consider it utterly absurd — and Niels, for one, would agree — were I to close my mind to the problems and ideas of earlier philosophers simply because they cannot be expressed in a more precise language. True, I often have great difficulty in grasping what these ideas are meant to convey, but when that happens, I always try to translate them into modern terminology and to discover whether they throw up fresh answers. But I have no principled objections to the re-examination of old questions, much as I have no objections to using the language of any of the old religions. We know that religions speak in images and parables and that these can never fully correspond to the meanings they are trying to express. But I believe that, in the final analysis, all the old religions try to express the same contents, the same relations, and all of these hinge around questions about values. The positivists may be right in thinking that it is difficult nowadays to assign a meaning to such parables. Nevertheless, we ought to make every effort to grasp their meaning, since it quite obviously refers to a crucial aspect of reality; or perhaps we ought to try putting it into modern language, if it can no longer be contained in the old."

"If you think about such problems in that way, then, quite

obviously, you cannot accept the equation of truth and predictive power. But what is your own criterion of truth in science?"

"We may find it more helpful to revert to our old comparison between Ptolemy's astronomy and Newton's conception of planetary motions. If predictive power were indeed the only criterion of truth, Ptolemy's astronomy would be no worse than Newton's. But if we compare Newton and Ptolemy in retrospect, we gain the clear impression that Newton's equations express the paths of the planets much more fully and correctly than Ptolemy's did, that Newton, so to speak, described the plan of nature's own construction. Or, to take an example from modern physics: when we learn that the principles of conservation of energy, charge, etc., have a quite universal character, that they apply in all branches of physics and that they result from the symmetry inherent in the fundamental laws, then we are tempted to say that symmetry is a decisive element in the plan on which nature has been created. In saying this I am fully aware that the words 'plan' and 'created' are once again taken from the realm of human experience and that they are metaphors at best. But it is quite easy to see that everyday language must necessarily fall short here. I suppose that is all I can say about my own conception of scientific truth."

"Quite so, but positivists will object that you are making obscure and meaningless noises, whereas they themselves are models of analytic clarity. But where must we seek for the truth, in obscurity or in clarity? Niels has quoted Schiller's 'Truth dwells in the deeps.' Are there such deeps and is there any truth? And may these deeps perhaps hold the meaning of life and death?"

A few hundred yards away, a large liner was gliding past, and its bright lights looked quite fabulous and unreal in the bright blue dusk. For a few moments, I speculated about the human destinies being played out behind the lit-up cabin windows, and suddenly Wolfgang's questions got mixed up with it all. What precisely was this steamer? Was it a mass of iron with a central power station and electric lights? Was it the expression of human intentions, a form resulting from interhuman relations? Or was it a consequence of biological laws, exerting their formative powers not merely on protein molecules but also on steel and electric currents? Did the word "intention" reflect the existence merely of these formative powers or of these biological laws in the human consciousness? And what did the word "merely" mean in this context?

My silent soliloquy now turned to more general questions. Was it utterly absurd to seek behind the ordering structures of this world a "consciousness" whose "intentions" were these very structures? Of course, even to put this question was an anthropomorphic lapse, since the word "consciousness" was, after all, based purely on human experience, and ought therefore to be restricted to the human realm. But in that case we would also be wrong to speak of animal consciousness, when we have a strong feeling that we can do so significantly. We sense that the meaning of "consciousness" becomes wider and at the same time vaguer if we try to apply it outside the human realm.

The positivists have a simple solution: the world must be divided into that which we can say clearly and the rest, which we had better pass over in silence. But can anyone conceive of a more pointless philosophy, seeing that what we can say clearly amounts to next to nothing? If we omitted all that is unclear, we would probably be left with completely uninteresting and trivial tautologies.

We walked on in silence and had soon reached the northern tip of the Langelinie, whence we continued along the jetty as far as the small beacon. In the north, we could still see a bright strip of red; in these latitudes the sun does not travel far beneath the horizon. The outlines of the harbor installations stood out sharply, and after we had been standing at the end of the jetty for a while, Wolfgang asked me quite unexpectedly:

"Do you believe in a personal God? I know, of course, how difficult it is to attach a clear meaning to this question, but you can probably appreciate its general purport."

"May I rephrase your question?" I asked. "I myself should prefer the following formulation: Can you, or anyone else, reach the central order of things or events, whose existence seems beyond doubt, as directly as you can reach the soul of another human being? I am using the term 'soul' quite deliberately so as not to be misunderstood. If you put your question like that, I would say yes. And because my own experiences do not matter so much, I might go on to remind you of Pascal's famous text, the one he kept sewn in his jacket. It was headed 'Fire' and began with the words: 'God of Abraham, Isaac and Jacob — not of the philosophers and sages.' "

"In other words, you think that you can become aware of the central order with the same intensity as of the soul of another person?"

"Perhaps."

"Why did you use the word 'soul' and not simply speak of another person?"

"Precisely because the word 'soul' refers to the central order, to the inner core of a being whose outer manifestations may be highly diverse and pass our understanding.

"If the magnetic force that has guided this particular compass — and what else was its source but the central order? — should ever become extinguished, terrible things may happen to mankind, far more terrible even than concentration camps and atom bombs. But we did not set out to look into such dark recesses; let's hope the central realm will light our way again, perhaps in quite unsuspected ways. As far as science is concerned, however, Niels is certainly right to underwrite the demands of pragmatists and positivists for meticulous attention to detail and for semantic clarity. It is only in respect to its taboos that we can object to positivism, for if we may no longer speak or even think about the wider connections, we are without a compass and hence in danger of losing our way."

Despite the late hour, a small boat made fast on the jetty and took us back to Kongens Nytorv, whence it was easy to reach Niels' house.

# ALBERT EINSTEIN

■

Albert Einstein's religious beliefs were as subtle as his scientific ones. Consequently he found it difficult to express them in terms that lesser mortals could readily understand. Part of the problem is that the same words mean different things to different listeners; when someone asked Einstein if he believed in God, the questioner usually was asking whether Einstein believed in *his* — the questioner's — God, about whom he had seldom reflected as carefully as had Einstein. In this essay, written in 1939, Einstein reminds his readers that science does not necessarily supply answers to ethical and theological questions. His modesty on this point serves as an enduring example to true believers less forthcoming about the limits of their own ideas.

# Science and Religion

## I

DURING THE LAST CENTURY, and part of the one before, it was widely held that there was an unreconcilable conflict between knowledge and belief. The opinion prevailed among advanced minds that it was time that belief should be replaced increasingly by knowledge; belief that did not itself rest on knowledge was superstition, and as such had to be opposed. According to this conception, the sole function of education was to open the way to thinking and knowing, and the school, as the outstanding organ for the people's education, must serve that end exclusively.

One will probably find but rarely, if at all, the rationalistic standpoint expressed in such crass form; for any sensible man would see at once how one-sided is such a statement of the position. But it is just as well to state a thesis starkly and nakedly, if one wants to clear up one's mind as to its nature.

It is true that convictions can best be supported with experience and clear thinking. On this point one must agree unreservedly with the extreme rationalist. The weak point of his conception is, however, this, that those convictions which are necessary and determinant for our conduct and judgments, cannot be found solely along this solid scientific way.

For the scientific method can teach us nothing else beyond how facts are related to, and conditioned by, each other. The aspiration toward such objective knowledge belongs to the highest of which man is capable, and you will certainly not suspect me of wishing to belittle the achievements and the heroic efforts of man in this sphere. Yet it is equally clear that knowledge of what *is* does not open the door directly to what *should be*. One can have the clearest and most complete knowledge of what *is*, and yet not be able to deduct from that what should be the *goal* of our human aspirations. Objective knowledge provides us with powerful instruments for the achievements of certain ends, but the ultimate goal itself and the longing to reach it must come from another source. And it is hardly necessary to argue for the view that our existence and our activity acquire meaning only by the setting up of such a goal and of corresponding values. The knowledge of truth as such is wonderful, but it is so little capable of acting as a guide that it cannot prove even the justification and the value of the aspiration towards that very knowledge of truth. Here we face, therefore, the limits of the purely rational conception of our existence.

But it must not be assumed that intelligent thinking can play no part in the formation of the goal and of ethical judgments. When someone realizes that for the achievement of an end certain means would be useful, the means itself becomes thereby an end. Intelligence makes clear to us the interrelation of means and ends. But mere thinking cannot give us a sense of the ultimate and fundamental ends. To make clear these fundamental ends and valuations, and to set them fast in the emotional life of the individual, seems to me precisely the most important function which religion has to perform in the social life of man. And if one asks whence derives the authority of such fundamental ends, since they cannot be stated and justified merely by reason, one can only answer: they exist in a healthy society as powerful traditions, which act upon the conduct and aspirations and judgments of the individuals; they are there, that is, as something living, without its being necessary to find justification for their existence. They come into being not through demonstration but through revelation, through the medium of powerful personalities. One must not attempt to justify them, but rather to sense their nature simply and clearly.

The highest principles for our aspirations and judgments are given to us in the Jewish-Christian religious tradition. It is a very high goal which, with our weak powers, we can reach only very inadequately, but which gives a sure foundation to our aspirations and valuations. If one were to take that goal out of its religious form and look merely at its purely human side, one might state it perhaps thus: free and responsible development of the individual, so that he may place his powers freely and gladly in the service of all mankind.

There is no room in this for the divinization of a nation, of a class, let alone of an individual. Are we not all children of one father, as it is said in religious language? Indeed, even the divinization of humanity, as an abstract totality, would not be in the spirit of that ideal. It is only to the individual that a soul is given. And the high destiny of the individual is to serve rather than to rule, or to impose himself in any other way.

If one looks at the substance rather than at the form, then one can take these words as expressing also the fundamental democratic position. The true democrat can worship his nation as little as can the man who is religious, in our sense of the term.

What, then, in all this, is the function of education and of the school? They should help the young person to grow up in such a spirit that these fundamental principles should be to him as the air which he breathes. Teaching alone cannot do that.

If one holds these high principles clearly before one's eyes, and compares them with the life and spirit of our times, then it appears glaringly that civilized mankind finds itself at present in grave danger. In the totalitarian states it is the rulers themselves who strive actually to destroy that spirit of humanity. In less threatened parts it is nationalism and intolerance, as well as the oppression of the individuals by economic means, which threaten to choke these most precious traditions.

A realization of how great is the danger is spreading, however, among thinking people, and there is much search for means with which to meet the danger — means in the field of national and international politics, of legislation, of organization in general. Such efforts are, no doubt, greatly needed. Yet the ancients knew something which we seem to have forgotten. All means prove but a blunt instrument, if they have not behind them a living spirit. But if the longing for the achievement of the goal is powerfully alive within us, then shall we not lack the strength to find the means for reaching the goal and for translating it into deeds.

## II

It would not be difficult to come to an agreement as to what we understand by science. Science is the century-old endeavor to bring together by means of systematic thought the perceptible phenomena of this world into as thoroughgoing an association as possible. To put it boldly, it is the attempt at the posterior reconstruction of existence by the process of conceptualization. But when asking myself what religion is I cannot think of the answer so easily. And even after finding an answer which may satisfy me at this particular moment I still remain convinced that I can never under any circumstances bring together, even to a slight extent, all those who have given this question serious consideration.

At first, then, instead of asking what religion is I should prefer to ask what characterizes the aspirations of a person who gives me the impression of being religious: A person who is religiously enlightened appears to me to be one who has, to the best of his ability, liberated himself from the fetters of his selfish desires and is preoccupied with thoughts, feelings, and aspirations to which he clings because of their super-personal value. It seems to me that what is important is the force of this super-personal content and the depth of the conviction concerning its overpowering meaningfulness, regardless of whether any attempt is made to unite this content with a divine Being, for otherwise it would not be possible to count Buddha and Spinoza as religious personalities. Accordingly, a religious person is devout in the sense that he has no doubt of the significance and loftiness of those super-personal objects and goals which neither require nor are capable of rational foundation. They exist with the same necessity and matter-of-factness as he himself. In this sense religion is the age-old endeavor of mankind to become clearly and completely conscious of these values and goals and constantly to strengthen and extend their effect. If one conceives of religion and science according to these definitions then a conflict between them appears impossible. For science can only ascertain what *is*, but not what *should be*, and outside of its domain value judgments of all kinds remain necessary. Religion, on the other hand, deals only with evaluations of human thought and action: it cannot justifiably speak of facts and relationships between facts. According

831

to this interpretation the well-known conflicts between religion and science in the past must all be ascribed to a misapprehension of the situation which has been described.

For example, a conflict arises when a religious community insists on the absolute truthfulness of all statements recorded in the Bible. This means an intervention on the part of religion into the sphere of science; this is where the struggle of the Church against the doctrines of Galileo and Darwin belongs. On the other hand, representatives of science have often made an attempt to arrive at fundamental judgments with respect to values and ends on the basis of scientific method, and in this way have set themselves in opposition to religion. These conflicts have all sprung from fatal errors.

Now, even though the realms of religion and science in themselves are clearly marked off from each other, nevertheless there exist between the two strong reciprocal relationships and dependencies. Though religion may be that which determines the goal, it has, nevertheless, learned from science, in the broadest sense, what means will contribute to the attainment of the goals it has set up. But science can only be created by those who are thoroughly imbued with the aspiration towards truth and understanding. This source of feeling, however, springs from the sphere of religion. To this there also belongs the faith in the possibility that the regulations valid for the world of existence are rational, that is, comprehensible to reason. I cannot conceive of a genuine scientist without that profound faith. The situation may be expressed by an image: Science without religion is lame, religion without science is blind.

Though I have asserted above that in truth a legitimate conflict between religion and science cannot exist I must nevertheless qualify this assertion once again on an essential point, with reference to the actual content of historical religions. This qualification has to do with the concept of God. During the youthful period of mankind's spiritual evolution human fantasy created gods in man's own image, who, by the operations of their will were supposed to determine, or at any rate to influence the phenomenal world. Man sought to alter the disposition of these gods in his own favor by means of magic and prayer. The idea of God in the religions taught at present is a sublimation of that old conception of the gods. Its anthropomorphic character is shown, for instance, by the fact that men appeal to the Divine Being in prayers and plead for the fulfilment of their wishes.

Nobody, certainly, will deny that the idea of the existence of an omnipotent, just and omnibeneficent personal God is able to accord man solace, help, and guidance; also, by virtue of its simplicity it is accessible to the most undeveloped mind. But, on the other hand, there are decisive weaknesses attached to this idea in itself, which have been painfully felt since the beginning of history. That is, if this being is omnipotent then every occurrence, including every human action, every human thought, and every human feeling and aspiration is also His work; how is it possible to think of holding men responsible for their deeds and thoughts before such an almighty Being? In giving out punishment and rewards He would to a certain extent be passing judgment on Himself. How can this be combined with the goodness and righteousness ascribed to Him?

The main source of the present-day conflicts between the spheres of religion and of science lies in this concept of a personal God. It is the aim of science to establish general rules which determine the reciprocal connection of objects and events in time and space. For these rules, or laws of nature, absolutely general validity is required — not proven. It is mainly a program, and faith in the possibility of its accomplishment in principle is only founded on partial successes. But hardly anyone could be found who would deny these partial successes and ascribe them to human self-deception. The fact that on the basis of such laws we are able to predict the temporal behavior of phenomena in certain domains with great precision and certainty is deeply embedded in the consciousness of the modern man, even though he may have grasped very little of the contents of those laws. He need only consider that planetary courses within the solar system may be calculated in advance with great exactitude on the basis of a limited number of simple laws. In a similar way, though not with the same precision, it is possible to calculate in advance the mode of operation of an electric motor, a transmission system, or of a wireless apparatus, even when dealing with a novel development.

To be sure, when the number of factors coming into play in a phenomenological complex is too large scientific method in most cases fails us. One need only think of the weather, in which case prediction even for a few days ahead is impossible. Nevertheless no one doubts that we are confronted with a causal connection whose causal components are in the main known to us. Occurrences in this domain are beyond the reach of exact prediction because of the variety of factors in operation, not because of any lack of order in nature.

We have penetrated far less deeply into the regularities obtaining within the realm of living things, but deeply enough nevertheless to sense at least the rule of fixed necessity. One need only think of the systematic order in heredity, and in the effect of poisons, as for instance alcohol, on the behavior of organic beings. What is still lacking here is a grasp of connections of profound generality, but not a knowledge of order in itself.

The more a man is imbued with the ordered regularity of all events the firmer becomes his conviction that there is no room left by the side of this ordered regularity for causes of a different nature. For him neither the rule of human nor the rule of divine will exists as an independent cause of natural events. To be sure, the doctrine of a personal God interfering with natural events could never be *refuted,* in the real sense, by science, for this doctrine can always take refuge in those domains in which scientific knowledge has not yet been able to set foot.

But I am persuaded that such behavior on the part of the representatives of religion would not only be unworthy but also fatal. For a doctrine which is able to maintain itself not in clear light but only in the dark, will of necessity lose its effect on mankind, with incalculable harm to human progress. In their struggle for the ethical good, teachers of religion must have the stature to give up the doctrine of a personal God, that is, give up that source of fear and hope which in the past placed such vast power in the hands of priests. In their labors they will have to avail themselves of those forces which are capable of cultivating the Good, the True, and the Beautiful in humanity itself. This is, to be sure, a more difficult but an incomparably more worthy task.[1] After religious teachers accomplish the refining process indicated they will surely recognize with joy that true religion has been ennobled and made more profound by scientific knowledge.

If it is one of the goals of religion to liberate mankind as far as possible from the bondage of egocentric cravings, desires, and fears, scientific reasoning can aid religion in yet another sense. Although it is true that it is the goal of science to discover rules which permit the association and foretelling of facts, this is not its only aim. It also seeks to reduce the connections discovered to the smallest possible number of mutually independent conceptual elements. It is in this striving after the rational unification of the manifold that it encounters its greatest successes, even though it is precisely this attempt

1. This thought is convincingly presented in Herbert Samuel's book *Belief and Action.*

which causes it to run the greatest risk of falling a prey to illusions. But whoever has undergone the intense experience of successful advances made in this domain, is moved by profound reverence for the rationality made manifest in existence. By way of the understanding he achieves a far-reaching emancipation from the shackles of personal hopes and desires, and thereby attains that humble attitude of mind towards the grandeur of reason incarnate in existence, and which, in its profoundest depths, is inaccessible to man. This attitude, however, appears to me to be religious, in the highest sense of the word. And so it seems to me that science not only purifies the religious impulse of the dross of its anthropomorphism but also contributes to a religious spiritualization of our understanding of life.

The further the spiritual evolution of mankind advances, the more certain it seems to me that the path to genuine religiosity does not lie through the fear of life, and the fear of death, and blind faith, but through striving after rational knowledge. In this sense I believe that the priest must become a teacher if he wishes to do justice to his lofty educational mission.

# ACKNOWLEDGMENTS

All possible care has been taken to trace the ownership of every work and to make full acknowledgment for its use. If any errors or omissions have accidentally occurred, they will be corrected in subsequent editions provided notification is sent to the publisher.

*One: The Realm of the Atom*

"Atoms in Motion" from *The Feynman Lectures on Physics*, vol. I, by R. Feynman, R. Leighton, and M. Sands. Copyright © 1970 by the California Institute of Technology. Reprinted with permission of Addison-Wesley Publishing Company, Inc.

"The Large and the Small" from *The World of Elementary Particles* by Kenneth W. Ford. Copyright © 1958 by The University Press. Reprinted with permission of Cambridge University Press.

"The Gay Tribe of Electrons" from *Mr. Tompkins Explores the Atom* by George Gamow. Copyright © 1958 by The University Press. Reprinted with the permission of Cambridge University Press.

"Radioactive Substances" by Pierre Curie, Nobel Prize in Physics Address, 1903. Copyright 1903.

"$E = mc^2$" from *Out of My Later Years* by Albert Einstein. Copyright © 1956, © 1984 by the Estate of Albert Einstein. Reprinted with permission of Carol Publishing Group.

"The Man Who Listened" from *The Second Creation* by Robert P. Crease and Charles C. Mann. Copyright © 1986 by Robert P. Crease and Charles C. Mann. Reprinted with permission of Macmillan Publishing Company.

*Acknowledgments*

"Theory of Electrons and Positrons" by Paul A. M. Dirac, Nobel Prize in Physics Address, 1933. Copyright LES PRIX NOBEL 1933. Reprinted with permission.

"The Copenhagen Interpretation of Quantum Theory" from *Physics and Philosophy: Volume Nineteen of World Perspectives* by Werner Heisenberg. Planned and edited by Ruth Nanda Anshen. Copyright © 1958 by Werner Heisenberg. Reprinted with permission of Harper & Row, Publishers, Inc.

"Uncertainty and Complementarity" and "Schrödinger's Cat," excerpts from *The Cosmic Code* by Heinz R. Pagels. Copyright © 1982. Reprinted with permission of Simon & Schuster, Inc.

"Unified Theories of Physics" by Timothy Ferris. Copyright © 1988 by Timothy Ferris. Originally published as "Grand Unification Theories: Faith in Ultimate Simplicity" in *Next: The Coming Era in Science*, edited by Holcomb B. Noble, Boston: Little, Brown and Company, 1988. Reprinted with permission of Timothy Ferris.

"Butterflies and Superstrings" from *Infinite in All Directions* by Freeman J. Dyson. Copyright © 1988 by Freeman J. Dyson. Reprinted with permission of Harper & Row, Publishers, Inc.

"The Distinction of Past and Future" from *The Character of Physical Law* by Richard P. Feynman. Copyright © 1965 by MIT Press. Reprinted with permission of MIT Press.

"The Second Law of Thermodynamics" from *Treatise on Thermodynamics* by Max Planck, translated by Alexander Ogg. Copyright 1930 by Cambridge University Press.

"The Age of the Elements" from *New Frontiers in Astronomy* by David N. Schramm. Copyright © 1974 by Scientific American, Inc. All rights reserved. Reprinted with permission.

"The Two Masses" from *The Subatomic Monster* by Isaac Asimov. Copyright © 1985 by Nightfall, Inc. Reprinted by permission of Doubleday, a division of Bantam Doubleday Dell Publishing Group, Inc.

"Einstein's Law of Gravitation" from *The ABC of Relativity* by Bertrand Russell. Copyright © 1958 by Signet.

of the Universe" by Allan Sandage from *The Universe*, edited by Byron Preiss. Copyright © 1987 by Byron Preiss. Reprinted with permission of Bantam Books, a division of Bantam Doubleday Dell Publishing Group, Inc.

"The Exploration of Space" from *The Realm of the Nebulae* by Edwin Hubble. Copyright 1936. Reprinted with permission of Yale University Press.

"Spherical Space" from *The Expanding Universe* by Arthur Stanley Eddington. Copyright 1933 by Macmillan Publishing Company. Reprinted in *The Realm of Science* edited by Stanley B. Brown. Reprinted with permission of Cambridge University Press.

"The Primeval Atom" by Georges Lemaître, translated by Betty H. Korff and Serge A. Korff. Copyright 1950. Reprinted with permission.

"The New Physics and the Universe" by James Trefil from *The Universe*, edited by Byron Preiss. Copyright © 1987 by Byron Preiss. Reprinted with permission of Bantam Books, a division of Bantam Doubleday Dell Publishing Group, Inc.

"Let There Be Light: Modern Cosmogony and Biblical Creation" by Owen Gingerich from *Is God a Creationist?*, edited by Roland Mushat Frye. Copyright © 1983 by Charles Scribner's Sons. Reprinted with permission of Charles Scribner's Sons, an imprint of Macmillan Publishing Company.

"The First Three Minutes" from *The First Three Minutes: A Modern View of the Origin of the Universe* by Steven Weinberg. Copyright © 1977 by Steven Weinberg. Reprinted with permission of Basic Books, Inc., Publishers, New York, and Andre Deutsch Ltd., London.

"The Magic Furnace" from *The Creation of Matter: The Universe from Beginning to End* by Harald Fritzsch, translated by Jean Steinberg. English translation copyright © 1984 by Basic Books, Inc. Originally published in German as *Vom Urknall zum Zerfall: Die Welt Zwischen Aufgang und Ende*. Copyright © 1984 by R. Piper Verlag, Munich. Reprinted with permission of Basic Books, Inc., Publishers, New York.

Acknowledgments

"The Origin of the Universe" from *A Modern Day Yankee in a Connecticut Court* by Alan Lightman. Copyright © 1986.

"How Will the World End?" from *The Left Hand of Creation: The Origin and Evolution of the Expanding Universe* by John D. Barrow and Joseph Silk. Copyright © 1983 by Basic Books, Inc. Reprinted with permission of Basic Books, Inc., Publishers, New York, and William Heinemann Ltd., London.

## Three: The Cosmos of Numbers

"A Mathematician's Apology" from *A Mathematician's Apology* by Godfrey Harold Hardy. Copyright © 1967. Reprinted with permission of Cambridge University Press.

"Mathematics and Creativity" by Alfred Adler. Copyright © 1972 by The New Yorker Magazine, Inc. Reprinted with permission.

"How Long Is the Coast of Britain?" from *The Fractal Geometry of Nature* by Benoit B. Mandelbrot. Copyright © 1983 by Benoit B. Mandelbrot.

"Chaos" from *Chaos* by James Gleick. Copyright © 1987 by Viking Penguin.

"Computers" from *The Lives of a Cell* by Lewis Thomas. Copyright © 1974 by Viking Penguin.

"The Computer and the Brain" from *The Computer and the Brain* by John von Neumann. Copyright © 1958. Reprinted with permission of Yale University Press.

"Can a Machine Think?" by Alan Turing from *Mind*, vol. 59. Copyright 1950. Reprinted with permission of Oxford University Press.

"The Loss of Certainty" from *Mathematics: The Loss of Certainty* by Morris Kline. Copyright © 1980 by Morris Kline. Reprinted with permission of Oxford University Press.

"The Unreasonable Effectiveness of Mathematics in the Natural Sciences" from *Communications in Pure and Applied Mathematics* by

Eugene P. Wigner. Copyright © 1960 by Eugene P. Wigner. Reprinted with permission of John Wiley & Sons, Inc.

"What Is Mathematics?" from *The World Within the World* by John D. Barrow. Copyright © 1988. Reprinted with permission of Oxford University Press.

"The Limits of Mathematics" from *The Mathematical Experience* by Philip J. Davis and Reuben Hersh. Copyright © 1981 by Birkhäuser Boston, Inc. Reprinted with permission.

## Four: The Ways of Science

"Albert Einstein" from *Albert Einstein: A Biographical Memoir*, vol. 52, by John Archibald Wheeler. Copyright © 1980. Reprinted with permission from National Academy Press.

"Autobiographical Notes" from *Albert Einstein: Autobiographical Notes* by Albert Einstein, translated by Paul A. Schilpp. Reprinted with permission of the publisher, Open Court Publishing Company. This book is itself reprinted from The Library of Living Philosophers, vol. VIII, *Albert Einstein: Philosopher-Scientist*, edited by Paul A. Schilpp, Open Court, 1970.

"A Tribute to Max Planck" by Albert Einstein. Excerpts from *Where Is Science Going?* by Max Planck. Reprinted with permission of Unwin Hyman Ltd.

"Rutherford" from *Variety of Men* by C. P. Snow. Copyright © 1967 by C. P. Snow. Reprinted with permission of Curtis Brown Ltd., London.

"Dirac" by Jagdish Mehra from *The Physicist's Conception of Nature*, edited by Jagdish Mehra. Copyright © 1973 by Jagdish Mehra. Reprinted with permission of Kluwer Academic Publishers.

"Hilbert" by Paul A. M. Dirac from *The Physicist's Conception of Nature*, edited by Jagdish Mehra. Copyright © 1973 by Jagdish Mehra. Reprinted with permission of Kluwer Academic Publishers.

"Quest for Order" by John Tierney from *A Passion to Know*, edited

843

"The Two Cultures" from *The New Statesman and Nation* by C. P. Snow. Copyright © 1956. Reprinted with permission of Cambridge University Press.

"Women in Science" from *Women in Science: Portraits from a World in Transition* by Vivian Gornick. Copyright © 1983, © 1990 by Vivian Gornick. Reprinted with permission of Simon and Schuster, Inc.

"Iron" from *The Periodic Table* by Primo Levi, published by Michael Joseph Ltd. English translation copyright © 1984 by Schocken Books, Inc. Italian text copyright © 1975 by Giulio Einaudi Editore s.p.a. Torino.

"When I Heard the Learn'd Astronomer" from *Leaves of Grass* by Walt Whitman. Copyright 1891 McKay.

" 'I Am Like a Slip of Comet . . .' " from *The Poems of Gerard Manley Hopkins* by Gerard Manley Hopkins. Fourth edition, copyright © 1967, Oxford University Press.

"Arcturus" by Emily Dickinson from *The Poems of Emily Dickinson*, edited by Thomas H. Johnson. Copyright © 1951, © 1955, © 1979, © 1983 by the President and Fellows of Harvard College. Reprinted with permission of the publishers and the Trustees of Amherst College. Cambridge, Mass.: The Belknap Press of Harvard University Press.

"Star-Swirls" from *The Beginning and the End and Other Poems* by Robinson Jeffers. Copyright © 1963 by Garth Jeffers and Donnan Jeffers. Reprinted with permission of Random House, Inc.

"Galaxy" from *Ravenswood* by Richard Ryan. Copyright © 1973 by Dolmen Press.

"Molecular Evolution" from *Poems of Science* by James Clerk Maxwell, edited by John Heath-Stubbs and Phillips Salman. Copyright © 1984.

"Atomic Architecture" from *Stars and Atoms Have No Size* by A. M. Sullivan. Copyright 1946 by Penguin, U.S.A.

"Seeing Things" from *The Collected Poems of Howard Nemerov* by

Howard Nemerov. Copyright © 1977 by The University of Chicago Press. Reprinted with permission of the author.

"Cosmic Gall" from *Telephone Poles and Other Poems* by John Updike. Copyright © 1960 by John Updike. Originally appeared in *The New Yorker*. Reprinted with permission of Alfred A. Knopf, Inc., and Andre Deutsch Ltd., London.

"Nomad" from *Anchor Dragging Poems* by Anthony Piccione. Copyright © 1977 by Anthony Piccione. Reprinted with permission of BOA Editions Ltd.

"Little Cosmic Dust Poem" by John Haines from *Songs from Unsung Worlds* by Bonnie Bilyeu Gordon. Copyright © 1985 by Birkhäuser Boston, Inc. Reprinted with permission of Birkhäuser Boston, Inc.

"Connoisseur of Chaos" from *The Collected Poems of Wallace Stevens*. Copyright 1942 by Wallace Stevens and renewed © 1970 by Holly Stevens. Reprinted with permission of Alfred A. Knopf, Inc., and Faber & Faber Ltd. of London.

" 'The Kings of the World . . .' " from *News of the Universe* by Rainer Maria Rilke, translated by Robert Bly. Copyright © 1980.

"The Windy Planet" by Annie Dillard. Reprinted with permission of Annie Dillard.

"Space Shuttle" from *Lady Faustus* by Diane Ackerman. Copyright © 1983 by Diane Ackerman. Reprinted with permission of William Morrow & Co.

"Epistemology" from *Ceremony and Other Poems* by Richard Wilbur. Copyright © 1978.

"What Science Says to Truth" by William Watson in *Songs of Science: An Anthology* edited by Virginia Shortridge. Marshall Jones copyright 1930 Golden Quill Press.

"Essay on Man" from *The Poetical Works of Alexander Pope* by Alexander Pope. Copyright 1897 Macmillan Publishing Company.

## Acknowledgments

"True Enough: To the Physicist" by Johann Wolfgang von Goethe from *Songs from Unsung Worlds* by Bonnie Bilyeu Gordon, translated by Michael Hamburger. Copyright © 1985 by Birkhäuser Boston, Inc. Reprinted with permission of Birkhäuser Boston, Inc.

"About Planck Time" from *Terms to Be Met* by George Bradley. Copyright © 1986 by George Bradley. Reprinted with permission of Yale University Press.

"The Observatory Ode" from *The Kiss: A Jambalaya* by John Frederick Nims. Copyright © 1982 by John Frederick Nims. Reprinted with permission of Houghton Mifflin Company.

"The Nature of Science" from *Adding a Dimension* by Isaac Asimov. Copyright © 1964 by Isaac Asimov. Reprinted with permission of Doubleday, a division of Bantam Doubleday Dell Publishing Group, Inc.

"The Art of Discovery" from *The Search for Solutions* by Horace Freeland Judson. Copyright © 1980 by Playback Associates. Reprinted with permission of Henry Holt and Company, Inc.

"The Structure of Scientific Revolutions" by Thomas S. Kuhn. Copyright © 1970. Reprinted with permission of University of Chicago Press.

"The Logic of Scientific Discovery" from *The Logic of Scientific Discovery* by Karl Popper. Copyright © 1959 by Karl Popper.

"Causality and Complementarity" from *Atomic Physics and Human Knowledge* by Niels Bohr. Copyright © 1963.

"Letters to Max Born" by Albert Einstein from *The Born-Einstein Letters* by Max Born, translated by Irene Born, 1971. Reprinted with permission of Walker & Company.

"Black Magic and White Magic" from *Magic, Science, and Civilization* by Jacob Bronowski. Copyright © 1978 by Columbia University Press. Reprinted with permission.

"Positivism, Metaphysics, and Religion" from *Physics and Beyond:*

Acknowledgments

*Encounters and Conversations by Werner Heisenberg* by Werner Heisenberg, translated by Arnold J. Pomerans. Planned and edited by Ruth Nanda Anshen. Copyright © 1971 by Harper & Row, Publishers, Inc. Reprinted with permission of the publisher.

"Science and Religion" from *Out of My Later Years* by Albert Einstein. Copyright © 1956, © 1984 by the Estate of Albert Einstein. Reprinted with permission of Carol Publishing Group.

# INDEX

absolute zero, 410–411
accelerator. *See* particle accelerator
Ackerman, Diane, 774
Adams, Henry, 423
Adams, John Couch, prediction of
  Neptune, 141–142
Adler, Alfred, on mathematics and
  creativity, 435–446
Agnew, Harold, 457
Ahlers, Günter, 470
Aiyar, V. Ramaswami, 650
Alexander, Hugh, 632, 633, 639
algorithms, iterative, and calculation of pi,
  655–659
Allende meteorite, 382
Alpha Centauri, 276, 296
Alpher, Ralph, 121, 331–332, 333
Alvarez, Luis W., 261–262, 263–267, 269;
  on Berkeley cyclotron, 730–740
Alvarez, Walter, 261, 266, 269, 270
Ambartsumian, Victor A., 313
American Astronomical Society, 1925
  meeting, 321
Anders, Edward, 270
Anderson, Carl D., 77–78, 79
Andromeda galaxy, 276, 322, 338, 340
angstrom, 5, 20
Ångström, Knut, 52
anthropic principle, 393–394
antimatter, 78, 122, 786
Antiphon, 652
Archimedes, 652–653
Argo, Harold and Mary, 714
Aristotle, mechanics of, 541
Arnett, W. David, 282
artificial intelligence, 475–519. *See also*
  computers
Asaro, Frank, 261
Asimov, Isaac: on relativity, 184–193; on
  science, 781–784
asteroids, 262–265
astrology, 816
astrophysics, 109, 262
Atkinson, Geoffrey S., 711
atom, 21–37, 366; primeval, 328, 360–
  364. *See also* atomic hypothesis; nucleus;
  particles, elementary

atomic bomb project, 456–457, 705–729
Atomic Energy Commission, Pakistan,
  672–673
Atomic Energy Commission, U.S., von
  Neumann on, 625
atomic hypothesis, 3–17, 117
Atoms for Peace Conference, Salam at,
  670

Baade, Walter, 329
Babbage, Charles, 498, 509
Bacher, Robert F., 712
Bacon, Francis, 784, 797–798, 818–819,
  820
Bahcall, John N., 304
Bailey, David H., 657
Balick, Bruce, 307
Bankoff, Leon, 661
Bardeen, Jim, 230
Baron, Edward A., 287–288, 290
Barrow, John D.: on end of world, 425–
  427; on mathematics, 541–558
Baxter, Rodney J., 649
Bayley, Don, 639, 640, 643
Becklin, Eric E., 307, 317
Becquerel, Antoine Henri, 50, 52
Bekenstein, Jacob, 230, 231
Bell's theorem, 550
Berkeley Radiation Laboratory, Alvarez
  on, 730–740
Berndt, Bruce C., 649, 659
Bessel functions, 712
beta-decay, 149, 174
Bethe, Hans A., 536, 690–691, 707, 734;
  at Los Alamos, 710, 711, 712–713, 714,
  717, 722; on supernovas, 277–291
Big Bang, 118, 121, 171, 173, 174, 217,
  220–221, 226, 333, 368, 369, 370, 374,
  375, 384, 385, 389, 563
biological sciences, 144, 540; and
  radioactivity, 55
Birkhoff, G. D., 623
black holes, 571; Chandrasekhar and,
  607–609, 611–612; entropy of, 135,
  228–231; Hawking on, 132–135, 221,
  226–238; Penrose on, 203–225; and
  white holes, 217